Numerical Modeling for Electromagnetic Non-Destructive Evaluation

Numerical Modeling for Electromagnetic Non-Destructive Evaluation

NATHAN IDA

Department of Electrical Engineering
University of Akron
Ohio, USA

CHAPMAN & HALL

London · Glasgow · Weinheim · New York · Tokyo · Melbourne · Madras

Published by Chapman & Hall, 2-6 Boundary Row, London SE1 8HN, UK

Chapman & Hall, 2-6 Boundary Row, London SE1 8HN, UK

Blackie Academic & Professional, Wester Cleddens Road, Bishopbriggs, Glasgow G64 2NZ, UK

Chapman & Hall GmbH, Pappelallee 3, 69469 Weinheim, Germany

Chapman & Hall USA, One Penn Plaza, 41st Floor, New York, NY10119, USA

Chapman & Hall Japan, ITP-Japan, Kyowa Building, 3F, 2-2-1 Hirakawacho, Chiyoda-ku, Tokyo 102, Japan

Chapman & Hall Australia, Thomas Nelson Australia, 102 Dodds Street, South Melbourne, Victoria 3205, Australia

Chapman & Hall India, R. Seshadri, 32 Second Main Road, CIT East, Madras 600 035, India

---

First edition 1995

Printed in Great Britain by T.J. Press, Padstow, Cornwall

ISBN 0 412 46830 1

A catalogue record for this book is available from the British Library

∞ Printed on permanent acid-free text paper, manufactured in accordance with ANSI/NISO Z39.48-1992 and ANSI/NISO Z39.48-1984 (Permanence of Paper).

Contents

2. Analytic methods of solution

3. The finite difference method

7. Finite element formulation

8. Boundary integral, volume integral and combined formulations

Foreword

This text on numerical methods applied to the analysis of electromagnetic nondestructive testing (NDT) phenomena is the first in a series devoted to all aspects of engineering nondestructive evaluation. The timing of this series is most appropriate as many university engineering/physics faculties around the world, recognizing the industrial significance of the subject, are organizing new courses and programs with engineering NDE as a theme. Additional texts in the series will cover electromagnetics for engineering NDE, microwave NDT methods, ultrasonic testing, radiographic methods and signal processing for NDE. It is the intended purpose of the series to provide senior—graduate level coverage of the material suitable for university curricula and to be generally useful to those in industry with engineering degrees who wish to upgrade their NDE skills beyond those needed for certification. This dual purpose for the series reflects the very applied nature of NDE and the need to develop suitable texts capable of bridging the gap between research laboratory studies of NDE phenomena and the real world of certification and industrial applications.

The reader might be tempted to question these assertions in light of the rather mathematical nature of this first text. However, the subject of numerical modeling is of critical importance to a thorough understanding of the field—defect interactions at the heart of all electromagnetic NDT phenomena.

Maxwell's equations, the material constitutive relationships and the Lorentz force equation describe all electromagnetic NDT field phenomena and also provide the theoretical basis from which the governing partial differential equations are derived, often in terms of appropriate potential functions.

Analytical solutions of the governing partial differential equations are only available for the simplest of NDT geometries and hence numerical methods must be used if reasonable solutions are to be obtained for practical electromagnetic NDT geometries. Such solutions are needed for test design purposes, for physical understanding of the field—defect interactions and for the generation of data to aid in the inverse or defect characterization problem.

This text describes different numerical approaches to the solution of all three types of partial differential equation governing electromagnetic NDT phenomena.

Elliptic partial differential equations, such as the Laplace and Poisson forms, govern static electromagnetic NDT phenomena associated respectively with potential drop and magnetic flux leakage methods of testing. Parabolic partial differential equations are shown to apply to all diffusive NDT phenomena such as those associated with eddy current testing. Finally, the wave phenomena of microwave testing can be categorized as hyperbolic in nature, requiring both spatial and temporal discretization for time and domain numerical solution.

A thorough understanding of the techniques described in this text will provide the reader with all the necessary background needed to develop a library of algorithms suitable for solving electromagnetic NDT problems.

W. Lord
Ames, Iowa
USA

Introduction

Modeling of electromagnetic *NDT* phenomena involves the solution of the electromagnetic field equations under specified conditions. While we often refer to this by using some descriptive aspect of the problem such as "active leakage fields", or "eddy current methods", it is not immediately evident from these expressions what exactly is involved in modeling of the given process. For example, the eddy current problem can be either a time dependent or a time-harmonic application. It can be transient, linear or nonlinear, and finally, it may be solved in various ways, under various restrictions. In fact, we may start by defining the general, time dependent wave equation and view the eddy current problem as a subset of this, more general equation. Under this assumption, the equation is an eddy current equation when the terms associated with displacement currents are small compared with terms associated with induced currents. Thus, the use of the term "eddy current method" does not necessarily imply a particular method of solution. Yet in modeling, this is perhaps the most important aspect of the problem since it defines the methods we need to use. If a problem is described by a wave equation, we must use methods that can handle propagation aspects. A time dependent problem may be solved more readily by, say, a finite difference technique while a time-harmonic problem can be solved using finite difference, finite element or, perhaps, a method of moments technique.

1 ELLIPTIC, PARABOLIC AND HYPERBOLIC PROCESSES

The approach in this work is to study first the various types of equations since the solution method depends to a large extent on the form of the equation. This allows simple classification of the various types of problems in nondestructive testing *(NDT)*. The techniques available for each type of equation are applicable to the equation in general, regardless of the actual testing configuration we describe.

The field equations fall under one of the following three types of partial differential equations:

Elliptic partial differential equations
Parabolic partial differential equations
Hyperbolic partial differential equations

This classification allows us to separate treatment of the various types of equations into three broad groups. However, the classification should only be viewed as a guideline. There are equations which do not necessarily fall under one group yet will be solved using techniques adopted for that group.

Classification of partial differential equations is based on the form of the equations and the relation of their coefficients. The structures of the various equations we treat is given in chapters *5*, *9*, and *10*, where both the general approach and exceptions to the general approach are given. For now we note the following general properties.

Elliptic equations are time independent or static equations. They describe an equilibrium or boundary value problem. As an example, the electrostatic and magnetostatic equations fall under this category. In terms of testing, both residual and active leakage fields are described by elliptic partial differential equations although the two methods are different in that residual leakage fields have no current densities as sources (these are generally described by Laplace's equation with the sources of the field as boundary conditions). Residual leakage fields are dominated by magnetization processes while active leakage fields include a source current (Poisson-type equations). In addition, residual leakage methods must use nonlinear methods while, for example, electrostatic solutions are almost always linear.

The Helmholtz equation, which is a wave equation in the frequency domain, is also viewed as an elliptic equation for the purpose of this work. This approach is taken here for a number of reasons, the primary reason being that similar methods are used for the solution of the problem. There are, however, major differences. One important difference is in the solution method. Boundary value problems are deterministic problems while the solution to Helmholtz's equation requires eigenvalue methods.

Parabolic partial differential equations are field equations which contain a first order time derivative. Thus they describe, for example, the time dependent, (transient) eddy current methods of testing. We note here that the time-harmonic eddy current problem is described by a Poisson-type equation in complex variables and therefore, it will be solved as an elliptic equation rather than as a parabolic equation. This simplification is afforded by the phasor transformation but it also indicates that we will only be able to treat linear applications of eddy currents. Nonlinear applications must be treated in the time domain as parabolic processes.

Hyperbolic partial differential equations, contain a second order time derivative with or without a first order time derivative. The general wave equation is of this type. As with the eddy current problem, time-harmonic wave propagation can often be solved in a much simpler way. One example is the modal solution for the sourceless (homogeneous) Helmholtz equation. The sources as well as other aspects of propagation such as scattering are taken into account by a general solution of the wave equation in the time domain.

2 GENERAL APPROACHES TO SOLUTION OF FIELD PROBLEMS

The analytic solution to any problem is the best solution we can obtain. It means that our solution is exact and therefore, an exact model of the problem. Also, because the solution is in the form of an exact expression, an analytic solutions allows simple parametric study of any aspect of the geometry, fields, and the interaction between them. This is obviously an ideal situation that can be met only in certain, simple situations. The solution of general, three dimensional field problems is either very complex or impossible to do analytically and very expensive in terms of resources and time if done numerically. Yet, in the field of nondestructive testing any realistic defect is three dimensional. The characterization of such defects requires a complete specification of the magnetic fields and current densities in the vicinity of defects, a task which can only be accomplished by the solution of the full three-dimensional field equations.

3 THE ANALYTIC APPROACH

The analytic approach to modeling is characterized by two major aspects. First, the solution is general and exact. A general solution is obtained, normally in the form of a mathematical relation which can then be used for analysis, parametric studies, calibration of test systems and the like. Second, derivation of the model involves rather standard (if not simple) mathematical techniques.

3.1 Range of solution methods

As mentioned above, the methods employed for analytic solution are standard mathematical methods. The best known methods are those based on the direct solution of partial differential equations by separation of variables. In these methods, the electromagnetic field equations are reformulated into the form of a second order partial differential equation by, essentially, a change in variables. Depending on the equation, solution may proceed by direct integration or by defining separable variables. Normally, this is followed by integration over a generic solution domain. Introduction of the boundary conditions of the problem completes the process.

The most general methods of solution for Laplace's and Poisson's equations are based on separation of variables. These include many applications in static and time dependent fields. The solution is normally in terms of harmonic functions (or the equivalent exponential functions) in Cartesian coordinates and in terms of Bessel functions or modified Bessel functions in cylindrical coordinates.

Other methods employed are the method of images and conformal mapping. The first is often employed for solution of Poisson's equation while the second applies to Laplace's equation in two-dimensional geometries.

Analytical methods, while extremely powerful when they can be employed, are limited to a relatively small class of problems. This includes one-dimensional problems, highly symmetric two-dimensional applications and some three-dimensional applications. While this is a rather restrictive set, the analytic solution of field problems is extremely important, especially for verification of numerical methods and for understanding the physical behavior of fields. Thus, it is sometimes more useful to solve a one-dimensional wave propagation problem analytically, because it is simple and provides all aspects of wave propagation, than to solve a full, three-dimensional, complicated problem which provides little additional fundamental information. However, if the details of the more general three-dimensional geometry are required, the one-dimensional solution is of little use.

3.2 Limitations

Most of the limitations of analytic models were pointed out above and stem from the limited applicability of the model. The following is a summary of some of the main limitations.

1. Only linear models can be considered.
2. Materials must be isotropic. Layered materials can sometimes be considered and, in the one-dimensional case, anisotropy can also be treated.
3. Only one- and two-dimensional applications in rather simple geometries can be solved. There are a few exceptions to this rule but in general only simple, highly symmetric problems can be treated.
4. In the context of *NDT*, defects and complex shapes can rarely be included.
5. Nonuniform fields (such as nonuniform plane waves or non-plane waves) cannot be treated.

4 THE NUMERICAL APPROACH IN NDT

In light of the fact that analytical methods are of limited scope in the modeling of *NDT*, the natural choice is to turn to numerical methods. Numerical approaches, both differential methods, based on the finite element or finite difference techniques and integral methods, serve to analyze the testing environment under most conditions, including those that do not allow a meaningful analytic solution.

The purpose of numerical methods can be easily summarized as serving as a test bed for design and analysis of testing parameters. Implementation of numerical techniques for the purpose of modeling *NDT* parameters should be viewed as a means of improving testing and reducing the costs involved in design and testing of equipment.

Numerical techniques allow analysis of most field/material interactions and as such they also promise to provide a means for solving the inverse problem in *NDT*.

4.1 Importance

We will view numerical approaches to modeling of electromagnetic fields as a direct consequence of the failure of analytic methods to provide adequate models. This approach seeks to find adequate, efficient models for realistic problems while, at the same time, acknowledging the fact that numerical models are only approximations to the analytic model. In this sense, the analytic model remains our goal.

In practical terms, numerical models can be, and often are, very accurate and quite flexible. They provide solutions to increasingly large and complex problems with moderate computer resources. Although it cannot be claimed that any problem has a numerical solution in practice, it does in principle. For example, nonlinear eddy current problems are rather difficult to solve because of the difficulty in describing the hysteresis curve. However, this is more of a practical difficulty than a fundamental flaw in numerical models.

Thus, the importance of numerical models is in their ability to approximate the solution to a wide range of practical problems and therefore provide a well established means of designing a test.

4.2 Limitations

There are of course many limitations associated with numerical methods. In fact, the advantages of analytical techniques are the limitations of numerical methods.

The first and foremost limitation is the fact that we do not obtain a closed form solution. This means that the solution process must be repeated for any set of input data even if one set is a minor perturbation on a previous set.

Second, a numerical approach requires both the availability of a computer and proficiency of the engineer to use it effectively. While computers are prevalent, good efficient software is not and, at least for the foreseeable future, the users of numerical methods will have to be reasonably proficient in their use, including programming. Also to be mentioned is the fact that numerical models can be extremely large, requiring considerable computer resources and the associated costs of access, and software maintenance.

5 NUMERICAL METHODS

Numerical methods can be viewed as a generic approach to approximation of the solution over the solution domain. As long as the approximation is compatible with the problem we solve, and satisfies the boundary conditions of the problem, it can form the basis of a successful numerical method. There are of course many other aspects that must be considered before adopting a particular method. Some have to do with the formulation we employ, some with the economics of solution and some with the type of results we need. We will not try to discuss this any further here since it would be difficult to do so before introducing the various methods of approximation and the requirements imposed on them. We will however point out to the various methods we use and outline their basic characteristics in general, simple terms.

All numerical methods used for field solution have the following common aspects:

1. Approximations: either the function or its derivatives are approximated.
2. Solution domain and approximation domain over which the solution is valid.
3. Boundary conditions that must be satisfied.
4. The approximation transforms the partial differential equation into a system of linear (or nonlinear) algebraic equations which can be solved by standard matrix methods.

5.1 The finite difference method (*FDM*)

The finite difference method is a general method of approximating the derivatives of a function in terms of the values of the function itself. When these approximations are introduced into a partial differential equation, and the derivatives are evaluated on a set of points (or grid), an approximate solution of the values of the function at the points of the grid is found. The solution obtained is only valid at the discretization points (or nodes). This forms a rather simple method of solving partial or ordinary differential equations. Both time and space derivatives can be approximated and the method is applicable to any order derivatives. The solution can take place in the time or frequency domain. A solution can be refined by simply increasing the number of points at which the derivatives are evaluated.

5.2 The finite element method (*FEM*)

The finite element method is different in that an approximation of the solution itself is sought over a specified domain: that of a finite size, geometrical volume called a finite element. This approximation is then introduced in the field equations or an equivalent form of the field equations. A method of minimizing the error due the approximation is also part of the method. The solution we obtain is again in terms of the actual function we seek and is obtained at discrete points but, unlike the finite difference technique, the solution is valid anywhere within the volume of the solution domain. Finite element methods, are in general more efficient in discretization of space, especially in awkward geometries, but cannot handle time. Approximations in time are normally handled by finite difference techniques either by themselves or in conjunction with space discretization using finite elements.

5.3 Boundary element methods (*BEM*)

Boundary element methods are modified finite element methods that use the integral forms of the field equations. As with any integral form, the solutions are obtained in terms of equivalent sources. In boundary element methods, the equivalent sources are defined on the boundary of the geometry under consideration. Standard finite elements are used and the solution is in terms of variables at the nodes of the finite elements but, because only the surface of the geometry is considered, the dimensionality of the problem is reduced by one. One of the major advantages of these methods is that three-dimensional problems can be treated as two-dimensional problems. However, from the *NDT* point of view, this approach has the disadvantage that only equivalent sources are available. A detailed solution in the volume is not obtained and defect characterization can only be done indirectly. Boundary element methods are very difficult to apply on complex surfaces, surfaces with corners, and to nonlinear materials. In addition, they result in fully populated systems of equations which are more expensive to solve. The trade-off between finite elements and boundary element methods is not clear cut and often a finite element method is less expensive to solve in spite of the higher dimensionality involved because of the sparse nature of the system of equations.

5.4 Coupled *FEM/BEM* methods

To take advantage of the basic features of the finite element method and the economy of boundary integral methods, the two methods can be combined. In this type of solution, a finite element approximation is used in the interior of the solution domain, and an integral method is used on the boundary of the solution domain. This has the combined effect of allowing detailed solution within materials and the reduction of problem size using the boundary element method. The two solutions (in terms of fields or potentials) must be coupled at the boundary using for example, potential functions. While the advantages of doing so are obvious, the limitations are the same as for boundary integral methods in that the problem must be linear and the surface "well behaved".

5.5 Volume integral methods

As with boundary integral methods, the principle is to find a solution in terms of a function which is indicative of variations in materials. In volume integral methods, the solution is in terms of volume variables such as current densities or magnetization in materials. This has the advantage of further reducing the problem size. However, this is done at the expense of resolution. While finite element methods provide detailed solutions in terms of the primary variables, volume integral methods provide solutions in terms of global variables. This is very useful if we are interested in global parameters such as impedance or energy but, in general, cannot completely replace differential methods. The main appeal of the method is its simplicity and, in particular, its economy. It is not unusual to obtain a solution using volume integral methods while only solving a system of equations with a few dozens equations. An equivalent differential equations approach using finite elements may require hundreds or thousands of variables.

CHAPTER 1

The electromagnetic field equations

1.1 INTRODUCTION

This chapter attempts to summarize the electromagnetic field theory. It should be understood that it is only possible to point out the various relations with minimal or no discussion. A discussion of any depth, will take more than this volume. Also, only the subjects considered important to this work are given. A few subjects, such as skin depth in tubular materials and magnetic properties of materials are discussed in some detail because they are particularly important in the context of *NDT*, and because they are seldom part of courses on electromagnetics. This chapter should only be used as a reference chapter. We assume that the reader is familiar with electromagnetic theory or that this theory is taught separate from this book. Some of the subjects which the reader should be familiar with are pointed out as problems at the end of the chapter. In this sense, the problems in this chapter are different than in subsequent chapters. They are more theoretical in nature and will require review of material in textbooks. Some of the sources that can be used for this purpose are given in the bibliography section.

Much of the missing material is discussed as it becomes necessary in subsequent chapters. Thus, for example, various forms of the field equations, which are given here without comment are described in detail in other chapters, including their derivation from Maxwell's equations. This approach is based on the principle that it is difficult to discuss a topic and motivate oneself to study it unless the utility of the subject is understood. In chapters dealing with the application of numerical techniques to solution of field problems it will become evident that a good understanding of the field equations is a prerequisite for meaningful analysis of the problem.

1.2 MAXWELL'S EQUATIONS IN DIFFERENTIAL FORM

The general, time dependent, Maxwell's equations in differential form (sometimes also called point or local form) are

$$\nabla\times\mathbf{E} = -\frac{\partial \mathbf{B}}{\partial t} \tag{1.1}$$

$$\nabla\times\mathbf{H} = \mathbf{J} + \frac{\partial \mathbf{D}}{\partial t} \tag{1.2}$$

$$\nabla\cdot\mathbf{D} = \rho \tag{1.3}$$

$$\nabla\cdot\mathbf{B} = 0 \tag{1.4}$$

where the various quantities involved are defined as:

E - the electric field intensity, given in *Volts/meter* [*V/m*]
H - the magnetic field intensity, given in *Amperes/meter* [*A/m*]
D - the electric flux density, given in *Coulombs/m*^{2} [C/m^2]
B - the magnetic flux density, given in Tesla [T] or *Webers/m*^{2} [Wb/m^2]
J - the current density, given in *Amperes/m*^{2} [A/m^2]
ρ - the volume electric charge density, given in *Coulombs/m*^{3} [C/m^3]

Equation (1.1) is known as Faraday's law of induction. This equation defines the basic laws of induction and, it is most often associated with eddy current applications. Eq. (1.2) is Ampere's law. Here we chose to write it in its general form, including the term $\partial\mathbf{D}/\partial t$ which is known as the displacement current density term as opposed to the conduction current density **J**. Eq. (1.3) is Gauss' law which we sometimes associate with electrostatic applications but which is also required for other applications as we will see later. Eq. (1.4) is not associated with a particular law and simply states the nonexistence of isolated magnetic poles. Because of the form of the expression it is sometimes referred to as the magnetic form of Gauss' law.

1.3 MAXWELL'S EQUATIONS IN INTEGRAL FORM

The differential forms of Maxwell's equations in Eqs. (1.1) through (1.4) will be used extensively in this work. However, we can also obtain an integral representation of the equations. While the differential form is more convenient in calculations using methods such as the finite element or finite difference methods, the integral form is more convenient in analytic calculation of fields and in the various integral methods of numerical calculation such as the method of moments and boundary element methods. The integral representation is obtained by simple integration of the four equations and use of vector theorems to simplify the resulting relations. We will not perform all steps here (see problem 1.1) but point out that we perform an open surface integration on both sides of Eqs (1.1) and (1.2) and then use Stoke's theorem to transform the volume integrals on the left hand side into line integrals. Eqs. (1.3) and (1.4) are integrated over a volume and then the left hand side is transformed into closed surface integrals by use of the divergence theorem. The integral form of Maxwell's equations is

$$\oint_C \mathbf{E}\cdot d\mathbf{l} = -\int_s \frac{\partial \mathbf{B}}{\partial t}\cdot d\mathbf{s} \tag{1.5}$$

$$\oint_C \mathbf{H}\cdot d\mathbf{l} = \int_s \left(\mathbf{J} + \frac{\partial \mathbf{D}}{\partial t}\right)\cdot d\mathbf{s} \tag{1.6}$$

$$\oint_s \mathbf{B}\cdot d\mathbf{s} = 0 \tag{1.7}$$

$$\oint_s \mathbf{D}\cdot d\mathbf{s} = \int_v \rho dv \tag{1.8}$$

The right hand side of these equations represent the sources of the fields. For this reason, these are sometimes integrated directly to simplify representation. For example, integration of flux density **B** over a surface S results in the total flux through the surface. Thus, Eq. (1.5) can be written as

$$\oint_C \mathbf{E}\cdot d\mathbf{l} = -\frac{\partial \Phi}{\partial t} \tag{1.9}$$

where Φ is the total flux through the surface enclosed by the contour C. This relation is immediately recognized as the "induction" equation. Similarly, in Eq. (1.6) we can integrate the current density over the surface to obtain the total current passing through the surface enclosed by the contour C:

$$\oint_C \mathbf{H}\cdot d\mathbf{l} = I_c + I_d \tag{1.10}$$

where I_c is the conduction current and I_d the displacement current. This division into source and displacement currents can also be done in terms of the current density in Eq. (1.6). In addition, the source current can be viewed as containing all currents (or current densities) except displacement current. This includes conduction, convection and eddy currents.

Finally, performing the integration on the right hand side of Eq. (1.8) we get

$$\oint_S \mathbf{D} \cdot \mathbf{ds} = Q \tag{1.11}$$

which is a familiar form of Gauss' law and Q is the total charge enclosed by the surface S.

1.4 CONSTITUTIVE RELATIONS

The field equations above are written as linear partial differential equations. This important property of the field equations is defined by the interaction of fields with materials and, therefore we will associate linearity or nonlinearity of field relations with material properties. We also note that since the field equations describe vector relations, and since there are four field quantities (**E**, **H**, **D**, and **B**), the set in Eqs. (1.1) through (1.4) seems to be equivalent to *12* scalar equations in *12* unknowns. This, however is not the case. The first two equations and the last two are not independent and, in fact, we can show that the last two equations can be obtained from the first two by simple vector algebra (see problem 1.2). Thus, we in fact only have *6* equations in *12* unknowns.

The two additional relations needed to complete the system of equations are the material constitutive relations:

$$\mathbf{B} = \mu \mathbf{H} \tag{1.12}$$

$$\mathbf{D} = \varepsilon \mathbf{E} \tag{1.13}$$

These two vector equations are equivalent to six scalar equations. We will discuss material properties in subsequent sections since the interaction of fields with materials is the basis of nondestructive testing and the most challenging aspect of computation. However, for now it suffices to note that these relations provide a link between the four fields. μ is the permeability of materials and ε the permittivity. In general these relations are nonlinear.

In addition, we define a constitutive relation involving current densities and the electric field intensity

$$\mathbf{J} = \sigma \mathbf{E} \tag{1.14}$$

1.5 ELECTROMAGNETIC INTERFACE CONDITIONS

The electromagnetic field undergoes changes as it passes from one material to another. As an example, the concentration of the magnetic field in gaps of ferromagnetic materials is well known. This change is based on the constitutive relations given above. If we take for example the relation $\mathbf{B}=\mu\mathbf{H}$, and assume two materials in contact, it is clear that the flux density **B** in each material is different. For this reason, a detailed knowledge of the behavior of fields at the interface between materials is important for computation and for our understanding of electromagnetic phenomena. This is done by applying Maxwell's equations at the interface between two general media. However, here we will simply give the relations that govern fields at material interfaces without giving any proof. The reader may refer to many of the references in the bibliography section for this purpose. Also, problems 1.3 through 1.8 deal with the subject of boundary or interface conditions. We will use the terms boundary and interface interchangeably in this section but, in later chapters, interface conditions will be used exclusively at material interfaces while boundary conditions will always refer to conditions on the boundary of the solution domain.

1.5.1 Interface conditions for the electric field

The electric field intensity is governed by the constitutive relation in Eq. (1.13). By rewriting this in therms of the electric field intensity alone we can write the interface conditions for the two components of the field as

$$E_{1t}=E_{2t}$$

$$D_{1n}-D_{2n}=\rho_s$$

where t indicates tangential components, n normal components, and ρ_S is the surface charge density at the interface, if any such charge density exists. Thus, the tangential component of the electric field intensity (the component along the interface) is continuous regardless of materials. The normal component is discontinuous and the discontinuity is equal to the surface charge density at the interface.

In the particular case where the surface charge density is zero, the normal component of the electric flux density is also continuous across the interface. For linear, isotropic materials, we can substitute the constitutive relations to obtain interface conditions in terms of **E** or **D**:

$$\frac{D_{1t}}{\varepsilon_1}=\frac{D_{2t}}{\varepsilon_2}$$

$$\varepsilon_1 E_{1n} - \varepsilon_2 E_{2n} = \rho_s$$

We can now summarize the interface conditions for the electric field as follows: At surfaces with surface charge densities

$$E_{1t} = E_{2t} \tag{1.15}$$

$$\varepsilon_1 E_{1n} - \varepsilon_2 E_{2n} = \rho_s \tag{1.16}$$

$$\frac{D_{1t}}{\varepsilon_1} = \frac{D_{2t}}{\varepsilon_2} \tag{1.17}$$

$$D_{1n} - D_{2n} = \rho_s \tag{1.18}$$

At surfaces without surface charge densities

$$E_{1t} = E_{2t} \tag{1.19}$$

$$\varepsilon_1 E_{1n} = \varepsilon_2 E_{2n} \tag{1.20}$$

$$\frac{D_{1t}}{\varepsilon_1} = \frac{D_{2t}}{\varepsilon_2} \tag{1.21}$$

$$D_{1n} = D_{2n} \tag{1.22}$$

1.5.2 Interface conditions for the magnetic field

The interface conditions for the magnetic field are obtained in a similar manner. However, in this case, the surfaces between the two materials may contain a surface current density instead of a surface charge density. The following summarizes the interface conditions for the magnetic field.
At interfaces with surface current densities

$$H_{1t} - H_{2t} = J_s \tag{1.23}$$

$$\mu_1 H_{1n} = \mu_2 H_{2n} \tag{1.24}$$

$$\frac{B_{1t}}{\mu_1} - \frac{B_{2t}}{\mu_2} = J_s \tag{1.25}$$

$$B_{1n} = B_{2n} \tag{1.26}$$

At interfaces without surface current densities

$$H_{1t} = H_{2t} \tag{1.27}$$

$$\mu_1 H_{1n} = \mu_2 H_{2n} \tag{1.28}$$

$$\frac{B_{1t}}{\mu_1} = \frac{B_{2t}}{\mu_2} \tag{1.29}$$

$$B_{1n} = B_{2n} \tag{1.30}$$

1.5.3 Other interface conditions

The relations above can be simplified under various conditions. If the materials are lossless (that is, they are very poor conductors), the following set of interface conditions may be used since lossless materials cannot support accumulation of charges or flow of currents at interfaces.

$$E_{1t} = E_{2t} \tag{1.31}$$

$$D_{1n} = D_{2n} \tag{1.32}$$

$$H_{1t} = H_{2t} \tag{1.33}$$

$$B_{1n} = B_{2n} \tag{1.34}$$

This is the case, for example, at the interface between two perfect dielectrics. Under these conditions, the components of the fields are continuous. The discontinuity in, for example, the normal component of **E** still exists since this is defined by the change in ε across the interface.

Another important type of interface is that between a conductor and a perfect dielectric. In this case, we can expect to have both surface current densities and surface charge densities at the interface. For most metals, the conductivity is high and we may assume they behave as perfect conductors for most practical purposes. The interface conditions between a dielectric and a conductor can be deduced from the general conditions in Eqs. (1.15) through (1.26) by setting the electric and magnetic field intensities inside the conductor to zero. These are:

$$E_{1t} = 0 \tag{1.35}$$

$$D_{1n} = \rho_s \tag{1.36}$$

$$H_{1t} = J_s \tag{1.37}$$

$$B_{1n} = 0 \tag{1.38}$$

This type of condition is quite common at high frequencies (e.g. in the microwave region) where it is normally referred to as an "electric wall condition" to indicate that the tangential component of the electric field intensity is zero.

Another type of interface condition is that between a perfect dielectric and an infinitely permeable material. Since such materials do not actually exist, this condition should be viewed only as an approximation. This approximate condition implies that the tangential component of **H** is zero at the interface. On the other hand, the tangential component of the electric field is now calculated from Eq (1.1) after adding the magnetic current density (magnetization) **M.** In effect we assume that a magnetic wall has a surface magnetic current density or surface magnetization. The conditions at the boundary are:

$$E_{1t} = -M_s \tag{1.39}$$

$$D_{1n} = 0 \tag{1.40}$$

$$H_{1t} = 0 \tag{1.41}$$

$$B_{1n} = 0 \tag{1.42}$$

This condition, in analogy with the electric wall condition is called a "magnetic wall condition" and is quite useful in analyzing microwave structures like waveguides, cavities and apertures.

Another aspect of boundary conditions is the behavior of fields at infinity. While we normally do not need to apply these conditions specifically, their existence is implied since we must assume that electromagnetic fields propagate throughout space. These conditions are called radiation or absorption boundary conditions. As an example, an absorption boundary condition at infinity, implies that none of the energy in the electromagnetic field is reflected back. In most cases this can be justified by observing that, in a lossy medium, the only way we might have reflections in an infinite domain is if the energy in the source producing the fields is infinite. For this reason, we will normally assume that only an outward propagating wave and an implicit radiation condition at infinity exist. In some cases, it is useful to create an artificial radiation boundary condition at a finite boundary. This is often done in computation of electromagnetic fields to ensure that artificially located boundaries do not create reflections which, in the actual infinite domain do not exist. We expand on this in chapters 9 and 10. The distinction between absorption and radiation boundary conditions is important. In an absorption boundary condition, the reflections are eliminated by dissipating the energy in the wave. As an example, a matched lossy material of sufficient thickness will ensure that no significant waves can reflect from the finite boundary. A radiation boundary condition does not dissipate energy but rather radiates it outward.

A related condition that is sometimes useful is an impedance boundary condition. In this type of condition the impedance on the boundary is taken into account. As an example, the resistivity of the conducting boundaries of waveguides and resonators can be used as impedance boundary conditions.

1.6 MATERIAL PROPERTIES

The constitutive relations mentioned in section 1.4 define the interaction between fields and materials. We will discuss these next in some detail since the electric and magnetic properties of materials are, perhaps, the most important factors in testing because of their effect on material behavior and on fields. In terms of modeling, material properties define not only what type of computation is needed, but also limit the model. Thus, a ferromagnetic sample may imply the need for nonlinear models while tests on a dielectric may require solution of a wave equation. The magnetic properties of a material are defined by the magnetic permeability. Permeability affects two important properties of the magnetic field and therefore of any application that uses the magnetic field. First, the flux density **B**, which is the quantity one is usually interested in, is larger for larger values of the permeability for a given source field intensity **H** [*A/m* or *Ampere turns/meter*]. Second, the permeability of many important engineering materials is nonlinear. In eddy current testing it influences the skin depth and through it the effective depth at which eddy current tests are useful. Permeability of free space is equal to $4\pi\times10^{-7}$ [*Henry/meter*]. Tables 1.1 to 1.3 list the relative permeabilities of some selected materials.

Electric fields are influenced by the permittivity or dielectric constant of dielectric materials. The dielectric constant is important in modeling static electric fields as well as at high frequencies. The dielectric constant of free space is equal to 8.85×10^{-12} [*Farad/meter*]. Both permittivity and permeability are, in general, complex. This is particularly important for permittivity, especially in the microwave domain. The three basic material properties: conductivity, permittivity, and permeability are discussed next.

1.6.1 Conductivity

Conductivity of a material can be broadly defined as its ability to conduct electric current. Under the influence of an electric field, free electrons move at various velocities. The electron velocity is proportional to the electric field and, therefore, the current density **J** in a conductor can be directly related to the applied electric field intensity **E** as

$$\mathbf{J} = \sigma\mathbf{E} \tag{1.43}$$

where σ is the material conductivity. Conductivity can be field dependent (that is, nonlinear), and it is almost always temperature dependent. The relation in Eq. (1.43) is defined for conductors but we also use it for low conductivity materials such as conducting composites. Eq. (1.43) is known as Ohm's law, written here in its vector form.

1.6.2 Permittivity

The electric properties of dielectric materials, are largely defined by polarization of charges within the material due to an applied external electric field. Since charges in these materials are bound, conduction is negligible, but polarization of charges (i.e. alignment of electric dipoles with the external electric field) may be significant. This polarization increases the electric flux density in the material. For this reason we can write the constitutive relation of Eq. (1.13) as

$$\mathbf{D} = \varepsilon_0\mathbf{E} + \mathbf{P} \tag{1.44}$$

where $\mathbf{P}$ is the polarization vector and ε_0 is the permittivity of free space. The polarization vector is proportional to the electric field. Since the polarization vector $\mathbf{P}$ is proportional to the external electric field intensity $\mathbf{E}$ we can write

$$\mathbf{D} = \varepsilon_0\mathbf{E} + \varepsilon_0\chi_e\mathbf{E} = \varepsilon_0(1+\chi_e)\mathbf{E}$$

where χ_e is a dimensionless quantity called electric susceptibility. The quantity

$$\varepsilon = \varepsilon_0(1+\chi_e)$$

is defined as the complex permittivity. This is often written as

$$\varepsilon = \varepsilon' + j\varepsilon'' \tag{1.45}$$

to signify the fact that permittivity is a complex quantity. By substituting the complex permittivity in Eq. (1.13) we get

$$\mathbf{D} = \varepsilon\mathbf{E}$$

This is identical in form to Eq. (1.13) but here ε is complex while in Eq. (1.13) ε is real. ε does not have to be linear or isotropic. The imaginary part of ε represents dielectric losses. Lossless materials have a real permittivity. In addition we define a relative permittivity through the relation

$$\varepsilon = \varepsilon_0\varepsilon_r$$

1.6.3 Permeability

Magnetic properties of materials are due to the interaction of external magnetic fields and moving charges in materials. Atomic scale magnetic fields are produced inside materials through orbiting electrons. Equivalent current loops are generated due to these electrons. These behave like small magnets (magnetic moments). With an assemblage of many such magnetic moments, the material volume contains a certain magnetic moment density. A net magnetic field is generated inside the material. This internal field is either aligned with the external field to increase the total field or opposes it to decrease the total field. If the internal magnetic moments are randomly oriented, as is often the case, the net internal field is zero and the material behaves like free space from the magnetic point of view. This is the case for nonmagnetic materials. Without discussing the details of magnetic materials here we indicate that a total magnetization vector **M** exists in the material, in the presence of an external magnetic field. Using this magnetization vector as analogous to the polarization vector in the electric field, the magnetic flux density **B** may be written as

$$\mathbf{B} = \mu_0 \mathbf{H} + \mathbf{M} \tag{1.46}$$

where **M** is the magnetization vector and μ_0 is the permeability of free space. Magnetization can be written in terms of the magnetic field intensity as

$$\mathbf{M} = \mu_0 \chi_m \mathbf{H} \tag{1.47}$$

where χ_m is the magnetic susceptibility of the material and is generally, complex. From Eq. (1.46) we get an expression for the complex permeability μ:

$$\mu = \mu_0(1+\chi_m) = \mu' + j\mu'' \tag{1.48}$$

Substituting this into Eq. (1.12) we get the third constitutive relation

$$\mathbf{B} = \mu \mathbf{H}$$

The imaginary part of the complex permeability represents losses due to damping forces in the magnetic material. Since there are no magnetic currents, we cannot view these losses as conduction losses. The real part represents materials without magnetic losses. In most cases there will be no need to consider permeability as a complex quantity since the real part for most materials is dominant. Permeability of many ferromagnetic materials is nonlinear as we will see shortly. For practical use, it is common to define a relative permeability μ_r through the relation

$$\mu = \mu_0 \mu_r$$

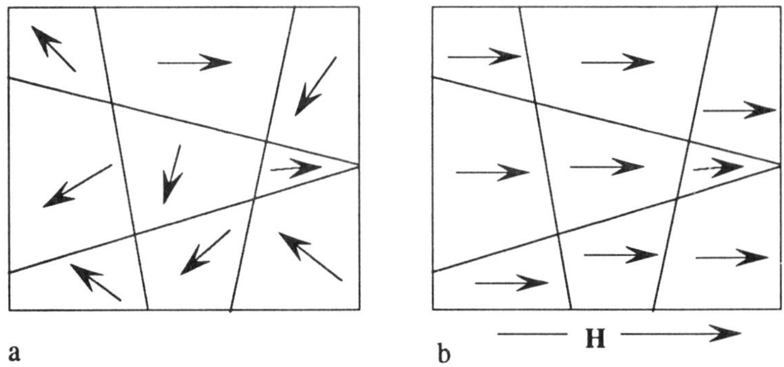

Figure 1.1. Orientation of magnetic domains in ferromagnetic materials. a. Random orientation in demagnetized materials. b. Magnetized ferromagnetic material.

Magnetic materials are divided into three groups based on their permeabilities. These are diamagnetic, paramagnetic and ferromagnetic materials.

Diamagnetic materials, are those for which the internal magnetic field is zero under normal conditions. When an external magnetic field is applied, a net internal field opposing the external field is produced. The relative permeabilities exhibited by diamagnetic materials are slightly smaller than 1.0. This group of materials includes many important materials like copper, lead and silver. Table 1.1 lists permeabilities of some diamagnetic materials.

Paramagnetic materials are similar to diamagnetic materials but the internal magnetic field is in the same direction as the external field. It increases the total field slightly due to alignment of the external and internal fields. The relative permeability of the material is slightly larger than 1.0. Examples of materials in this group are air and aluminum as well as some stainless steels. Table 1.2 lists a few paramagnetic materials and their permeabilities.

Ferromagnetic materials vary from diamagnetic and paramagnetic materials considerably. Their permeability is very large and the internal structure of their magnetic moments is different. Many of the moments are aligned in a certain direction within a very small region called a magnetic domain. Neighboring domains have similar structure, with the net magnetic domain directed in one direction. Under excitation free conditions, the magnetic domains tend to be aligned randomly. The internal field is very small or zero (with the exception of natural or artificial permanent magnets). This is shown schematically in Figure 1.1. Under an applied external field, those domains that have a net field aligned in the direction of the applied field grow in size while other domains shrink. The total field is larger because of the alignment of the external and internal fields. This behavior is related to the hysteresis curve of a given ferromagnetic material. As the external field grows, more domains are aligned, further increasing the total field. Beyond a certain field, most of the magnetic domains are aligned with the field and an increase in the field does not increase the internal field. Thus, a saturation in the magnetic field has been reached. Materials typical to this group are iron, steels, nickel and some stainless steels.

Ferrimagnetic materials are ferromagnetic materials but are essentially insulating as far as conductivity is concerned. In practice this means that their conductivities are very low. Ferrites and powdered materials like magnetite are part of this group of materials. Table 1.3 lists some typical ferromagnetic and ferrimagnetic materials and their permeabilities.

As is evident from any hysteresis curve, the permeability of ferromagnetic materials is not constant but varies with the field. This is exhibited through the slope of the initial magnetization curve to which the permeability is tangent. Thus, most ferromagnetic materials are highly nonlinear materials.

Table 1.1 Relative permeabilities for some diamagnetic materials.

Material	μ_r
Silver	*0.99998*
Copper	*0.999991*
Water	*0.999991*
Mercury	*0.999968*
Bismuth	*0.99983*
Lead	*0.999983*

Table 1.2 Relative permeabilities for some paramagnetic materials.

Material	μ_r
Vacuum	*1.0*
Air	*1.00000036*
Aluminum	*1.000021*
Palladium	*1.0008*

Table 1.3 Relative permeabilities for some ferromagnetic materials. These figures represent maximum values and depend on composition and treatment of material.

Material	μ_r
Cobalt	*250*
Nickel	*600*
Iron	*6,000*
Supermalloy	*1,000,000*
Carbon Steel (*0.9%C*)	*100*
Silicon Iron (*4%Si*)	*7,000*
Permalloy (*78.5% Ni*)	*100,000*
Fe_3O_4 (Ferrimagnetic)	*100*
Ferrites (Ferrimagnetic)	*5,000*

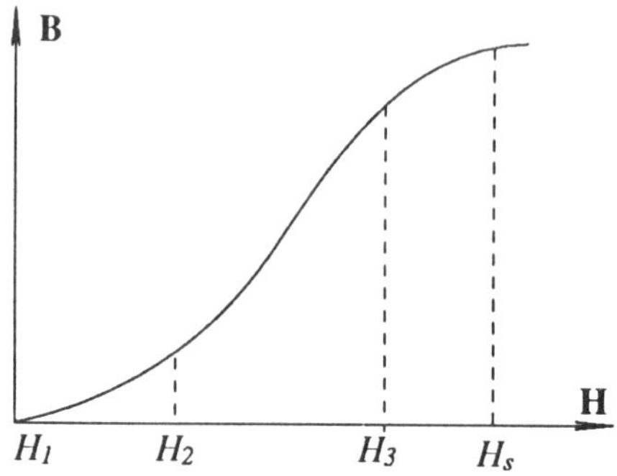

Figure 1.2 The initial magnetization curve of a ferromagnetic material.

1.7 HYSTERESIS

The behavior of magnetic materials described above can only explain the fact that permeability is larger or smaller than that of free space but cannot account for all magnetic phenomena that exist within materials. While the constitutive relation for permeability in Eq. (1.12) or Eq. (1.46) is generally correct, the permeability of ferromagnetic material is field dependent, with a very complex dependency relation. Thus, in general, Eq. (1.12) must be written as a nonlinear relation:

$$\mathbf{B} = \mu(H)\mathbf{H} \tag{1.49}$$

This relation can also be explained qualitatively from the domain behavior of a ferromagnetic material. Initially, the domains are randomly oriented (Figure 1.1a). As the applied field increases, domains begin to grow by displacing other domains and eventually occupying most of the material volume (Figure 1.1b). Any increase of the field will have little effect on the domains and therefore on the flux density in the material.

A plot of the relation in Eq. (1.49) describing the flux density as a function of the field intensity is useful in looking at magnetic materials. For linear materials (materials for which the permeability is constant at any field value) this curve is a straight line whose slope is equal to the permeability. The curve in Figure 1.2 describes the behavior of iron. When starting with a non-magnetized specimen, the initial applied field and the flux density are both zero. As the applied field is increased, the flux density also increases but, unlike linear materials, the curve is not linear. At some field value, H_3, the curve starts leveling off and the slope of the curve is reduced significantly. Any increase beyond the field H_3, increases the flux density but not at the same rate as at lower points on the curve. In fact, the slope in this section of the curve approaches unity, meaning that the relative permeability approaches one. This is a saturation region. The level of field at which saturation occurs depends on the material tested. The whole curve described in Figure 1.2 is a magnetization curve. Since it starts with zero applied field it is also called an initial magnetization curve.

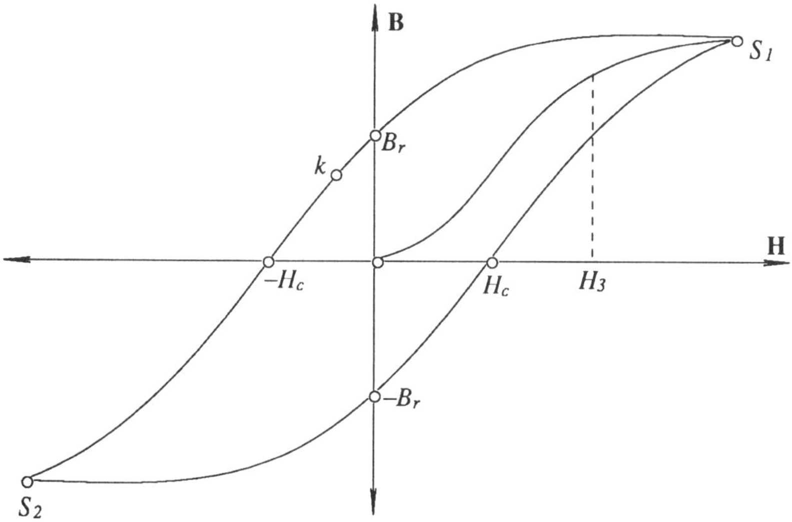

Figure 1.3 The hysteresis or magnetization curve of a ferromagnetic material.

If the applied field is reduced, the curve, instead of retracing the initial magnetization curve, moves to the left. This is shown in Figure 1.3. The flux density is reduced, up to point B_r where the applied field is zero. This residual flux is called remnance or retentivity. It is typical of magnetic materials, but cannot be explained from normal field relations. The fact that a remnant flux density may exist without a source to sustain it is very important in permanent magnets where the magnitude of this coercive force is an important parameter of the magnet. Applying a reverse magnetic field reduces the flux density up to the point $-H_c$ where an applied field intensity exists without an associated flux density. This field intensity is called the coercivity or the coercive force of the material. Further increase in the negative field intensity will trace the magnetization curve through point S_2 where a saturation point has again been reached except that the field intensity and the flux density are negative. If we now decrease the applied field to zero a point symmetric to B_r is reached. Similarly, by increasing the applied field intensity to a value equal (but positive) to H_c, the flux density is again zero. Further increase in the field intensity retraces the curve to point S_1. Repeating the process described above will result in a retracing of the outer curve but not that of the initial magnetization curve. This unique magnetization curve is called the hysteresis curve and is typical of all magnetic materials although hysteresis curves of different materials including their coercive forces and remnance, are markedly different.

The slope to this curve, at any point, is the ratio of the magnetic flux density and the magnetic field intensity (B/H). This is the magnetic permeability. The slope is relatively high in the lower portions of the initial magnetization curve and is gradually reduced to unity. At this point (i.e. beyond point H_3) the material has reached magnetic saturation. The magnetic domains change but little and the material behaves linearly.

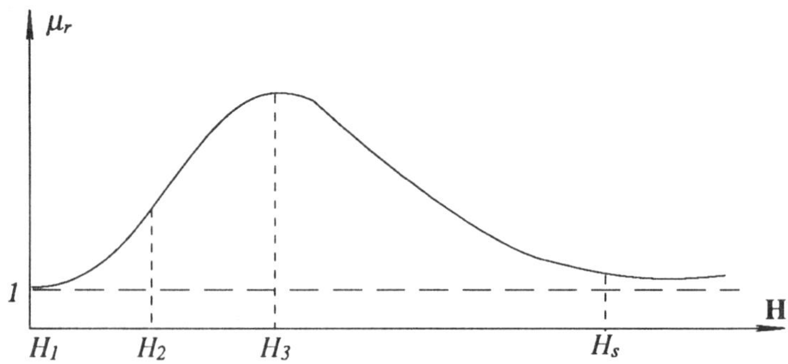

Figure 1.4 Permeability curve for an initial magnetization curve.

A curve, describing the slope to the initial magnetization curve of Figure 1.2 is shown in Figure 1.4. It shows, that for this material (iron) the initial relative permeability is low, increases gradually and then, as the field approaches saturation decreases, approaching one.

The hysteresis curve in Figure 1.3 is typical of ferromagnetic materials such as iron. It has four sections bounded by the four quadrants of the coordinate system. The first and second quadrants are of particular importance for nondestructive testing methods. The curve in the first quadrant, is created by an applied field or source and is therefore called a magnetization curve. The initial magnetization curve is created by starting with a demagnetized specimen of the material and then increasing the field within the material. In testing, this section of the curve is often referred to as the active part of the curve. All dc applications of magnetic testing that depend on active magnetization are governed by this section of the curve.

The second quadrant, with the limits at B_r and H_c is the demagnetization quadrant. All magnetic materials after magnetization, will relax to the point B_r or to a point in the second quadrant such as point k. This is the region in which permanent magnets operate. The coercivity (H_c) and remnance (B_r) of materials are very different from each other and serve to distinguish between different magnetic materials. Table 1.4 lists the coercivity and remnance of some soft magnetic materials and Table 1.5 lists the coercivity and remnance of some hard magnetic materials.

Table 1.4 Coercivity and remnance of some soft magnetic materials. Figures for H_c and B_r are approximate and are strongly dependent on material treatment.

Material	H_c [A/m]	B_r [T]	B_s [T]
Iron, pure, annealed	100	1.2	2.16
Supermalloy	0.2	10^{-4}	0.8
Steel (0.9%C),	4,000	1.0	2.0
Silicon Iron (4%Si)	20	0.5	1.95
Ni-Zn ferrite	16	0	0.34

Table 1.5 Coercivity and remnance of some hard magnetic materials.

Material	H_c [A/m]	B_r [T]
Carbon Steel (0.9%C)	4,000	1.0
Alnico 2	43,000	0.73
Alnico 2	44,000	1.2
Alnico 2	83,500	0.7
Alnico 2	126,000	1.04
Samarium-Cobalt	560,000	0.84
Ferroxdur 2	143,000	0.34

Different materials have different magnetization curves. This is related to the material structure, its magnetic domains and material treatment. The shape and magnitude of the magnetization curves defines the use of materials. Since the area under the hysteresis curve represents energy, one may need to consider this in design of devices. In many devices, such as transformers, this is a detrimental property since this energy is dissipated in the magnetic core. In other cases, this property is useful as in use of magnetic cores for switching applications. Permanent magnets rely on this effect to store energy.

1.8 MAGNETIZATION

Magnetization of a specimen is often necessary in *NDT* applications. This may be called for in the case of active leakage fields to bring the magnetic field to a required level and generate leakage fields or in the residual case to set up the initial field from which the specimen is allowed to relax. In the case of permanent magnets, magnetization is used to set up the field in the specimen. Magnetization of materials finds application in eddy current testing as well. To magnetize a specimen, it is necessary to apply a magnetic field to that specimen. The way in which this field is applied varies depending on practical considerations but the same basic effect must be obtained. The field in the specimen must be increased to a required level (active field). In general, if a specimen is initially demagnetized, the field is gradually increased through the initial magnetization curve to a required value. For residual field testing, the field is left to relax to a point in the second quadrant. For previously magnetized specimens, it is usually better to demagnetize the specimen first and then to magnetize it to the required level. In testing applications, the magnetization process is used to bias the material. With leakage field methods, the magnetization sets up the leakage fields required for testing. In eddy current methods, magnetization is sometimes used to saturate ferromagnetic materials. This reduces the relative permeability of the material to nearly *1* and allows a deeper penetration of the field in the test material. For many magnetic materials, once they have been magnetized, there is a remnant field (i.e. B_r) or otherwise a coercive force must be applied to cancel this field (i.e. H_c). This remnant or residual field is useful for testing purposes but in most cases it is necessary to first demagnetize the specimen before a controlled field is applied. Demagnetization is also required after testing to ensure that the specimen is demagnetized. For a uniformly magnetized specimen, demagnetization can be performed by simply creating a field opposing the original source field. This will, in effect, bring the operating point of the material to point H_c on the curve in Figure 1.3. This approach works well with permanent magnets which have a uniform field inside but is seldom practical in testing because of different operating points in different sections of the material. Simple and effective demagnetization of materials can also be achieved by heating the material beyond the Curie temperature and then cooling it in a zero field environment. However, this method is seldom practical.

The most common approach to demagnetization of materials is to cycle the material through the hysteresis curve while gradually reducing the field intensity to zero. The effect is that shown in Figure 1.5. This method, if started with a high enough field intensity and reduced slowly results in a properly demagnetized specimen. In practice, demagnetization is performed by applying an *ac* field and gradually reducing its amplitude to zero. Complete demagnetization is usually a very time consuming process. In practical situations, it is usually limited to reducing the flux density to some acceptable level. If the specimen is then magnetized this low magnetic level is not important. In other situations, where zero field is required, the cycling may be continued at a slower rate to ensure lower levels of residual magnetization

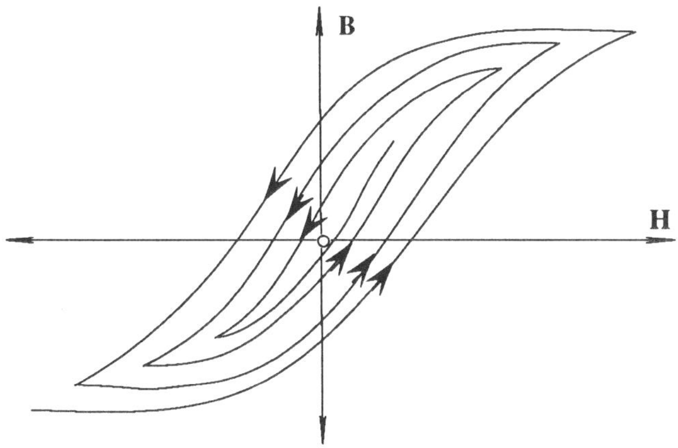

Figure 1.5 Demagnetization sequence for a magnetized ferromagnetic sample.

1.8.1 Minor hysteresis loops

If a sample is already magnetized and a small change in the field occurs, a small magnetization curve, superimposed on the main magnetization curve is generated. An example would be a large *dc* current corresponding to a point on the hysteresis curve and a small *ac* current superimposed on it. This would be the situation if an eddy current test is performed on a magnetized sample. Alternatively, if the magnetizing current is suddenly decreased and then increased again to the previous level the same effect is created. This situation results in a change in the hysteresis curve as shown in Figure 1.6. Thus, a small, oval-shaped curve, similar to the hysteresis curve is described at the given point on the magnetization curve. These loops are called minor hysteresis loops to distinguish them from the normal (or major) hysteresis loop. They are important for nondestructive testing since the operating point may be on one of these curves rather than on the major hysteresis curve. Since permeability is defined as the ratio of B and H, the permeability of a minor loop is $\Delta B/\Delta H$:

$$\mu_{inc} = \frac{\Delta B}{\Delta H} \tag{1.50}$$

This permeability, called an incremental permeability, depends on the location of the minor curve on the hysteresis loop and decreases as the normal magnetization increases. The slope of minor loops is always smaller than that of the major loop at a given point ($\mu_{inc}<\mu$). Incremental permeability can be defined either on the hysteresis curve or on the initial magnetization curve. In most cases, only the upper part of the initial magnetization curve will exhibit this effect since the lower part of the curve is reversible.

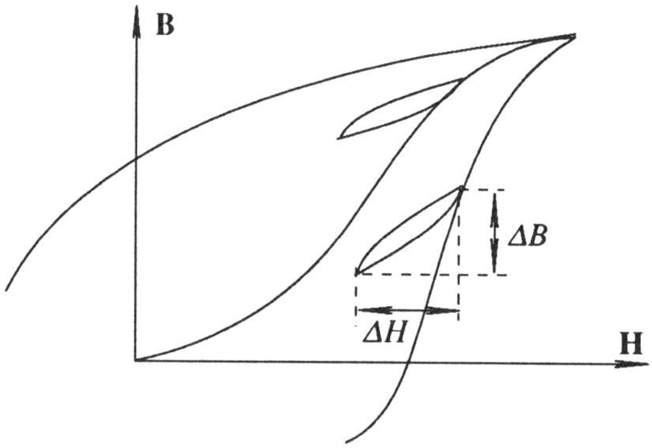

Figure 1.6 Minor hysteresis loops in a magnetized ferromagnetic sample.

1.9 PERMANENT MAGNETS

Permanent magnets are materials that retain a significant remnant flux density when the external field passing through the material is removed. Permanent magnetic materials are also called hard magnetic materials. The magnetic properties of a permanent magnet are defined by the fact that the operating point of a permanent magnet is in the second quadrant of the hysteresis loop of the material. Thus, the coercive field intensity H_c of the magnet and its flux density are opposite in direction anywhere in the magnet. In other words, permeability of a permanent magnet is negative.

The energy associated with the field of a permanent magnet is calculated as for any field and is given by

$$W = \int_v \frac{1}{2} HB dv \qquad (1.51)$$

where v is the volume of the magnet.

Thus, for a given volume of the magnet, energy depends only on the product BH. There is a point on the characteristic curve $B(H)$ of the magnet at which this product is maximum, representing the operating point of the magnet at which it can produce maximum energy. At points $B=B_r$ (and therefore $H=0$) or $B=0$ (and therefore $H=H_c$), the BH product is zero and the magnet does not produce any external energy.

In terms of design characteristics, a good permanent magnet is one which has a high coercive field intensity H_c as well as high remnant flux density B_r. A high coercive field intensity is important because it does not allow the magnet to be demagnetized and a high B_r is normally associated with the capacity of the magnet to produce high magnetic fields in magnetic circuits in which it is inserted. Examples of some permanent magnetic materials with their characteristics are given in Table 1.6.

Our interest in permanent magnets in the context of computation is two fold. First, they afford an example of calculation of static fields but, more importantly, they are sometimes used to provide a *dc* bias field to probes, magnetization yokes and in leakage field methods. Correct modeling of their fields is therefore essential for a good model of the test in which they are used. We will treat magnets as equivalent coils as shown in Figure 1.7. This model of the permanent magnet has the particular advantage of being compatible with other electromagnetic applications. A more correct model (in terms of the physical principles involved) is to use the magnetization of the magnet as the source of its field as shown in Figure 1.8.

Table 1.6 Properties of permanent magnets.

	B_r [T]	H_c [KA/m]	$(BH)_{max}$ [KJ/m^3]	μ_r
ALNICO	1.25	60	50	3-5
FERRITE	0.38	240	25	1.1
Sm-Co	0.9	700	150	1.05
Ne-Fe-B (20°C)	1.15	800	230	1.05

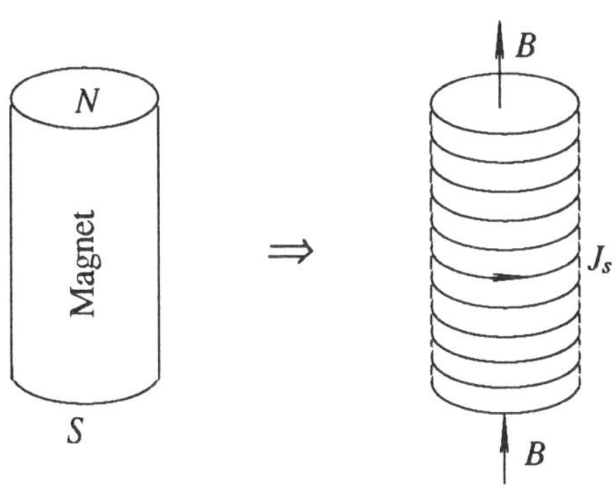

Figure 1.7 A permanent magnet and its equivalent coil.

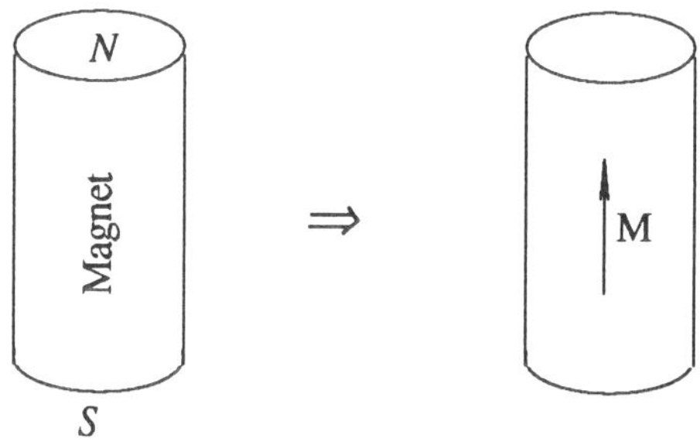

Figure 1.8 A permanent magnet described in terms of magnetization

1.10 THE POYNTING THEOREM, ENERGY AND FORCES

Energy in the electromagnetic field is governed by the Poynting theorem. The Poynting theorem states that the time rate of change in the stored energy (electric and magnetic) minus the power dissipated in a volume is equal to the power leaving that volume. To put this in more concrete terms, the Poynting vector is:

$$\mathbf{P} = \mathbf{E}\times\mathbf{H} \tag{1.52}$$

This represents power density but, in addition, it also shows the direction of flow of energy. Using the Poynting vector, and integrating it over a closed surface (enclosing a volume v), gives an expression for the Poynting theorem as

$$\oint_s (\mathbf{E}\times\mathbf{H})\cdot d\mathbf{s} = -\frac{\partial}{\partial t}\int_v \left(\frac{\mathbf{H}\cdot\mathbf{B}}{2} + \frac{\mathbf{E}\cdot\mathbf{D}}{2}\right)dv - \int_v \mathbf{E}\cdot\mathbf{J}dv \tag{1.53}$$

The left side of the expression represents the total flow of power through the area bounding the volume. The first two terms on the right-hand side are the rates of change in the stored magnetic and electric energy (power) while the third term is the dissipated power through ohmic losses or, if sources exist within the volume, this represents the sources as well.

The Poynting theorem describes all energy relations in a system be they electrostatic, magnetostatic or time dependent. Because the vector product between the electric field intensity and the magnetic field intensity is taken, these two quantities must be related (i.e. they must be generated by the same sources) otherwise the results obtained will have no meaning.

The expression in Eq. (1.53) is an instantaneous quantity. For practical purposes, a time averaged quantity is normally required. For periodic time variation of fields, this can be done by averaging over a time T (usually a cycle of the field), giving the time average Poynting vector:

$$\mathbf{P}_{av} = \frac{1}{T}\int_0^T \mathbf{P}(t)dt \tag{1.54}$$

1.10.1 The complex Poynting vector

In some instances as, for example, when using time-harmonic fields, we will require the complex equivalent of the Poynting vector and Poynting theorem. The complex Poynting vector is defined as

$$\mathbf{P} = \frac{1}{2}(\mathbf{E}\times\mathbf{H}^*) \tag{1.55}$$

where * indicates a complex conjugate and **P**, **E** and **H** are phasors. The time averaged complex Poynting vector is

$$\mathbf{P}_{av} = \frac{1}{2}Re(\mathbf{E}\times\mathbf{H}^*) \tag{1.56}$$

where *Re* indicates the real part of the complex Poynting vector. The equivalent form of Eq. (1.53) is

$$\oint_s \mathbf{P}\cdot d\mathbf{s} = j2\omega\int_v\left(\frac{1}{4}\mu\mathbf{E}\cdot\mathbf{E}^* - \frac{1}{4}\mu\mathbf{H}\cdot\mathbf{H}^*\right)dv + \int_v \frac{1}{2}\mathbf{E}\cdot\mathbf{J}_0^* dv - \int_v \frac{1}{2}\sigma|\mathbf{E}|^2 dv \tag{1.57}$$

where the power due to sources and dissipated power are written explicitly. The term on the left hand side is the total power flowing out of the volume v. The first term on the right hand side is the total electric and magnetic power while the second term on the right hand side is the total (complex) power of the source, if such a source exists. The last term is the dissipated power in the volume.

1.10.2 Forces

Forces are defined in general through the use of the Lorentz force equation. This defines the force on a moving charge as

$$\mathbf{F} = q\mathbf{E} + q\mathbf{v}\times\mathbf{B} \tag{1.58}$$

where q is a charge and $\mathbf{v}$ is the velocity of the charge. The first term is the electric force on the charge and exists whenever an electric field exists. The second is the force in the magnetic field and only exists if the charge moves.

Calculation of forces in conjunction with field analysis is performed either by the relation above or an equivalent relation. In some cases, the use of the Maxwell stress tensor is preferred (see bibliography).

1.11 POTENTIAL FUNCTIONS

Potential functions are viewed here as alternative representation of the electromagnetic field. In other words, we may find it simpler, or more useful to use an auxiliary function to describe the field properties rather than to use the field variables described in section 1.2. This may allow simplification in computation of fields, better understanding of the results we obtain or, in many cases, perform computation more economically. Potential functions are defined based on the properties of the magnetic field.

1.11.1 The electric scalar potential for static fields

The electric scalar potential is one of the most widely used potential functions. The potential is defined based on the irrotational property of the static electric field. That is

$$\nabla \times \mathbf{E} = 0 \tag{1.59}$$

This is a direct consequence of setting the time derivative to zero in Eq. (1.1). Any function that is irrotational (that is, its curl is zero), can be written as the gradient of a scalar potential. In the case of the electric field intensity we choose the following

$$\mathbf{E} = -\nabla V \tag{1.60}$$

This definition is consistent with our conventions of flow of current. It simply states that if we can calculate the electric scalar potential (or voltage), then we can also calculate the electric field intensity. The advantages of calculating the voltage V instead of $\mathbf{E}$ will be discussed later but for now we can intuitively say that the calculation of a single scalar function must be easier than the calculation of a vector function. If nothing else, the scalar function is single-valued while the vector function requires three components.

1.11.2 The magnetic scalar potential

If we can define an electric scalar potential solely on the basis of the irrotational nature of **E**, then it also follows that we should be able to define a magnetic scalar potential if the magnetic field intensity is irrotational:

$$\nabla\times\mathbf{H} = 0 \tag{1.61}$$

If this is satisfied, the magnetic field intensity can be written in terms of a magnetic scalar potential as

$$\mathbf{H} = -\nabla\Phi \tag{1.62}$$

We note from Eq. (1.2) that this can only be done if both the conduction and displacement currents are zero. Also, note that our definition here is purely mathematical and we have not even insisted that the scalar potential have a physical meaning. In fact, the physical meaning (which actually exists and will be discussed) is not important at this stage: a scalar potential can be defined if the field is irrotational. The magnetic scalar potential is often used for calculation of fields in magnetic materials, especially for static fields and for dynamic fields in nonconducting materials because under these conditions there are no currents in materials and the condition above is satisfied.

1.11.3 The magnetic vector potential

Scalar potentials are defined based on the irrotational properties of the fields. We can also define potentials based on the solenoidal nature of fields (that is on the divergence free condition). Because the magnetic flux density **B** is solenoidal in nature (i.e. $\nabla\cdot\mathbf{B}=0$), it can be derived as the curl of another vector (again, based on the vector algebra property)

$$\mathbf{B} = \nabla\times\mathbf{A} \tag{1.63}$$

A is a vector and because it defines the magnetic flux density we call it a magnetic vector potential. We will not attempt to explain the physical meaning of the magnetic vector potential or its usefulness at this stage. We note however, that if **A** is also solenoidal (that is if $\nabla\cdot\mathbf{A}=0$), we can define a second order vector potential as

$$\mathbf{A} = \nabla\times\mathbf{W} \tag{1.64}$$

and the magnetic flux density can be written in terms of the second order magnetic vector potential as

$$\mathbf{B} = \nabla\times(\nabla\times\mathbf{W})$$

The question wether we should use this or any other potential function is a separate issue. At this point we merely comment on the fact that this can be done. The answer, will lie in the usefulness of a particular relation. In general, the magnetic vector potential, or any other potential function will be used if it is of some advantage to do so.

We should note here that any vector must be defined in terms of its curl and divergence (this is called the Helmholtz's theorem, see problem 1.9). The curl and divergence of **E** and **H** are defined in Eqs. (1.1) through (1.4). For the magnetic vector potential, this must be done separately. One choice is to make the magnetic vector potential solenoidal ($\nabla\cdot\mathbf{A}=0$) as above. However, there are cases when this is not necessarily the best choice.

1.11.4 The electric vector potential

Another vector function formulation employs the electric vector potential (**T**) to describe the current distribution in conducting regions. The electric vector potential can be regarded as a variation for **H** as it has the same curl relation but, generally, a different divergence. The magnetic vector potential is defined as

$$\nabla\times\mathbf{T} = \mathbf{J} \tag{1.65}$$

Since the divergence of **H** is zero it can be redefined as

$$\mathbf{H} = \mathbf{T} - \nabla\Omega \tag{1.66}$$

where Ω is a magnetic scalar potential.

1.12 GAGE CONDITIONS

The use of vector potential functions requires the specification of the curl and the divergence of the potential functions for unique representation of field quantities. The curl is normally specified based on the properties of the field (e.g. in the case of the magnetic vector potential the solenoidal nature of **B** allows the definition of **A** from $\mathbf{B}=\nabla\times\mathbf{A}$). The divergence must then be specified to be consistent with the field equations. We will normally use one of two gage conditions. One is the condition mentioned above

$$\nabla\cdot\mathbf{A} = 0 \tag{1.67}$$

and is known as the Coulomb gage condition. The second is

$$\nabla \cdot \mathbf{A} + \mu\varepsilon \frac{\partial V}{\partial t} = 0 \tag{1.68}$$

and is known as the Lorentz gage condition. We say that this relation is consistent with the field equations because it leads to the continuity equation. The gage conditions may be used as above or in the frequency domain.

1.13 FIELD EQUATIONS IN TERMS OF POTENTIAL FUNCTIONS

1.13.1 Derivation for the magnetic vector potential

After defining various potential functions, we show here a few of the methods in which the potential functions can be used. Later, while discussing numerical methods we will refer to this process as a formulation process. At this stage we simply substitute the potentials defined above into the field equations and discuss the resulting equations. Also, we will point to a few of the main advantages of each of these derived equations. It should be pointed out again, that unless there is an advantage to using a potential function the field variables should be used directly. We start with the vector potential functions because they are more general and then discuss scalar potentials.

Substituting the magnetic vector potential from Eq. (1.63) into Eq. (1.1), gives:

$$\nabla \times \mathbf{E} = -\frac{\partial(\nabla \times \mathbf{A})}{\partial t}$$

and therefore, the electric field intensity can be written as

$$\mathbf{E} = -\frac{\partial \mathbf{A}}{\partial t}$$

This electric field intensity is only the time dependent part of the electric field intensity. In addition, the static term given in Eq. (1.60) also exists

$$\mathbf{E} = -\nabla V$$

The total electric field intensity is the sum of the two electric field intensities

$$\mathbf{E} = -\frac{\partial \mathbf{A}}{\partial t} - \nabla V \tag{1.69}$$

In the case of the magnetic vector potential, substituting Eq. (1.63) into the second of Maxwell's equations (Eq. (1.2)) and using the electric field **E** as represented in Eq. (1.69) gives

$$\nabla\times\left(\frac{1}{\mu}\nabla\times\mathbf{A}\right) = \mathbf{J}_s + \mathbf{J}_e + \varepsilon\frac{\partial}{\partial t}\left(-\nabla V - \frac{\partial\mathbf{A}}{\partial t}\right) \tag{1.70}$$

where the relations $\mathbf{B}=\mu\mathbf{H}$ and $\mathbf{D}=\varepsilon\mathbf{E}$ were used in their scalar form, and the current density **J** was separated into $\mathbf{J}_S$ (applied or source current density) and $\mathbf{J}_e$ (induced or eddy current density). Assuming linear media, and using the vector identity

$$\nabla\times(\nabla\times\mathbf{A}) = \nabla(\nabla\cdot\mathbf{A}) - \nabla^2\mathbf{A} \tag{1.71}$$

Eq. (1.70) becomes:

$$\nabla(\nabla\cdot\mathbf{A}) - \nabla^2\mathbf{A} = \mu\mathbf{J} - \mu\varepsilon\nabla\left(\frac{\partial V}{\partial t}\right) - \mu\varepsilon\left(\frac{\partial^2\mathbf{A}}{\partial t^2}\right) \tag{1.72}$$

This equation represents Maxwell's equations in terms of the magnetic vector potential **A** and the electric scalar potential *V*. If we require an equation in terms of the vector potential only, the scalar potential must be removed. This is done observing that the divergence of **A** has not been specified. As pointed out earlier, we are free to choose the divergence in any way we wish as long as our choice is consistent with the field equations (see problem 1.10). From Eq. (1.72) we note that if we rewrite the equation as

$$\nabla\left[(\nabla\cdot\mathbf{A}) + \mu\varepsilon\left(\frac{\partial V}{\partial t}\right)\right] - \nabla^2\mathbf{A} = \mu\mathbf{J} - \mu\varepsilon\left(\frac{\partial^2\mathbf{A}}{\partial t^2}\right)$$

an immediate choice for the divergence suggests itself as

$$\nabla\cdot\mathbf{A} + \mu\varepsilon\frac{\partial V}{\partial t} = 0 \tag{1.73}$$

Substitution of this expression in Eq. (1.72) yields the following:

$$\nabla^2\mathbf{A} + \mu\mathbf{J} - \mu\varepsilon\left(\frac{\partial^2\mathbf{A}}{\partial t^2}\right) = 0 \tag{1.74}$$

This is the nonhomogeneous wave equation for the magnetic vector potential but, for the purpose of this discussion, the most important point is that we have obtained an equation in terms of a single potential function, representing Maxwell's equations. If this equation is chosen for solution, the magnetic vector potential is obtained. This auxiliary function, although fundamental in electromagnetics is seldom of interest by itself. To calculate field variables, we use the relations in Eqs. (1.63), (1.69) and (1.73) to calculate the magnetic field intensity, the electric field intensity and the electric scalar potential.

1.13.2 Derivation for the electric vector potential

The curl-curl relation in Eq. (1.70) can also be written in terms of the electric vector potential **T** and Ω instead of **A** and V

$$\nabla\times(\nabla\times\mathbf{T}) = \mu\sigma\frac{\partial}{\partial t}(\mathbf{T} - \nabla\Omega) \tag{1.75}$$

where constant conductivity and permeability are assumed. Using the identity in Eq. (1.71) this can be written as

$$\nabla^2\mathbf{T} = -\mu\sigma\left(\frac{\partial\mathbf{T}}{\partial t}\right) + \mu\sigma\,\nabla\Omega = 0 \tag{1.76}$$

The electric vector potential equation has several advantages over the magnetic vector potential formulation. Since **T** is used to describe the current distribution, many *3-D* problems can be solved by computing only a single component of the electric vector potential such as in the case of planar flow of currents. In other cases two or three components may be necessary. An important feature of the method is that in non-conducting regions, **T** is either zero or constant and thus need be calculated only in conducting regions. There are two significant problems associated with this approach. One is the need to specify the interface conditions at the conducting boundaries. The second is associated with the calculation of **T** and Ω in conducting regions. Since $\nabla\Omega$ and **T** are of the same order of magnitude and $\mathbf{H}=\mathbf{T}-\nabla\Omega$, large percentage errors may occur in the calculation of the magnitude of **H** in areas of small skin depth.

1.14 DERIVATION IN TERMS OF SCALAR POTENTIALS

A representation in terms of the electric scalar potential, is obtained starting with Maxwell's third equation:

$$\nabla\cdot\varepsilon\mathbf{E} = \rho \tag{1.77}$$

where the relation $\mathbf{D}=\varepsilon\mathbf{E}$ was again used. Substitution of the value for **E** from Eq. (1.69) gives

$$\nabla \cdot \varepsilon\left(-\nabla V - \frac{\partial \mathbf{A}}{\partial t}\right) = \rho$$

Assuming ε is constant and can be taken outside the divergence gives

$$\nabla \cdot \nabla V + \frac{\partial}{\partial t}(\nabla \cdot \mathbf{A}) = -\frac{\rho}{\varepsilon}$$

Since $\nabla . \nabla V \equiv \nabla^2 V$, and using again Eq. (1.73) for the divergence of **A** we get

$$\nabla^2 V - \mu\varepsilon \frac{\partial^2 V}{\partial t^2} = -\frac{\rho}{\varepsilon} \qquad (1.78)$$

This is again a nonhomogeneous wave equation in terms of the electric scalar potential V. Note however that there is a fundamental difference between Eq. (1.74) and Eq. (1.78). The first contains current densities. The current density is a vector and therefore, the potential function required for solution must also be a vector. In the second case, we have only scalar sources (charge density) and a scalar potential is sufficient. In this sense, our choice of potential function is dictated by sources. We are free to choose any potential function as long as it represents the sources properly.

Simpler representations are possible for static fields. These are all based on the fact that the electric or magnetic field are curl-free under certain conditions. For the curl-free electric field we can write the following

$$\mathbf{E} = -\nabla V \qquad (1.79)$$

To define the electric field completely, we must also define the divergence of the electric field using Eq. (1.3)

$$\nabla \cdot \mathbf{E} = \frac{\rho}{\varepsilon} \qquad (1.80)$$

Calculating the divergence of Eq. (1.79) and equating with Eq. (1.80) gives

$$\nabla \cdot \mathbf{E} = \frac{\rho}{\varepsilon} = -\nabla \cdot \nabla V$$

and, since $\nabla . \nabla V \equiv \nabla^2 V$, we get

$$\nabla^2 V = -\frac{\rho}{\varepsilon}$$

or, in a more familiar form, in Cartesian coordinates

$$\frac{\partial^2 V}{\partial x^2} + \frac{\partial^2 V}{\partial y^2} + \frac{\partial^2 V}{\partial z^2} = -\frac{\rho}{\varepsilon} \tag{1.81}$$

This is the well-known Poisson equation which, in this case describes electrostatic phenomena. If the physical system has no sources, we may be able to simplify the equation (and its solution) by removing the source term. Under this condition we get the Laplace form of the equation

$$\frac{\partial^2 V}{\partial x^2} + \frac{\partial^2 V}{\partial y^2} + \frac{\partial^2 V}{\partial z^2} = 0 \tag{1.82}$$

Using identical steps, we can write the magnetic scalar potential representation.

If in a region of space the current density is zero, Eq. (1.2) becomes

$$\nabla \times \mathbf{H} = 0$$

In this case the field intensity can be defined as the gradient of a magnetic scalar potential as

$$\mathbf{H} = -\nabla \Phi \tag{1.83}$$

This is analogous to the definition of the electrostatic field as $\mathbf{E}=-\nabla V$, and therefore, the equivalent form of Eq. (1.82) for the magnetic scalar potential is

$$\frac{\partial^2 \Phi}{\partial x^2} + \frac{\partial^2 \Phi}{\partial y^2} + \frac{\partial^2 \Phi}{\partial z^2} = 0 \tag{1.84}$$

We note however that in this particular case, a Poisson-type equation cannot be used since the divergence-free condition for the magnetic field intensity in Eq. (1.4) is used explicitly. This of course is not surprising since we know that the magnetic field does not have individual poles (the equivalent to charge in the magnetic case) as sources.

A few scalar and vector potential representation are presented above. There are many others and some are presented in later chapters. The basic question of why we resort to this representation has not yet been answered directly. We give here a simple, but incomplete answer. The reason it is incomplete is that sometimes, a particular choice is dictated by the numerical technique we use and has little to do with field theory or any other consideration.

Why a scalar potential should be used whenever possible is obvious; instead of solving a vector equation (with it's three components in three dimensions or two components in two dimensions) we solve a single scalar equation. The economy of solution is obvious. There are however some disadvantages associated with this approach. First, if the field variables are needed, we must calculate gradients of potentials. This means calculating derivatives as in Eqs.

(1.62) or (1.79). Since the calculations are numerical, we must calculate numerical derivatives. This can lead to large errors.

As for vector potential representation, an example of how this may be useful is given now.

Consider a two-dimensional case in which the magnetic field intensity varies in the x and y direction but does not vary in the z direction. By writing the magnetic flux density explicitly, we have

$$\mathbf{B} = \hat{\mathbf{x}}\left(\frac{\partial A_z}{\partial y} - \frac{\partial A_y}{\partial z}\right) + \hat{\mathbf{y}}\left(\frac{\partial A_x}{\partial z} - \frac{\partial A_z}{\partial x}\right) + \hat{\mathbf{z}}\left(\frac{\partial A_y}{\partial x} - \frac{\partial A_x}{\partial y}\right)$$

If $\mathbf{B}$ is two-dimensional (in the x-y plane), the last term must be zero. This means that A_y and A_x are constant in the x and y directions and, therefore, their derivatives with respect to any coordinate are zero. This gives:

$$\mathbf{B} = \hat{\mathbf{x}}\frac{\partial A_z}{\partial y} - \hat{\mathbf{y}}\frac{\partial A_z}{\partial x} \tag{1.85}$$

Thus, instead of requiring a two component vector ($\mathbf{B}$), a single component vector (A_z) is sufficient for solution in the two-dimensional case. Solution of a field problem in two dimensions in terms of the magnetic vector potential can be done as a scalar solution rather than a vector solution, considerably simplifying the process. There are other advantages as well but we defer those to sections on the numerical solution of fields where, before choosing a particular potential (or combination of potentials) we will have to justify our choice. We note again that the original field variables may always be chosen rather than introducing potentials.

We complete this section by pointing out that in some cases a representation in terms of more than one potential function everywhere in the solution domain may be required. To show an example, consider the relation in Eq. (1.72)

$$\nabla(\nabla\cdot\mathbf{A}) - \nabla^2\mathbf{A} = \mu\mathbf{J} - \mu\varepsilon\nabla\left(\frac{\partial V}{\partial t}\right) - \mu\varepsilon\left(\frac{\partial^2\mathbf{A}}{\partial t^2}\right) \tag{1.86}$$

Instead of using the Lorentz gage, we can use Coulomb's gage and write

$$\nabla^2\mathbf{A} + \mu\mathbf{J} - \mu\varepsilon\nabla\left(\frac{\partial V}{\partial t}\right) - \mu\varepsilon\left(\frac{\partial^2\mathbf{A}}{\partial t^2}\right) = 0 \tag{1.87}$$

If we choose this representation, another equation must be added since Eq. (1.87) can only represent three relations for the three components of $\mathbf{A}$. The fourth equation is needed to represent V. A simple choice is to impose the divergence-free condition on the total current in the solution domain. Since currents must flow in closed paths, their divergence is always zero. This condition may be written as

$$\nabla.\left[\mu\mathbf{J} - \mu\varepsilon\nabla\left(\frac{\partial V}{\partial t}\right) - \mu\varepsilon\left(\frac{\partial^2\mathbf{A}}{\partial t^2}\right)\right] = 0 \qquad (1.88)$$

Thus, a coupled system of equations consisting of Eqs. (1.87) and (1.88) must now be solved. The reasons why we may choose to do so will become apparent in chapter 7.

In addition, note that the current density given in Eqs. (1.87) and (1.88), contains all current densities except for displacement currents. In low frequency *NDT* applications only conduction current densities are important and these include source current densities and eddy current densities

$$\mathbf{J} = \mathbf{J}_s + \mathbf{J}_e$$

The eddy current density can be written in terms of the electric field intensity and conductivity as

$$\mathbf{J}_e = \sigma\mathbf{E} = \sigma(-\nabla V - \frac{\partial\mathbf{A}}{\partial t})$$

Thus, Eq. (1.87) can be written as

$$\nabla^2\mathbf{A} + \mu\mathbf{J}_s + \mu\sigma\left(-\nabla V - \frac{\partial\mathbf{A}}{\partial\tau}\right) - \mu\varepsilon\nabla\left(\frac{\partial V}{\partial t}\right) - \mu\varepsilon\left(\frac{\partial^2\mathbf{A}}{\partial t^2}\right) = 0 \qquad (1.89)$$

while Eq. (1.88) is

$$\nabla.\left[\mu\mathbf{J}_s + \mu\sigma\left(-\nabla V - \frac{\partial\mathbf{A}}{\partial\tau}\right) - \mu\varepsilon\nabla\left(\frac{\partial V}{\partial t}\right) - \mu\varepsilon\left(\frac{\partial^2\mathbf{A}}{\partial t^2}\right)\right] = 0 \qquad (1.90)$$

To generalize the discussion in sections 1.13 and 1.14 the following can be stated:

1. A scalar potential can be used in a source-free domain.
2. A scalar potential may be used in a domain which only contains a scalar source such as charge densities.
3. A vector potential must be used in regions that contain current densities.

The field equations defined above assume that one of the potentials is sufficient for solution and that we may represent the solution by that potential alone. This is true in general but not always the most economical method. In most cases, the region that includes currents is small while most of the solution domain contains no currents. Based on the conclusions above, we must still use a vector potential for representation everywhere.

Alternatively we may use a vector potential in current regions and a scalar potential in regions devoid of currents. This can be done provided that the two solutions can be correctly matched on the interfaces between the two domains.

1.14.1 Other scalar potential representations

In Eq. (1.84) we obtained a representation of the magnetic field in terms of the magnetic scalar potential. Since the gradient of Φ defines the total field in the whole region it is called the "total scalar potential".

For regions containing current distributions the scalar potential cannot be used as in Eq. (1.83). In this case the field is written as the sum of two parts, one arising from the currents in the region and the other representing the rest of the field.

$$\mathbf{H} = \mathbf{H}_c + \mathbf{H}_m$$

$\mathbf{H}_m$ satisfies the curl-free condition $\nabla\times\mathbf{H}_m=0$ and therefore can be represented by a scalar potential. This scalar potential which we denote Ψ, does not define the total field $\mathbf{H}$ and is therefore called the "reduced scalar potential".

$$\mathbf{H}_m = -\nabla\Psi$$

The second term, $\mathbf{H}_c$ is due to current distributions and cannot be represented by a scalar potential. It has to be determined analytically from the current distributions, by direct integration before the scalar potential ψ can be calculated. The main advantage of this method is that it allows scalar representation of current regions but, in addition to the need for analytical calculation of the fields due to currents, the method suffers from severe problems of cancellation. That is, the two terms $\mathbf{H}_m$ and $\mathbf{H}_c$ are of the same order of magnitude in magnetic regions with the net result that the total field, calculated from the two terms tends to be inaccurate. This happens since the two terms are calculated separately and then added.

One approach to removing this difficulty is to use the reduced scalar potential Ψ in current regions and the total scalar potential Φ in current-free regions. This however, requires the coupling of the two potentials at the interfaces of the two regions. In a current region of constant permeability, the following can be used:

$$-\nabla\mu\nabla\Psi + \nabla\mu\mathbf{H}_c = 0 \qquad (1.91)$$

and in current-free regions containing magnetic materials, the total scalar potential Φ is used:

$$-\nabla\mu\nabla\Phi + \nabla\mu\mathbf{H}_m = 0 \tag{1.92}$$

The coupling between Φ and Ψ is imposed by requiring continuity of the normal components of **B** and the tangential components of **H** at the interfaces of the regions.

Other approaches to the magnetostatic field problem using scalar potentials, related to the above two methods exist. One is the use of the total scalar potential outside current regions and a vector potential in current regions avoiding the need for analytic integration of the current contribution.

The main disadvantage in the use of the scalar potentials is in their inability to specify current distributions. The coupling of the scalar potential with analytic calculations in current regions seems to be adequate for magnetostatic problems but for eddy current calculations the current has to be specified by a vector function. These methods are therefore generally limited to magnetostatic calculations.

1.15 TIME-HARMONIC FIELDS

In many applications, the sources of the electromagnetic field are sinusoidal. While all considerations given above apply to these fields as well, the sinusoidal form offers some advantages in terms of formulation and solution of the field equations. The main reason for this is that under the phasor transformation (see problem 1.13), the equations look like static equations. We therefore call these equations quasi-static. If the electromagnetic field vectors are characterized by sinusoidal variation in time, we can place their phasor equivalents in Eqs. (1.1) through (1.4). As an example, the electric field intensity is written as

$$\mathbf{E} = \mathbf{E}_0 e^{j\omega t}$$

where the vector notation has been retained but now **E** is a phasor. ω is the angular frequency of the field and j is the complex number. The derivative with respect to time is therefore

$$\frac{\partial \mathbf{E}}{\partial t} = j\omega\mathbf{E}_0 e^{j\omega t} = j\omega\mathbf{E} \tag{1.93}$$

By substituting $j\omega\mathbf{E}$ for $\partial\mathbf{E}/\partial t$ and $j\omega\mathbf{B}$ for $\partial\mathbf{B}/\partial t$, and using the notation above we get the time harmonic form of Maxwell's equations:

$$\nabla\times\mathbf{E} = -j\omega\mathbf{B} \tag{1.94}$$

$$\nabla\times\mathbf{H} = \mathbf{J} + j\omega\mathbf{D} \tag{1.95}$$

$$\nabla \cdot \mathbf{D} = \rho \tag{1.96}$$

$$\nabla \cdot \mathbf{B} = 0 \tag{1.97}$$

Now, any of the time dependent equations we used in previous sections may be written in time-harmonic form by replacing $\partial/\partial t$ with $j\omega$. As an example, the general time dependent equation for the magnetic vector potential (Eq. (1.74)) becomes

$$\nabla^2 \mathbf{A} + \mu \mathbf{J} + \omega^2 \mu\varepsilon\, \mathbf{A} = 0$$

While the fields are still time dependent, the solution will be in terms of **A** which now is a phasor and is only frequency dependent. The equation is therefore "static-like" in that no time derivatives are present. The solution however, has nothing to do with static fields. The phasor is

$$\mathbf{A} = \mathbf{A}_0 e^{j\omega t} \tag{1.98}$$

and, if the time domain representation is needed we must use the phasor transformation to do so. This is accomplished by writing

$$\mathbf{A}(t) = Re\left[\mathbf{A}_0 e^{j\omega t}\right] \tag{1.99}$$

In other words, the real part of the phasor must be evaluated.

This transformation also points out to a very important aspect of solution in the frequency domain. While in the time domain the quantities we deal with are all real, in the frequency domain they are all complex. Thus, we have exchanged a solution in the time domain for a "static-like" solution in complex variables. As with all previous formulations, we should only do so if there is an advantage in computation.

1.16 NONLINEAR FIELDS

Nonlinear magnetic fields are often used in *NDT*, primarily in static methods of testing. These include active and residual leakage field methods. From the field equations point of view, the only difference between linear and nonlinear fields is in the constitutive relations for the magnetic field. The constitutive relation in Eq. (1.12) is now written as

$$\mathbf{B} = \mu(H)\mathbf{H} \tag{1.100}$$

where the dependence can be simple such as that defined by the initial magnetization curve in Figure 1.2 or the full hysteresis curve in Figure 1.3.

Even though the change in equations seems to be minor, the practical aspects of this additional complexity are major. However, most difficulties are computational, not theoretical and therefore there is no need to modify any of the equations given in previous sections.

Nonlinear methods are used for calculation of static fields but not eddy currents and wave propagation since nonlinearities are not often encountered in these domains.

1.17 PLANE WAVES AND SCATTERING

1.17.1 The wave equation

In section 1.2 we wrote Maxwell's equations in terms of the magnetic vector potential as a general time dependent partial differential equation. In section 1.15 the same equations were written in the frequency domain. Nothing was said about their solution or indeed, about the properties of the equations. Also, we noted that the choice of the magnetic vector potential was, in a way, arbitrary and that we could have written the equations in terms of **E**, **H**, **B**, or **D**, or in terms of any other "properly" defined vector or scalar function. In this section we discuss, briefly, the so-called wave equations. They are called wave equations because their solutions lead to propagating waves. To obtain a wave equation in any of the vector field variables, all that is needed is to use Maxwell's equations and manipulate them so that the resulting partial differential equation is an equation in a single variable. For example, if a wave equation in terms of **E** is needed, we first take the curl on both sides of Eq. (1.1) and substitute Eq. (1.2) to eliminate **H** (see problem 1.14). The result is

$$\nabla^2\mathbf{E} - \mu\varepsilon\frac{\partial^2\mathbf{E}}{\partial t^2} = \mu\frac{\partial\mathbf{J}}{\partial t} + \nabla\left(\frac{\rho}{\varepsilon}\right) \tag{1.101}$$

Similarly, we obtain for **H** and **A** (see problem 1.15 and 1.16):

$$\nabla^2\mathbf{H} - \mu\varepsilon\frac{\partial^2\mathbf{H}}{\partial t^2} = \nabla\times\mathbf{J} \tag{1.102}$$

$$\nabla^2\mathbf{A} - \mu\varepsilon\left(\frac{\partial^2\mathbf{A}}{\partial t^2}\right) = -\mu\mathbf{J} \tag{1.103}$$

or, we can write a wave equation in terms of the electric scalar potential (see problem 1.17) as:

$$\nabla^2 V - \mu\varepsilon \frac{\partial^2 V}{\partial t^2} = -\frac{\rho}{\varepsilon} \tag{1.104}$$

In these equations, and others like these, the right hand side represents the sources of the fields. The solution to these equations is a propagating wave. It is the interaction of waves and materials that is often our interest in nondestructive testing.

Equations (1.101), (1.102), (1.103), and (1.104) are second order, partial differential equations. Therefore, they have standard solutions. This does not imply that the solution to a particular physical situation is simple. This solution requires imposition of boundary conditions which may not be possible to do analytically. However, the equations can certainly be solved numerically.

1.17.2 The time-harmonic wave equation

In the time-harmonic domain, the phasor notation is again adopted (i.e. $\partial/\partial t = j\omega$) The wave equations given above are:

$$\nabla^2 \mathbf{E} + \omega^2 \mu\varepsilon \mathbf{E} = j\omega\mu \mathbf{J} + \nabla\left(\frac{\rho}{\varepsilon}\right) \tag{1.105}$$

$$\nabla^2 \mathbf{H} + \omega^2 \mu\varepsilon \mathbf{H} = \nabla\times\mathbf{J} \tag{1.106}$$

$$\nabla^2 \mathbf{A} + \omega^2 \mu\varepsilon \mathbf{A} = -\mu\mathbf{J} \tag{1.107}$$

$$\nabla^2 \mathrm{V} + \omega^2 \mu\varepsilon \mathrm{V} = -\frac{\rho}{\varepsilon} \tag{1.108}$$

In these equations the sources are left essentially as in Eqs. (1.101) through (1.104), although, sometimes it will become convenient to write the sources in different forms. As an example, in Eq. (1.105), the term $j\omega\mu\mathbf{J}$ is often written as $j\omega\mu\sigma\mathbf{E}$ by using the constitutive relation for **J**. The main advantage in doing so is to show that this term is a loss term by incorporation of the conductivity explicitly. Also, any of these sources may not exist in a particular situation.

1.17.3 The Helmholtz equations

Particularly useful forms of the wave equations are the source-free (or homogeneous) wave equations. These are obtained by removing the sources from Eqs. (1.105) through (1.108). The implication is that now we solve the problem without sources and therefore the solutions are characteristic of geometry rather than a particular source:

$$\nabla^2\mathbf{E} + \omega^2\mu\varepsilon\mathbf{E} = 0 \tag{1.109}$$

$$\nabla^2\mathbf{H} + \omega^2\mu\varepsilon\mathbf{H} = 0 \tag{1.110}$$

$$\nabla^2\mathbf{A} + \omega^2\mu\varepsilon\mathbf{A} = 0 \tag{1.111}$$

$$\nabla^2\mathrm{V} + \omega^2\mu\varepsilon\mathrm{V} = 0 \tag{1.112}$$

These are known as the homogeneous or Helmholtz type equations. Note that the general permittivity and permeability instead of free space permittivity and permeability are used to indicate that, under the conditions of no sources and no losses, these apply in general media. If we wish to include lossy materials, these must be included through the concepts of complex permittivity and complex permeability. We note, however, that the equations are linear. Their solution is relatively simple compared to the general wave equations and we shall make considerable use of these in subsequent chapters. It is also possible to write similar equations in terms of **D** or **B**.

1.18 PROPAGATION OF WAVES: PLANE WAVES IN LOSSLESS DIELECTRICS

The solution to any of the equations above is necessary to describe behavior of electromagnetic fields in materials if propagation effects are important. This becomes particularly useful in testing in the microwave domain where propagation effects are essential. We will describe next some of the fundamental properties of waves as obtained from the solution of the time-harmonic wave equation. For this purpose we assume that the waves are uniform plane waves and, for simplicity, look at the properties of a wave, propagating in a given direction with a single, transverse component. A uniform plane wave is a wave (that is: a solution to a wave equation), in which the electric and the magnetic field intensities are constant in magnitude and phase and each is in a constant direction on planes perpendicular to the direction of propagation. The assumption of plane waves simplifies discussion but does not, in any way, restrict it.

We only present here the time-harmonic case because it is more often used. However, the same can be done, if required, in the time domain. Also, since any time dependent wave can be transformed into the frequency domain, the solution is rather general. We start with the one-dimensional wave equation with the electric field in the x direction and propagating in the z direction. The wave equation is

$$\frac{d^2E_x}{dz^2} + \omega^2\mu\varepsilon E_x = 0 \tag{1.113}$$

where regular derivatives are used instead of partial derivatives since E_x depends on z only. Before writing the solution we observe the following:

1. The term $\omega^2\mu\varepsilon$ has units of $[rad/m]^2$. For this reason, we normally define a quantity

$$k = \omega\sqrt{\mu\varepsilon} \tag{1.114}$$

and call it the "wave number" or a "phase constant". The term wave number indicates the number of wavelengths in one cycle of the wave and will become apparent shortly but, for now, we note that as a phase constant, it defines how the wave propagates. Analogously, in free space, we define the "free space wave number" as

$$k_0 = \omega\sqrt{\mu_0\varepsilon_0}$$

where μ_0 and ε_0 are the permeability and permittivity of vacuum. k_0 is a real number in vacuum, but, for a general material k is normally a complex number.

2. The term $\mu\varepsilon$ has units of $[sec/m]^2$. Therefore, a velocity can be defined as

$$v = \frac{1}{\sqrt{\mu\varepsilon}} \tag{1.115}$$

This velocity, is known as the velocity of propagation of the wave or phase velocity. In free space this velocity is the speed of light

$$c = \frac{1}{\sqrt{\mu_0\varepsilon_0}}$$

The wave number can now be written as:

$$k = \omega\sqrt{\mu\varepsilon} = \frac{\omega}{v} \qquad or \qquad k_0 = \omega\sqrt{\mu_0\varepsilon_0} = \frac{\omega}{c}$$

3. The square root of the ratio between the product of permeability and permittivity of any material and that of the product of permeability and permittivity of free space is the index of refraction of the material.

$$n = \sqrt{\frac{\mu\varepsilon}{\mu_0\varepsilon_0}} = \sqrt{\mu_r\varepsilon_r} \tag{1.116}$$

   where μ_r and ε_r are the relative permeability and relative permittivity respectively ($\mu=\mu_0\mu_r$, $\varepsilon=\varepsilon_0\varepsilon_r$).

The solution to Eq. (1.113) is

$$E_x(z) = E_0^+ e^{-jkz} + E_0^- e^{jkz} \tag{1.117}$$

or, in the time domain

$$E_x(z,t) = E_0^+ \cos(\omega t - kz + \phi) + E_0^- \cos(\omega t + kz + \phi) \tag{1.118}$$

where E_0^+ and E_0^- are constants to be determined from the boundary conditions of the problem. The notation $(+)$ and $(-)$ indicates that the first term is a propagating wave in the positive z direction and the second a propagating wave in the negative z direction. The constants can be complex and are arbitrary. An arbitrary phase angle ϕ is added for completeness.

For a wave that only propagates outward from the source (no reflection from boundaries) we have

$$E_x(z) = E_0^+ e^{-jkz} \tag{1.119}$$

or, in the time domain

$$E_x(z,t) = E_0^+ \cos(\omega t - kz + \phi) \tag{1.120}$$

This wave propagates with a phase velocity equal to

$$v_p = \frac{\omega}{k} = \frac{1}{\sqrt{\mu\varepsilon}} \quad [m/sec]$$

This velocity is the same velocity calculated in Eq. (1.115). Thus phase velocity in free space is equal to the speed of light. Phase velocity is the speed of propagation of the electromagnetic wave. Now we can also define the wavelength λ (in meters) which is that distance the wave travels in one cycle

$$\lambda = \frac{v}{f} = \frac{2\pi}{k} \qquad [m] \tag{1.121}$$

The name "wave number" now becomes obvious. If we write

$$k = \frac{2\pi}{\lambda} \tag{1.122}$$

k is the number of wavelengths in one cycle.

All this was a consequence of the solution we found for the electric field. However, an electromagnetic wave is always composed of an electric and a magnetic field. To find the magnetic field we use Maxwell's first equation (Eq. (1.94)). For the conditions described above (single component electric field intensity), Eq. (1.94) becomes

$$\frac{\partial E_x}{\partial z} = -j\omega\mu H_y$$

By writing this for a forward propagating wave, we have for H_y

$$H_y^+ = \frac{j}{\omega\mu}\frac{\partial E_x^+}{\partial z}$$

and, substituting the solution for E_x^+ from Eq. (1.117), gives

$$H_y^+(z) = \frac{k}{\omega\mu}E_x^+(z) \tag{1.123}$$

The ratio between $H(z)$ and $E(z)$, has units of $(1/\Omega)$. We define the quantity $\omega\mu/k$ as

$$\eta = \frac{\omega\mu}{k} = \sqrt{\frac{\mu}{\varepsilon}} \qquad [\Omega] \tag{1.124}$$

This quantity is an impedance and is called the intrinsic impedance of the material. The intrinsic impedance of free space is

$$\eta_0 = \sqrt{\frac{\mu_0}{\varepsilon_0}} = 377 \qquad [\Omega] \tag{1.125}$$

The intrinsic impedance is a complex number in general media. Eq. (1.123) is now

$$H_y^+(z) = \frac{1}{\eta} E_x^+(z) \tag{1.126}$$

Both **H** and **E** propagate in the same direction. They are orthogonal to each other and to the direction of propagation. This means that **E** and **H** are transverse electromagnetic (*TEM*) waves.

The discussion above is restricted to a single component of the electric and magnetic field intensities as well as to linear media. However, we can do the same with any other component of the electric or magnetic field, and use any other direction of propagation. The restriction to lossless media, is also minor in that all that is necessary is replacement of the permittivity with the complex permittivity. This and the effect of losses are discussed next. As for the assumption of linearity, it is normally a good approximation at high frequencies.

1.19 PROPAGATION OF PLANE WAVES IN LOSSY MEDIA

1.19.1 Losses in materials

The loss mechanism in materials is most easily described through the use of complex permittivity since it includes two loss components; one loss component is due to conduction currents, the other due to dielectric, or polarization losses. The complex permittivity is defined as

$$\varepsilon = \varepsilon' - j\left(\frac{\omega\varepsilon'' + \sigma}{\omega}\right) \tag{1.127}$$

The real part of the complex permittivity is the term we normally associate with dielectrics as the dielectric constant. In lossless materials, the imaginary part is zero and our definition is correct. In lossy materials, the imaginary part is nonzero and has two parts; the first, due to ε'' is the dielectric loss while the second due to σ is the conduction or Ohmic loss, although, in practice the two types of losses are indistinguishable.

The degree to which a dielectric is lossy is quantified by the ratio of the imaginary (lossy) and the real (lossless) parts of Eq. (1.127). This is written as

$$\tan\delta = \frac{\omega\varepsilon'' + \sigma}{\omega\varepsilon'} \tag{1.128}$$

and is known as the loss tangent or the loss factor of the material. In most dielectrics, Ohmic losses are relatively small while polarization losses are small in "good" dielectrics. In lossy dielectrics, at microwave frequencies, polarization losses dominate. In high conductivity materials, Ohmic losses are dominant.

1.19.2 Propagation of waves in lossy dielectrics

A lossy dielectric is defined as a material with small but not negligible losses regardless of the sources of losses (conduction as well as dielectric losses).

To analyze the effect of losses in dielectrics, we substitute the complex permittivity into the field equations and, based on this redefine the basic properties of the wave solution we outlined in section 1.18. With these steps, Eq. (1.109) becomes

$$\nabla^2 \mathbf{E} + \omega^2 \mu\varepsilon \left(1 - j\,\frac{\omega\varepsilon'' + \sigma}{\omega\varepsilon} \right) \mathbf{E} = 0$$

Since$(\omega\varepsilon''+\sigma)$ can be viewed as an equivalent conductivity and $\varepsilon=\varepsilon'$, we replace $\omega\varepsilon''+\sigma$ by σ and ε' by ε. Now, the wave number can be written as:

$$k = \omega\sqrt{\mu\varepsilon}\sqrt{\left(1 - \frac{j\sigma}{\omega\varepsilon}\right)}$$

A propagation constant is defined through the following relation

$$-\gamma^2 = k^2$$

or

$$\gamma = j\omega\sqrt{\mu\varepsilon}\sqrt{\left(1 - \frac{j\sigma}{\omega\varepsilon}\right)} \tag{1.129}$$

This term can now be substituted for jk in Eq. (1.113), using the same assumptions made for the lossless case. The one-dimensional wave equation can now be written as

$$\frac{d^2 E_x}{dz^2} - \gamma^2 E_x = 0 \tag{1.130}$$

and the general solution has the same two wave components discussed in the previous section: one travelling in the positive z direction, the other in the negative z direction

$$E_x(z) = E_0^+ e^{-\gamma z} + E_0^- e^{+\gamma z} \tag{1.131}$$

For practical purposes, we assume only an outgoing wave exists:

$$E_x(z) = E_0^+ e^{-\gamma z} \tag{1.132}$$

Denoting

$$\gamma = \alpha + j\beta$$

we get

$$E_x(z) = E_0^+ e^{-(\alpha + j\beta)z} = E_0^+ e^{-\alpha z} e^{-j\beta z} \tag{1.133}$$

The propagating wave has the same form as Eq. (1.119) if β is replaced by k and the exponential term $e^{-\alpha z}$ multiplies the expression. This is therefore a wave, propagating in the positive z direction, with a phase velocity v and with an exponentially decaying amplitude. The phase velocity is now

$$v = \frac{\omega}{\beta}$$

and the wavelength is

$$\lambda = \frac{2\pi}{\beta}$$

The magnetic field intensity is written directly, using Eq. (1.124), as

$$H_y^+(z) = \frac{1}{\eta} E_x^+(z)$$

where the intrinsic impedance is

$$\eta = \frac{j\omega\mu}{\gamma} \tag{1.134}$$

and is a complex number. The imaginary part of the intrinsic impedance is due to losses

$$\eta = \frac{j\omega\mu}{\gamma} = \sqrt{\frac{\mu}{\varepsilon\left(1 - \frac{j\sigma}{\omega\varepsilon}\right)}} = \eta_n\left(1 + \frac{j\sigma}{\omega\varepsilon}\right) \tag{1.135}$$

where η_n is the lossless intrinsic impedance. The imaginary part (reactive) is normally small and is often neglected for low loss dielectrics.

The exponential term $e^{-\alpha z}$ is an attenuation term. The wave is attenuated as it propagates in space. The constant α is therefore an attenuation constant. The attenuation constant is measured in *Nepers/meter*. The *Neper* is a dimensionless constant. A *Neper* is defined as that attenuation constant for which the wave is attenuated to $1/e$ of its amplitude in one meter.

1.19.3 Propagation of waves in low loss dielectrics

A low loss dielectric is a material for which $\sigma/\omega\varepsilon << 1$. The propagation constant can now be approximated by

$$\gamma = j\omega\sqrt{\mu\varepsilon}\sqrt{\left(1-\frac{j\sigma}{\omega\varepsilon}\right)} \approx j\omega\sqrt{\mu\varepsilon}\left(1-\frac{j\sigma}{2\omega\varepsilon}\right) \tag{1.136}$$

The attenuation constant is therefore

$$\alpha \approx \frac{\sigma}{2}\sqrt{\frac{\mu}{\varepsilon}} \tag{1.137}$$

and the phase constant is

$$\beta \approx \omega\sqrt{\mu\varepsilon} \tag{1.138}$$

The phase constant for low loss dielectrics is essentially unchanged from that for lossless dielectrics.

The intrinsic impedance in low loss dielectrics is complex and can be approximated as

$$\eta = \eta_n\left(1+\frac{j\sigma}{\omega\varepsilon}\right) \tag{1.139}$$

where η_n is the no-loss intrinsic impedance for the same material (i.e. with the same μ and ε' but $\varepsilon''=0$). The phase velocity remains essentially unchanged from that in the lossless case since β changes very little $(v=\omega/\beta)$.

1.19.4 Propagation of waves in conductors

In highly conductive materials the losses are high; that is: $\sigma >> \omega\varepsilon$ or, $\varepsilon'' >> \varepsilon'$. This is equivalent to saying that displacement currents in Maxwell's equations are low compared to conduction currents. Note that high conductivity does not mean perfect conductivity. In the case of high conductivity materials, the complex propagation constant can be written as

$$\gamma \approx j\omega\sqrt{\mu\varepsilon}\sqrt{-\frac{j\sigma}{\omega\varepsilon}} = (1+j)\sqrt{\frac{\omega\mu\sigma}{2}} \tag{1.140}$$

where the following relation was used

$$\sqrt{j} = (1+j)/\sqrt{2} \tag{1.141}$$

From this we get

$$\alpha = \beta = \sqrt{\frac{\omega\mu\sigma}{2}} = \frac{1}{\delta} \tag{1.142}$$

The attenuation and phase constants are equal and are very large. The wave is attenuated rapidly to the point where propagation in conducting media can only exist within short distances. The propagating wave can now be written as

$$E_x(z) = E_0^+ e^{-z/\delta} e^{-jz/\delta} \tag{1.143}$$

where the term

$$\delta = \sqrt{\frac{2}{\omega\mu\sigma}} \tag{1.144}$$

was used here and in Eq. (1.141). This constant is known as the skin depth or depth of penetration. It is defined as that distance through which the amplitude of a plane wave is attenuated to $1/e$ of its original amplitude. The skin depth in conductors is normally small. In the microwave range it can be of the order of a few microns (depending on material and frequency).

The phase velocity in good conductors is

$$v = \frac{\omega}{\beta} = \omega\delta = \sqrt{\frac{2\pi}{\mu\sigma}} \tag{1.145}$$

and is obviously small compared to the phase velocity in dielectrics or free space.

The intrinsic impedance is

$$\eta = \sqrt{\frac{\mu}{\varepsilon}} \approx \sqrt{\frac{j\omega\mu}{\sigma}} = (1+j)\frac{1}{\sigma\delta} \tag{1.146}$$

The phase angle of the intrinsic impedance is $45°$. This is characteristic of good conductors for which the magnetic field intensity lags behind the electric field intensity by $45°$.

The wavelength also changes dramatically compared to free space or lossless dielectrics. It is given by

$$\lambda = \frac{2\pi}{\beta} = 2\pi\delta \qquad (1.147)$$

Since δ is very small, the wavelength is also very short.

The skin depth is very useful in low frequency applications since it defines the range in which currents or fields can be assumed to exist. In high frequency applications, it defines the depth to which waves propagate. Materials with small skin depths cannot support a propagating wave because of the large attenuation. Also, the skin depth defines the range of many tests with particular difficulties in eddy current testing in ferromagnetic materials.

An important point with regard to skin depth, one that is often overlooked is that skin depth is only properly defined for plane waves and only for sinusoidal excitations. We will take up this point later on in this chapter and in chapter 2.

1.20 MICROWAVES, WAVEGUIDES AND RESONANT CAVITIES

1.20.1 Waveguides

The propagation of uniform plane waves as described in the previous sections are assumed to occur in unbounded space. Also, the fields were transverse to the direction of propagation (that is, none of the components of the electric or magnetic field intensity was in the direction of propagation). We now discuss the propagation of waves in guiding structures. The purpose of these structures is to guide waves, with particular characteristics, along the structure. These are called waveguides. The idea of a waveguide is not limited to microwaves but in practice, waveguides are only used in the microwave domain and therefore, we only use these ideas for modeling of fields in this domain. As a direct extension to waveguides we also discuss microwave cavity resonators. These are essentially sections of waveguides shorted at one or both ends to create an open or a closed cavity resonator.

A wave with an electric field perpendicular to the direction of propagation, but with a component of the magnetic field in the direction of propagation is called a transverse electric *(TE)* wave. Similarly, if the magnetic field is perpendicular to the direction of propagation, and the electric field has a component in the direction of propagation, this is a transverse magnetic *(TM)* wave. While we do not define here the conditions under which one mode of propagation or the other may exist, we note that both modes can be excited in waveguides. In terms of solutions in the waveguide or cavity resonator the two modes are obtained by solving similar equations. Since *TM* modes are characterized by transverse magnetic fields and have a longitudinal component (in the direction of propagation) of the electric field, we write the equations for this mode in terms of the longitudinal component. Similarly for *TE* modes, the equation will be written in terms of the longitudinal magnetic field.

To define directions, we assume a rectangular waveguide as shown in Figure 1.9. The waveguide is directed in the z direction. This is the direction along which waves are guided. The transverse directions are x and y. Figure 1.9a also shows the necessary boundary conditions on the conducting surfaces for *TE* propagation and Figure 1.9b shows the conditions for *TM* propagation.

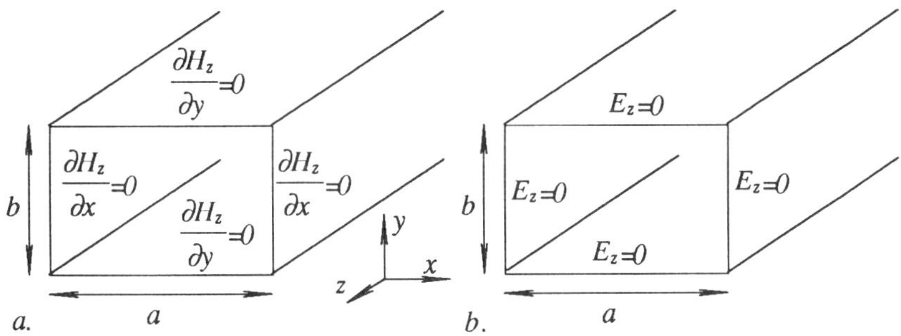

Figure 1.9. Rectangular waveguide and boundary conditions. a. Boundary conditions for TE waves. b. Boundary conditions for TM waves.

The necessary equation for *TE* waves is

$$\frac{\partial^2 H_z}{\partial x^2} + \frac{\partial^2 H_z}{\partial y^2} + (\gamma^2 + k^2)H_z = 0 \qquad (1.148)$$

This is a wave equation in H_z alone and will be taken from now on as the defining equation for *TE* waves. It is obtained by assuming that the component E_z is zero (since the fields are transverse electric). Similarly, the equation for *TM* waves is

$$\frac{\partial^2 E_z}{\partial x^2} + \frac{\partial^2 E_z}{\partial y^2} + (\gamma^2 + k^2)E_z = 0 \qquad (1.149)$$

in which γ and k are as defined above. These equations are Helmholtz equations and, therefore have characteristic or modal solutions. A solution is obtained for each characteristic value (eigenvalue) and each eigenvalue corresponds to a cutoff frequency. A cutoff frequency is a frequency below which waves cannot propagate in the waveguide while above it, propagation takes place. Thus, analysis of a waveguide consists of finding the cutoff frequencies of the waveguide and the corresponding fields within the cavity for each frequency. Each solution is called a mode. We only show here the modes and fields for rectangular waveguides as shown in Figure 1.9 but a similar approach can be followed for cylindrical waveguides or, for that matter (and often with considerable more difficulty), for any cross-section waveguide.

As an example of modes we give here the *TE* and *TM* fields and modes for rectangular cross-section waveguides.

For *TM* modes, we solve Eq. (2.149) subject to the boundary conditions given in Figure (1.9b). The solution is performed using standard separation of variables (see problems 1.18). The resulting fields are

$$E_x(x,y) = \frac{-\gamma}{\gamma^2 + k^2} E_0 \frac{m\pi}{a} \cos\left(\frac{m\pi}{a}\right)x \sin\left(\frac{n\pi}{b}\right)y \qquad (1.150)$$

$$E_y(x,y) = \frac{-\gamma}{\gamma^2 + k^2} E_0 \frac{n\pi}{b} \sin\left(\frac{m\pi}{a}\right)x \cos\left(\frac{n\pi}{b}\right)y \qquad (1.151)$$

$$H_x(x,y) = \frac{j\omega\varepsilon}{\gamma^2 + k^2} E_0 \frac{n\pi}{b} \sin\left(\frac{m\pi}{a}\right)x \cos\left(\frac{n\pi}{b}\right)y \qquad (1.152)$$

$$H_z(x,y) = \frac{-j\omega\varepsilon}{\gamma^2 + k^2} E_0 \frac{m\pi}{a} \cos\left(\frac{m\pi}{a}\right)x \sin\left(\frac{n\pi}{b}\right)y \qquad (1.153)$$

where

$$\gamma^2 + k^2 = \left(\frac{m\pi}{a}\right)^2 + \left(\frac{n\pi}{b}\right)^2$$

is called the dispersion relation and m,n, are integers that define the possible modes present in the waveguide. We now rewrite this as

$$\gamma = j\beta = j\sqrt{\omega^2\mu\varepsilon - \left(\frac{m\pi}{a}\right)^2 - \left(\frac{n\pi}{b}\right)^2} \qquad (1.154)$$

based on the definition in Eq. (1.140). To define the modes in terms of cutoff frequencies we set $\gamma = 0$ since below that value, γ is real and the wave cannot propagate. This gives a relation between frequency, dimensions of the waveguide and the mode indices m,n as

$$f_c = \frac{1}{2\pi\sqrt{\mu\varepsilon}} \sqrt{\left(\frac{m\pi}{a}\right)^2 + \left(\frac{n\pi}{b}\right)^2} \qquad (1.155)$$

The mode indices can be any integers as long as the fields in Eq. (1.150) through (1.153) do not vanish. Modes in a waveguide are defined by these integers and the integer defines the number of half-cycle variations in the fields in each dimension of the waveguide's cross section. Thus, if $m=1$, $n=3$, this defines a TM_{13} mode with one half-cycle variation in the x direction and 3 half-cycles variation in the y direction. We note however that neither m nor n can be zero since that would force all field components to be zero. The lowest possible mode (that is: the one with lowest cutoff frequency) is the TM_{11} mode.

For TE modes we must solve Eq. (1.148) with the boundary conditions in Figure 1.9a. After similar steps as for the TM modes we get

$$E_x(x,y) = \frac{j\omega\mu}{\gamma^2 + k^2} H_0 \frac{n\pi}{b} \cos\left(\frac{m\pi}{a}\right)x \sin\left(\frac{n\pi}{b}\right)y \tag{1.156}$$

$$E_y(x,y) = \frac{-j\omega\mu}{\gamma^2 + k^2} H_0 \frac{m\pi}{a} \sin\left(\frac{m\pi}{a}\right)x \cos\left(\frac{n\pi}{b}\right)y \tag{1.157}$$

$$H_x(x,y) = \frac{\gamma}{\gamma^2 + k^2} H_0 \frac{m\pi}{a} \sin\left(\frac{m\pi}{a}\right)x \cos\left(\frac{n\pi}{b}\right)y \tag{1.158}$$

$$H_y(x,y) = \frac{\gamma}{\gamma^2 + k^2} H_0 \frac{n\pi}{b} \cos\left(\frac{m\pi}{a}\right)x \sin\left(\frac{n\pi}{b}\right)y \tag{1.159}$$

The dispersion relation and cutoff frequencies for TE modes are the same as for TM modes.

However, in TE modes either m or n can be zero (but not both). This indicates that the lower resonant mode is a TE_{0n} or TE_{m0}, depending on the dimensions a and b. If $a>b$, the lowest cutoff frequency is for a TE_{10} mode. TM and TE modes with identical indices have identical frequencies.

1.20.2 Cavity resonators

A cavity resonator is a shorted waveguide, such that it forms a closed structure of some well-defined shape. A rectangular cavity is built by adding two conducting walls at $z=0$ and $z=d$ as shown in Figure 1.10. The main difference between cavities and waveguides is that in cavities, the z direction imposes additional boundary conditions and there is no propagation of waves as in waveguides. The cavity acts as a resonant structure with the associated exchange of energy between the electric and magnetic field at resonant frequencies. A cavity resonator is equivalent to an LC circuit (or an RLC circuit in the case of a lossy cavity).

Analysis of fields in a cavity is very similar to that for waveguides except for the obvious additional term since now we have a three-dimensional structure. The equations we need to solve are simple extensions of Eqs. (1.148) and (1.149). The equation for TM fields is

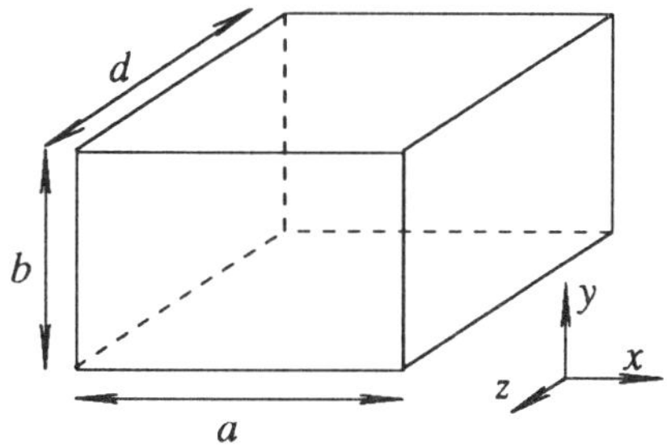

Figure 1.10 A rectangular cavity resonator.

$$\frac{\partial^2 E_z}{\partial x^2} + \frac{\partial^2 E_z}{\partial y^2} + \frac{\partial^2 E_z}{\partial z^2} + (\gamma^2 + k^2)E_z = 0 \qquad (1.160)$$

while for *TE* fields we must solve the following

$$\frac{\partial^2 H_z}{\partial x^2} + \frac{\partial^2 H_z}{\partial y^2} + \frac{\partial^2 H_z}{\partial z^2} + (\gamma^2 + k^2)H_z = 0 \qquad (1.161)$$

where the term $(\gamma^2 + k^2)$ is the cutoff wave number and is given in Eq. (1.166) below.

(a) *TM* and *TE* modes in cavity resonators

Thus, following steps identical to those for *TM* fields in waveguides, we get the fields for *TM* modes in (rectangular) cavity resonators as

$$E_x(x,y,z) = \frac{-1}{\gamma^2 + k^2} E_0 \frac{p\pi}{d}\frac{m\pi}{a} \cos\left(\frac{m\pi}{a}\right)x \sin\left(\frac{n\pi}{b}\right)y \sin\left(\frac{p\pi}{d}\right)z \qquad (1.162)$$

$$E_y(x,y,z) = \frac{-1}{\gamma^2 + k^2} E_0 \frac{p\pi}{d}\frac{n\pi}{b} \sin\left(\frac{m\pi}{a}\right)x \cos\left(\frac{n\pi}{b}\right)y \sin\left(\frac{p\pi}{d}\right)z \qquad (1.163)$$

$$H_x(x,y,z) = \frac{j\omega\varepsilon}{\gamma^2 + k^2} E_0 \frac{n\pi}{b} \sin\left(\frac{m\pi}{a}\right)x \cos\left(\frac{n\pi}{b}\right)y \cos\left(\frac{p\pi}{d}\right)z \qquad (1.164)$$

$$H_y(x,y,z) = \frac{-j\omega\varepsilon}{\gamma^2 + k^2} E_0 \frac{m\pi}{a} \cos\left(\frac{m\pi}{a}\right)x \sin\left(\frac{n\pi}{b}\right)y \cos\left(\frac{p\pi}{d}\right)z \qquad (1.165)$$

where the dispersion relation is

$$\gamma^2 + k^2 = \left(\frac{m\pi}{a}\right)^2 + \left(\frac{n\pi}{b}\right)^2 + \left(\frac{p\pi}{d}\right)^2 \qquad (1.166)$$

and the resonant frequencies (equivalent to cutoff frequencies in waveguides) are

$$f_{mnp} = \frac{1}{2\pi\sqrt{\mu\varepsilon}}\sqrt{\left(\frac{m\pi}{a}\right)^2 + \left(\frac{n\pi}{b}\right)^2 + \left(\frac{p\pi}{d}\right)^2} \qquad (1.167)$$

where the indices *m,n,p*, indicate the mode at which the cavity resonates.

The *TE* fields are similarly given as

$$E_x(x,y,z) = \frac{j\omega\mu}{\gamma^2 + k^2} H_0 \frac{n\pi}{b} \cos\left(\frac{m\pi}{a}\right)x \sin\left(\frac{n\pi}{b}\right)y \sin\left(\frac{p\pi}{d}\right)z \qquad (1.168)$$

$$E_y(x,y,z) = \frac{-j\omega\mu}{\gamma^2 + k^2} H_0 \frac{m\pi}{a} \sin\left(\frac{m\pi}{a}\right)x \cos\left(\frac{n\pi}{b}\right)y \sin\left(\frac{p\pi}{d}\right)z \qquad (1.169)$$

$$H_x(x,y,z) = \frac{\gamma}{\gamma^2 + k^2} H_0 \frac{m\pi}{a} \sin\left(\frac{m\pi}{a}\right)x \cos\left(\frac{n\pi}{b}\right)y \cos\left(\frac{p\pi}{d}\right)z \qquad (1.170)$$

$$H_y(x,y,z) = \frac{\gamma}{\gamma^2 + k^2} H_0 \frac{n\pi}{b} \cos\left(\frac{m\pi}{a}\right)x \sin\left(\frac{n\pi}{b}\right)y \cos\left(\frac{p\pi}{d}\right)z \qquad (1.171)$$

with the same dispersion relation and resonant frequencies as for the *TM* modes. From the equations for *TM* and *TE* modes we note that for *TM* modes, neither *m* nor *n* can be zero while *p* can be zero. For *TE* modes, either *m* or *n* can be zero (but not both) while *p* must be nonzero.

The various modes possible are defined by *m,n*, and *p*. Some of the modes may have the same resonant frequency even though they are different modes. As an example, for a cubic cavity *(a=b=d)*, TM_{110}, TE_{011}, TE_{101} all have the same resonant frequency. These are called degenerate modes. The mode with lowest frequency (which depends on *a*, *b*, and *d*) is the dominant mode.

1.20.3 Energy relations in waveguides and cavity resonators

Energy relations in waveguides and cavity resonators are defined by the Poynting theorem as given in section 1.10. In particular, since most relations in these devices are for time harmonic fields, the time harmonic form of the Poynting theorem is normally used to define energy relations in waveguides and cavity resonators. Equation (1.57) defines these relations.

The stored electric and magnetic energy in a volume (such as a cavity or a section of a waveguide) is given as

$$W_0 = \int_v \left(\frac{\varepsilon |E|^2}{2} + \frac{\mu |H|^2}{2} \right) dv \tag{1.172}$$

where E and H are the fields in the volume v. This relation is correct at any frequency regardless of resonance. The dissipated power in the volume, either due to presence of lossy materials or due to losses in the volume walls is

$$P_d = \frac{1}{2} \int_v \sigma |E|^2 dv \tag{1.173}$$

where σ may represent conducting materials in the cavity, including lossy dielectrics or, more often, conductivity of the walls.

The energy relations in Eqs. (1.172) and (1.173) also define the quality factor of cavity resonators as the ratio between the stored energy in the cavity and the dissipated power per unit time

$$Q = \frac{\omega_0 W_0}{P_d} \tag{1.174}$$

where ω_0 is the resonant frequency of the cavity. Since the higher the Q factor, the more selective the cavity is, Q is a measure of the bandwidth of the cavity. It also defines, indirectly, the amount of energy needed to couple into the cavity to maintain an energy balance. Ideal cavities have an infinite quality factor. For testing purposes, measurement of Q-factors is a simple method of testing lossy materials and monitoring processes such as curing and drying in materials such as ceramics and composites.

Energy relations are also useful in determining shifts in resonant frequencies of cavity resonators.

1.21 SKIN DEPTH

One of the most important aspects of electromagnetic nondestructive testing at low frequencies is the concept of skin depth. We discuss this concept in some detail here since it is often misunderstood and almost never properly defined. To do so, we first give the standard definition and point to the source of this definition. In particular, it is important to understand that the definition, as most commonly used, applies only to planar materials, which are good conductors, and under plane wave excitation. We often "extend" the use of skin depth to cylindrical conductors and arbitrary excitation without proper justification. As an example, the definition of skin depth for a coil enclosing a poor cylindrical conductor does not satisfy any of the three conditions used for the definition of the skin depth and yet, it is common to still use the standard definition.

1.21.1 Skin depth in planar materials

Skin depth or standard depth of penetration in *NDT* is normally defined as that distance in which the amplitude of eddy currents is reduced to $1/e$, although, as shown above, skin depth is not necessarily restricted to eddy currents. Any propagating field quantity (**B**, **H**, **E**, etc. will behave in identical fashion). In most instances, this would be related to the amplitude at the surface of the conductor. In this case skin depth may be defined as that distance in which the amplitude is reduced to $1/e$ of the magnitude at the surface. However, skin depth can be, and sometimes is used at any location in a conductor, not only at the surface. In eddy current testing, the skin depth is a basic parameter. It defines, although in inexact terms, the depth within which an eddy current test may be useful. One must realize, however, that the amplitude is never reduced to zero and eddy currents exist at considerable depths. Their amplitudes are very low and may require special techniques for detection. As an example, the amplitude at *5* skin depths below the surface of a good conductor is reduced to about *0.6%* of the amplitude at the surface.

The skin depth is defined as

$$\delta = \frac{1}{\sqrt{\pi f \mu \sigma}} \quad [m] \tag{1.175}$$

where: δ = skin depth (or depth of penetration) $[m]$
$\pi = 3.141592654$
f = frequency $[Hz]$
μ = magnetic permeability $[H/m]$
σ = electric conductivity $[S/m]$

Although a very useful parameter, it must be used with care for the following reasons:

1. Skin depth is defined from the attenuation of a uniform plane wave as it penetrates into a flat, semi-infinite conductor. For any other geometry or for non-plane waves this is only an approximation to the actual current distribution. A coil, for example, has a non-planar wavefront.
2. The form shown in Eq. (1.175) only applies to uniform plane wave penetration in good conductors. Poor conductors such as composite materials alter this expression.

3. It is only applicable to time-harmonic fields. Transient eddy currents contain more than one frequency and their penetration in conducting materials is different.
4. The expression assumes linear, isotropic materials.

Under these conditions, the eddy current density can be evaluated anywhere in the semi-infinite material as

$$J_r = J_0 e^{(-r/\delta)} \tag{1.176}$$

where J_0 is the current density at the surface, r is the depth at which the current density is calculated and δ is the skin depth as calculated in Eq. (1.175). Similarly, the phase angle can be evaluated as

$$\theta = r\sqrt{\pi f \mu \sigma} = r/\delta \tag{1.177}$$

The magnitude of the eddy currents varies exponentially from the surface and the phase angle lags the phase angle at the surface by *1 radian* per skin depth (57°/skin depth).

The equation for skin depth as presented above assumes that a plane wave impinges on a semi-infinite plane of a high conductivity material. In practice, the plane wave condition is rarely satisfied. We normally use the same approximation for the skin depth regardless of frequency, conductivity, permeability or geometry. In many ways, this is appropriate since the whole purpose of the skin depth in eddy current testing is to estimate the amplitude and phase of the eddy currents with depth. On the other hand, this estimate is not very good unless the conditions under which the skin depth was calculated are actually met.

1.21.2 Skin depth in tubular materials

For cylindrical conductors, the skin depth in Eq. (1.175) is a good approximation only if the radius of the conductor is large compared with the skin depth. To see how large the error in estimating the skin depth is, the eddy current density in cylindrical conductors can be calculated exactly and compared with that predicted by the skin depth equation in Eq. (1.175).

The solution to the eddy current density in a cylindrical conductor is obtained by solving Maxwell's equations in cylindrical coordinates. The method and details of solution are given in chapter 2. We however use the results here for comparison.

(a) Skin depth in hollow cylindrical conductors

To define a skin depth in a cylindrical conductor, one proceeds in exactly the same manner as for the planar case except for the use of cylindrical coordinates. To calculate the behavior of eddy currents with depth in the material, we assume that the only variation in eddy currents is with depth *(r)*. Eddy currents are assumed to be circumferential (in the ϕ direction) and constant at any given depth r. The problem is illustrated in Figure 1.11. This, in effect, allows the solution of a one-dimensional problem in cylindrical coordinates in a manner equivalent to the method used to calculate the skin depth in a semi-infinite planar material.

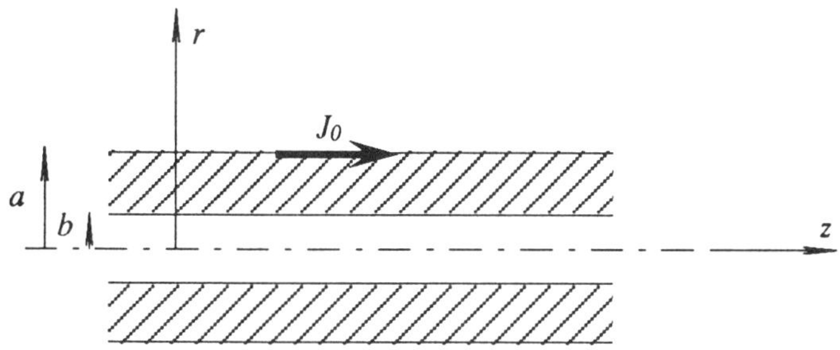

Figure 1.11 Geometry for calculation of skin depth in cylindrical conductors.

The solution for eddy currents in a hollow cylinder is

$$J(r) = CI_0 \frac{j\omega\sigma\mu}{2} r + DK_0 \frac{j\omega\sigma\mu}{2} r \tag{1.178}$$

where I_0 and K_0 are the modified Bessel functions of order zero and can be found in mathematical tables. The argument for the functions is complex and therefore, the current density is a complex variable.

The constants C and D are evaluated from the boundary conditions on the two surfaces of the conductor and are discussed in chapter 2. The phase angle of the eddy currents is found from the real and imaginary parts of Eq. (1.178) as

$$\alpha = tg^{-1}\left(\frac{Im(J(r))}{Re(J(r))}\right) \tag{1.179}$$

Thus, Eq. (1.178) describes the variation of eddy currents with depth. This equation clearly shows the difference between eddy current distribution in an infinite planar material and in a hollow conductor. Only if the skin depth happens to be small compared with the outer radius of the cylinder is the planar form of the definition a good approximation.

(b) Skin depth in solid cylindrical conductors

In a solid cylindrical conductor, the eddy currents behave exactly in the same way as in a hollow cylinder for which the inner radius is zero. This is given as

$$J(r) = \frac{I_0(P(r))}{I_0(P(a))} J_0 e^{j\omega t} \qquad (1.180)$$

This expression is again different than either a hollow cylinder or an infinite plane. Note that unlike the definition of skin depth in planar geometries, one cannot talk about skin depth or standard depth of penetration in cylindrical geometries. Comparison of Eqs. (1.176), (1.178) and (1.180) shows considerable variations in the magnitude of eddy currents for a given depth in the material. Similarly, their phase angles show considerable variations when Eqs. (1.177) and (1.179) are compared.

In Figure 1.12, the estimated eddy current density for a hollow cylinder of inner radius $b=2.25mm$, outer radius $a=4.25mm$ are plotted using the expressions in Eqs. (1.176), (1.178) and (1.180). The variations are quite significant. In this case, the skin depth equation predicts a current density close to the inner surface of the conductor which is about half its actual value. Similarly, the phase angle of the eddy current is completely different as can be seen from Figure 1.13. While the phase angle must be linear if the skin depth equation is used, this is clearly not the case for the cylindrical conductor using either the hollow cylinder equation (exact in this case) or the solid cylinder equation.

Thus, the correct estimate for the skin depth depends on the parameters of the test. Particular attention should be paid to testing at low frequencies, and low conductivity materials. In these cases, it is often likely that the conditions for the skin depth equation are not satisfied and the estimate would be in error. On the other hand, since the skin depth equation underestimates the actual eddy current density, in many practical situations it may not be important to calculate the eddy current density exactly.

The skin depth is also an important factor to consider for *ac* magnetic particle inspection because, even at $60\ Hz$, the skin depth can be quite small. For example, for a ferromagnetic material with conductivity of $\sigma= 0.5\times10^7$, a relative permeability $\mu_r=100$ and at a frequency of $60\ Hz$, the skin depth is $3mm$. This is probably an overestimate because a linear material was assumed in the derivation of the skin depth. For this reason, ac magnetic particle methods, such as the so-called "swinging field" methods, generally detect only defects which are open to the surface. On the other hand, these methods can be more sensitive to surface breaking defects because the applied field is concentrated at the surface.

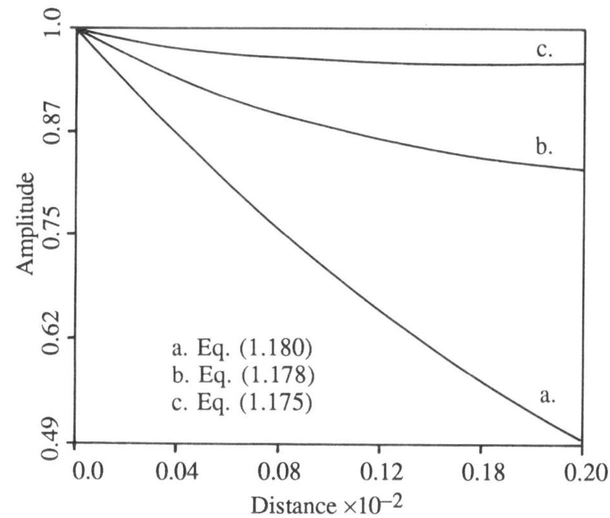

Figure 1.12 Eddy currents for a hollow cylindrical conductor calculated with the standard skin depth formula (Eq. (1.176)), Eq. (1.178) and Eq. (1.180) for $\delta<a$.

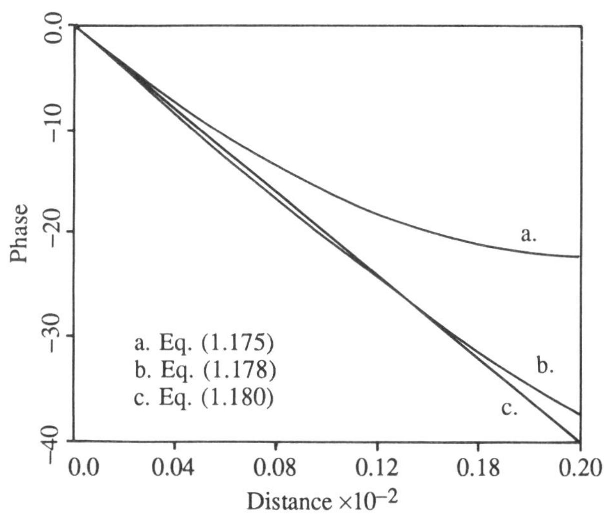

Figure 1.13 Eddy currents phase angle for a hollow cylindrical conductor calculated with the standard skin depth formula (Eq. (1.176)), Eq. (1.179) and Eq. (1.180) for $\delta<a$.

1.22 CLASSIFICATION OF FIELD EQUATIONS

From the discussion of the electromagnetic field equations it is obvious that the type of equation that needs to be solved depends on the problem at hand. Regardless of the variable used, there are three classes of problems that we need to treat in the context of modeling of *NDT* applications. These are:

1. Static or quasi-static problems. These are time independent and are exemplified by the Laplace and Poisson equations mentioned above. The equations can be linear or nonlinear although, for the quasi-static equation (the eddy current equation) we will almost always assume linearity. In terms of variables, almost every variable discussed in sections 1.11 through 1.14 will be used, depending on application. These types of equations are treated separately in chapters *4* through *7* as elliptical partial differential equations.
2. Time dependent or transient problems. These are time dependent, non-propagating problems and are generally described by a diffusion equation not a wave equation. Eddy current problems with arbitrary excitation, and skin effect applications are examples of this type. The most commonly used variable for this purpose is the magnetic vector potential and the electric vector potential. These aspects of modeling are discussed in chapter 9 where we discuss the solution of parabolic partial differential equations.
3. Wave propagation applications. This type of application is characterized by a wave equation either in the time domain or the frequency domain. Examples are the Helmholtz equation for mode analysis in waveguides and cavity resonators, and the general, time dependent wave equation. The formulation in terms of field variables as well as potentials is given in chapter *10*.

1.23 PROBLEMS

1.1 Derive the integral form of Maxwell's equations. Use open surface integration on both sides of Eqs. (1.1) and (1.2) and then use Stokes' theorem to convert the left hand side into closed contour integrals. Similarly use volume integration on both sides of Eqs. (1.3) and (1.4) and the divergence theorem to convert the left hand side into closed surface integrals to obtain Eqs. (1.7) and (1.8).

1.2 Show that Eqs. (1.3) and (1.4) can be obtained from Eqs. (1.1) and (1.2) if the continuity equation is postulated and therefore, that only two of

Maxwell's equations are independent equations. The continuity equation is:

$$\nabla \cdot \mathbf{J} = -\frac{\partial \rho}{\partial t}$$

and is a statement of the law of conservation of charge.

1.3 Derive the interface conditions at the boundary between two perfect dielectrics for the electric field. Assume tangential and normal components of the electric fields on each side of the interface and use Gauss' and Faraday's laws to derive the relations between the components in the two materials. Which of the components are continuous across the interface? How does this change if a surface charge density exists on the interface between the two dielectrics?

1.4 Derive the interface conditions at the boundary between two materials with different permeabilities for the magnetic field. Assume tangential and normal components of the magnetic fields on each side of the interface and use Gauss' and Ampere's laws to derive the relations between the components in the two materials. Which of the components are continuous across the interface. How does this change if a surface current density exists on the interface between the two materials?

1.5 Derive the interface conditions for the electric field at the interface between a perfect dielectric and a perfect conductor. Show that the electric field must be perpendicular to the surface of the conductor.

1.6 Derive the interface conditions for the magnetic field between free space and a perfectly permeable material (material which has infinite permeability). Show that the magnetic field must be perpendicular to the surface of the material. How does this change if permeability is finite?

1.7 Derive the interface conditions for the electric and magnetic fields for two general materials. Assume that both surface current densities and surface charge densities may exist and discuss the implications. Show that the interface conditions derived in problems 1.3 through 1.6 can be derived from these general interface conditions by simple approximations on material properties.

1.8 Derive interface conditions for the current density **J** between two, general, lossy dielectrics. Show that a surface charge density must exist at the interface if both conductivity and permittivity must be considered. Calculate the surface current density.

1.9 Helmholtz's theorem states that a vector field is uniquely determined to within an additive constant if its divergence and its curl are specified. Show that a vector field **P** can be written as the sum of the gradient of a scalar function U and the curl of a vector function **F**:

$$\mathbf{P} = -\Delta U + \nabla \times \mathbf{F}$$

To do so write the vector **P** as the sum of an irrotational (curl-free) and a solenoidal (divergence-free) part.

1.10 Show that imposing the Lorentz gage condition is equivalent to specifying the current continuity equation. That is, by using the Lorentz gage, the divergence of fields is properly specified (see also problem 1.2).

1.11 Show that under static condition, the Lorenz gage condition reduces to that of Coulomb's gage condition and that it properly defines the divergence for static fields.

1.12 Derive the interface conditions for the magnetic vector potential **A** and the electric vector potential **T** following the methods used in problems 1.3 through 1.8.

1.13 Derive the various relations needed for use of phasors in field relations. In particular show that $\partial/\partial t$ can be replaced by $j\omega$. How is phase represented? Given a phasor, calculate the time dependent form and vice versa.

1.14 Derive the nonhomogeneous wave equation in the time domain in terms of variable **B**, following the method in section 1.13.1. Rewrite the equation in the frequency domain and as a homogeneous wave equation.

1.15 Derive the nonhomogeneous wave equation in the time domain in terms of variable **D**, following the method in section 1.13.1. Rewrite the equation in the frequency domain and as a homogeneous wave equation.

1.16 Derive the nonhomogeneous wave equation in the time domain in terms of variable **H**, following the method in section 1.13.1. Rewrite the equation in the frequency domain and as a homogeneous wave equation.

1.17 Derive the nonhomogeneous wave equation in the time domain in terms of variable V, following the method in section 1.13.1. Rewrite the equation in the frequency domain and as a homogeneous wave equation.

1.18 Perform separation of variables on Eq. (1.149) and find a general solution for the two-dimensional geometry in Figure 1.9b. By substituting the boundary conditions in Figure 1.9b show that Eqs. (1.150) through (1.153) represent the fields in a rectangular waveguide. The amplitude E_0 is an arbitrary constant. Explain the meaning of this constant.

1.19 Repeat problem 1.18 for *TE* modes using Eq. (1.148) and the geometry in Figure 1.9a. Show that the general solution for *TE* modes leads to the fields in Eqs. (1.156) through (1.159).

1.20 Using the results in problems 1.18 and 1.19, derive the *TE* and *TM* solutions for a rectangular cavity resonator. Start with Eqs. (1.160) and (1.161).

1.21 Define skin depth for a low loss dielectric using Eq. (1.137) using the relation $\delta=1/\alpha$. Compare this relation with that in Eq. (1.144). Using these two relations compare the field in sea water ($\sigma= 4\ S/m$) at a depth of $1m$ as calculated using the two relations. What is the error in calculation of skin depth?

1.24 BIBLIOGRAPHY

The material covered in this chapter is mostly simple field theory as studied at the undergraduate level. The choice of topics was dictated by those aspects that are required for modeling and therefore, it is an incomplete treatment of fields. The references given below can be used to supplement or expand subjects treated here.

The first three references are recommended for an initial study while references [4] through [10] can be used for more advanced treatment and topics. Smythe's book [11] is particularly interesting in its approach to problem solving and contains an extensive discussion of skin depth. There are many other texts on the subject and almost any of these includes most of the topics discussed here.

[1] Plonus, M.A. (1978), *Applied Electromagnetics*, McGraw-Hill, New York.
[2] Lorain, P and Corson, D.R. (1990), *Electromagnetism Principles and Applications*, 2nd ed., W. H. Freeman and Company, San Francisco.
[3] Hayt, W.H. (1989), *Engineering Electromagnetics*, 5th ed., McGraw-Hill, New York.
[4] Plosney, P and Collin, R.E. (1961), *Principles and Applications of Electromagnetic Fields*, McGraw-Hill, New York.
[5] Collin, R.E. (1991), *Electromagnetic Theory of Guided Waves*, McGraw-Hill, NY.
[9] Kong, J.A. (1975), *Theory of Electromagnetic Waves*, John Wiley & Sons, New York.
[10] Harrington, R.F. (1961), *Time-harmonic Electromagnetic Fields*, McGraw-Hill, New York.
[11] Smythe, W.R. (1991), *Static and Dynamic Electricity*, McGraw-Hill, New York.
[12] Maxwell, J.C. (1954), *A Treatise on Electricity and Magnetism*, Vol. 1 and 2, Dover Publications, New York.

CHAPTER 2

Analytic methods of solution

2.1 INTRODUCTION

Analytic solutions for electromagnetic *NDT* problems are rather difficult to obtain. There are many reasons for this but perhaps the main reason is the complex interaction of the electromagnetic field with materials and in particular, with material variations such as defects.

However, there are a number of methods that can be applied and very significant results can be obtained for a number of problems in *NDT*. The value of analytic methods of solution cannot be overemphasized: only analytic methods provide exact solutions to a whole class of problems. Because of this, analytic solutions are often used to verify numerical solutions.

All classical methods of solution can be employed in *NDT*. These include separation of variables, conformal mapping, method of images and other related methods. Perhaps the most general is the method of separation of variables by which, a partial differential equation is solved directly. This aspect of solution is very tempting because, after all, there can be no better solution than a direct, exact solution of the original partial differential equation. There are however severe limitations to analytic methods. The most obvious is the assumption of linearity which is required for separation of variables to apply. Similarly, for an analytic solution to be practical, the geometry must be rather simple and in most cases, highly symmetric.

Not the least limitation from a practical point of view is the considerable mathematical prowess required for many analytical solutions. These skills are not easily and quickly imparted and, considering the limited scope of analytical methods, not always justified for this purpose.

2.2 ANALYTIC METHODS

The analytic approach to modeling is characterized by the following general aspects:

1. The solution is general and exact. Parametric studies of the test geometry are possible and there is no need for approximations. Although this is an ideal case, we will often consider analytic models which contain various degrees of approximations and simplification of the exact problem. This approach is taken because an analytic model is always easier to apply, less expensive to compute, and more accurate (under the assumptions introduced in the model) than a numerical solution.
2. The model is described in a closed form, normally as a mathematical relation which can then be solved for any problem that satisfies the assumptions implicit in the model.
3. Derivation of the model is almost always done by solving the partial differential equations exactly by a separation of variables approach. In some cases, other methods may be used such as conformal mapping or method of images but these are merely indirect methods of solving the same partial differential equations.

2.2.1 Range of solution methods

The process of finding an analytic solution to a field problem is not always obvious to the user and often requires considerable mathematical skills. There are however a few general techniques which can be employed for this purpose.

Perhaps the best known methods are those based on the direct solution of partial differential equations by separation of variables. In these methods, the electromagnetic field equations are cast into the form of a second order partial differential equation by essentially, change in variables (although we may actually call this process a "formulation in terms of potential functions", or "formulation in terms of field variables"). Then, depending on their form, the equations are decoupled by defining separable variables. Normally, this is followed by integration over a generic solution domain. Introduction of the boundary conditions of the problem completes the process.

Thus, the general solution of the static field equations in the Cartesian coordinate system is in terms of sine, cosine, hyperbolic sine and hyperbolic cosine functions. This is known from the separation of variables solution to any elliptic partial differential equation. The solution of the same problem in cylindrical coordinates leads to a solution in terms of Bessel functions or modified Bessel functions.

Another powerful technique often employed is the general method of images. Although this method is best known for its application to simple problems in electrostatics, it is quite general. It is equally applicable in magnetostatic and magnetodynamic problems, with or without current sources. The main idea of the method, and the reason for its simplicity, is the ability to find an equivalent system of sources which produce the same solution in the domain of interest. If the equivalent source distribution can be easily solved, or if it has a solution based on other methods, then the original problem has the same solution in the domain of interest.

The main difficulty with this method is that the equivalent sources are not necessarily easy to find and, in general, the method can only be applied with highly symmetric geometries. There are however a number of rather interesting problems that can be solved using this technique either exactly, or approximately.

A third technique of considerable importance is conformal mapping. The principle involved is rather simple: find a transformation from the solution domain into an auxiliary domain in which the solution can be obtained easily. Then perform the inverse transformation to obtain the solution in the physical domain. This process is similar to any transform method. For successful application, the transformation must be unique and must be easily inverted. However, the method is a two-dimensional method and its application can be rather difficult. On the other hand, a large number of problems were solved using this method and whenever the need arises for a solution, it is quite likely that an existing solution can be found or modified to apply to the problem at hand.

2.2.2 Range of solvable problems

Analytic solutions can be found to a number of classes of problems. Even though it is not possible to include all aspects of solution, the following types of problems normally have an analytic solution

1. One-dimensional problems. In many cases, a one-dimensional problem can be solved by direct integration. While these are of little help in solving the general *NDT* problem they offer a means of verifying formulations and solutions.
2. Highly symmetric two-dimensional problems.
3. Some three-dimensional problems which can be viewed as two-dimensional problems by transformation of the system of coordinates.

2.2.3 Importance of analytic solutions for verification

It was mentioned above that analytic solutions are our ultimate goal. If we could hope to find analytic models for the general *NDT* problem (or for that matter, in any other aspect of science), this should be our only effort. Unfortunately, it is sufficient to look at the list in the previous paragraph to see that this is not the case. However, our considerable interest in analytic solutions, be they limited as they are, is in the fact that analytic solutions can be used to verify much more complex and general numerical models.

As a rule, we can always obtain a numerical solution to a given problem. However, because the solution is numerical rather than in closed form, we must find an independent way of confirming its accuracy and indeed, its correctness. There are only two ways of doing so; one is experimental, the other by comparison to analytic solutions.

It is in this capacity that we will consider analytic solutions in this work. Thus, although our approach will be to look at numerical modeling, we will also resort to analytic solution for verification of numerical results whenever possible and appropriate. Experimental verification is much more difficult to apply because of the errors involved in experimentation and because often it is too difficult or too expensive to obtain good experimental data.

2.2.4 Limitations

Most of the limitations of analytic models were pointed out above and stem from the assumptions required to define the model. The following is a summary of some of the main limitations.

1. Only linear models can be considered.
2. Materials must be isotropic. Layered materials can sometimes be considered and, in the one-dimensional case, anisotropy can be treated.
3. Only one- and two-dimensional applications in rather simple geometries can be solved. There are a few exceptions to this rule but in general only simple, highly symmetric problems can be considered.
4. In the context of *NDT*, defects and complex shapes cannot be included.
5. Nonuniform fields (such as nonuniform plane waves or non plane waves) cannot be treated.

In the following sections, a number of general methods and their application to *NDT* are considered. These include applications from the whole spectrum of field problems with emphasis on those that provide some useful information rather than just a technique. In the process of defining the problem a number of assumptions will be necessary. These assumptions are in fact the limitations of the analytic model but, without these assumptions the solution cannot be obtained.

2.3 SEPARATION OF VARIABLES: SOLUTION TO LAPLACE'S EQUATION

Separation of variables for Laplace's equation can be accomplished relatively easily. A number of important problems in electromagnetics and therefore, in *NDT* can be treated using the general method. Also, the use of the method is rather instructive because of the principles involved.

After giving the general principles we will demonstrate its application to calculation of the skin depth in tubular materials.

2.3.1 Separation of variables in Cartesian coordinates

To outline the method of separation of variables consider first Laplace's equation in Cartesian coordinates

$$\frac{\partial^2 \Phi}{\partial x^2} + \frac{\partial^2 \Phi}{\partial y^2} + \frac{\partial^2 \Phi}{\partial z^2} = 0 \tag{2.1}$$

where Φ indicates a scalar function such as the electric or magnetic scalar potentials. Assuming linear behavior, the solution can be written as the product of three functions

$$\Phi(x,y,z) = X(x)Y(y)Z(z) \tag{2.2}$$

where $X(x)$ is only dependent on the x variable, $Y(y)$ on y and $Z(z)$ only depends on z.

Substitution of this general solution into Eq. (2.1) gives

$$Y(y)Z(z)\frac{d^2X(x)}{dx^2} + X(x)Z(z)\frac{d^2Y(y)}{dy^2} + X(x)Y(y)\frac{d^2Z(z)}{dz^2} = 0$$

where the partial derivatives were replaced by ordinary derivatives because of dependency on a single variable. Dividing both sides by $\Phi(x,y,z)$ gives

$$\frac{1}{X(x)}\frac{d^2X(x)}{dx^2} + \frac{1}{Y(y)}\frac{d^2Y(y)}{dy^2} + \frac{1}{Z(z)}\frac{d^2Z(z)}{dz^2} = 0$$

In this form each term depends on a single variable and, therefore, can be separated. To do so, each term must be equal to a constant to be determined. The equation can be written as

$$\frac{1}{X(x)}\frac{d^2X(x)}{dx^2} = -k_x^2 \tag{2.3}$$

$$\frac{1}{Y(y)}\frac{d^2Y(y)}{dy^2} = -k_y^2 \tag{2.4}$$

$$\frac{1}{Z(z)}\frac{d^2Z(z)}{dz^2} = k_z^2 \tag{2.5}$$

where

$$k_x^2 + k_y^2 + k_z^2 = 0 \tag{2.6}$$

and the negative sign in Eqs. (2.3) and (2.4) as well as the use of the square in Eq. (2.6) are arbitrary. This choice will be convenient later when we evaluate the constants for specific applications. Thus, the following three differential equations are obtained

$$\frac{d^2X(x)}{dx^2} + k_x^2X(x) = 0 \tag{2.7}$$

$$\frac{d^2Y(y)}{dy^2} + k_y^2Y(y) = 0 \tag{2.8}$$

$$\frac{d^2Z(z)}{dz^2} - k_z^2Z(z) = 0 \tag{2.9}$$

These three equations have standard solutions which we can now use. The first two equations are identical in form and therefore, have the same solution. The general solution for $X(x)$ and $Y(y)$ are:

$$X(x) = A_1 sin(k_x x) + A_2 cos(k_x x) = B_1 e^{jk_x x} + B_2 e^{-jk_x x} \tag{2.10}$$

$$Y(y) = A_3 sin(k_y y) + A_4 cos(k_y y) = B_3 e^{jk_y y} + B_4 e^{-jk_y y} \tag{2.11}$$

The third term has the following solution

$$Z(z) = A_5 sinh(k_z z) + A_6 cosh(k_z z) = B_5 e^{k_z z} + B_6 e^{-k_z z} \tag{2.12}$$

where the constants A_i or B_i must be evaluated to obtain a solution. These are evaluated from the boundary conditions of the problem. Either the harmonic form of solution or the exponential form may be used, depending on which form is most appropriate.

Thus, the general solution for the three-dimensional Laplace equation is written as

$$\Phi(x,y,z) = X(x)Y(y)Z(z) = \left[A_1 sin(k_x x) + A_2 cos(k_x x)\right]. \\ .\left[A_3 sin(k_y y) + A_4 cos(k_y y)\right].\left[A_5 sinh(k_z z) + A_6 cosh(k_z z)\right] \tag{2.13}$$

or

$$\Phi(x,y,z) = X(x)Y(y)Z(z) = \left[B_1 e^{jk_x x} + B_2 e^{-jk_x x}\right]. \\ .\left[B_3 e^{jk_y y} + B_4 e^{-jk_y y}\right].\left[B_5 e^{k_z z} + B_6 e^{-k_z z}\right] \tag{2.14}$$

In two dimensions, we simply assume that only the x and z components exist and that $k_y=0$. Thus, the solutions are essentially identical with the middle term (the term $Y(y)$) removed and with $k_z^2=-k_x^2$.

We also note that if any of the constants k_x, k_y or k_z are zero, the corresponding equation has a simple linear solution. If, for example, $k_x=0$, the solution for $X(x)$ is $X(x)=A_1x+A_2$.

2.3.2 Separation of variables in cylindrical coordinates

An essentially identical process can be followed in cylindrical coordinates. We start with Helmholtz's equation in cylindrical coordinates

$$\frac{1}{r}\frac{\partial}{\partial r}\left(r\frac{\partial\Phi}{\partial r}\right) + \frac{1}{r^2}\frac{\partial^2\Phi}{\partial\phi^2} + \frac{\partial^2\Phi}{\partial z^2} + k^2\Phi = 0 \tag{2.15}$$

or

$$\frac{\partial^2\Phi}{\partial r^2} + \frac{1}{r}\frac{\partial\Phi}{\partial r} + \frac{1}{r^2}\frac{\partial^2\Phi}{\partial\phi^2} + \frac{\partial^2\Phi}{\partial z^2} + k^2\Phi = 0 \tag{2.16}$$

The general solution is written as

$$\Phi(r,\phi,z) = F(r)G(\phi)H(z) \tag{2.17}$$

Substituting this in Eq. (2.16) we get

$$G(\phi)H(z)\frac{\partial^2 F(r)}{\partial r^2}+G(\phi)H(z)\frac{1}{r}\frac{\partial F(r)}{\partial r}+F(r)H(z)\frac{1}{r^2}\frac{\partial^2 G(\phi)}{\partial \phi^2}$$
$$+F(r)G(\phi)\frac{\partial^2 H(z)}{\partial z^2}+k^2F(r)G(\phi)H(z)=0$$

Dividing the equation by $\Phi(r,\phi,z)$ gives

$$\frac{1}{F(r)}\frac{d^2F(r)}{dr^2}+\frac{1}{F(r)}\frac{1}{r}\frac{dF(r)}{dr}+\frac{1}{G(\phi)}\frac{1}{r^2}\frac{d^2G(\phi)}{d\phi^2}+\frac{1}{H(z)}\frac{d^2H(z)}{dz^2}+k^2=0 \quad (2.18)$$

First we write for the last term

$$\frac{d^2H(z)}{dz^2}+k_z^2H(z)=0 \quad (2.19)$$

where k_z is a constant. Now, Eq. (2.18) is written as

$$\frac{r^2}{F(r)}\frac{d^2F(r)}{dr^2}+\frac{r}{F(r)}\frac{dF(r)}{dr}+\frac{1}{G(\phi)}\frac{d^2G(\phi)}{d\phi^2}+r^2\left(k^2-k_z^2\right)=0 \quad (2.20)$$

The third term can be written as

$$\frac{d^2G(\phi)}{d\phi^2}+k_\phi^2G(\phi)=0 \quad (2.21)$$

The remaining term is, therefore

$$r^2\frac{d^2F(r)}{dr^2}+r\frac{dF(r)}{dr}+\left[-k_\phi^2+r^2\left(k^2-k_z^2\right)\right]F(r)=0 \quad (2.22)$$

To write this in a more compact form we write $k_r^2=k^2-k_z^2$. This gives the following form:

$$r^2\frac{d^2F(r)}{dr^2}+r\frac{dF(r)}{dr}+\left[(k_rr)^2-k_\phi^2\right]F(r)=0 \quad (2.23)$$

Equations (2.19) and (2.21) are similar to Eqs. (2.7) and (2.8) and, therefore, have similar solutions. These are

$$G(\phi)=A_1sin(k_\phi\phi)+A_2cos(k_\phi\phi)=B_1e^{jk_\phi\phi}+B_2e^{-jk_\phi\phi} \quad (2.24)$$

$$H(z) = A_3 sin(k_z z) + A_4 cos(k_z z) = B_3 e^{jk_z z} + B_4 e^{-jk_z z} \tag{2.25}$$

Equation (2.23) is the Bessel differential equation if we write $k_\phi^2 = m^2$. This is permissible since k_ϕ^2 is a constant. Thus, its solution is

$$F(r) = A_5 J_m(k_r r) + A_6 Y_m(k_r r) \tag{2.26}$$

where $J_m(k_r r)$ is the Bessel function of the first kind and mth order while $Y_m(k_r r)$ is the Bessel function of the second kind and mth order. The argument of the two Bessel functions is $k_r r$. While we will not discuss Bessel functions in any detail, these are common and available in many of the references in the bibliography section.

The solution is, therefore,

$$\Phi(r,\phi,z) = \left[A_5 J_m(k_r r) + A_6 I_m(k_r r)\right]. \\ .\left[A_1 sin(k_\phi \phi) + A_2 cos(k_\phi \phi)\right].\left[A_3 sin(k_z z) + A_4 cos(k_z z)\right] \tag{2.27}$$

or:

$$\Phi(r,\phi,z) = \left[A_5 J_m(k_r r) + A_6 I_m(k_r r)\right]\left[B_1 e^{jk_\phi \phi} + B_2 e^{-jk_\phi \phi}\right]. \\ .\left[B_3 e^{jk_z z} + B_4 e^{-jk_z z}\right] \tag{2.28}$$

The axisymmetric case, even though separation of variables is performed in cylindrical coordinates, is somewhat different. The potential is independent of the ϕ coordinate. The equation to solve is:

$$\frac{1}{r}\frac{\partial}{\partial r}\left(r\frac{\partial \Phi}{\partial r}\right) + \frac{\partial^2 \Phi}{\partial z^2} + k^2 \Phi = 0 \tag{2.29}$$

Therefore, only the terms $F(r)$ and $H(z)$ exist. The general solution is:

$$\Phi(r,z) = \left[A_5 J_m(k_r r) + A_6 I_m(k_r r)\right]\left[A_3 sin(k_z z) + A_4 cos(k_z z)\right] \tag{2.30}$$

A special case is the solution of Laplace's equation in the r-ϕ plane. Since the potential Φ is independent of z, the problem is two-dimensional. In this case, we have

$$\frac{1}{r}\frac{\partial}{\partial r}\left(r\frac{\partial \Phi}{\partial r}\right) + \frac{1}{r^2}\frac{\partial^2 \Phi}{\partial \phi^2} = 0 \tag{2.31}$$

Using the following expression for the solution:

$$\Phi(r,\phi) = F(r)G(\phi) \tag{2.32}$$

The equations we get are

$$\frac{r}{F(r)}\frac{d}{dr}\left(r\frac{dF(r)}{dr}\right) - k_r^2 = 0 \tag{2.33}$$

$$\frac{1}{G(\phi)}\frac{d^2G(\phi)}{d\phi^2} + k_\phi^2 = 0 \tag{2.34}$$

and

$$k_\phi^2 + k_r^2 = 0 \tag{2.35}$$

The choice here was to use a positive constant in the second expression since then the solution is identical to that in Eq. (2.24). The solution for $G(\phi)$ is:

$$G(\phi) = A_1 sin(k\phi) + A_2 cos(k\phi) = B_1 e^{jk\phi} + B_2 e^{-jk\phi} \tag{2.36}$$

The equation for $F(r)$ can now be written as

$$r^2\frac{d^2F(r)}{dr^2} + r\frac{dF(r)}{dr} - k^2F(r) = 0 \tag{2.37}$$

where the term k was used since $k_\phi = -k_r = k$. Also, because the solution for $G(\phi)$ is periodic, k is an integer. The solution of this equation is

$$F(r) = B_3 r^k + B_4 r^{-k} \tag{2.38}$$

and the solution for $\Phi(r,\phi)$ is

$$\Phi(r,\phi) = F(r)G(\phi) = \left[A_1 sin(k\phi) + A_2 cos(k\phi)\right]\left[B_3 r^k + B_4 r^{-k}\right] \tag{2.39}$$

2.4 EXAMPLE: SKIN EFFECT

Before demonstrating the use of separation of variables in a two-dimensional geometry we will use the results above to discuss the very important issue of skin depth in planar and cylindrical geometries. The skin depth was defined in chapter 1 from the point of view of a plane wave propagating in conducting materials. Here we show that the same can be obtained by solving for the current density directly. This calculation does not actually require the method of separation of variables but we give it here because the results obtained above are required for the solution. In addition, the calculation of skin depth in cylindrical conductors is shown.

2.4.1 Skin effect in planar conductors

The geometry used for computation is shown in Figure 2.1. The conductor is assumed to be semi-infinite in size and any current density on the surface of the conductor is uniform and does not depend on the y or z coordinate. To solve for the current density directly we write from Maxwell's equations

$$\nabla\times\mathbf{E} = -j\omega\mu\mathbf{H} \tag{2.40}$$

$$\nabla\times\mathbf{H} = \mathbf{J} \tag{2.41}$$

From the constitutive relation $\mathbf{J} = \sigma\mathbf{E}$ and using Eq. (2.40) we can write

$$\nabla\times\frac{\mathbf{J}}{\sigma} = -j\omega\mu\mathbf{H}$$

Taking the curl on both sides of this relation gives

$$\nabla\times(\nabla\times\mathbf{J}) = -j\omega\mu\sigma(\nabla\times\mathbf{H})$$

and using the vector identity in Eq. (1.71) and Eq. (2.41) we get

$$-\nabla^2\mathbf{J} + \nabla(\nabla\cdot\mathbf{J}) = -j\omega\mu\sigma\mathbf{J}$$

Using the zero divergence of currents we finally get

$$\nabla^2\mathbf{J} - j\omega\mu\sigma\mathbf{J} = 0 \tag{2.42}$$

In the case given above, only one component of current density need be considered. If the current density is in the z direction but only varies with the x direction (depth in material) we can write

$$\frac{d^2J_z}{dx^2} - j\omega\mu\sigma J_z = 0 \tag{2.43}$$

This equation is identical in form to Eq. (2.9) and its solution is given in Eq. (2.12). We will use the exponential form of the equation as

$$J(x) = B_1 e^{k_x x} + B_2 e^{-k_x x} \tag{2.44}$$

The constants are now evaluated based on boundary conditions. The boundary conditions are

$$J(x=0) = J_0 e^{j\omega t} \qquad J(x \to \infty) = 0$$

The second of these relations is really not necessary since the constant B_1 must be zero; otherwise, the current density would not be finite as x tends to infinity. Thus, the only condition necessary is the first. We get

$$J(x=0) = J_0 e^{j\omega t} = B_2 e^0 = B_2 \qquad or: \qquad B_2 = J_0 e^{j\omega t}$$

From Eqs. (2.9) and (2.43) we write

$$k_x = \sqrt{j\omega\mu\sigma} = (1 + j)\sqrt{\frac{\omega\mu\sigma}{2}}$$

where the relation

$$\sqrt{j} = \frac{(1 + j)}{\sqrt{2}}$$

was used. Substituting k_x and B_2 in Eq. (2.44) we get

$$J(x) = J_0 e^{j\omega t} e^{-\sqrt{\omega\mu\sigma/2}\,x} e^{-j\sqrt{\omega\mu\sigma/2}\,x}$$

Or, in a more common form:

$$J(x) = J_0 e^{-\sqrt{\omega\mu\sigma/2}\,x} \cos\left[\omega t - \sqrt{\frac{\omega\mu\sigma}{2}}x\right]$$

Thus, the current density decays exponentially from the surface and the phase angle changes linearly from the surface. The term in the exponent preceding the variable x is the skin depth as defined in Eq. (1.143). The skin depth and phase of the current density are

$$\delta = \sqrt{\frac{2}{\omega\mu\sigma}} \tag{2.45}$$

$$\varphi = \omega t - \sqrt{\frac{\omega\mu\sigma}{2}}x \tag{2.46}$$

The current density can be written in terms of the skin depth as

$$J(x) = \left[J_0 e^{-x/\delta}\right] e^{-jx/\delta} e^{j\omega t}$$

or:

$$J(x) = \left[J_0 e^{-x/\delta}\right] \cos\left[\omega t - x/\delta\right]$$

where the term in the first bracket is the magnitude of the current density. Both magnitude and phase are strongly dependent on skin depth as is well known.

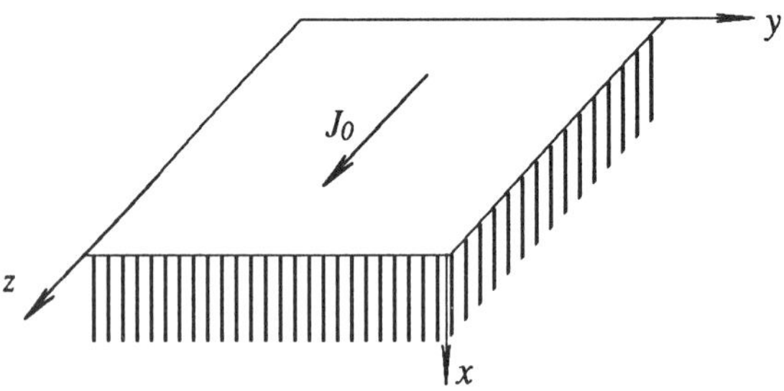

Figure 2.1 Geometry used for calculation of skin depth in planar conductors. The conductor fills the lower half-space.

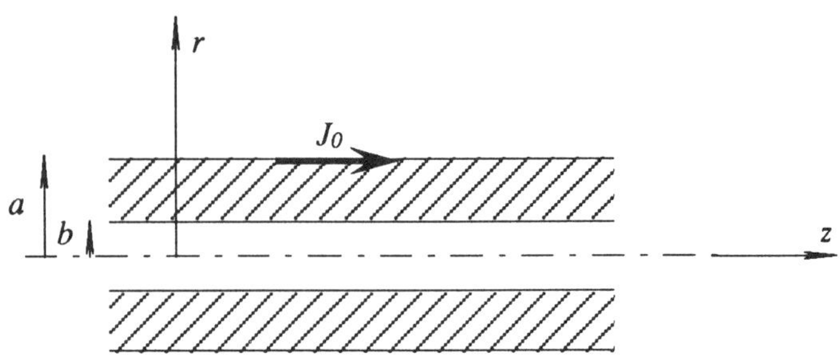

Figure 2.2 Geometry used for calculation of skin depth in cylindrical conductors. A hollow conductor is shown. For solid cylinders, b=0.

2.4.2 Skin effect in cylindrical conductors

The equivalent form of Eq. (2.43) in cylindrical coordinates is

$$\frac{d^2 J_z}{dr^2} + \frac{1}{r}\frac{dJ_z}{dr} - j\omega\mu\sigma J_z = 0 \tag{2.47}$$

This equation describes the current density in a cylindrical conductor assuming that the current density is known at the surface and it only changes in the radial direction. The basic geometry considered is shown in Figure 2.2.

Rewriting the equation above as

$$r^2\frac{d^2J_z}{dr^2}+r\frac{dJ_z}{dr}-(k_r r)^2J_z=0 \tag{2.48}$$

brings this equation to the form of Bessel's equation given in Eq. (2.23). However, because of the negative sign in the last term this is called the modified Bessel equation. Its solution is similar to the solution in Eq. (2.26) but in terms of what are called modified Bessel functions. The solution is

$$J(r)=A_1I_0(k_r r)+A_2K_0(k_r r) \tag{2.49}$$

where I_0 is the modified Bessel function of the first kind and zeroeth order and K_0 is the modified Bessel function of the second kind and zeroeth order. As with J_m and Y_m, these functions can be found either tabulated or in various analytic forms in mathematical handbooks. In addition we note that k_r is given as

$$k_r=\sqrt{j\omega\mu\sigma}=(1+j)\sqrt{\frac{\omega\mu\sigma}{2}}$$

and, therefore, the argument of the modified Bessel functions is

$$k_r r=(1+j)r\sqrt{\frac{\omega\mu\sigma}{2}}$$

To evaluate the current density everywhere in a cylindrical conductor we must use the boundary conditions of the problem. These are a finite current density at the surface (at $r=a$) and zero at $r=0$. This latter condition means that $A_2=0$. Thus, at $r=a$ we have

$$J(r{=}a)=J_0=A_1I_0(k_r a) \qquad \text{or:} \qquad A_1=\frac{J_0}{I_0(k_r a)}$$

The current density in the solid cylindrical conductor is given as

$$J(r)=J_0\frac{I_0(k_r r)}{I_0(k_r a)} \tag{2.50}$$

and the phase angle is given as

$$\alpha = tan^{-1}\left(\frac{(Im(J(r))}{Re(J(r))}\right) \tag{2.51}$$

Similar calculations can be done for hollow cylindrical conductors by imposing the condition at the outer surface of the cylinder as above and requiring that the current density be zero at the inner surface. The formula for the current density in a hollow cylindrical conductor with inner radius a_0 and outer radius a is given by

$$J(r) = J_0 e^{j\omega t}\left[\frac{K_1(k_r a_0)I_0(k_r r)}{I_0(k_r a)K_1(k_r a_0) + I_1(k_r a_0)K_0(k_r a)} + \frac{I_1(k_r a_0)K_0(k_r r)}{I_0(k_r a)K_1(k_r a_0) + I_1(k_r a_0)K_0(k_r a)}\right] \tag{2.52}$$

where the index (1) indicates modified Bessel functions of the first kind and first order.

The results above are quite different than in a planar conductor as is shown in Figures 1.12 and 1.13. These two figures compare the estimated eddy current density and phase angle at *60Hz* for a hollow cylinder of inner radius $b = 2.25$ *mm*, outer radius $a = 4.25mm$, a solid cylinder of radius $a = 4.25mm$ and the skin depth and phase angle as calculated from the standard formula for flat materials. The variations are quite significant and are discussed briefly in section 1.21. The reason for the large variations is that the skin depth (as calculated from the standard formula) is $2.25mm$. This is comparable with the radius of the cylinder and, therefore, the standard formula cannot be used to estimate skin depth.

In Figures 2.3 and 2.4, a similar calculation is performed but now the parameters are such that the skin depth δ is smaller than the radius (in this case b=$2.25mm$ and a=$10mm$). Clearly, the error in using the flat material calculation is very small and the normal practice of using the skin depth equation universally is justified.

The calculation above demonstrates the value of an analytic model: a parametric study is easily performed and results can be compared on a wide range of values. The same problem could be solved numerically but, each calculation requires a new solution and, depending on the range of parameters, may require a different set of approximations.

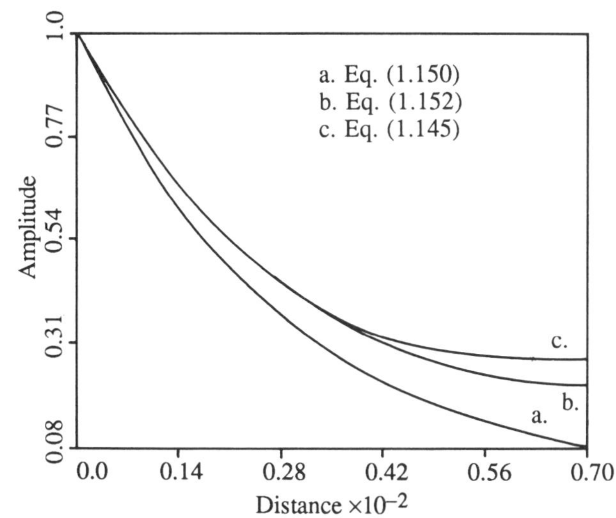

Figure 2.3 Eddy currents for a hollow cylindrical conductor, calculated with standard skin depth formula (Eq. (2.45)), Eq. (2.50), and Eq. (2.52) for $\delta > a$.

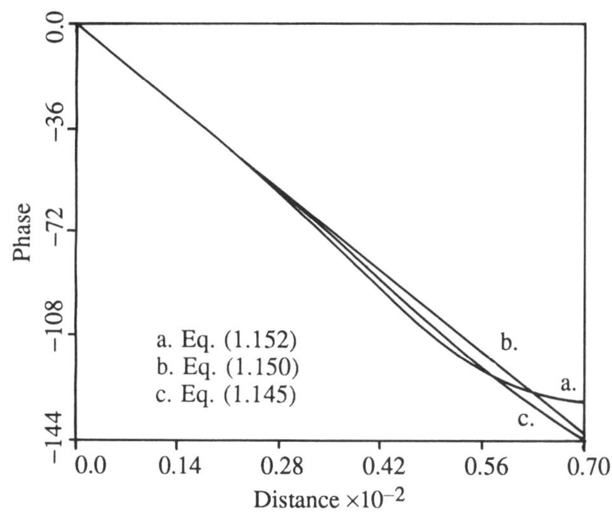

Figure 2.4 Eddy currents phase angle for a hollow cylindrical conductor calculated with standard skin depth formula (Eq. (2.45)), Eq. (2.50), and Eq. (2.52) for $\delta > a$.

2.5 EXAMPLE: TM MODES IN A RECTANGULAR WAVEGUIDE

As an example of the use of separation of variables in two dimensions we will calculate *TM* modes in a rectangular waveguide. Similar calculations can be done for cylindrical waveguides but will not be pursued here.

To define directions, we assume a rectangular waveguide as shown in Figure 2.5a. The waveguide is directed in the z direction. This is the direction along which waves are guided. The transverse directions are x and y. Figure 2.5b also shows the necessary boundary conditions on the conducting surfaces for *TM* propagation.

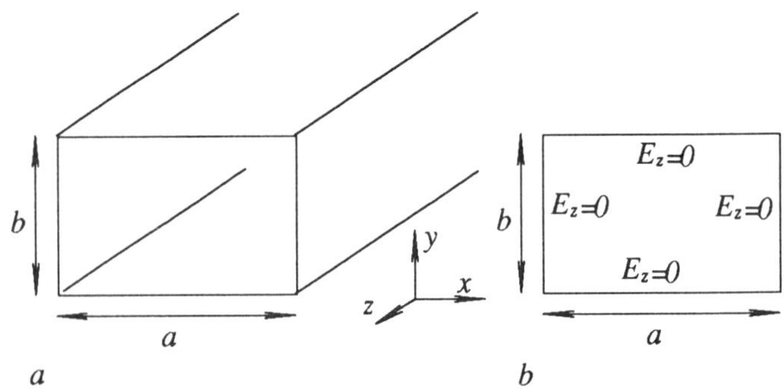

Figure 2.5. a. Structure of a rectangular waveguide. b. Boundary conditions for TM waves.

The equation for *TM* waves is given in Eq. (1.149) as

$$\frac{\partial^2 E_z}{\partial x^2} + \frac{\partial^2 E_z}{\partial y^2} + (\gamma^2 + k^2)E_z = 0 \qquad (2.53)$$

in which γ and k are defined in Eqs. (1.154). To solve this using separation of variables we write the equation as

$$\frac{\partial^2 E_z}{\partial x^2} + \frac{\partial^2 E_z}{\partial y^2} + p^2 E_z = 0 \qquad (2.54)$$

where $p^2 = \gamma^2 + k^2$. First we must perform the separation of variables as for Eq. (2.15). This is done by assuming a solution for E_z as

$$E_z(x,y) = X(x)Y(y)$$

Substitution of this general solution into Eq. (2.54) gives

$$\frac{1}{X(x)}\frac{d^2X(x)}{dx^2}+\frac{1}{Y(y)}\frac{d^2Y(y)}{dy^2}+p^2=0$$

The equation can be separated as

$$\frac{d^2X(x)}{dx^2}+k_x^2X(x)=0$$

and

$$\frac{d^2Y(y)}{dy^2}+k_y^2Y(y)=0$$

with the added condition that

$$k_x^2+k_y^2=p^2$$

The general solution is, therefore

$$E_z(x,y)=X(x)Y(y)=\left[A_1 sin(k_x x)+A_2 cos(k_x x)\right]\left[A_3 sin(k_y y)+A_4 cos(k_y y)\right] \quad (2.55)$$

where the solutions in Eqs. (2.24) and (2.25) are used in their harmonic form.

To evaluate the solution we must use the boundary conditions of the problem. These are shown in Figure 2.5b and can be summarized as

$$E(x{=}0,y)=0, \qquad E(x{=}a,y)=0$$

$$E(x,y{=}0)=0, \qquad E(x,y{=}b)=0$$

Thus, since the solution must be zero at $x=0, x=a, y=0, y=b$, only the sine functions exist and the periods of the sine functions must be such that they are zero at $x=0$ and $x=a$ for the first sine function in Eq. (2.55) and at $y=0$ and $y=b$ for the second sine function. This happens for the following conditions

$$k_x=\frac{m\pi}{a} \qquad k_y=\frac{n\pi}{b}$$

for any integers m,n, (except for $m=0, n=0$, for which only a trivial solution is obtained). Thus we can write the general solution as

$$E_z(x,y) = E_0 \frac{m\pi}{a} \sin\left(\frac{m\pi}{a}\right)x \sin\left(\frac{n\pi}{b}\right)y \qquad (2.56)$$

where the constant E_0 is an arbitrary scaling amplitude. From this we can now calculate individual terms of **E** or **H**. Since this is a *TM* mode, only transverse components of the electric and magnetic field can exist. The transverse fields are evaluated from the general solution using the source-free Maxwell's equations. The transverse components are evaluated from the following two relations:

$$\nabla \times \mathbf{E} = -j\omega\mu\mathbf{H}$$

$$\nabla \times \mathbf{H} = j\omega\varepsilon\mathbf{H}$$

The transverse components of fields for *TM* propagation are:

$$E_x(x,y) = \frac{-\gamma}{\gamma^2 + k^2} E_0 \frac{m\pi}{a} \cos\left(\frac{m\pi}{a}\right)x \sin\left(\frac{n\pi}{b}\right)y \qquad (2.57)$$

$$E_y(x,y) = \frac{-\gamma}{\gamma^2 + k^2} E_0 \frac{n\pi}{b} \sin\left(\frac{m\pi}{a}\right)x \cos\left(\frac{n\pi}{b}\right)y \qquad (2.58)$$

$$H_x(x,y) = \frac{j\omega\varepsilon}{\gamma^2 + k^2} E_0 \frac{n\pi}{b} \sin\left(\frac{m\pi}{a}\right)x \cos\left(\frac{n\pi}{b}\right)y \qquad (2.59)$$

$$H_z(x,y) = \frac{-j\omega\varepsilon}{\gamma^2 + k^2} E_0 \frac{m\pi}{a} \cos\left(\frac{m\pi}{a}\right)x \sin\left(\frac{n\pi}{b}\right)y \qquad (2.60)$$

where

$$\gamma^2 + k^2 = \left(\frac{m\pi}{a}\right)^2 + \left(\frac{n\pi}{b}\right)^2$$

This term is called a dispersion relation in (see chapter 1) and is a characteristic term for the waveguide. Chapter 1 also gives a discussion of calculation of modes based on this relation but this is not repeated here since the purpose is to show how the original equation is solved to obtain the modes.

2.6 GREEN'S FUNCTION METHOD

The idea of a Green's function method is rather general and can be applied to a number of problems. The main requirement is that the problem be linear and that the Green's function for the geometrical configuration can be found or can be synthesized from simpler geometrical configurations.

The Green's function is the field of a source. The source is assumed to be of unit strength and, therefore, the Green's function can be viewed as the general solution of a field problem due to a unit source. From this rather special solution, more general results can be obtained as will be shown shortly.

To outline the general method in its simplest form we will consider here two simple problems. The first is the solution for the electric scalar potential due to a charge distribution. The second is the solution for the magnetic vector potential due to a current distribution. Both problems are solved in unbounded space to avoid the need for satisfying boundary conditions. Considerably more complex problems can be solved and some of these are discussed in chapter 8, in the context of defining formulations for boundary element and volume element methods.

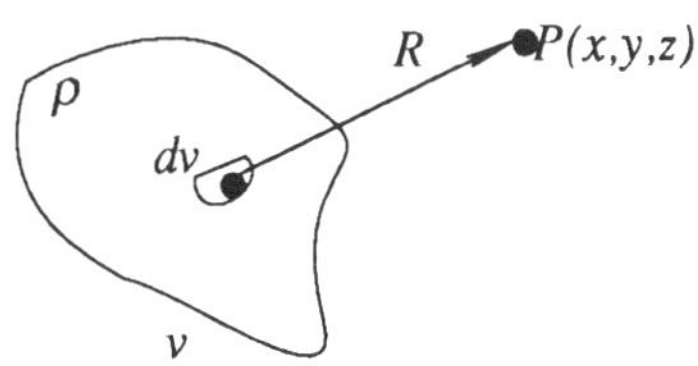

Figure 2.6 General geometry for calculation of electric fields from volume charge distributions.

Consider the problem in Figure 2.6 where a charge density ρ is distributed uniformly in a volume v. The problem is solved by solving the following Poisson equation

$$\nabla^2 V(x,y,z) = -\frac{\rho}{\varepsilon_0} \qquad (2.61)$$

However, from basic field theory, we know that the electric potential due to a point charge in free space is given by

$$V(x,y,z) = \frac{q}{4\pi\varepsilon_0 R} \qquad (2.62)$$

where R is the distance between the point charge and the location at which the field is calculated. If the charge is distributed over a volume, the solution is given by

$$V(x,y,z) = \int_v \frac{\rho}{4\pi\varepsilon_0 R}\, dv \tag{2.63}$$

This can be written in the following form:

$$V(x,y,z) = \int_v \left[\frac{\rho}{\varepsilon_0}\right]\left[\frac{1}{4\pi R}\right] dv \tag{2.64}$$

The first term in the integral is due to the source while the second is geometrical. Comparing this with Eq. (2.62), it can be viewed as the potential due to a point charge of magnitude *1*. This is therefore the Green's function for this problem. Thus, the solution is written as

$$V(x,y,z) = \int_v \frac{\rho}{\varepsilon_0} G\, dv \tag{2.65}$$

The advantage of this method is obvious: If a Green's function can be found for the problem, then a general solution is immediately available.

The second problem we wish to emphasize here is shown in Figure 2.7. It consists of a two-dimensional geometry with a current density in the cross-section as shown. Again, the problem is unbounded.

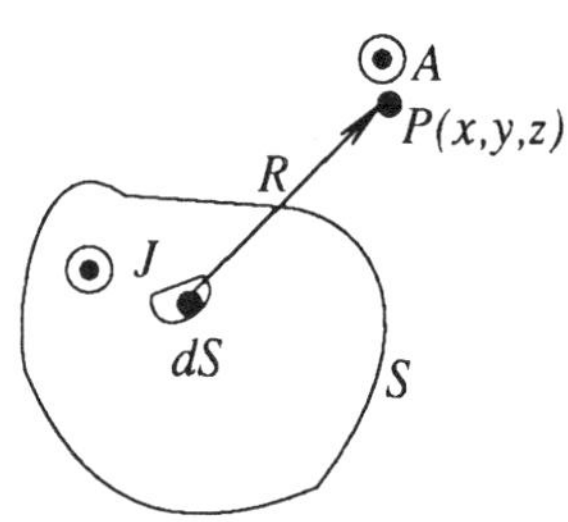

Figure 2.7 General geometry for calculation of magnetic field from current distributions. Geometry shown is a cross-section of a very long conductor.

To solve the problem, the following Poisson equation must be solved

$$\nabla^2 A_z(x,y) = -\mu J_s \tag{2.66}$$

Again, from basic field theory, the magnetic vector potential due to an infinitesimally thin wire, carrying a current *I* is given as

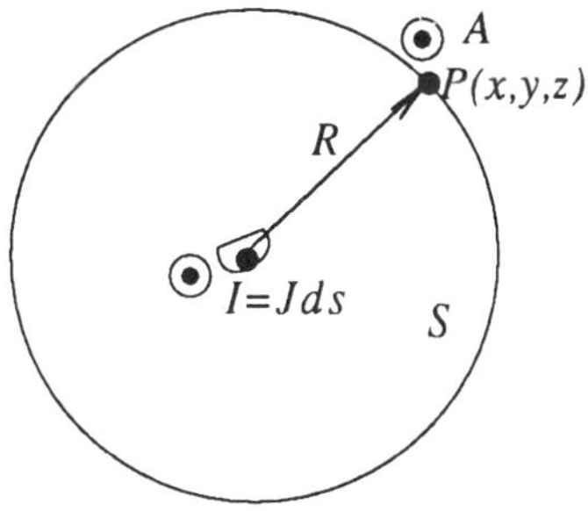

Figure 2.8. Geometry used to calculate the magnetic vector potential due to a wire with cross-sectional area ds. Biot-Savart's or Ampere's law may be used.

$$A_z(x,y) = \frac{\mu_0 I}{2\pi} \ln r \tag{2.67}$$

For an infinitesimally thick conductor shown in Figure 2.8, with current density J, the solution for the magnetic vector potential is

$$dA_z(x,y) = \frac{\mu_0 J ds}{2\pi} \ln r$$

Thus, in complete analogy with the previous problem, the solution can be written as

$$A_z(x,y) = \int_s \frac{\mu_0 J ds}{2\pi} \ln r$$

or:

$$A_z(x,y) = \int_s \mu_0 J G ds \tag{2.68}$$

where the Green's function is

$$G = \frac{\ln r}{2\pi} \tag{2.69}$$

and represents the solution for a point unit current.

Both solutions above were obtained in an unbounded domain. If we introduce boundary conditions into the problem, the solution will change. However, since the sources have not changed, we must conclude that the Green's function must be modified. Similarly, if we were to modify the basic problems above in any way (other that the trivial case of changing the source strength) the corresponding Green's function would be different. This immediately points to the difficulty in solving problems using the Green's function approach: the need to find the Green's function which, very often, is difficult to obtain. The main difficulty is, therefore, the construction of the Green's function for the problem at hand since this includes both geometrical considerations and boundary conditions for the problem.

Nevertheless, there are a fair number of rather complex problems that can be solved using this technique, one of which is shown below.

2.6.1 Example: fields and voltages in the remote field effect

In this example, the analytic solution for a coil moving inside a conductive tube is considered. This is a typical geometry in eddy current testing as well as in testing with the remote field effect. The solution takes into account the velocity of motion of the coil but, as with other analytical methods, only linear materials are considered.

The geometry of the problem is shown in Figure 2.9. A coil is moving with velocity v in the z direction inside a conductive tube of infinite length. The characteristics of the tube are measured by conductivity σ and permeability μ (permeability of air is denoted by μ_0). The goal is to construct an analytic solution to the field induced by the exciting current density J in the coil. With this field it is then possible to calculate induced currents in the tube as well as induced voltage in a pickup coil. In testing with the remote field effect a second coil is typically moving together with the inducing coil and the induced voltage in this coil is often the measured quantity. In eddy current calculations, there is normally only a driven coil but sometimes there are pickup coils as well.

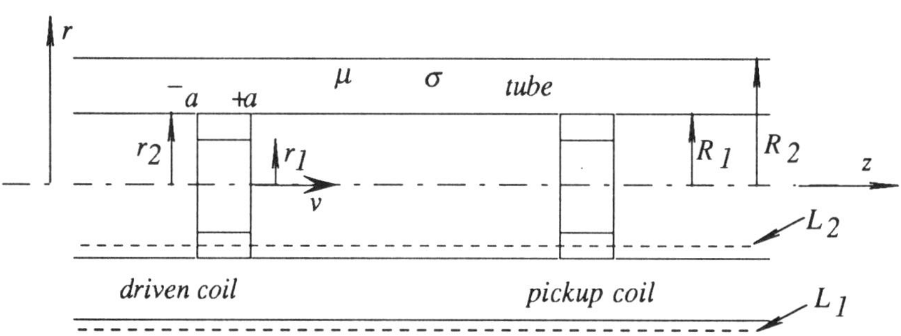

Figure 2.9. Remote field effect testing geometry as used for computation. Tube material is ferromagnetic but linear.

To solve the problem we establish a coordinate system which is moving with the coil. Under this system,

$$\mathbf{B} = \mathbf{B}' \tag{2.70}$$

$$\mathbf{E} = \mathbf{E}' + \mathbf{v}\times\mathbf{B} \tag{2.71}$$

where **B**' and **E**' are the fields in a stationary reference frame. Using the magnetic vector potential **A** together with Coulomb's gage, namely, $\nabla\times\mathbf{A}=\mathbf{B}$, $\nabla\cdot\mathbf{A}=0$, and substituting these into Maxwell's equations, we obtain the following governing equations:

$$\nabla^2\mathbf{A} + \mu_0\mathbf{J} = 0 \qquad \textit{in air} \tag{2.72}$$

$$\nabla^2\mathbf{A} - \left(\mu\sigma v\frac{\partial A}{\partial z} + \frac{\partial A}{\partial t}\right) = 0 \qquad \textit{in tube} \tag{2.73}$$

Because the geometry is cylindrical we write Eqs. (2.72) and (2.73) in cylindrical coordinates as

$$\frac{1}{r}\frac{\partial}{\partial r}\left(r\frac{\partial A}{\partial r}\right) + \frac{\partial^2 A}{\partial z^2} - \frac{1}{r^2}A = -\mu_0 J \qquad \textit{in air} \tag{2.74}$$

$$\frac{1}{r}\frac{\partial}{\partial r}\left(r\frac{\partial A}{\partial r}\right) + \frac{\partial^2 A}{\partial z^2} - \mu\sigma v\frac{\partial A}{\partial z} - \mu\sigma\frac{\partial A}{\partial t} - \frac{1}{r^2}A = 0 \qquad \textit{in tube} \tag{2.75}$$

where A denotes the ϕ component of the magnetic vector potential, and v is the velocity of the coil. This approach is typical of what we call axisymmetric geometries: the magnetic flux density is always in the direction of the current density in coils and therefore in the ϕ direction. This causes a magnetic flux density in the two directions perpendicular to the current in the coil.

Denoting Λ to be the operator

$$\Lambda = \frac{1}{r}\frac{\partial}{\partial r}\left(r\frac{\partial}{\partial r}\right) + \frac{\partial^2}{\partial z^2} - \mu\sigma v\frac{\partial}{\partial z} - \mu\sigma\frac{\partial}{\partial t} - \frac{1}{r^2}$$

the problem is modeled as

$$\Lambda A = -\mu_0 J$$

where A is bounded at $r = r_0$ and $r = \infty$.

To solve the problem above, the following Green's function problem is considered:

$$\Lambda G = -\mu_0 \delta(r - r_0)\, \delta(z - z_0) \tag{2.76}$$

and G is also bounded at $r=r_0$ and $r=\infty$. This Green's function represents the solution due to a unit current density at a point (r_0,z_0).

If J is uniformly distributed over a domain Ω and G can be found, we may construct the solution to Eq. (2.76) as

$$A = \int_\Omega JG((r - r_0),(z - z_0))dr_0 dz_0 \tag{2.77}$$

This solution satisfies Eqs. (2.74) and (2.75) as can be verified by direct substitution.

To find the Green's function, we first assume that the current density is restricted to be inside the tube (that is, the coil is a through coil). Thus, to find the Green's function the following problem is considered:

$$\frac{1}{r}\frac{\partial}{\partial r}\left(r\frac{\partial A}{\partial r}\right) + \frac{\partial^2 A}{\partial z^2} - \frac{1}{r^2} A = -\mu_0 \delta(r - r_0)\delta(z_0) \qquad \text{in air} \tag{2.78}$$

$$\frac{1}{r}\frac{\partial}{\partial r}\left(r\frac{\partial A}{\partial r}\right) + \frac{\partial^2 A}{\partial z^2} - \mu\sigma v\frac{\partial A}{\partial z} - \mu\sigma\frac{\partial A}{\partial t} - \frac{1}{r^2} A = 0 \qquad \text{in tube} \tag{2.79}$$

which describes a loop (single turn coil), carrying a unity current density, inside the tube at a radius r_0, and located at a point z_0, as shown in Figure 2.10.

We also assume that the current in the coil is sinusoidal and that the frequency is ω. This implies that a phasor notation can be used and time dependency is $e^{j\omega t}$. Denoting

$$A = \hat{A}_e e^{j\omega t}$$

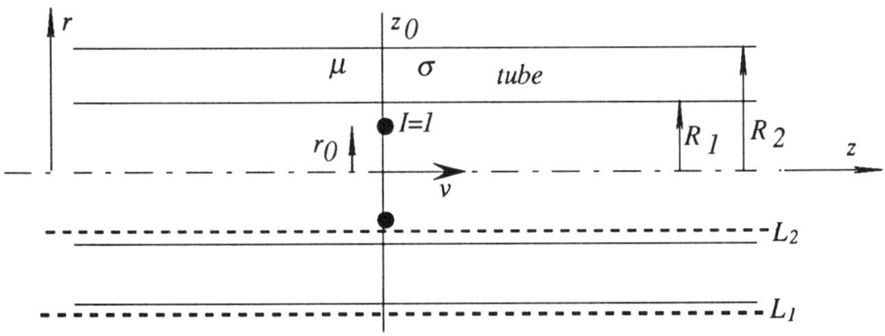

Figure 2.10 Location and dimensions of the unit current loop used for construction of Green's function.

and substituting this into Eqs. (2.78) and (2.79), gives

$$\frac{1}{r}\frac{\partial}{\partial r}\left(r\frac{\partial \hat{A}}{\partial r}\right)+\frac{\partial^2 \hat{A}}{\partial z^2}-\frac{1}{r^2}\hat{A}=-\mu_0\delta(r-r_0)\delta(z_0) \qquad \text{in air} \qquad (2.80)$$

$$\frac{1}{r}\frac{\partial}{\partial r}\left(r\frac{\partial \hat{A}}{\partial r}\right)+\frac{\partial^2 \hat{A}}{\partial z^2}-\mu\sigma v\frac{\partial \hat{A}}{\partial z}-j\omega\mu\sigma\hat{A}-\frac{1}{r^2}\hat{A}=0 \qquad \text{in tube} \qquad (2.81)$$

To simplify the solution we take the Fourier transform with respect to *z*, of both equations. This allows solution in the frequency domain followed by a transform back into the time domain. We obtain

$$\frac{1}{r}\frac{\partial}{\partial r}\left(r\frac{\partial \tilde{A}}{\partial r}\right)-\left(\xi^2+\frac{1}{r^2}\right)\tilde{A}=-\mu_0\delta(r-r_0) \qquad \text{in air} \qquad (2.82)$$

$$\frac{1}{r}\frac{\partial}{\partial r}\left(r\frac{\partial \tilde{A}}{\partial r}\right)-\left(\xi^2+j\mu\sigma(\xi v+\omega)+\frac{1}{r^2}\right)\tilde{A}=0 \qquad \text{in tube} \qquad (2.83)$$

where ξ is the transformed *z* coordinate and $\tilde{A}$ is the transformed vector potential

$$\hat{A}=\frac{1}{2\pi}\int_{-a}^{+a}\tilde{A}e^{jz\xi}\,d\xi$$

The general solutions to Eqs. (2.82) and (2.83) are:

$$\tilde{A}=aI_1(\xi r)+bK_1(\xi r) \qquad \text{in air} \qquad (2.84)$$

and

$$\tilde{A}=cI_1(\zeta r)+dK_1(\zeta r) \qquad \text{in tube} \qquad (2.85)$$

where $a,b,c,$ and d are appropriate constants to be determined from boundary conditions and

$$\zeta^2=\xi^2+j\mu\sigma(\xi v+\omega)$$

Now the magnetic vector potential inside the tube can be written as

$$\tilde{A}_1=a_1I_1(\xi r) \qquad r\le r_0 \qquad (2.86)$$

$$\tilde{A}_2=c_1I_1(\zeta r)+a_2K_1(\zeta r) \qquad R_1\ge r>r_0 \qquad (2.87)$$

where r is the general variable, r_0 the location of the loop and R_1 the inner radius of the tube as shown in Figure 2.10.
Since the magnetic vector potential changes at $r = r_0$, the continuity and jump conditions at $r = r_0$ are:

$$\tilde{A}_1 = \tilde{A}_2$$

$$\frac{\partial \tilde{A}_2}{\partial r} - \frac{\partial \tilde{A}_1}{\partial r} = -\mu_0$$

From the two conditions, the two constants a_1 and a_2 in Eqs. (2.86) and (2.87) are calculated as:

$$a_1 = \mu_0 r_0 K_1(\xi r_0) + c_1$$

$$a_2 = \mu_0 r_0 I_1(\xi r_0)$$

where c_1 is a constant.

In a similar fashion, the magnetic vector potential inside the tube wall and outside the tube can be written as

$$\tilde{A}_3 = c_2 I_1(\zeta r) + c_3 K_1(\zeta r) \qquad R_2 \geq r \geq R_1 \tag{2.88}$$

$$\tilde{A}_4 = c_4 K_1(\xi r) \qquad r > R_2 \tag{2.89}$$

To evaluate the constants we use the following interface conditions between air and material, from the property of the electromagnetic field,

$$\tilde{A}_2 = \tilde{A}_3, \qquad \frac{1}{\mu_0}\frac{\partial}{\partial r}(r\tilde{A}_2) = \frac{1}{\mu}\frac{\partial}{\partial r}(r\tilde{A}_3) \qquad at \qquad r = R_1$$

$$\tilde{A}_3 = \tilde{A}_4, \qquad \frac{1}{\mu_0}\frac{\partial}{\partial r}(r\tilde{A}_4) = \frac{1}{\mu}\frac{\partial}{\partial r}(r\tilde{A}_3) \qquad at \qquad r = R_2$$

Now a system of equations is established as

$$\begin{bmatrix} I_{IR_1} & -CI_{IR_1} & -CK_{IR_1} & 0 \\ \mu\xi R_1 I_{OR_1} & -\mu_0\zeta R_1 CI_{OR_1} & \mu_0\zeta R_1 CK_{OR_1} & 0 \\ 0 & CI_{IR_2} & CK_{IR_2} & -K_{IR_2} \\ 0 & \mu_0\zeta R_2 CI_{OR_2} & -\mu_0\zeta R_2 CK_{OR_2} & \mu\xi R_2 K_{OR_2} \end{bmatrix} \begin{bmatrix} c_1 \\ c_2 \\ c_3 \\ c_4 \end{bmatrix} = \begin{bmatrix} -\mu_0 r_0 I_{Ir_0} K_{IR_1} \\ \mu\mu_0 r_0 \xi R_1 I_{Ir_0} K_{OR_1} \\ 0 \\ 0 \end{bmatrix} \tag{2.90}$$

where the following short form notation was used

$$I_{nz} = I_n(\xi z), \qquad K_{nz} = K_n(\xi z), \qquad CI_{nz} = CI_n(\zeta z), \qquad CK_{nz} = CK_n(\zeta z)$$

Eq. (2.90) leads to the solution of Eq. (2.82) and (2.83). The solutions for the magnetic vector potential in the four domains are:

$$\widetilde{A_1} = \mu_o r_o\left(K_1(\xi r_o) + \frac{P_1}{Q} I_1(\xi r_o)\right) I_1(\xi r) \qquad r_o \ge r \tag{2.91}$$

$$\widetilde{A_2} = \mu_o r_o I_1(\xi r_o)\left(\frac{P_1}{Q} I_1(\xi r) + K_1(\xi r)\right) \qquad R_1 \ge r > r_o \tag{2.92}$$

$$\widetilde{A_3} = \mu_o r_o I_1(\xi r_o)\left(\frac{P_2}{Q} I_1(\zeta r) + \frac{P_3}{Q} K_1(\zeta r)\right) \qquad R_2 \ge r > R_1 \tag{2.93}$$

$$\widetilde{A_4} = \mu_o r_o \frac{P_4}{Q} I_1(\xi r_o) K_1(\xi r) \qquad r > R_2 \tag{2.94}$$

where

$$P_1 = \left(\mu_r CK_{IR_2}K_{OR_2} - \frac{\zeta}{\xi} K_{IR_2}CK_{OR_2}\right)\left(\mu_r CI_{IR_1}K_{OR_1} + \frac{\zeta}{\xi} K_{IR_1}CI_{OR_1}\right)$$
$$- \left(\mu_r CI_{IR_2}K_{OR_2} + \frac{\zeta}{\xi} K_{IR_2}CI_{OR_2}\right)\left(\mu_r CK_{IR_1}K_{OR_1} - \frac{\zeta}{x} K_{IR_1}CK_{OR_1}\right)$$

$$P_2 = \frac{\mu_r}{\xi R_1}\left(\mu_r CK_{IR_2}K_{OR_2} - \frac{\zeta}{\xi} K_{IR_2}CK_{OR_2}\right)$$

$$P_3 = \frac{\mu_r}{\xi R_1}\left(\mu_r CI_{IR_2}K_{OR_2} + \frac{\zeta}{\xi} K_{IR_2}CI_{OR_2}\right)$$

$$P_4 = -\frac{\mu_r}{\xi^2 R_1 R_2}$$

$$Q = \left(\mu_r CK_{IR_2}K_{OR_2} - \frac{\zeta}{\xi} K_{IR_2}CK_{OR_2}\right)\left(\mu_r CI_{IR_1}I_{OR_1} - \frac{\zeta}{\xi} I_{IR_1}CI_{OR_1}\right)$$
$$- \left(\mu_r CI_{IR_2}K_{OR_2} + \frac{\zeta}{\xi} K_{IR_2}CI_{OR_2}\right)\left(\mu_r CK_{IR_1}I_{OR_1} + \frac{\zeta}{\xi} I_{IR_1}CK_{OR_1}\right)$$

and $\mu_r = \mu/\mu_0$ is the relative permeability of the material.

Green's function for the problem defined in Eq. (2.76), is, therefore

$$G = \frac{\mu_0 r_0 e^{j\omega t}}{2\pi} \int_{-\infty}^{+\infty} [FI_1(\chi r) + HK_1(\chi r)] e^{j\xi(z - z_0)} d\xi \qquad (2.95)$$

where:

$$F = K_1(\xi r_0) + \frac{P_1}{Q} I_1(\xi r_0) \qquad r_0 \geq r$$

$$F = \frac{P_1}{Q} I_1(\xi r_0) \qquad R_1 \geq r > r_0$$

$$F = \frac{P_2}{Q} I_1(\xi r_0) \qquad R_2 \geq r > R_1$$

$$F = 0 \qquad r > R_2$$

$$H = 0 \qquad r_0 \geq r$$

$$H = I_1(\xi r_0) \qquad R_1 \geq r > r_0$$

$$H = \frac{P_3}{Q} I_1(\xi r_0) \qquad R_2 \geq r > R_1$$

$$H = \frac{P_3}{Q} I_1(\xi r_0) \qquad R_2 \geq r > R_1$$

and

$$\chi = \begin{cases} \zeta & R_2 \geq r > R_1 \\ \xi & otherwise \end{cases}$$

From Eq. (2.77) and Eq. (2.95), the solution to Eqs. (2.74) and (2.75) may be written as

$$A(r,z) = Je^{j\omega t} \int_{r_1}^{r_2} dr_0 \int_{z-a}^{z+a} dz_0 \frac{1}{2\pi} \int_{-\infty}^{+\infty} \mu_0 r_0 [FI_1(\chi r) + HK_1(\chi r)] e^{i\xi(z-z_0)} d\xi \qquad (2.96)$$

In this solution, the coil is assumed to occupy the region between $-a$ and $+a$ and between r_1 and r_2 (not to be confused with R_1 and R_2, the inner and outer radii of the tube in Figure 2.9.

The solution in Eq. (2.96) although certainly not a simple solution, can be evaluated using software libraries and the magnetic vector potential can be found anywhere in the *r-z* plane. The notation dr_0 and dz_0 indicates integration over the cross section of the coil as opposed to the general coordinates *x,y*.

From the definition of the magnetic vector potential in Eq. (1.63), the *r* and *z* components of the magnetic flux density can be calculated as

$$B_r = \frac{\partial A}{\partial z} \tag{2.97}$$

$$B_z = r\frac{\partial}{\partial r}(rA) \tag{2.98}$$

where B_r, B_z are the radial and axial components of the flux density.

To calculate the induced voltage *V* in an identical coil at $z=z_1$, we write the following

$$V_s = \frac{\partial}{\partial t}\int_s \mathbf{B}\cdot d\mathbf{s} = -j\omega\int_0^d r\,dr\int_0^{2\pi} B_z d\theta = -j2\pi\omega\int_0^d B_z r\,dr \tag{2.99}$$

for a loop of radius *d* inside the tube. Performing integration over the wire area, *V* is obtained as

$$V = -j2\pi n\omega\int_{r_1}^{r_2} dr\int_{z_1-a}^{z_1+a} dz\int_0^r B_z r\,dr \tag{2.100}$$

These formulae can now be used to obtain any analytic data required. It is also possible to include axisymmetric defects in the solution but with considerable more difficulty.

A sample of numerical results possible with this type of solution is shown in Figures 2.11 through 2.13. Figures 2.11a and 2.11b show the tangential flux density (*z* component, tangential to the tube) on lines L_1 and L_2 shown in Figure 2.9. Figure 2.11a is for zero coil velocity and Figure 2.11b for a coil velocity of *10 m/sec*. Figures 2.12a and 2.12b show the normal flux density (*r* component, normal to the tube) on lines L_1 and L_2. Figure 2.12a is for zero coil velocity and Figure 2.12b for a velocity of *10 m/sec*. Figure 2.13 shows the induced voltage in the pickup coil due to velocity of the coil assembly. For the calculations presented here, outer tube diameter is *40 mm*, inner diameter is *28 mm*, coil width is *10 mm*, coil's inner diameter is *20 mm* and the two coils are separated a distance of *112 mm*. Relative permeability of the tube is *50*, conductivity is 5.0×10^6 *S/m* and frequency is *50 Hz*.

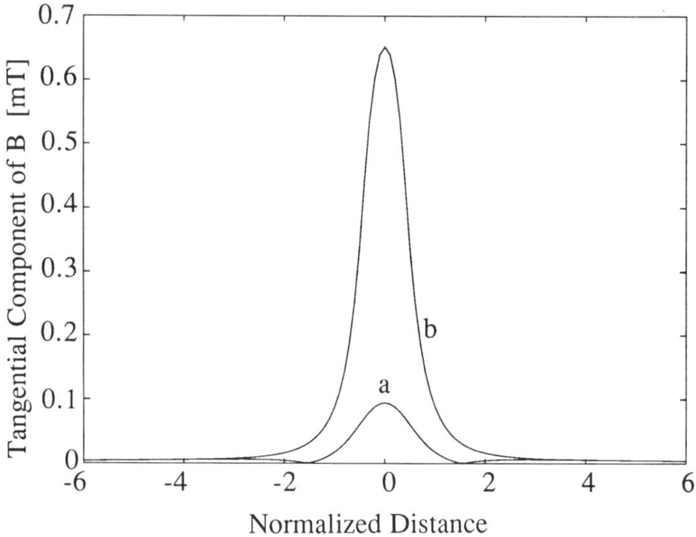

Figure 2.11 Tangential component of flux density inside and outside a ferromagnetic tube. a. Flux density on line L_1. b. Flux density on line L_2. Distance is in meters starting at center of coil, flux density is in Tesla.

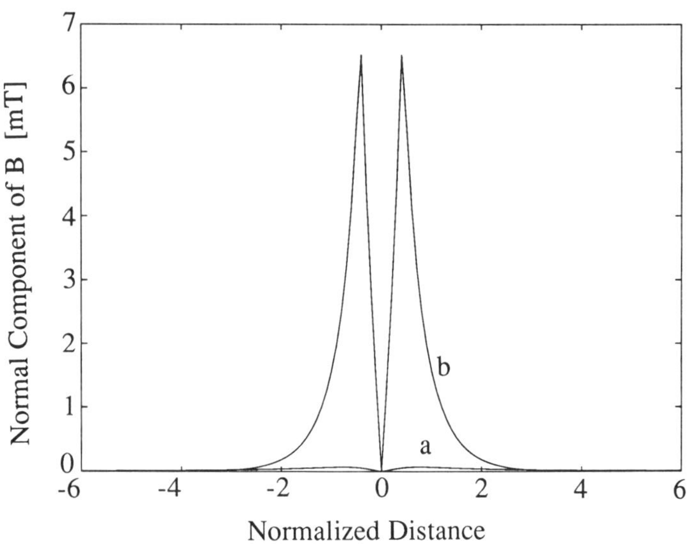

Figure 2.12 Normal component of flux density inside and outside a ferromagnetic tube. a. Flux density on line L_1. b. Flux density on line L_2. Distance is in meters starting at center of coil, flux density is in Tesla.

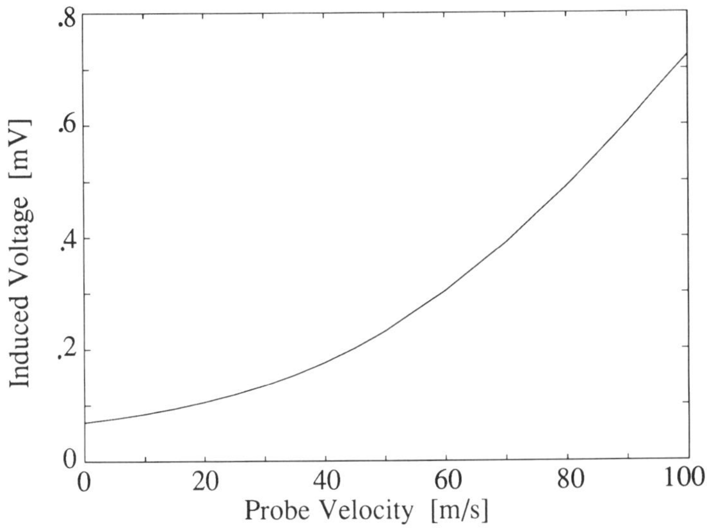

Figure 2.13 Induced voltage in a pickup coil due to probe velocity. Velocity is in m/sec.

An argument can be made that the solution obtained here is quite complex, and perhaps not every engineer can be expected to have the skill to obtain a solution of this kind. The counter argument is that, even though the solution is complex, it is in closed form and applies equally well to other problems of the same kind. Once the solution is obtained, and the required computer routines for evaluation of the various integrals are set up, a parametric study of any variable can be carried out with very little effort. The same study, if done numerically, would be much more tedious in evaluation and then only numerical results are available.

2.7 CONFORMAL MAPPING

2.7.1 General

The method of conformal mapping relies on a very simple principle: if the problem (geometry and boundary conditions) can be mapped into a much simpler geometry, the transformed problem can be solved and then the solution mapped back into the physical domain. Further, for a transformation to be conformal, we require that angles are preserved in the transformation such that both magnitude and sense of angles remain unchanged.

Thus, the question of solution by conformal transformation is one of finding an appropriate transformation and then solving the transformed problem. This of course is not always possible but there are a number of problems that can and have been solved using this method.

To understand the power of the method we will first consider an example in terms of its geometrical interpretation. Consider the calculation of potential distribution in the end region of a parallel plate capacitor. As one would expect, the fringing effect at the end region is rather complex. However, the solution for potential inside a parallel plate capacitor without fringing is rather simple: the potential is uniformly distributed since we assume the plates are infinite in extent (no fringing). If we could physically take a parallel plate capacitor (infinitely large plates) and fold the two plates at the center, we would obtain a capacitor, semi-infinite in size. The end at which folding takes place shows the fringing effect. This process is shown in Figure 2.14, and the potential (as well as flux) lines shown in Figure 2.15. Between the plates, to the left of the end zone, the potential is essentially uniform while outside, the potential distribution is quite complex. If a unique transformation can be found between the two geometries in Figure 2.14, the solution in Figure 2.15 can be obtained from that of a parallel plate capacitor by use of this transformation.

The question is, how can we find an appropriate transformation and, perhaps even more important, what are the limitations of the method. The answer to the first question is not very encouraging: while there are some standard methods used to derive the transformation for certain classes of problems, there is no single method that applies to all geometries. More often than not, it is a matter of modifying an existing transformation or finding it in a "dictionary" of conformal transformations. The latter approach should always be considered first since there are many transformations that have been formulated and can be made use of. The bibliography section lists some of the more common sources as well as sources on applications of conformal mapping.

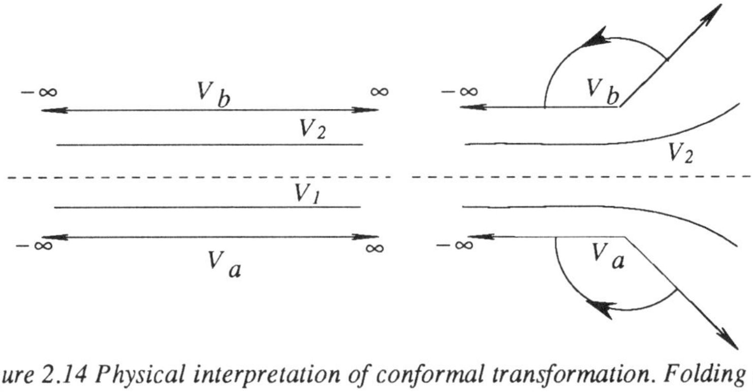

Figure 2.14 Physical interpretation of conformal transformation. Folding of the plates of the capacitor transforms the field of an infinite capacitor to that of a semi-infinite capacitor.

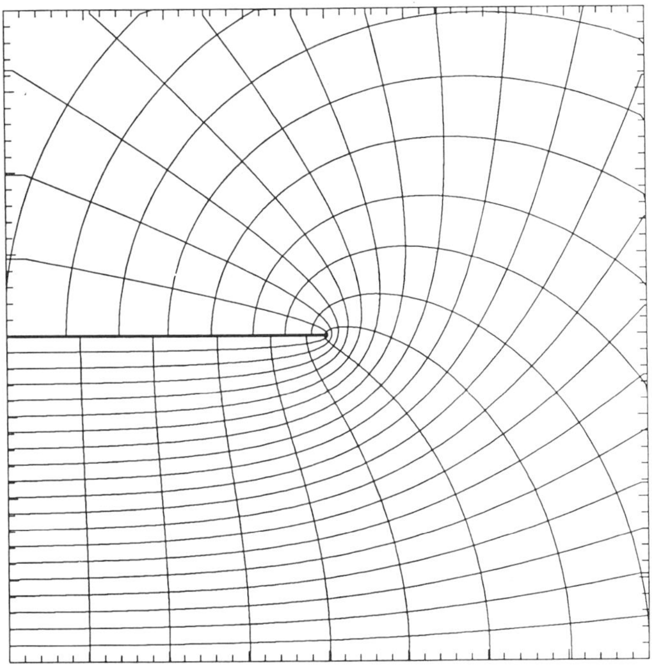

Figure 2.15 Potential and flux lines for a semi-infinite parallel plate capacitor. Only the upper half of the capacitor in the vicinity of the edge is shown.

As to limitations, conformal mapping applies mostly to solution of Laplace's equation in two-dimensional geometries. It is also possible to solve Poisson-type equations as well as eigenvalue problems but there is only a handful of problems of this type that has been attempted.

The transformation that we seek is between two complex planes: $z(x,y)$ and $w(u,v)$. More specifically, we seek a transformation of the form

$$x + jy = f(u,v) \tag{2.101}$$

The transformation is shown schematically in Figure 2.16. We require that every point in the z plane have a corresponding point in the w plane. Thus, a segment a-b corresponds to a segment a'-b' in the complex w plane.

Before considering the transformation itself, we assume that the transformation exists and look at how it may be used to solve Laplace's equation. Consider the uniform potential distribution shown in Figure 2.17. This potential represents the potential of a charged, semi-infinite conductor. The electric field due to this arrangement is constant throughout the upper half plane and the potential increases linearly from the conductor onwards towards $v=j\infty$.

The solution in Figure 2.17 can be readily obtained using Gauss' law and therefore, is a solution to Laplace's equation.

Now consider the transformation

$$x + jy = \sqrt{u + jv} \tag{2.102}$$

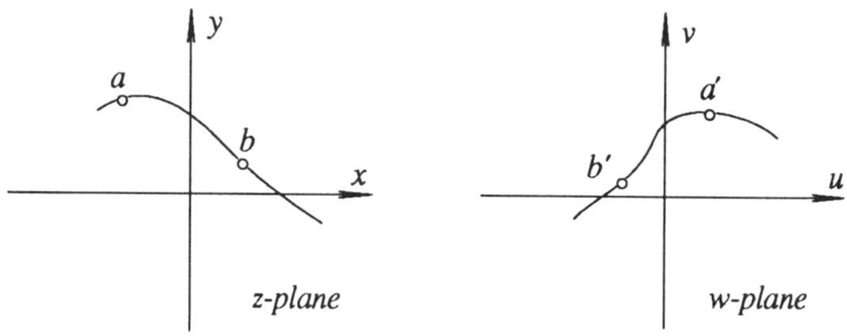

Figure 2.16 General conformal transformation. Correspondence between points in the two planes must be unique and angles must be preserved.

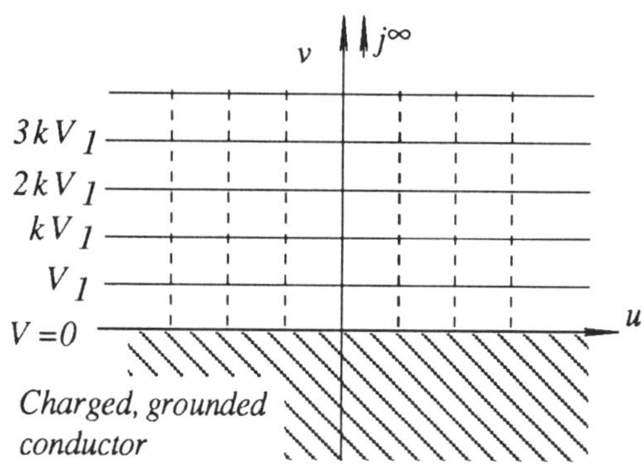

Figure 2.17 Field of a semi-infinite charged, grounded conductor. Potential and flux are uniform.

By squaring both sides we get

$$x^2 + j2xy - y^2 = u + jv$$

and the correspondence between the *w* plane and the *z* plane in Figure 2.17 is

$$2xy = v \qquad x^2 - y^2 = u \tag{2.103}$$

Thus, the negative real axis in the w plane is mapped into the imaginary axis in the z plane. The planar, semi-infinite conductor of Figure 2.17 is mapped into the geometry shown in Figure 2.18. A solution is therefore obtained for the charged conductor in Figure 2.18 which is a much more complicated geometry than that in Figure 2.17. This is shown in Figure 2.19 where some of the mapped potential and flux lines are shown.

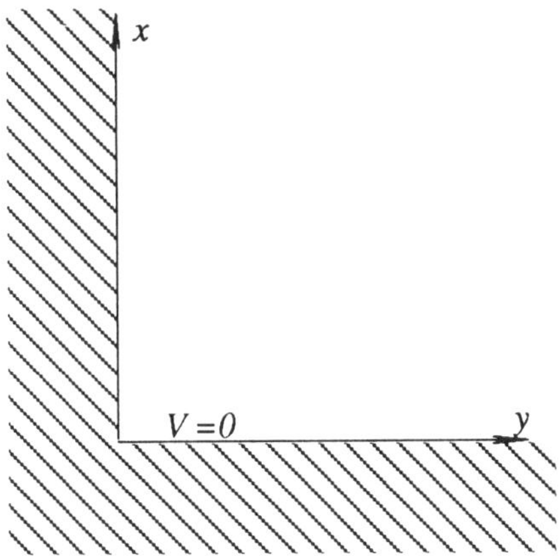

Figure 2.18 Geometry transformation.

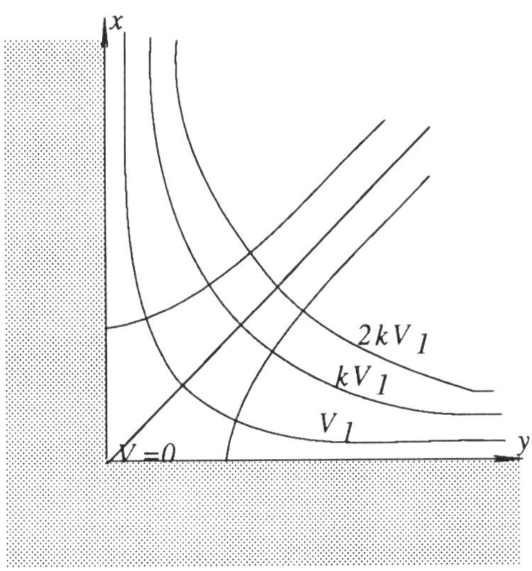

Figure 2.19 Field transformation.

To find a specific solution we note that any line parallel to the v axis is mapped into a hyperbolic line in the w plane. These are the flux lines. Thus we can write

$$k(x^2 - y^2) = \varphi_u \tag{2.104}$$

where the index u indicates that this is the flux in the w plane (parallel lines to the v axis). Since this flux is known, the flux in the z plane can be plotted directly. Similarly, for lines parallel to the u axis (potential lines) we can write

$$k2xy = V_v \tag{2.105}$$

The solution is shown in Figure 2.19.

The solution process is rather simple once the correct transformation is known. It is therefore important to look into methods of finding the transformation to a given problem. This, unfortunately is not simple. As mentioned above, systematic methods exist only for certain classes of problems while others must be solved either by "guessing" a transformation or by consulting a known solution.

2.7.2 The Schwarz-Christoffel transformation

We demonstrate here the use of the Schwarz-Christoffel transformation to define mappings of general polygonal geometries. The transformation is a general method that transforms the upper half-plane in one plane into the interior of a polygon in the second plane. In formal terms, and considering the two planes in Figure 2.20, the transformation is written as a differential equation

$$\frac{dz}{dw} = K(w - a')^{\alpha/\pi - 1}(w - b')^{\beta/\pi - 1}(w - c')^{\gamma/\pi - 1} \cdots\cdots \tag{2.106}$$

where the points $a', b', c', \ldots$ are points in the w plane (on the real axis) and the angles $\alpha, \beta, \gamma, \ldots$ are the angles of the polygon in the z plane. In this expression K is a constant and remains to be evaluated. To find the transformation between the z and w planes we integrate the expression in Eq. (2.106) and evaluate the constant K as well as any constants of integration.

We return now to the parallel plate capacitor discussed briefly above, whose solution is given in Figure 2.15. To find a solution to the field in the end zone, we first note that the field and potential distributions about the center line of the capacitor are symmetric. If we were to connect the capacitor in Figure 2.21 to a potential as shown, or the capacitor in Figure 2.11 to half the potential as shown, the solutions in Figure 2.22 or in the upper half of Figure 2.21 would be identical. We therefore take the geometry in Figure 2.22 as the z plane geometry. This can be accomplished by assuming the lower plate to be an infinite plate, while the upper plate to be a folded-over plate as shown in Figure 2.23. This geometry is viewed as an open polygon with angles $\alpha = 0$ at $z = -\infty$ and $\beta = 2\pi$ at $z = 0 + jd/2$. Thus, the mapping is quite simple: the left part of the real axis in the w plane (between points c' and a') is folded over and mapped onto the upper plate of the capacitor. The right segment (between points d' and e') is mapped onto the lower plate. Thus, the point $w = c'$ corresponds to $z = -\infty + jd/2$, $w = b'$ to $z = jd/2$, $w = d'$ to $z = -\infty$ and $w = e'$ to $z = \infty$.

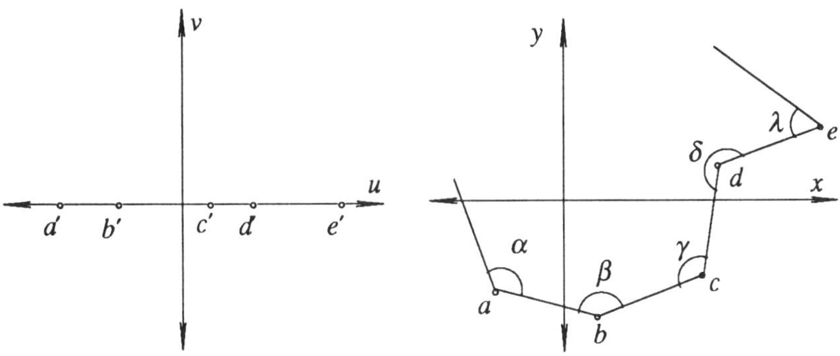

Figure 2.20 The Schwarz-Christoffel transformation.

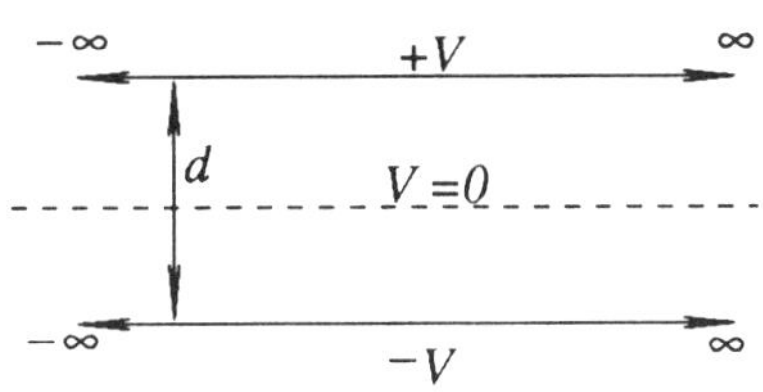

Figure 2.21 The starting point of the transformation is a parallel plate capacitor as in Figure 2. 14.

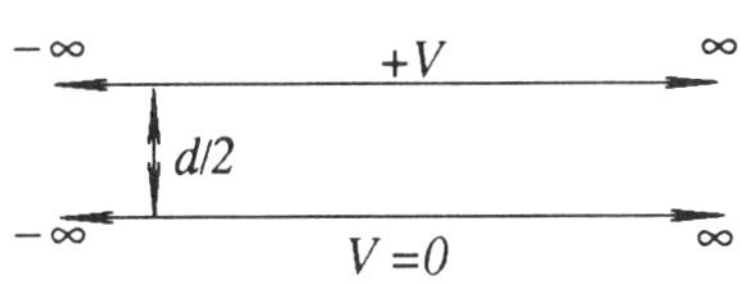

Figure 2.22 Equivalent geometry to that in Figure 2.21. This geometry is used for calculation.

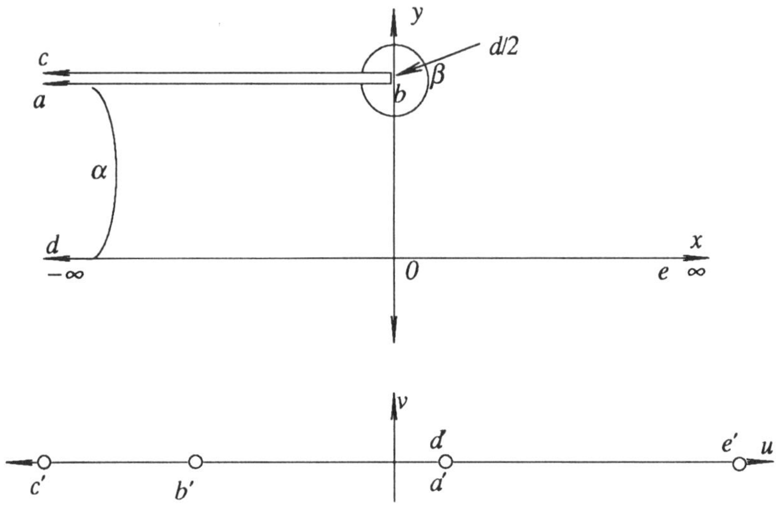

Figure 2.23 Conformal mapping for the semi-infinite capacitor. Correspondence between the w and z planes is shown.

The Schwarz-Christoffel transformation is therefore,

$$\frac{dz}{dw} = K(w - a')(w - b')^{-1} \tag{2.107}$$

The points a and b are arbitrary and we choose them as $a' = -1$ and $b' = 0$. This gives a convenient expression

$$\frac{dz}{dw} = K\left[1 + \frac{1}{w}\right]$$

Integrating we get the following

$$z = K\,(w + ln(w) + p) \tag{2.108}$$

where K and p are constants and Kp is the constant of integration. To determine the constants, consider point a. At this point $z = jd/2$ and $w = -1$. Substituting these in Eq. (2.108) gives

$$j\frac{d}{2} = K\,(-1 + ln(-1) + p) = K\,(-1 + j\pi + p) \tag{2.109}$$

where $ln(-1) = j\pi$ was used. From this

$$K = \frac{d}{2\pi} \tag{2.110}$$

by equating the imaginary parts on each side of Eq. (2.109). Using the fact that at $z = -\infty$, $w = 0$, gives

$$-\infty = K\,(0 + ln(0) + p) \tag{2.111}$$

where K is given above. Since $K>0$ and p can be chosen independently, we will choose $p=1$. Substituting these in Eq. (2.108) gives

$$z = \frac{d}{2\pi}(w + ln(w) + 1) \tag{2.112}$$

This is then the transformation and, by using the potential and flux lines of the parallel plate capacitor (which is uniform and known everywhere) we obtain the field of the finite capacitor as shown in Figure 2.15. As mentioned above, only the right hand, upper corner of the capacitor is shown.

While the discussion above dealt with electrostatic solutions, the method applies equally well to magnetic problems. In particular, the use of the magnetic scalar potential in Eq. (1.84) is identical in form and solution to the electrostatic Laplace's equation. The equivalent problem in magnetostatics is the magnetic potential between two perfectly permeable surfaces. This is satisfied (at least approximately) in air gaps of electric machines and in winding slots.

Other applications of the method are pointed out in the bibliography. However, we should mention here that the method as given above is in its simplest form. Much more complex geometries can be treated including curved geometries as well as the solution to Poisson's equation.

2.8 OTHER METHODS

While we cannot attempt to cover all analytic methods, there are a few methods that must be mentioned here. Although not as general as the methods described above these are often used and, in some cases, easier to evaluate. Among these we mention the direct evaluation from basic principles such as evaluation of the magnetic field intensity from Biot-Savart's or Ampere's laws. In combination with superposition, relatively complex geometries can be analyzed and, unlike the previous methods, their application is straightforward. Similarly, in electrostatic applications, Gauss' or Coulomb's laws are often used directly.

Another method that can sometimes be employed is the method of images which, unlike previous methods attempts to modify the problem into a new problem which, while much simpler to calculate, has the same solution as the original problem, at least in a portion of the domain.

2.8.1 Direct integration

To demonstrate the method, consider the geometry in Figure 2.24. It consists of a finite, rectangular bar carrying a current density J. We are interested in calculating the magnetic field intensity everywhere in space.

The solution is straightforward: find the solution for a very thin "wire" in the bar and view the bar as being composed of thin wires. Then, the solution for the bar is merely an integration over the bar, not unlike the Green's function solution we presented above. In this case it is easier to evaluate the magnetic vector potential of a segment. This is defined as a wire with cross section $dx'dz'$ and length $2L$. Using Figure 2.25 as a reference, the magnetic vector potential at a general point x,y,z, due to the segment at x',y',z' is

$$d\mathbf{A} = \hat{\mathbf{y}}\frac{\mu_0 dI}{4\pi}\int_{y'=-L}^{y'=+L}\frac{dy'}{R} \tag{2.113}$$

where $dI=Jds=Jdx'dz'$. Calculating R, gives

$$d\mathbf{A} = \hat{\mathbf{y}}\frac{\mu_0 J dx'dz'}{4\pi}\int_{y'=-L}^{y'=+L}\frac{dy'}{\left[(x-x')^2+(y-y')^2+(z-z')^2\right]^{1/2}}$$

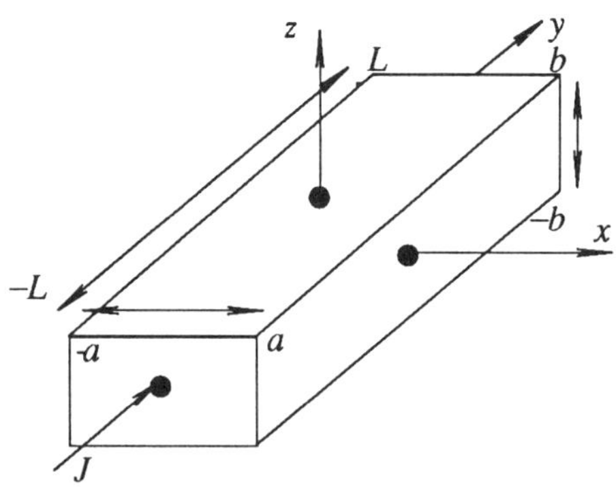

Figure 2.24 A rectangular, current carrying bar of length 2L. Origin of system of coordinates is located at the center of the bar for convenience.

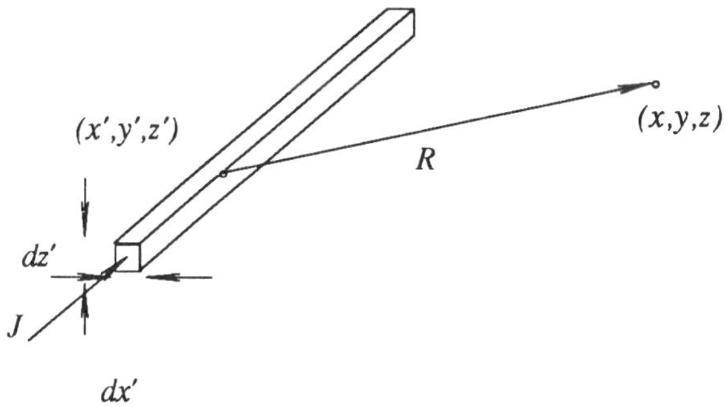

Figure 2.25 The elemental current segment used to build the solution.

The magnetic vector potential at any point in space is calculated by integrating this result over the cross section of the bar. This gives:

$$\mathbf{A} = \hat{\mathbf{y}}\frac{\mu_0 J}{4\pi}\int_{z'=-b}^{z'=+b}\left[\int_{x'=-a}^{x'=+a}\left[\int_{y'=-L}^{y'=+L}\frac{dy'}{\left[(x-x')^2+(y-y')^2+(z-z')^2\right]^{1/2}}\right]dx'\right]dz' \tag{2.114}$$

This expression can be directly evaluated rather easily but, for purposes of illustrating the general principle, is left as given above. In general, if a closed form solution to the problem cannot be found, numerical integration can be used. In spite of possible use of numerical integration, a solution of this sort should be viewed as analytic.

Now that the magnetic vector potential is known, the magnetic field intensity is calculated from the terms of the relation $\mathbf{B}=\nabla\times\mathbf{A}$ using the expression in Eqs. (1.63) and (1.12).

We note here that the only difficulty in this calculation is in integration. If the cross sectional shape were to be of a much more complex geometry, the integration would be much more difficult and might have to be done numerically. Also, the cross section is assumed to be uniform throughout the geometry and current density to be uniform in any cross section. However, it is sometimes possible to divide the cross section into smaller, simpler sections, integrate each separately and add the solutions up.

If instead of the geometry in Figure 2.24 the bar were infinitely long, we could use Ampere's law to calculate the magnetic field intensity directly. From Ampere's law, the magnetic field intensity at a distance R from the segment in Figure 2.26 (but assuming the segment is infinitely long) can be written as

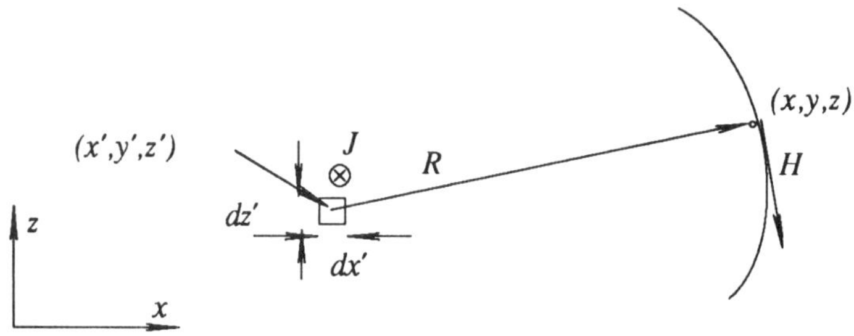

Figure 2.26 Use of Ampere's law to calculate the magnetic field intensity of an elemental current segment (shown in cross-section).

$$\oint_C \mathbf{H} \cdot d\mathbf{l} = I = Jdx'dz'$$

Using a contour at radius R we can write

$$Jdx'dz' = 2\pi RH_\phi$$

where H_ϕ is the ϕ component of the magnetic field intensity. This is shown in Figure 2.26. The ϕ component of the flux density due to the thin wire is

$$dH_\phi = \frac{Jdx'dz'}{2\pi r}$$

The two components of **H** in the x-z plane can now be written using Figure 2.26 as

$$dH_x = \frac{Jdx'dz'}{2\pi} \frac{(x-x')}{\left[(x-x')^2 + (z-z')^2\right]^{1/2}}$$

$$dH_z = -\frac{Jdx'dz'}{2\pi} \frac{(z-z')}{\left[(x-x')^2 + (z-z')^2\right]^{1/2}}$$

where we have used the same result as for a round conductor, and the distance r is written in terms of the general coordinates (x,z) and the coordinates of the current element (x',z'). This introduces no error since the segment is infinitesimally small in cross section. Now the solution is in two dimensions and the components of the magnetic field intensity are evaluated as follows

$$H_x = \frac{J}{2\pi}\int_{x'=-a}^{x'=+a}\left[\int_{z'=-b}^{z'=+b}\frac{(x-x')dz'}{\left[(x-x')^2+(z-z')^2\right]^{1/2}}dz'\right]dx' \qquad (2.115)$$

$$H_z = -\frac{J}{2\pi}\int_{x'=-a}^{x'=+a}\left[\int_{z'=-b}^{z'=+b}\frac{(z-z')dz'}{\left[(x-x')^2+(z-z')^2\right]^{1/2}}dz'\right]dx' \qquad (2.116)$$

and again, the expressions are rather easy to evaluate.

The two simple examples above can be extended to other geometries but, as a rule, only simple geometries can be treated in this fashion. Interfaces between materials, complex three-dimensional objects and nonlinear materials can seldom be evaluated.

Biot-Savart's law is often used for geometries other than planar, including the evaluation of fields of coils.

2.8.2 Method of images

The method of images is based on the principle that a field produced by a set of sources can be produced by other sets of sources. This is particularly true when we only require the evaluation of a field in a limited space.

The method of images is best known from electrostatics where the basic idea is that a charge (point charge or charge distribution) reflects in a conductor much the same as light reflects in a mirror. Under these conditions, the conductor is replaced by the image charge and the field due to the two charges is calculated. The method is usually associated with perfect conductors but this is not a necessary condition. Image methods can be used with conductors or dielectrics.

We will not discuss the method of images for charges (see bibliography) but an analogous method applies to currents over conductors. As with charges, we normally use the method with perfectly permeable materials but this is not a necessary condition. The general method of images is illustrated in Figure 2.27. As a simple illustration, consider the calculation of the magnetic field of the short bar over a permeable plane. The geometry is given in Figure 2.28. We assume the plane to be perfectly permeable. The equivalent image problem is shown in Figure 2.29. Now we can superimpose the solution for Figure 2.24 with an identical solution with negative current, displaced a distance $2d$.

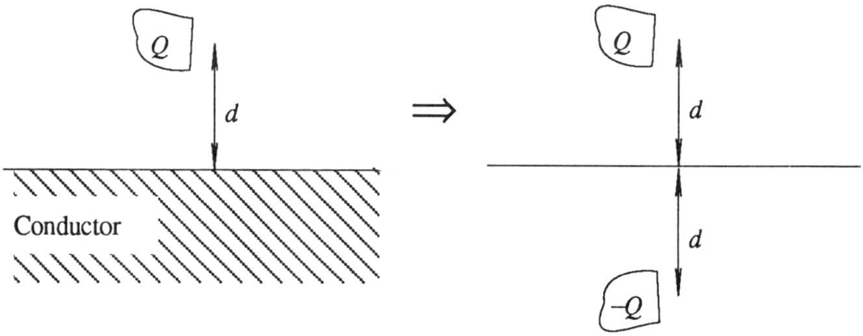

Figure 2.27 The general principle of the method of images. The sources shown as charges can be other sources and may have arbitrary shapes.

The solution for the upper conductor is given in Eq. (2.114), but since the conductor is displaced a distance *d* in the positive z direction we have

$$\mathbf{A} = \hat{\mathbf{y}}\frac{\mu_0 J}{4\pi}\int_{z'=-b+d}^{z'=+b+d}\left[\int_{x'=-a}^{x'=+a}\left[\int_{y'=-L}^{y'=+L}\frac{dy'}{\left[(x-x')^2+(y-y')^2+(z-z')^2\right]^{1/2}}\right]dx'\right]dz' \tag{2.117}$$

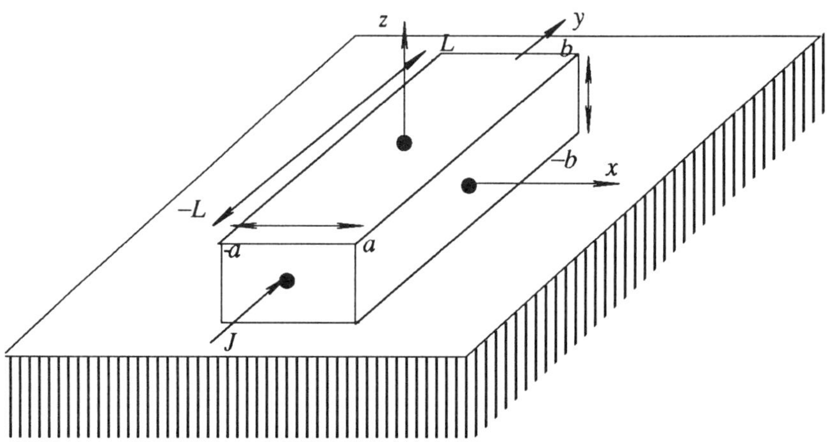

Figure 2.28 Rectangular bar over permeable planar conductor.

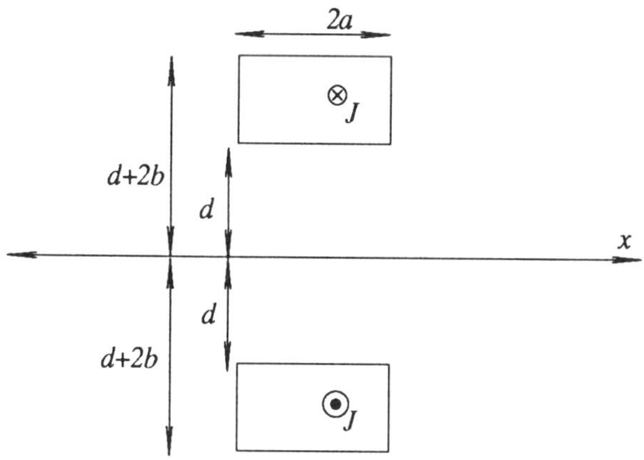

Figure 2.29 Equivalent image representation of Figure 2.28.

Similarly, for the lower conductor, the solution is displaced a distance $-d$ and in addition, the current density is negative. The solution is

$$\mathbf{A} = -\hat{\mathbf{y}}\frac{\mu_0 J}{4\pi}\int_{z'=-b-d}^{z'=+b-d}\left[\int_{x'=-a}^{x'=+a}\left[\int_{y'=-L}^{y'=+L}\frac{dy'}{\left[(x-x')^2+(y-y')^2+(z-z')^2\right]^{1/2}}\right]dx'\right]dz' \quad (2.118)$$

The field everywhere in space is now the sum of the solutions in Eqs. (2.117) and (2.118). We note however that this solution is only valid above the permeable plane in Figure 2.28. Below the plane the solution is zero as it should be in an infinitely permeable material. As with the previous solutions, the magnetic field intensity and magnetic flux densities are calculated using Eq. (1.63) and Eq. (1.12).

Variations on this type of problem can be solved using the same ideas used to solve for the electric field in dielectrics. For example, if the conducting half space in Figure 2.28 is not perfectly permeable, the solution is obtained by scaling the image current density in Figure 2.29 by the ratio $(\mu_r-1)/(\mu_r+1)$ where μ_r is the relative permeability of the bar. The solution now is rather similar to that in Eqs. (2.118) and (2.119) but the field due to the lower bar is scaled by the ratio $(\mu_r-1)/(\mu_r+1)$. This approach is a direct extension of calculation of electric fields due to point charges over dielectric materials. While the problems discussed here are rather simple, their numerical solution is, in fact more difficult than the analytic solution because of complications with boundary conditions and discretization. However, the extension of image methods to nonlinear problems and complex geometries is not possible. Calculation in general curved geometries is very difficult and anisotropic solutions are not possible. With all that the method has a value in being a very simple and mostly "physical" solution to Poisson's equation.

2.9 PROBLEMS

2.1 Consider the two-dimensional geometry shown in Figure 2.30. It consists of an infinite channel with conducting boundaries. Assuming three sides of the geometry are at zero potential and the upper side at *10V*, calculate the general solution using separation of variables and then substitute the values given to obtain a particular solution. Plot or sketch the solution. Assume the top and sides are electrically insulated from each other.

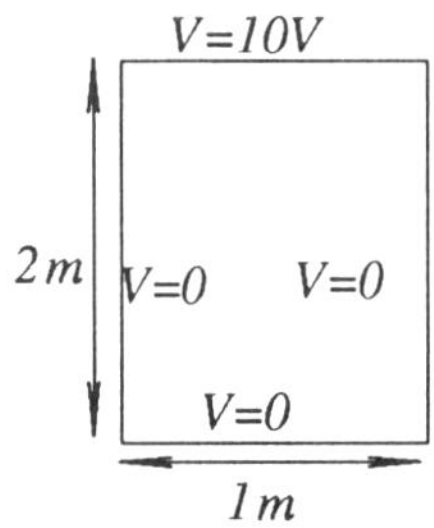

Figure 2.30.

2.2 A direct extension of problem 1.1 to a three dimensional geometry is given in Figure 2.31. For the given geometry and boundary conditions, calculate the potential everywhere inside the conducting box. Use $V = 0$ on all sides except $V = 10V$ on the top side (on the plane $y = b$).

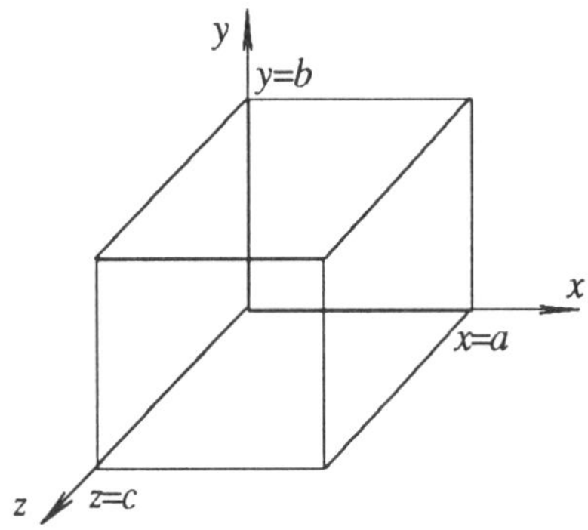

Figure 2.31.

2.3 Calculate the potential inside the hollow, cylindrical shell shown in Figure 2.32. The top of the cylinder is at *10V* while the bottom and side surfaces are at zero potential.

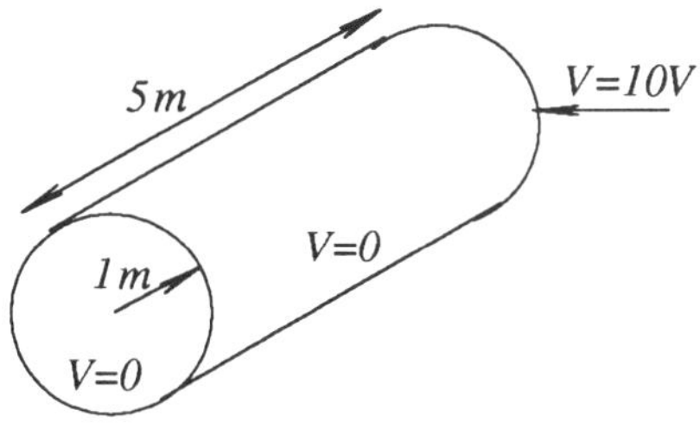

Figure 2.32.

2.4 Calculate the skin depth in sea water ($\sigma = 4S/m$, $\mu = \mu_0$) at *100Hz* and at *1GHz* ($1GHz=10^9Hz$). Compare the results using the standard formula in Eq. (2.45) and the attenuation constant for low loss materials given in Eq. (1.137). (Skin depth is defined as $\delta = 1/\alpha$). Which result is more accurate?. Does the answer change at different frequencies?

2.5 Using the method in section 2.5 find the modes and fields in a cubic resonant cavity with dimensions $a \times a \times a$. Calculate *TE* modes.

2.6 Using the Green's function method, find the field of a coil in free space. Assume the coil has inner radius a, outer radius b, and thickness c. Current density in the coil is given as J.

2.7 Repeat problem 2.6 if instead of current density, the number of turns is given as N and the current in the coil is I, keeping the same dimensions.

2.8 A charged cloud is assumed to be a sphere of radius a and contains a uniform charge density ρ. If the cloud is at a height d from the surface of the earth (and neglecting earth's effects) calculate the electric field at the earth's surface. Use the Green's function method and show that in this case, the same result is obtained using Gauss' law. Assume permittivity of free space everywhere.

2.9 Calculate the potential and electric field everywhere in space for two planes, separated a distance $2a$ as shown in Figure 2.33. The planes extend to infinity in all other directions. The left plate is grounded, while the right plate is connected to a potential V_0.

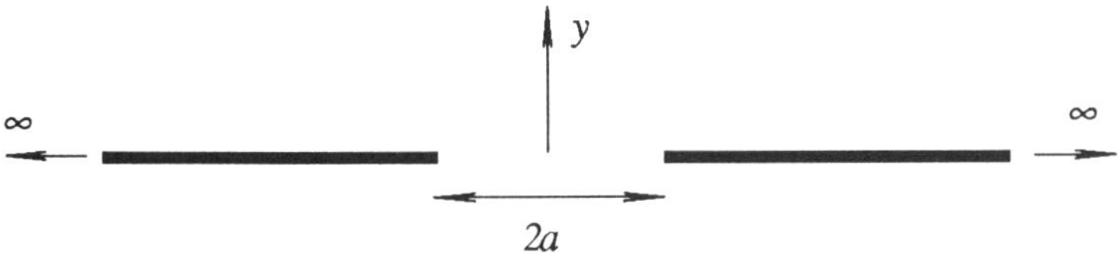

Figure 2.33.

2.10 Calculate the electric field intensity between the two surfaces shown in Figure 2.34. The surfaces are conducting and extend to infinity in the directions perpendicular to the page. Sketch the solution.

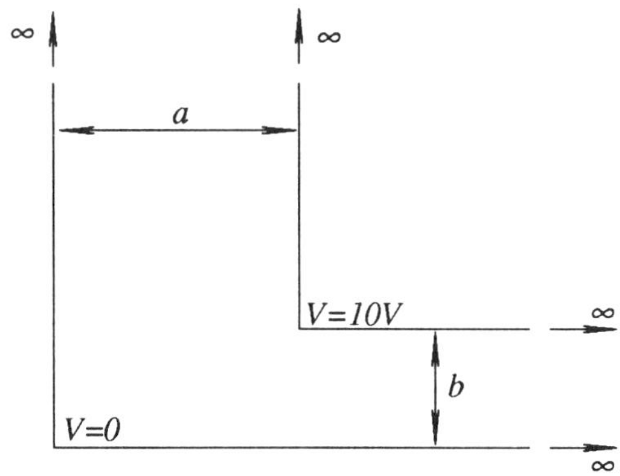

Figure 2.34.

2.11 Consider the L-shaped conductor in Figure 2.35. A total current I flows in the conductor and the conductor can be assumed to be very long. Calculate the magnetic field intensity at a general point $P(x,y,z)$.

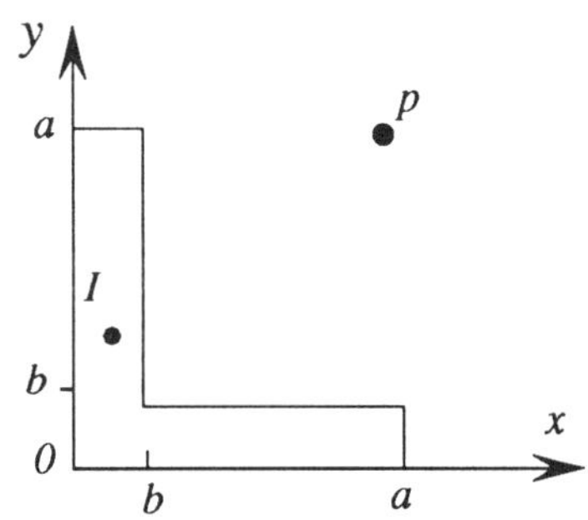

Figure 2.35.

2.12 Consider the geometry shown in Figure 2.36. Calculate the electric field intensity everywhere in space and the charge density on the surface of the conductor. The conductor fills the second, third, and fourth quadrants of space. Conductor can be assumed to be grounded.

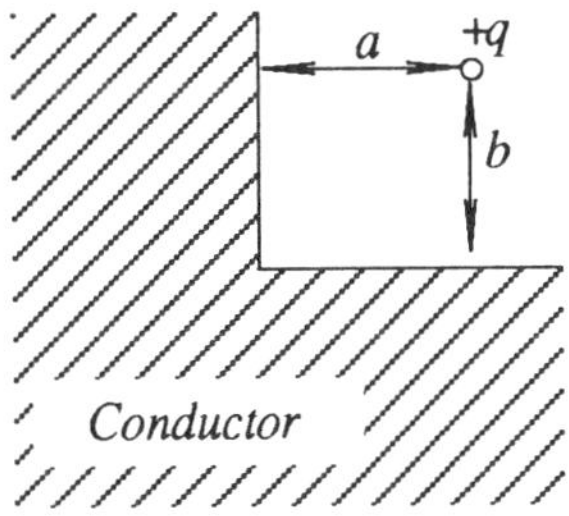

Figure 2.36.

2.13 Consider the geometry in Figure 2.37 where the two plates are conducting. A point charge is placed midway between the plates. Calculate the electric potential everywhere in space. You can assume one plate is connected to a zero potential. Hint: there will be an infinite number of image charges, similar to standing between two parallel mirrors.

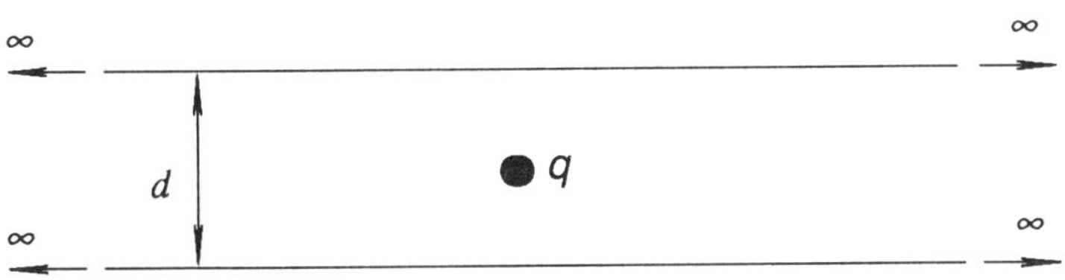

Figure 2.37.

2.14 An overhead transmission line carries a current of *1,000 Amperes*. The line is made of two wires spaced a distance *3 m* apart, at a height of *10 m* above ground. Current in the two wires is in opposite directions. Assume that the ground is perfectly conducting. Calculate the maximum magnetic field intensity at ground level. Where does this maximum occur? Assume that the wires are thin, straight, and infinitely long. What happens if the distance between the wires is reduced to *1 cm*?

2.10 BIBLIOGRAPHY

The number of problems that have been solved by application of analytical methods to field problems is extremely large. Only a few are referred to here and mostly those that are related to *NDT*. Almost all textbooks on electromagnetics discuss some of the methods with the exception of conformal transformation methods. One of the most extensive treatments of analytic methods is given in

[1] with an emphasis on separation of variables. Bessel's functions are also treated extensively. Reference [2] treats a variety of methods but excludes Green's function methods. A short but rather complete treatment of the Green's function approach is given in chapter 2 in reference [3]. Conformal mapping can be found in [4-6] as well as in [2] which also includes a short "dictionary" of conformal transformations. Many other applications of conformal mapping exist and some date from the earliest periods of field analysis. A selection of papers on analytic methods related to *NDT* are references [7-21]. In particular, [7] and [8] discuss integral solutions (separation of variables) in a variety of cylindrical geometries. A number of general methods are discussed in [22-28].

[1] Smythe, W.R, (1968), *Static and Dynamic Electricity*, McGraw-Hill, New York.

[2] Binns, K.J. and Lawrenson, P.J. (1991), *Analysis and Computation of Electric and Magnetic Field Problems*, Pergamon Press, 2nd edition, 1975.

[3] Collin R. E. , *Electromagnetic Theory of Guided Waves*, McGraw-Hill, New York.

[4] Gibbs, W.J. (1958), *Conformal Transformations in Electrical Engineering*, Chapman and Hall, London.

[5] Kober, H. (1952), *Dictionary of Conformal Representation*, Dover, New York.

[6] Bewley, L.V. (1963), *Two-Dimensional Fields in Electrical Engineering*, Dover, New York.

[7] Dodd, C.V. (1967), "*Solutions to electromagnetic induction problems*," Ph. D. Dissertation, University of Tennessee.

[8] Dodd, C.V., Deeds, W.E. and Luquire, J.W. (1969/1970), "Integral solutions to some eddy current problems", *International Journal of NDT*, Vol. 1, pp. 29-90.

[9] Waidelich, D.L. and Renken, C.J., (1956), "*The impedance of a coil near a conductor*", Proceedings of the National Electronics Conference,Vol. 12, pp. 188-196.

[10] Vine, J. (1964), "Impedance of a coil placed near to a conducting sheet", *Journal of Electronics and Control*, Vol. 16, pp. 596-577.

[11] Cheng, D.H.S. (1965), "The reflected impedance of a circular coil in the proximity of a semi-infinite medium", *IEEE Transactions on Instrumentation and Measurement*, Vol. 14, No. 3, pp.107-116.

[12] Graneau, P. and Swann, S.A. (1960), "Electromagnetic fault detection in non-ferrous pipes", *Journal of Electronics and Control*, Vol. 8, pp. 127-147.

[13] Graneau, P. (1961), "Frequency dependence of induced currents", *Journal of Electronics and Control*, Vol. 10, pp. 383-401.

[14] Vein, P.R. (1962), "Inductance between two loops in the presence of solid conducting bodies", *Journal of Electronics and Control*, Vol. 13, pp. 471-494.

[15] Burrows, M.L. (1964), "An examination of the coupled-circuit theory of eddy currents", *Journal of Electronics and Control*, Vol. 16, pp. 659-668.

[16] Freeman, E.M. and El-Markabi, M.H.S. (1979), "2-dimensional field analysis and input impedance of a finite length solenoidal coil", *Proceedings of the IEE*, Vol. 126, No. 1, pp. 135-139.

[17] Zatsepin, N.N. and Scherbinin, V.E. (1966), "Calculation of the magnetostatic field of surface defects, I. Field topography of defect models", *Defektoskopiya*, No. 5, pp. 50-59.

[18] Vlasov, V.V. and Komarov, V.A. (1971), "The magnetic field of eddy currents above a surface crack in metal with excitation of them by an applied inductor", *Defektoskopiya*, No. 6, pp. 63-76.

[19] Kahn, A.H., Spal, R. and Feldman, A. (1977), "Eddy current losses due to a surface crack in conducting material", *Journal of Applied Physics*, Vol. 48, No. 11, pp. 4454-4459.

[20] Shcherbinin, V.E. and Shur, M.L. (1976), "Calculating the effect of the boundaries of a product of the field of a cylindrical defect", *Defektoskopiya*, No. 6, pp. 30-36 (translation, 606-612).

[21] Sandovskii, V.A. (1971), "Calculating the tangential component of the magnetic field intensity due to eddy currents generated by an inductive transducer," Defektoskopiya, No. 3, pp. 90-94 (translation,331-334).

[22] Kriezis, E.E. and Tegopoulos, J.A. (1975), "Transient eddy current distribution in cylindrical shells," *IEEE Transactions on Magnetics*, Vol. MAG-11, pp. 1529-1531.

[23] Hammond, P. (1967), "Roth's method for the solution of boundary value problems in electrical engineering," *IEE Proceedings*, Vol. 114, No. 12, pp. 1969-1976.

[24] Kriezis, E.E. and Xypteras, I.E. (1979), "Eddy current distribution and loss in a semi-infinite conducting space due to a vertical current loop," *Archiv fur Electrische Energietechnik der Zeitung (ETZ)*, Vol. 7, pp. 201-207.

[25] Tegopoulos, J.A. and Kriezis, E.E. (1973), "Eddy-current distribution in cylindrical shells of finite length of one cylindrical boundary due to axial currents", *IEEE Transactions on Power Apparatus and Systems*, Vol. PAS-92, pp. 742-750.

[26] Lawrenson, P.J. and Ralph, M.C. (1970), "General 3-dimensional solution of eddy-currents and Laplacian fields in cylindrical structures", *IEE Proceedings*, Pt. A, Vol. 117, No. 2, pp. 467-472.

[27] Buyler, G.A. (1972), "Complete solution of the problem of determining the magnetic field in a plate moving at constant velocity in a plane alternating field," *Defektoskopiya*, No. 3, pp. 134-136 (translation, 371-373).

[28] Urankar, L., "Vector potential and magnetic field of current carrying circular finite arc segment in analytical form - Part I, *IEEE Transactions on Magnetics*, Vol. 16, pp. 1283-1288, 1980. Part II, *IEEE Transactions on Magnetics*, Vol. 18, pp. 911-917, 1982. Part III, *IEEE Transactions on Magnetics*, Vol. 18, pp. 1860-1867, 1982. Part IV, *IEEE Transactions on Magnetics*, Vol. 20, pp. 2145-2150, 1984. Part V: Polygonal cross section," *IEEE Transactions on Magnetics*, Vol. 26, No. 3, May 1990, pp. 1171-1180.

CHAPTER 3

The finite difference method

3.1 INTRODUCTION

The finite difference method *(FDM)* is, perhaps, the oldest numerical method for solution of differential and partial equations (the method dates from the time of Gauss). There are many reasons why the method is appealing and why it has been and is still being used so extensively. The method is relatively easy to apply as well as being general. It is equally applicable to dc fields, quasi static or transient fields and to linear and nonlinear problems. Thus, it can be applied to the solution of any field problem. The finite difference technique consists of replacing the partial derivatives of the partial differential equation by a finite difference formula which approximates the partial derivatives. The question is how are these approximations found and, from a more fundamental point of view, why are the approximations "good" representations of the partial derivatives. Once appropriate difference formulae have been derived, a solution can be obtained for the dependent variable at discrete points within the solution region either by an iterative process or by solution of a system of algebraic equations, depending on the finite difference formula chosen and the method of application of the approximation.

Of the numerical methods for partial differential equations, the finite difference method is the easiest to set up and solve. Discretization of space or time consists of defining a set of points (or grid) at which the unknown variables are calculated. Normally, the grids are regular, and result in a simple, systematic method of solving the resulting system of equations or the explicit evaluation of the unknowns. In addition, the approximation process is closely linked to the definition of the derivative and the errors involved in the approximation are well defined.

On the other hand, the implementation of the finite difference method is complicated by problems of convergence and stability of the solution in the case of time dependent applications, as well as by restrictions on the discretization process. Simple, regular grids are rather easy to define and solve but irregular grids are not. Discretization of complex geometries into regular grids is not always practical and irregular grids may in some cases render the solution nonconvergent. In particular, the discretization of awkward boundaries, small defects, and complex, irregular sources is difficult to obtain while still maintaining the basic simplicity of the method. In addition, the finite difference method is a nodal method and cannot take into account distributed parameters such as current densities, conductivities and permeabilities. These have to be described as equivalent nodal quantities with all the associated errors. The solution obtained is only valid at grid nodes. Most of these difficulties can be overcome but it is important to realize the limitations of the method at the outset.

Perhaps the main advantage of the finite difference method for field computation is its ability to handle time derivatives rather easily while other methods either cannot handle time derivatives or do so inefficiently. Therefore, even if the method is not used for discretization of space, it is often used in conjunction with other methods (such as the finite element method) when time dependent applications are called for.

3.2 THE FINITE DIFFERENCE APPROXIMATION

As mentioned above, the finite difference approximation consists of replacing derivatives (or, more commonly in this work, partial derivatives). Therefore, we need a systematic method to approximate first and second order derivatives. While higher order derivatives can also be approximated we will find no use for these in this work and will only discuss them briefly. However, the methods shown here apply to higher order derivatives as well.

Before discussing the general finite difference formulae, we will resort to a straightforward, direct method of approximation based on the definition of the derivative.

3.2.1 Finite difference formulae: an intuitive approach

Consider Figure 3.1, where a general function is described. The derivative dy/dx at a point x is the tangent to the curve at this point (y'). An approximation to the derivative can be found by taking two points, say on both sides of the point x and passing a straight line through them. If the two points are chosen to be equally spaced about the point x, as in Figure 3.1, (line y'_i) the following expression for the slope of the line is obtained

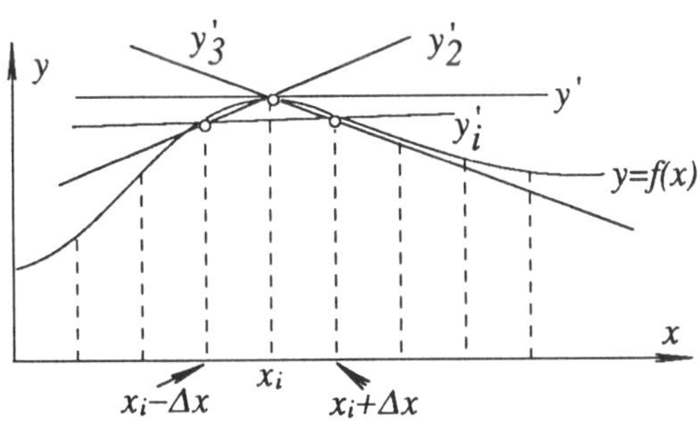

Figure 3.1. Exact derivative and approximations to the exact derivative.

$$\frac{dy(x_i)}{dx} \approx \frac{y(x_i + \Delta x) - y(x_i - \Delta x)}{2\Delta x} \tag{3.1}$$

This expression is immediately recognized as the definition of the derivative if we allow Δx to tend to zero

$$\frac{dy(x_i)}{dx} = lim_{\Delta x \to 0} \frac{y(x_i + \Delta x) - y(x_i - \Delta x)}{2\Delta x} \tag{3.2}$$

The expression in Eq. (3.1) is a difference formula and, while not the only possible approximation, it is a valid approximation to the derivative. Examples of other valid approximations, using the same points are (line y'_3):

$$\frac{dy(x_i)}{dx} \approx \frac{y(x_i + \Delta x) - y(x_i)}{\Delta x} \tag{3.3}$$

and (line y'_2)

$$\frac{dy(x_i)}{dx} \approx \frac{y(x_i) - y(x_i - \Delta x)}{\Delta x} \tag{3.4}$$

The three approximations are shown schematically in Figure 3.1, together with the exact derivative. We will not comment on these any further at this stage other than to note that the three approximations are not identical and therefore, one may be a better approximation than the other.

An approximation to the second order derivative is obtained using the first order derivatives. Suppose that we used the formula in Eq. (3.4) to calculate the first order derivative at all points of the function in Figure 3.1. The result is the function in Figure 3.2. This function is drawn as a continuous function to indicate that the number of points, and therefore Δx can be as small as we wish and therefore, the function describing the derivative can be as close to a continuous function as is practical. Now we start with the function in Figure 3.2 and calculate the first order derivative at a point x_i using Eq. (3.3). This gives

$$\frac{dy'(x_i)}{dx} \approx \frac{y'(x_i + \Delta x) - y'(x_i)}{\Delta x} \tag{3.5}$$

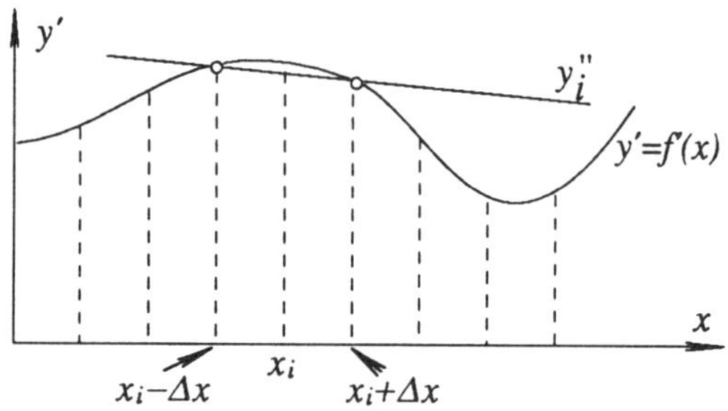

Figure 3.2 Approximation to the second order derivative.

From Figure 3.2, the derivatives $y'(x_i{+}\Delta X)$ and $y'(x_i)$ are given in terms of the original function as

$$y'(x_i + \Delta x) = \frac{y(x_i + \Delta x) - y(x_i)}{\Delta x}$$

$$y'(x_i) = \frac{y(x_i) - y(x_i - \Delta x)}{\Delta x}$$

Substitution of these two in Eq. (3.5) gives

$$\frac{dy'(x_i)}{dx} = \frac{d^2 y x_i)}{dx^2} \approx \frac{\dfrac{y(x_i + \Delta x) - y(x_i)}{\Delta x} - \dfrac{y(x_i) - y(x_i - \Delta x)}{\Delta x}}{\Delta x}$$

and, collecting terms, we get the approximation for the second order derivative as

$$\frac{d^2 y(x_i)}{dx^2} \approx \frac{y(x_i + \Delta x) - 2y(x_i) + y(x_i - \Delta x)}{(\Delta x)^2} \tag{3.6}$$

Again, as for the first order derivative, by using different combinations of points and formulae, we obtain different approximations.

The approximations given above were defined in terms of the definition of the derivative and clearly show the physical meaning of the approximation. Also to be noted is the fact that the approximation becomes better as Δx becomes smaller based on Eq. (3.2).

Before continuing, we rewrite Eqs. (3.1) and (3.6) in short form by denoting x_i as i, $x_i+\Delta x$ as x_{i+1} and $x_i-\Delta x$ as x_{i-1} and Δx as h:

$$\frac{dy_i}{dx} \approx \frac{y_{i+1} - y_{i-1}}{2h} \tag{3.7}$$

$$\frac{d^2 y_i}{dx^2} \approx \frac{y_{i+1} - 2y_i + y_{i-1}}{h^2} \tag{3.8}$$

The approximations above are given for ordinary derivatives of the function. We use the same approximations for partial derivatives. This method can be easily extended to higher order derivatives (see problem 3.2).

3.2.2 Finite difference formulae: a systematic approach

While the formulae derived in the previous section has the advantage of being derived from the definition of the derivative, we have no systematic way of evaluating the error involved in the approximation. To do so we define the same difference formulae (and others) using truncated Taylor series. This method allows a standard method of approximation and evaluation of errors.

By expanding the function $y=f(x)$ about the point x for $x=x_i-\Delta x$ and $x=x_i+\Delta x$ the following Taylor series are obtained

$$y(x_i + \Delta x) = y_i + \frac{dy_i}{dx}\Delta x + \frac{d^2 y_i}{dx^2}\frac{(\Delta x)^2}{2!} + \frac{d^3 y_i}{dx^3}\frac{(\Delta x)^3}{3!} + \ldots \tag{3.9}$$

$$y(x_i - \Delta x) = y_i - \frac{dy_i}{dx}\Delta x + \frac{d^2 y_i}{dx^2}\frac{(\Delta x)^2}{2!} - \frac{d^3 y_i}{dx^3}\frac{(\Delta x)^3}{3!} + \ldots \tag{3.10}$$

Subtracting Eq. (3.10) from Eq. (3.9) and rearranging the terms, an approximation to the first order derivative can be written as

$$\frac{dy_i}{dx} = \frac{y(x_i+\Delta x) - y(x_i-\Delta x)}{2\Delta x} - \left[\frac{1}{6}\frac{d^3 y_i}{dx^3}(\Delta x)^2\right] + \ldots \tag{3.11}$$

If derivatives of order two or higher are neglected we get the approximation in Eq. (3.1). Also, the dominant term we neglect is of the order of $(\Delta x)^2$ and, therefore, the error involved in the approximation is of this order.

By appropriate choice of expansions and combinations of expansions it is possible to derive other finite difference approximations. As an example, adding Eqs. (3.10) and (3.9) we get an approximation to the second order derivative:

$$\frac{d^2y_i}{dx^2} = \frac{y(x_i+\Delta x) - 2y(x_i) + y(x_i-\Delta x)}{(\Delta x)^2} - \left[\frac{1}{12}\frac{d^4y_i}{dx^3}(\Delta x)^2 + ...\right] \tag{3.12}$$

The error in this approximation happens to also be of the order of $(\Delta x)^2$ since the dominant term neglected is of this order.

3.2.3 Higher order formulae

The method used in the previous section can be used to approximate higher order derivatives as well. As mentioned above, this is done by an appropriate choice of expansions. As an example, an approximation to the third order derivative can be written by using a total of five points and the following expansions

$$\begin{aligned} y(x_i+2\Delta x) = y_i + \frac{dy_i}{dx}2\Delta x + \frac{d^2y_i}{dx^2}\frac{(2\Delta x)^2}{2!} + \frac{d^3y_i}{dx^3}\frac{(2\Delta x)^3}{3!} \\ + \frac{d^4y_i}{dx^4}\frac{(2\Delta x)^4}{4!} + ... \end{aligned} \tag{3.13}$$

$$\begin{aligned} y(x_i-2\Delta x) = y_i - \frac{dy_i}{dx}2\Delta x + \frac{d^2y_i}{dx^2}\frac{(2\Delta x)^2}{2!} - \frac{d^3y_i}{dx^3}\frac{(2\Delta x)^3}{3!} \\ + \frac{d^4y_i}{dx^4}\frac{(2\Delta x)^4}{4!} - ... \end{aligned} \tag{3.14}$$

To obtain an expression for the third order derivative, Eq. (3.14) is subtracted from Eq. (3.13)

$$y(x_i+2\Delta x) - y(x_i-2\Delta x) = 4\frac{dy_i}{dx}\Delta x + 2\frac{d^3y_i}{dx^3}\frac{(2\Delta x)^3}{3!} + 2\frac{d^5y_i}{dx^5}\frac{(2\Delta x)^5}{5!} + ...$$

or, after rearranging, and using the short form notation above,

$$y(i+2) - y(i-2) = 4\frac{dy_i}{dx}\Delta x + 2\frac{d^3y_i}{dx^3}\frac{(2\Delta x)^3}{3!} + 2\frac{d^5y_i}{dx^5}\frac{(2\Delta x)^5}{5!} + ...$$

Similarly, subtracting Eq. (3.10) from Eq. (3.9) we get

$$y(x_i+\Delta x) - y(x_i-\Delta x) = 2\frac{dy_i}{dx}\Delta x + 2\frac{d^3y_i}{dx^3}\frac{(\Delta x)^3}{3!} + 2\frac{d^5y_i}{dx^5}\frac{(\Delta x)^5}{5!} + ...$$

Solving for d^3y_i/dx^3 gives (after neglecting all terms with $(\Delta x)^5$ and higher

$$\frac{d^3y_i}{dx^3} = \frac{y(x_i+2\Delta x) - 2y(x_i-\Delta x) + 2y(x_i-\Delta x) + y(x_i-2\Delta x)}{2(\Delta x)^3} \tag{3.15}$$

The approximation here is also of the order of $(\Delta x)^2$ (see problem 3.5).

3.2.4 Forward, backward and central difference formulae

The finite difference expressions derived above use values at points on both sides of the point at which the derivatives are calculated. They are therefore called central difference expressions. On the other hand, the approximations in Eq. (3.3) and (3.4) use the point x_i and either a point in front of x_i $(x_i+\Delta x)$ or a point behind x_i $(x_i-\Delta x)$. The approximation in Eq. (3.3) is, therefore a forward difference formula while that in Eq. (3.4) is a backward difference formula. We define a forward difference formula as any formula that uses only values at the current level and future levels (for time derivatives) or values in "front" of the current value for spatial derivatives. Thus, in the context of the definitions above, only values of the function at point i, $i+1$, $i+1$... in any combination are allowed. This type of approximation has obvious applications in situations where values of the function at previous levels are not available. This is the case at a boundary where central difference schemes cannot be used. In the time domain, a forward difference formula means that only values at a time t and $t+\Delta t$ (or $t+n\Delta t$) are used (see Figure 3.3).

A backward difference formula uses only values at the present and past levels. These difference schemes are useful in cases where central differences are not applicable. In particular, backward difference formulae are useful in the time domain since normally we only know (or can evaluate) a function at present and past times. Backward difference formulae only require knowledge of "past" values as shown in Figure 3.3.

The simplest forward difference formula can be obtained by using Eqs. (3.9) and (3.10) and replacing Δx by zero in Eq. (3.10). The two equations become

$$y(x_i + \Delta x) = y_i + \frac{dy_i}{dx}\Delta x + ...$$

$$y(x_i) = y_i$$

Adding the two gives

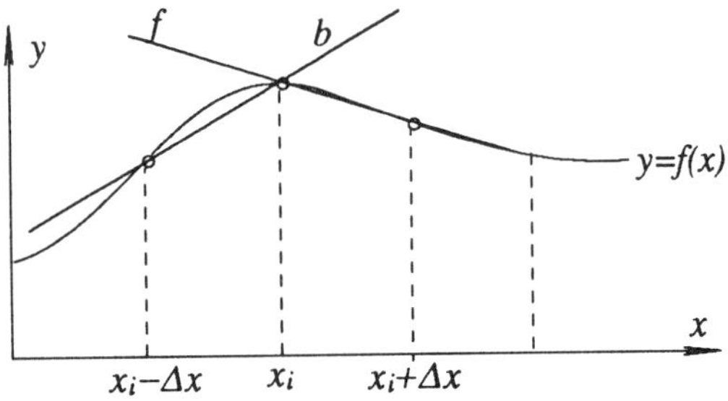

Figure 3.3 Forward (f) and backward (b) finite difference approximations.

$$y(x_i + \Delta x) + y(x_i) = 2y_i + \frac{dy_i}{dx}\Delta x + \ldots$$

and, rearranging and neglecting higher order terms gives

$$\frac{dy_i}{dx} = \frac{y(x_i + \Delta x) - y(x_i)}{\Delta x} \tag{3.16}$$

Thus, for example, the first order forward finite difference approximation (in time) can be written as

$$\frac{dy_i}{dt} = \frac{y_{i+1} - y_i}{\Delta t} \tag{3.17}$$

Similar steps lead to backward difference formulae. A backward difference formula for the first order time derivative can be written as

$$\frac{dy_i}{dt} = \frac{y_i - y_{i-1}}{\Delta t} \tag{3.18}$$

Unlike the central difference formula in Eq. (3.1), these forward and backward difference formulae have an accuracy of the order of (Δt) instead of $(\Delta t)^2$. This means that the use of these particular formulae is less advantageous than central differences although it is also possible to define forward difference formulas with higher accuracy by including more points in the derivation of the formulae (see problems 3.7 and 3.8).

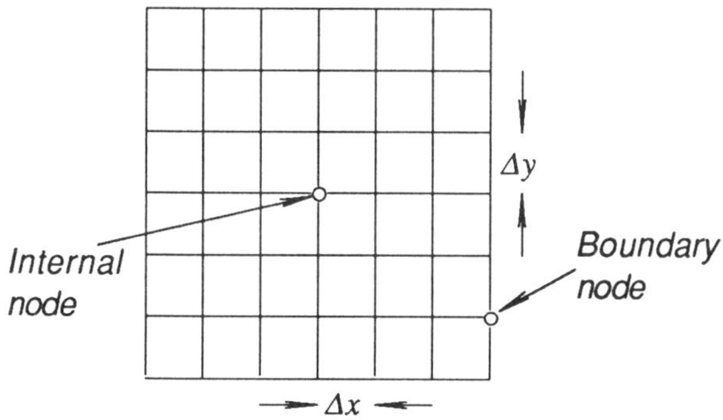

Figure 3.4 *A two-dimensional finite difference grid. Normally, $\Delta x=\Delta y$.*

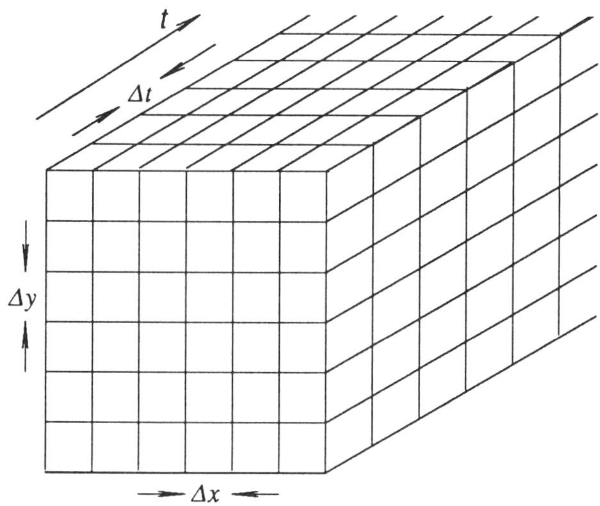

Figure 3.5 *A three dimensional finite difference grid. This mesh is suitable for two-dimensional time dependent applications.*

3.3 THE FINITE DIFFERENCE GRID

Before a problem can be formulated and solved, a finite difference grid must be defined. As mentioned above, the grid is a simple collection of nodes. In principle, the nodes can be arbitrarily located but, since we define the finite difference schemes based on equally spaced nodes, the grid should also have equally spaced nodes.

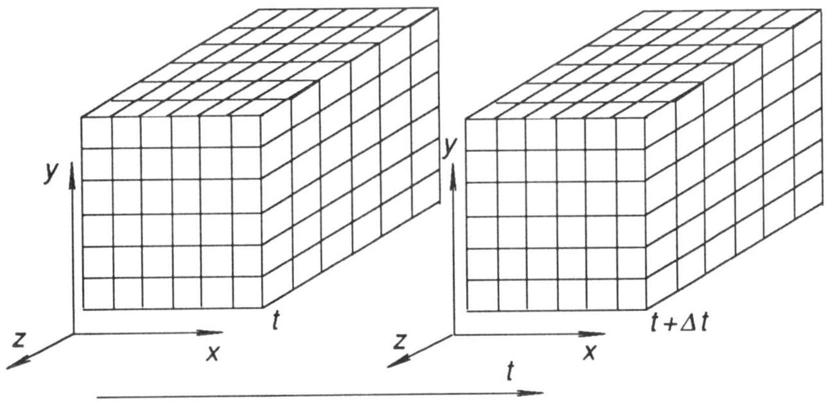

Figure 3.6 A three dimensional mesh for time dependent applications.

Figure 3.4 shows a two-dimensional grid as is normally used for solution of elliptic equations. Figure 3.5 shows a three-dimensional grid where two dimensions in space are shown, together with time discretization. Obviously, while Δx is taken to be equal to Δy, the spacing in time, Δt, is different. We will see shortly that the three spacings cannot be chosen independently. Also in this figure, time is shown as an open direction as is normally the case for time dependent problems. Figure 3.6 shows a grid for a three-dimensional, time dependent application.

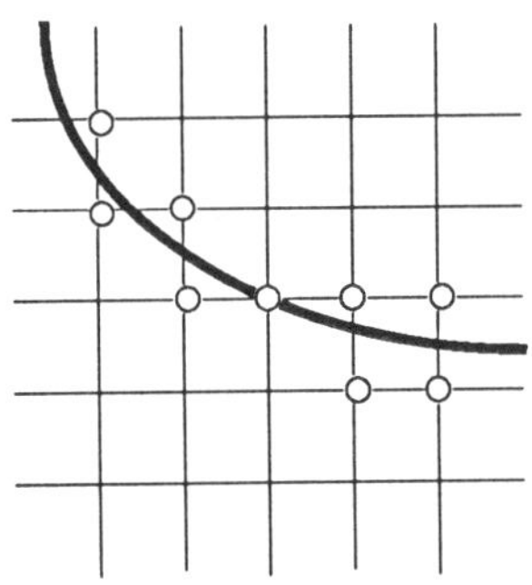

Figure 3.7 Curved boundary in a finite difference grid.

3.3.1 Curved boundaries

Uniform grids are easy to implement on rectangular geometries. Any deviation from a rectangular boundary, as for example a curved or slanted boundary causes some of the nodes to be off the boundary as shown in Figure 3.7.

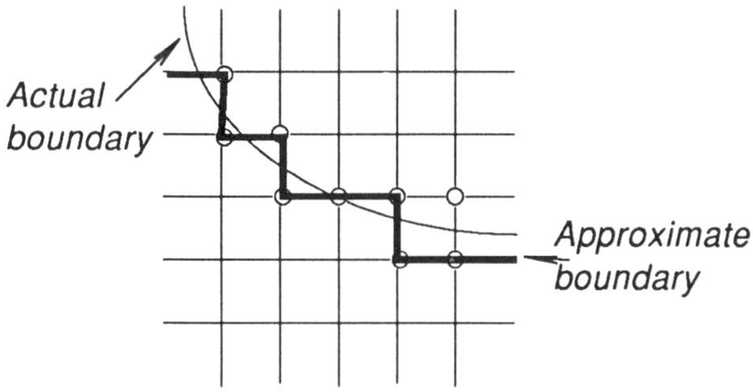

Figure 3.8 Approximation of a curved boundary by nearest node.

There are two approaches that can be adopted to overcome this difficulty. One is to use the nearest node as the boundary node for the mesh as shown in Figure 3.8. The approximate boundary is now the thick line. This method relies on the simple fact that, if we need (or if the solution requires) a more accurate approximation, all that needs to be done is to use a larger number of nodes. Of course, this makes the solution more expensive but it justifies the approximation. Figure 3.9 shows three approximations for a circular geometry. Each subsequent approximation subdivides the grid once. Thus if the grid spacing in Figure 3.9a is h, in Figure 3.9b it is $h/2$ and in Figure 3.9c it is $h/4$. The improvement is obvious but at a price: the number of nodes describing the geometry is 4, 9 and 37 in the three figures.

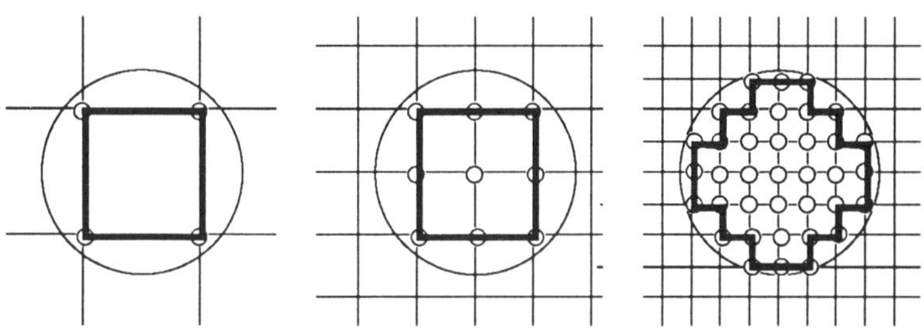

Figure 3.9 Improving curved boundary representation by mesh refinement.

An alternative method is shown in Figure 3.10. In this method, any grid line (lines in a grid are artifices: only the nodes actually exist and are used) that cuts the boundary, creates a node on the boundary at the point of intersection. The result is a grid, with nodes on the boundary but the uniformity of the mesh has been lost. To use this grid we perform interpolation between the available points and find a modified expression for the nodes at the boundary. Using Figure 3.10, the two dark points on the boundaries can be approximated by linear interpolation as

$$V_{11} = V_5\frac{\Delta x + a}{\Delta x} - V_4\frac{a}{\Delta x}$$

$$V_{10} = V_5\frac{\Delta y + b}{\Delta y} - V_8\frac{b}{\Delta y}$$

These equations may be used, for example, to calculate the potential at node *5* in terms of potentials at nodes *4*, *8*, and the two boundary values V_{10} and V_{11}.

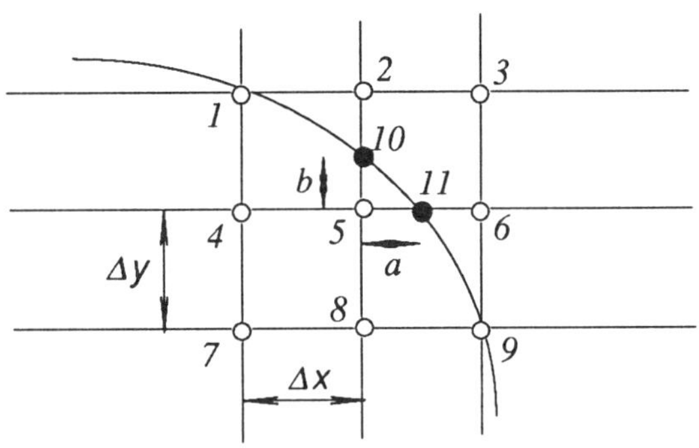

Figure 3.10 Use of interpolation to approximate nodes on curved boundaries.

3.4 EXPLICIT AND IMPLICIT FINITE DIFFERENCE METHODS

To introduce the ideas of implicit and explicit methods, we use a very simple electrostatic example. The problem is shown in Figure 3.11. It consists of an infinitely long, square box with boundary conditions as shown. This particular problem has an analytic solution, obtained by separation of variables (see problem 2.1), so that the *FDM* solution may be checked if required.

The implicit method of solution is introduced first. The first step is to define a finite difference grid over the solution domain. Since at this point we have no basis for any particular grid spacing we simply divide the solution domain into the grid shown in Figure 3.12. This forms a grid with *36* nodes, out of which, *20* nodes are on the boundary (and therefore known) and *16* are interior nodes and are unknown.

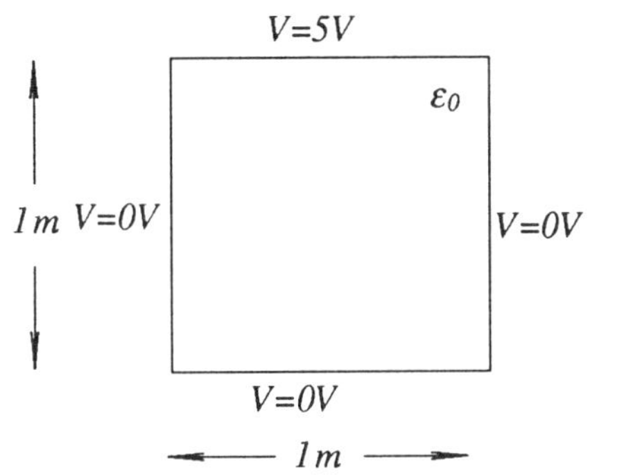

Figure 3.11 A simple geometry used to demonstrate explicit and implicit solutions for electrostatic fields (Laplace's equation).

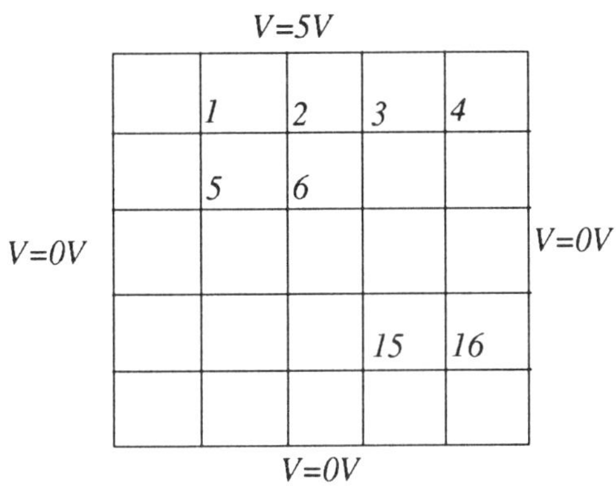

Figure 3.12 Finite difference grid and node numbering for internal nodes.

The second step is to define the finite difference approximation. The problem is described by Laplace's equation:

$$\frac{\partial^2 V}{\partial x^2} + \frac{\partial^2 V}{\partial y^2} = 0 \tag{3.19}$$

Next we define the approximations for the partial derivatives, using Eq. (3.8). Considering any interior point (i,j) in Figure 3.13 we can write for the second order partial derivative with respect to x:

$$\frac{d^2 V_{i,j}}{dx^2} \approx \frac{V_{i+1,j} - 2V_{i,j} + V_{i-1,j}}{h^2}$$

and for the second order partial derivative with respect to y:

$$\frac{d^2 V_{i,j}}{dy^2} \approx \frac{V_{i,j+1} - 2V_{i,j} + V_{i,j-1}}{h^2}$$

Substituting these expressions in Eq. (3.19), gives

$$\frac{V_{i+1,j} - 2V_{i,j} + V_{i-1,j}}{h^2} + \frac{V_{i,j+1} - 2V_{i,j} + V_{i,j-1}}{h^2} = 0$$

or, after collecting terms

$$V_{i+1,j} + V_{i-1,j} + V_{i,j+1} + V_{i,j-1} - 4V_{i,j} = 0 \qquad (3.20)$$

This approximation is independent of h but this is only a consequence of the zero right hand term and does not happen with other equations (such as Poisson's equation).

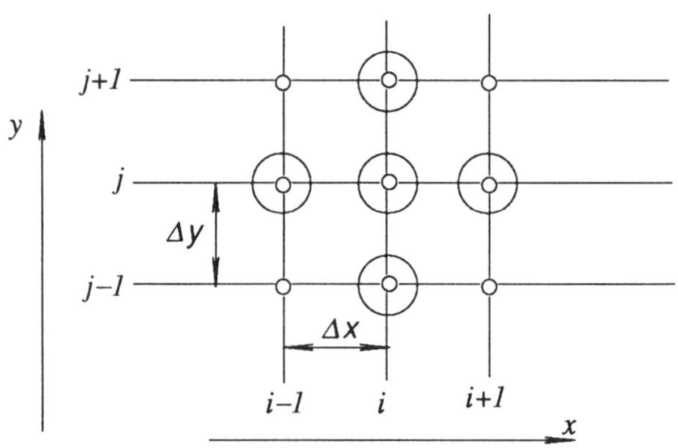

Figure 3.13 Notation used for finite difference solution. The five nodes marked are those used to approximate the Laplacian.

3.4.1 Implicit finite difference method

The potentials at all interior nodes are assumed to be unknowns. The approximation in Eq. (3.20) is now applied to any of the unknown potentials at nodes *1* through *16* in Figure 3.12. This results in *16* equations in *16* unknowns. The equations for node *1* and *7* are given here as an example

$$V_2 + 0 + V_5 + 5 - 4V_1 = 0$$

$$V_{11} + V_3 + V_8 + V_6 - 4V_7 = 0$$

An equation of this type is now established for each node and results in a system of equations as follows

$$\begin{bmatrix}
-4 & 1 & 0 & 0 & 1 & 0 & 0 & 0 & 0 & 0 & 0 & 0 & 0 & 0 & 0 & 0 \\
1 & -4 & 1 & 0 & 0 & 1 & 0 & 0 & 0 & 0 & 0 & 0 & 0 & 0 & 0 & 0 \\
0 & 1 & -4 & 1 & 0 & 0 & 1 & 0 & 0 & 0 & 0 & 0 & 0 & 0 & 0 & 0 \\
0 & 0 & 1 & -4 & 0 & 0 & 0 & 1 & 0 & 0 & 0 & 0 & 0 & 0 & 0 & 0 \\
1 & 0 & 0 & 0 & -4 & 1 & 0 & 0 & 1 & 0 & 0 & 0 & 0 & 0 & 0 & 0 \\
0 & 1 & 0 & 0 & 1 & -4 & 1 & 0 & 0 & 1 & 0 & 0 & 0 & 0 & 0 & 0 \\
0 & 0 & 1 & 0 & 0 & 1 & -4 & 1 & 0 & 0 & 1 & 0 & 0 & 0 & 0 & 0 \\
0 & 0 & 0 & 1 & 0 & 0 & 1 & -4 & 0 & 0 & 0 & 1 & 0 & 0 & 0 & 0 \\
0 & 0 & 0 & 0 & 1 & 0 & 0 & 0 & -4 & 1 & 0 & 0 & 1 & 0 & 0 & 0 \\
0 & 0 & 0 & 0 & 0 & 1 & 0 & 0 & 1 & -4 & 1 & 0 & 0 & 1 & 0 & 0 \\
0 & 0 & 0 & 0 & 0 & 0 & 1 & 0 & 0 & 1 & -4 & 1 & 0 & 0 & 1 & 0 \\
0 & 0 & 0 & 0 & 0 & 0 & 0 & 1 & 0 & 0 & 1 & -4 & 0 & 0 & 0 & 1 \\
0 & 0 & 0 & 0 & 0 & 0 & 0 & 0 & 1 & 0 & 0 & 0 & -4 & 1 & 0 & 0 \\
0 & 0 & 0 & 0 & 0 & 0 & 0 & 0 & 0 & 1 & 0 & 0 & 1 & -4 & 1 & 0 \\
0 & 0 & 0 & 0 & 0 & 0 & 0 & 0 & 0 & 0 & 1 & 0 & 0 & 1 & -4 & 1 \\
0 & 0 & 0 & 0 & 0 & 0 & 0 & 0 & 0 & 0 & 0 & 1 & 0 & 0 & 1 & -4
\end{bmatrix}
\begin{Bmatrix} V_1 \\ V_2 \\ V_3 \\ V_4 \\ V_5 \\ V_6 \\ V_7 \\ V_8 \\ V_9 \\ V_{10} \\ V_{11} \\ V_{12} \\ V_{13} \\ V_{14} \\ V_{15} \\ V_{16} \end{Bmatrix}
=
\begin{Bmatrix} -5 \\ -5 \\ -5 \\ -5 \\ 0 \\ 0 \\ 0 \\ 0 \\ 0 \\ 0 \\ 0 \\ 0 \\ 0 \\ 0 \\ 0 \\ 0 \end{Bmatrix}$$

This is a very simple, regular, and sparse structure which is typical of finite difference techniques. This structure is discussed again in chapter 11 in conjunction with solution methods. At this stage we also note that the system is symmetric, positive definite and diagonally dominant.

To obtain a solution to this system, any solution method can be used. Assuming Gaussian elimination is used, the solution is as shown in Figure 3.14. The values of the potential at the nodes are known but nowhere else. To obtain more useful data, the results can be plotted either as a contour plot (shown in Figure 3.15) or as a *3-D* field plot, shown in Figure 3.16. The results in Figures 3.15 and 3.16 were obtained for a similar geometry but the height of the box is *2 m*, the potential on the upper boundary is *10 V* and the figure is rotated by $90°$ for easier representation (see inset in Figure 3.15). These solutions required a larger number of points to obtain a smoother solution.

5.0	*5.0*	*5.0*	*5.0*	*5.0*	*5.0*
0.0	*2.9735*	*2.9735*	*2.9735*	*2.2727*	*0.0*
0.0	*1.1174*	*1.6477*	*1.6477*	*1.1174*	*0.0*
0.0	*0.5492*	*0.8523*	*0.8523*	*0.5492*	*0.0*
0.0	*0.2273*	*0.3598*	*0.3598*	*0.2273*	*0.0*
0.0	*0.0*	*0.0*	*0.0*	*0.0*	*0.0*

Figure 3.14 Solution for the potential in the geometry in Figure 3.11. Values are shown in the same order as the node numbers in Figure 3.12.

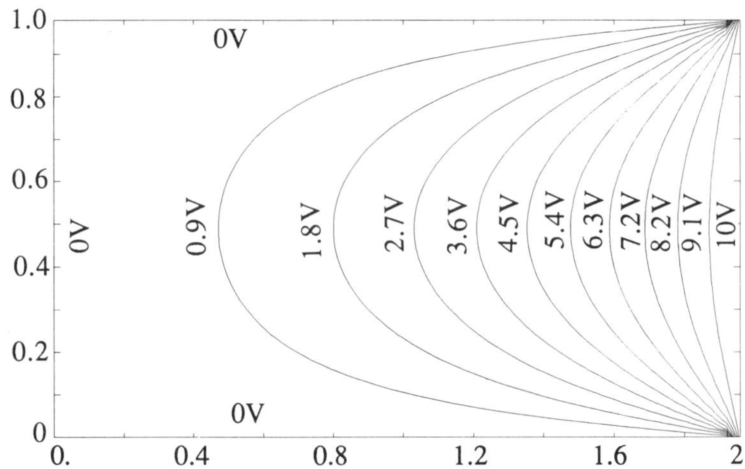

Figure 3.15 Contour lines for the geometry shown in inset.

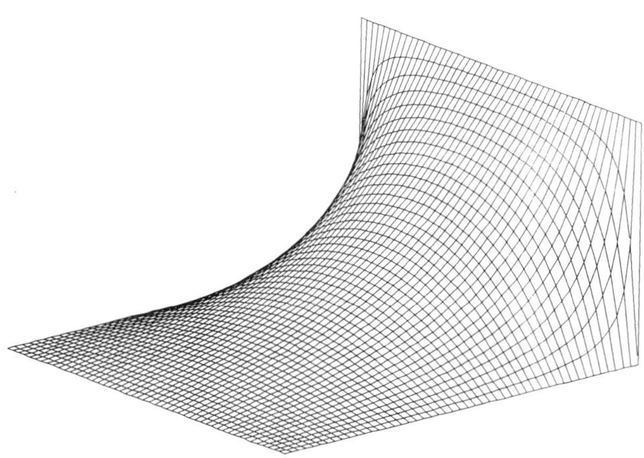

Figure 3.16 3-D plot of the solution in Figure 3.15.

The process described above is called an implicit process; the unknowns are implicit functions of the other unknowns and can only be evaluated by solving a system of equations.

3.4.2 Explicit finite difference method

In an explicit scheme, we start with the same approximation as for the implicit solution (Eq. (3.20)), but assume that all potentials are known, including at interior nodes, at any step of the solution. Before applying the approximation in Eq. (3.20) we set all interior potentials to zero (for lack of a better choice). For convenience Eq. (3.20) is rewritten as

$$V_{i,j} = \frac{V_{i+1,j} + V_{i-1,j} + V_{j,i+1} + V_{j,i-1}}{4} \tag{3.21}$$

and the question of evaluating a potential at node (i,j) is that of calculating the average of the four potentials above, below, to its left and to its right. The only additional point is that any potential that is updated, is then used for the evaluation of subsequent potentials.

As examples, the potentials at points *1*, *2*, *3*, *4*, *5*, and *6* (in Figure 3.12) are given as

$$V_1 = \frac{5+0+0+0}{4} = 1.25V \qquad V_2 = \frac{1.25+5+0+0}{4} = 1.5625V$$

$$V_3 = \frac{1.5625+5+0+0}{4} = 1.640625V$$

$$V_4 = \frac{1.640625+5+0+0}{4} = 1.66015625V$$

$$V_5 = \frac{0+1.25+0+0}{4} = 0.3125V$$

$$V_6 = \frac{0.3125+1.5625+0+0}{4} = 0.46875V$$

The solution proceeds similarly with all other nodes. At each step, a nodal value is calculated as an average of the previously known nodes. After sixteen steps, a new solution is obtained. This solution is shown in Figure 3.17. Comparing this solution with that obtained for the implicit solution in Figure 3.14 (which is exact within the round-off errors of computation), or with the analytic solution, this is obviously not the "correct" solution. Thus, we repeat the process, again starting with node *1*. We assume that the value at node *1* is unknown and evaluate it in terms of the neighboring nodes. Repeating the solution three more times, we obtain the solution in Figure 3.18. This solution is closer but still not "correct". To obtain an "accurate" solution we repeat this process, until the change in solution is lower than a pre-determined tolerance.

To see how the solution progresses consider Figure 3.19 which shows the solution at point *5* as a function of the iterations employed. After about twenty iterations, the solution is within *0.1%* of the exact solution, while after thirty iterations it is within *0.01%* of the exact solution. The solution therefore, converges to the exact solution (in this plot, $V_5=1.1174\ V$) as the number of iterations increases and, provided the number of iterations is sufficiently large, an accurate solution will be obtained.

5.0	5.0	5.0	5.0	5.0	5.0
0.0	1.25	1.5625	1.6406	1.6602	0.0
0.0	0.3125	0.4687	0.5273	0.5469	0.0
0.0	0.0781	0.1367	0.1660	0.1782	0.0
0.0	0.0195	0.0391	0.0513	0.0574	0.0
0.0	0.0	0.0	0.0	0.0	0.0

Figure 3.17 Solution to geometry in Figure 3.11 after one iteration.

5.0	5.0	5.0	5.0	5.0	5.0
0.0	2.0709	2.7094	2.7609	2.1677	0.0
0.0	0.8732	1.3229	1.3828	0.9849	0.0
0.0	0.3613	0.5997	0.6445	0.4446	0.0
0.0	0.1346	0.2351	0.2571	0.1754	0.0
0.0	0.0	0.0	0.0	0.0	0.0

Figure 3.18 Solution to geometry in Figure 3.11 after four iterations.

The only remaining question is how to stop the iteration procedure so that a good solution is obtained. We could obviously choose one node and follow the solution at this node. Since the correct solution is not known, all we can do is look at the change in solution and when the change in the solution between subsequent iterations is lower than a given error, we may assume the solution is acceptable. This can be written as

$$\left|V_i^n - V_i^{n-1}\right| \le \varepsilon$$

This is very simple but has the disadvantage that we infer the correctness of the solution based on a single node. If the solution also converges at all other nodes at the same rate (a fact we normally do not know and have no reason to assume) this approach is very good.

More appropriate would be a measure of convergence that takes into account the change at all nodes. A possible error criterion is

$$\frac{1}{N}\sum_{i=1}^{N}\left|V_i^n - V_i^{n-1}\right| \le \varepsilon \tag{3.22}$$

which is an average error per node.

The process described here is an iterative process and is explicit in that the solution is known at each iteration of the process. In general, explicit procedures are preferred over implicit methods. This is particularly important when very large grids are required and the implicit procedure would be prohibitively expensive. There are cases in which only an explicit method is possible. These occur in transient and propagation problems. In these types of calculations the solution in time is open ended and in general, we cannot assume a given time for solution. In addition, the time step is often minute, resulting in a very large number of time steps. Even if the solution could be bound in time, the number of unknowns may be so large as to render the solution impractical.

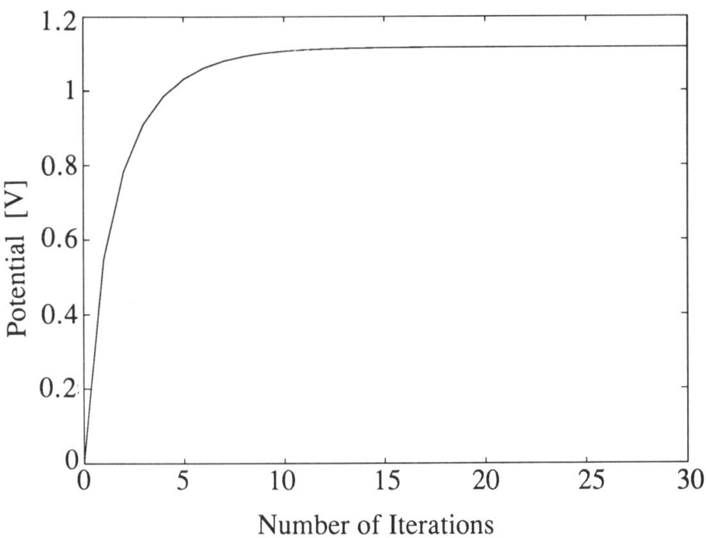

Figure 3.19 Solution at point five as a function of number of iterations.

3.5 FINITE DIFFERENCE APPROXIMATION FOR TIME DEPENDENT EQUATIONS

Consider the following two equations which, at this stage we take as merely partial differential equations without discussing their physical meaning or applications. These aspects are amply discussed in the following chapters.

$$k\left[\frac{\partial^2 \phi}{\partial x^2} + \frac{\partial^2 \phi}{\partial y^2}\right] = \frac{\partial \phi}{\partial t} \tag{3.23}$$

and

$$p\left[\frac{\partial^2 \phi}{\partial x^2} + \frac{\partial^2 \phi}{\partial y^2}\right] = \frac{\partial^2 \phi}{\partial t^2} \tag{3.24}$$

The finite difference approximations for Eq. (3.23) is

$$k\,\frac{\phi_{i+1,\,j,\,t} + \phi_{i-1,\,j,\,t} + \phi_{j,\,i+1,\,t} + \phi_{j,\,i-1,\,t} - 4\phi_{i,\,j,\,t}}{(\Delta x)^2} = \frac{\phi_{i,\,j,\,t+1} - \phi_{i,\,j,\,t}}{\Delta t}$$

where $\Delta x = \Delta y$ and a forward difference approximation in time is used. After rearranging, we get an expression for the function at time $t+\Delta t$ in terms of the solution at time t.

$$\phi_{i,\,j,t+1} = k\left[\frac{\Delta t}{(\Delta x)^2}\right]\left(\phi_{i+1,\,j,t} + \phi_{i-1,\,j,t} + \phi_{j,\,i+1,t} + \phi_{j,\,i-1,t} - 4\phi_{i,\,j,t}\right) + \phi_{i,\,j,t} \tag{3.25}$$

This equation can be solved if a grid is defined since the solution at point (i, j) at time $t+\Delta t$ depends only on values of the function at the previous time step. However, for the purpose of this section we do not need to solve the problem. This will be done in chapters 6 and 10.

Following almost identical steps, we get for the expression in Eq. (3.24)

$$\phi_{i,\,j,t+1} = p\left[\frac{(\Delta t)^2}{(\Delta x)^2}\right]\left(\phi_{i+1,\,j,t} + \phi_{i-1,\,j,t} + \phi_{j,\,i+1,t} + \phi_{j,\,i-1,t} - 4\phi_{i,\,j,t}\right) + 2\phi_{i,\,j,t} - \phi_{i,\,j,t-1} \tag{3.26}$$

where central difference formulae are used for the second order derivatives in space and time.

Equation (3.25) is characteristic of the solution of transient problems (parabolic equations) while Eq. (3.26) is characteristic of wave solutions (hyperbolic equations).

The solution of these equations is different than in the case of the time independent equation in that the choice of the time step Δt is dependent on the choice of the space step Δx.

Equation (3.25) is representative of parabolic equations and for these the required value for Δt is given as

$$k\left[\frac{\Delta t}{(\Delta x)^2}\right] \leq 0.25 \tag{3.27}$$

If Δt is chosen to satisfy this stability condition the solution will be stable and nonoscillatory. If an oscillatory solution (oscillating around the correct solution) is acceptable, the stability criterion can be written as

$$k\left[\frac{\Delta t}{(\Delta x)^2}\right] \leq 0.5 \tag{3.28}$$

Values above 0.5 result in unstable solutions.

For the hyperbolic equation in Eq. (3.26) the stability criterion is given as

$$p\left[\frac{(\Delta t)^2}{(\Delta x)^2}\right] \leq 1 \tag{3.29}$$

However, the value must be as close as possible to unity for a correct solution. The farther the value is from unity the worse the solution.

The stability criteria given above impose severe restrictions on the choice of both Δx and Δt. This aspect of computation will be discussed again in chapters 9 and 10 but it is instructive to consider the following example.

Consider a wave propagating in free space at $1GHz$. For the purpose of calculation, we assume that the spatial distance between two points is $1\ mm$. This is a very small spacing since at $1GHz$, the wavelength is $0.3\ m$ and the number of points in each direction needed to model a single wavelength would be 300. For a three-dimensional problem this translates into 27×10^6 nodes. As for the time spacing we note that

$$p=\frac{1}{c^2}=\frac{1}{9\times10^{16}}=0.11111\times10^{-16}$$

Therefore, for the condition in Eq. (3.29) to be satisfied we must have

$$9\times10^{16}\left(\frac{\Delta t}{\Delta x}\right)^2=1$$

and, for the $\Delta x=0.001$ we get:

$$\Delta t = 3.3334 \text{ picoseconds}$$

This is a rather small time step, equivalent to some *300 time steps per cycle* at the given frequency. A longer or shorter time step may be required depending on the resolution needed. The only way to achieve a change in the time step is by changing grid spacing. If grid spacing cannot be reduced because of problem restrictions (such as required resolution) a very large number of time steps may be required to satisfy the stability criterion. We discuss means of overcoming this problem in chapter 9.

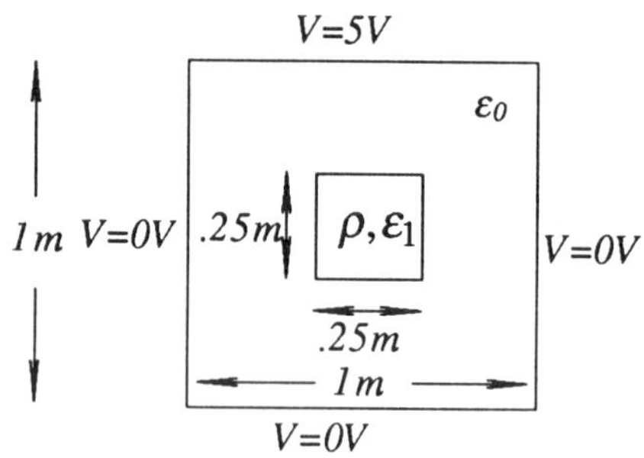

Figure 3.20 A variation of Figure 3.11 to which a dielectric material is added.

3.6 INCLUSION OF MATERIAL PROPERTIES

In the discussions that preceded, and in all examples given, we have not mentioned material properties. This was done to simplify the discussion. However, material properties are an integral part of the solution and must be considered when solving a problem.

In the context of finite differences, material properties must be associated with nodes of the mesh rather than with volumes or surfaces. Consider again the electrostatic problem in section 3.4. However, now a charge density is included as shown in Figure 3.20. The dimensions of the outer boundaries and the charge distribution are shown. The equation describing this situation is

$$\frac{\partial^2 V}{\partial x^2} + \frac{\partial^2 V}{\partial y^2} = -\frac{\rho}{\varepsilon} \tag{3.30}$$

and the finite difference representation is

$$\frac{V_{i+1,j} + V_{i-1,j} + V_{i,j+1} + V_{i,j-1} - 4V_{i,j}}{(\Delta x)^2} = -\frac{\rho_{i,j}}{\varepsilon_{i,j}} \qquad (3.31)$$

where both charge density and permittivity are those at point (i, j). We do not actually solve this problem but, to show how materials are taken into account, consider the two grids in Figure 3.21.

In Figure 3.21a, the grid spacing is larger than the charge distribution and no point of the grid is located within the charge area. When applying the *FD* formula in Eq. (3.31), no point in the grid contributes any charge density and the solution would be identical to that in section 3.4. Thus, the charge density does not exist as far as this grid is concerned.

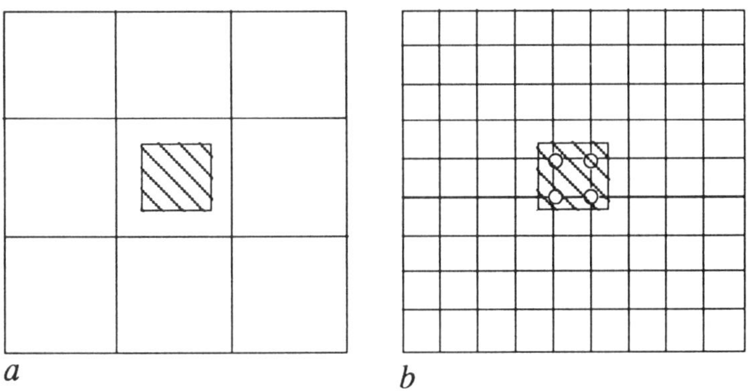

Figure 3.21 Two finite difference grids used to discretize Figure 3.20. a. No grid points in the dielectric. b. Dielectric includes four grid points.

In the grid in Figure 3.21b, the charge distribution contains four grid points. These four points contribute to the right hand side of Eq. (3.31) but no other points. Thus, the solution is equivalent to that of a charge density of size h^2. In this case, $h^2=0.0123\ m^2$ while the charge density is distributed on an area $0.0625\ m^2$. Since the actual distribution is almost 5 times larger than what the grid takes into account, the solution will be in error. The obvious answer is to use a fine enough grid spacing so that the charge density is properly modeled. However, this does not solve the problem on the edges of the charge distribution. As an example, consider the grid points shown in Figure 3.22.

Since the potential must be calculated at a point within the charge distribution, but very close to the edge, one node used for the calculation is outside the charge distribution. It is therefore reasonable to assume that the contribution of this approximation is different than that of a node for which all four neighboring nodes are within the charge distribution. Similarly, if a node at a corner of the interface is evaluated (see Figure 3.22, white dots), the contribution of this node should be smaller yet.

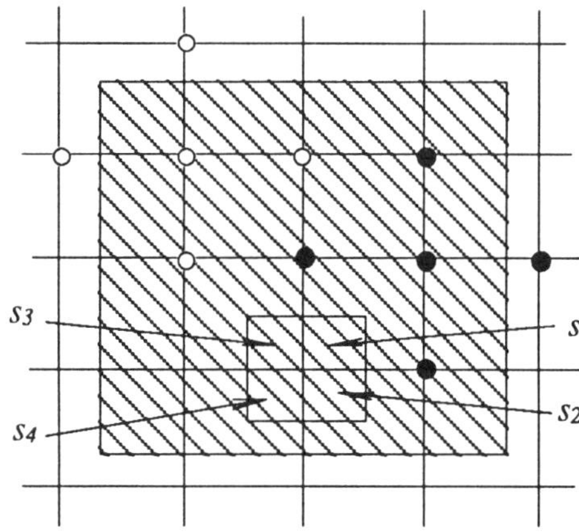

Figure 3.22 Finite difference approximation at edges and corners of the charge distribution.

This particular difficulty can be overcome by defining a domain associated with the node (i, j) as shown in Figure 3.22 for an arbitrary node. Four distinct regions around the node are then considered. Each domain may have associated with it different charge density as well as material properties. The charge density at the node is then an average of the charge densities in the four domains around the node. The charge density is calculated as

$$\rho_{i,j} = \frac{1}{4(h)^2}(S_1\rho_1 + S_2\rho_2 + S_3\rho_3 + S_4\rho_4)$$

and the permittivity at point (i,j) is

$$\varepsilon_{i,j} = \frac{1}{4(h)^2}(S_1\varepsilon_1 + S_2\varepsilon_2 + S_3\varepsilon_3 + S_4\varepsilon_4)$$

In these relations, S_1 through S_4 are the surfaces of the four domains. This procedure has the net effect of averaging material properties and sources at the edges to account for the nodal characteristic of finite difference approximations.

The same approach can be used to modify magnetic permeabilities at the interface between magnetic materials or the electric conductivity at the interface between two conductors.

Another aspect of computation with finite differences is shown in Figure 3.23a. Here, the source of the field is a point charge rather than a charge density. Since the approximation can only be in terms of charge densities, the point charge cannot be modeled directly. Instead, a small area around the charge is taken and the charge is assumed to be distributed over the small area. This will introduce a large error at the point charge (where the electric field tends to infinity) but, at some distance away, this error is minimal. The equivalent problem is shown in Figure 3.23b.

In addition, if the function we evaluate is discontinuous across material interfaces, a continuity or interface condition must also be included in the formulation. This point (and others) is discussed in chapter 6 together with other aspects of implementation.

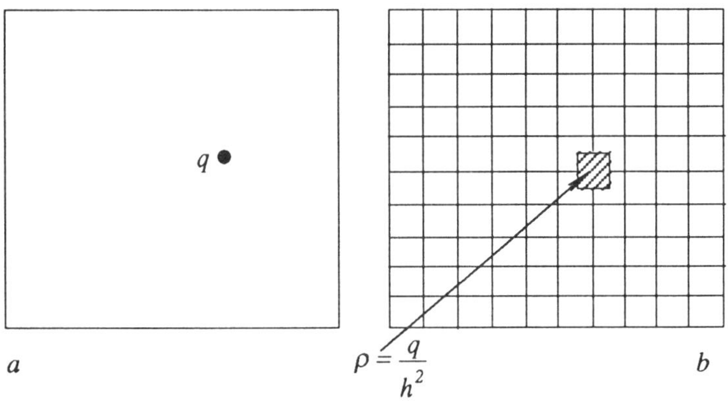

Figure 3.23 Modeling of point sources as equivalent distributed sources. The denser the mesh the more accurate the representation.

3.7 PROBLEMS

3.1 Show that the finite difference expression in Eq. (3.6) has error of the order of $(\Delta x)^2$ while those in Eqs. (3.7) and (3.8) have errors of the order of (Δx). What are the implications of this property for calculation?

3.2 Using the method in section 3.2.1 calculate a backward and a forward finite difference approximation to the second order derivative.

3.3 Using the method in section 3.2.1 extend the method to calculation of fourth order derivatives. Can it also be extended to calculation of third order derivatives?.

3.4 Using the Taylor series approach, write forward and central difference formulae for the second order derivative in time. What are the errors involved in the approximation? Compare these formulae to that in Eq. (3.12).

3.5 Show that the central difference formula for the third order derivative given in Eq. (3.15) has error of the order of $(\Delta x)^2$.

3.6 Consider the parallel plate capacitor in Figure 3.24. Neglecting fringing at edges, calculate the potential and electric field inside the capacitor. Show that solution is independent of discretization (number of nodes) by solving the problem with three nodes: one at the center and one on each plate. Explain.

3.7 Obtain a forward difference formula for first order derivatives with error of the order of $(\Delta x)^2$.

3.8 Repeat problem 3.7 for a backward difference formula.

3.9 Consider again the parallel plate capacitor in Figure 3.24 but now, a charge density ρ_0 exists everywhere inside the capacitor. Calculate the potential and electric field inside. Compare with the result in problem 3.6.

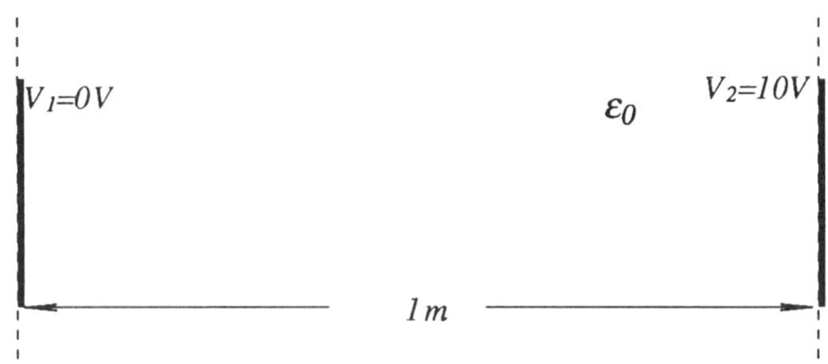

Figure 3.24.

3.10 Solve for the electric potential in the geometry given in Figure 3.25. This is a two-dimensional problem. Use both explicit and implicit methods. The solution to this problem is given in Figures 3.15 and 3.16. It was also found analytically in problem 2.1. You will have to write a small computer program to solve this problem.

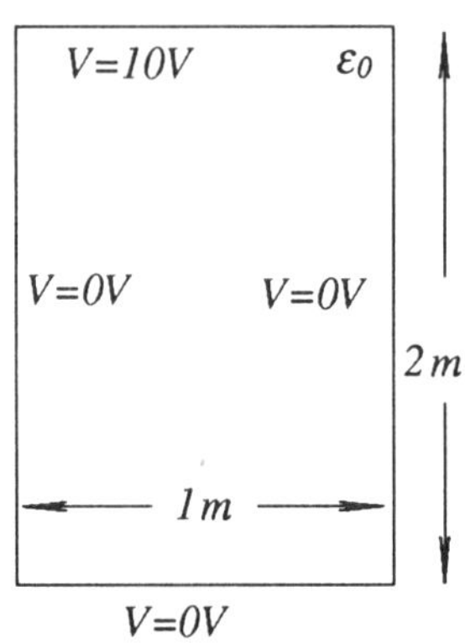

Figure 3.25.

3.8 BIBLIOGRAPHY

The finite difference method is discussed in many texts on numerical methods. While it is difficult to select those texts which are most useful, references [1] through [5] should prove to be a good introduction. The idea of weighting contributions of sources at edges can be found in [7,8] while some other topics including stability criteria can be found in [8,9].

[1] Morgan, M (ed.), (1990), *Finite Element and Finite Difference Methods in Electromagnetic Scattering*, Elsevier, New York.

[2] Richtmeyer, R.D. and Morton, K.W. (1967), *Differential Methods for Initial Value Problems*, John Wiley New York.

[3] Forsythe, G.E. and Wasow W.R. (1967), *Finite Difference Methods of Partial Differential Equations*, John Wiley, New York.

[4] Southwell, R.V. (1946), *Relaxation Methods in Theoretical Physics*, Oxford, London.

[5] James, M.L. Smith, G.M. and Walford, J.C. (1977), *Applied Numerical Methods for Digital Computation*, Harper and Row, New York.

[6] Erdelyi, E.A. and Fuchs E.F. (1970), "Nonlinear magnetic field analysis of D.C. machines - Part I: theoretical fundamentals," *IEEE Transactions on Power Apparatus and Systems*, Vol. PAS-89, pp. 1546-1554.

[7] Erdelyi, E.A. and Fuchs E.F. (1970), "Nonlinear magnetic field analysis of D.C. machines - Part II: application of the improved treatment," *IEEE Transactions on Power Apparatus and Systems*, Vol. PAS-90, pp. 1555-1564.

[8] Sieminieniuch, J.L. and Gladwell, I. (1978), "Analysis of explicit difference methods for a diffusion-convection equation," *International Journal for Numerical Methods in Engineering*, Vol. 12, No. 6, pp. 899-916.

[9] Dufort, E.C. and Frankel, S.P. (1953), "Stability conditions in the numerical treatment of parabolic differential equations," in *Mathematical Tables and Aids to Computation*, Vol. 7, pp. 135-152.

[10] K.J. Binns, K.J. and P.J. Lawrenson, P.J. (1973), *Analysis and Computation of Electric and Magnetic Field Problems*, Pergamon Press, London.

CHAPTER 4

The finite element method

4.1 INTRODUCTION

The finite element method attempts to approximate the continuous problem in a rather straightforward way. The steps involved are the following:

1. Discretization: the solution space is discretized into finite elements. Finite elements are either lines, surfaces or volumes.
2. Approximation: an approximation over a finite element is defined which has some required properties as we shall see shortly.
3. Minimization: a method of minimizing the error due to the discretization must be included.
4. Solution: the previous steps result in a system of equations (linear or nonlinear) which must be solved.
5. Post-processing of data may be required, depending on the function calculated in the previous steps.

The finite element method *(FEM)* approximates the solution rather than the partial derivatives in the differential equation. It is a volumetric method in that the approximation is valid at any point in the solution domain, not only at discrete points. The process calls for discretization of the continuum into any number of finite volume subspaces over which the approximation is valid. These subspaces, or finite elements, are of any well-defined geometrical shape while the approximation is normally a polynomial interpolated over a finite number of points that define the shape of the finite element. Because the restrictions on their shape and size are minimal, finite elements are well suited for discretization of awkward geometries. The elements may be nonuniformly distributed and can be of any size. Furthermore, it is possible to mix different types of elements to increase the *FEM* ability to handle a given geometry.

The *FEM* is based on spatial discretization of the solution domain. A given finite element has associated with it physical dimensions of space. Therefore, it is not suited for solution in the time domain. While it is possible to define a finite element to approximate a solution in time, this is not normally done since finite differences are much more efficient for this purpose. When time-dependent problems must be solved, the time variable is handled either by a combined finite element/finite difference technique or, if possible, by transforming the time dependent variables into the frequency domain (using the phasor transformation or, in more general terms, using the Fourier transform) and solving the problem in the frequency domain.

We introduce the finite element method using a very simple one-dimensional problem. This approach has two main benefits. First, since the solution is simple, it allows us to concentrate on the steps involved in the finite element approximation rather than the details of solving a complex problem. Second, all problems of integration, calculation of partial derivatives, and solving the resulting system of equations are simplified. From this we will get a good feel as to the approximation, assembly of the system of equations in terms of finite elements and to what a finite element is. In addition, we will be able to see the structure of finite element matrices, elemental and global systems of equations, and compare the approximate finite element results with the analytic solution to the same one-dimensional problem.

In subsequent sections and chapters we will use the same principles although the resulting finite elements, matrices, etc. are much more complex.

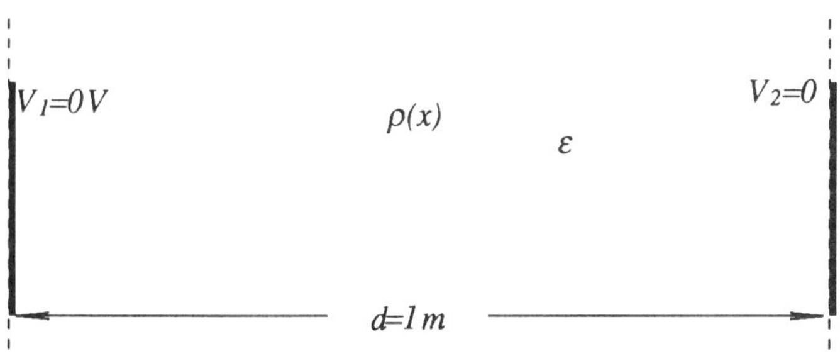

Figure 4.1 Geometry of a parallel plate capacitor with both plates grounded and a charge density distributed between the plates.

4.2 THE FINITE ELEMENT APPROXIMATION

4.2.1 A rudimentary finite element approximation: the one-dimensional Poisson's equation

Before attempting the general finite element solution to field problems we look first at a very simple problem which involves all steps necessary for solution. We consider here the solution of Poisson's equation for a simple electrostatic problem: that of a parallel plate capacitor as outlined in Figure 4.1. The general form for Poisson's equation for electrostatic fields is

$$\nabla^2 V(x,y,z) = -\frac{\rho(x,y,z)}{\varepsilon} \tag{4.1}$$

The equation as written refers to electrostatic solutions for the electric scalar potential but this is not important for the purpose of this chapter.

For the purpose of generality, the source term will be considered as being coordinate-dependent and in this case, independent of time (static). A simple, one-dimensional source is a source that varies parabolically with distance from the plates:

$$\rho(x) = \rho_0 x(x-d) \tag{4.2}$$

where d is the distance between the plates of the capacitor. While this type of source is not easily achievable in practice it is used here for reasons that will become apparent shortly. Note that we could have used a constant charge density, or no charge density at all, both of which are rather common situations.

Eq. (4.1) is now written in one dimension as

$$\frac{\partial^2 V}{\partial x^2} = -\frac{\rho_0 x(x-d)}{\varepsilon} \tag{4.3}$$

with boundary conditions

$$V(x=0) = 0. \qquad V(x=d) = 0. \tag{4.4}$$

In Eq. (4.3), the ordinary derivative may be used instead of the partial derivative since for the one dimensional case the two are identical

$$\frac{d^2 V}{dx^2} = -\frac{\rho_0 x(x-d)}{\varepsilon} \tag{4.5}$$

4.2.2 The analytic solution

The differential equation in Eq. (4.5) can be integrated directly to obtain the analytic solution. We do this first so that we may compare subsequent approximate solutions to the exact solution. This step, which is called a "validation step" is quite important. In this case, an analytic solution of the problem exists. Often this is not the case and we must use experimental data, approximate or simplified solutions, or, in extreme cases, even independently obtained numerical solutions.

Integrating twice the general form of the solution is obtained as

$$V(x) = -\frac{\rho_0 x^4}{12\varepsilon} + \frac{\rho_0 x^3 d}{6\varepsilon} + C_1 x + C_2 \tag{4.6}$$

The constants are evaluated from application of boundary conditions. With the given boundary conditions we get:

$$C_1 = -\frac{\rho_0 d^3}{12\varepsilon} \qquad C_2 = 0 \tag{4.7}$$

Thus, the solution to this boundary value problem is

$$V(x) = -\frac{\rho_0 x^4}{12\varepsilon} + \frac{\rho_0 x^3 d}{6\varepsilon} - \frac{\rho_0 x d^3}{12\varepsilon} \tag{4.8}$$

The reason we were able to obtain an analytical solution is the fact that the problem is one-dimensional and the general partial differential equation becomes an ordinary differential equation. Direct integration and evaluation of the constants of integration, based on given boundary conditions is normally not possible for other types of problems.

4.2.3 Approximate solution

The exact solution in Eq. (4.8) satisfies the differential equations everywhere in the solution domain as well as the specified boundary conditions. That is, when the solution is substituted into the differential equation we get:

$$R(x) = \frac{d^2V(x)}{dx^2} + \frac{\rho_0 x(x-d)}{\varepsilon} = 0 \tag{4.9}$$

If the solution is not known exactly, assuming a solution can be found which approximately satisfies the equation, we may be able to obtain an acceptable solution to the problem. Consider for example, the following function:

$$V(x) \approx Kx(x-d) \tag{4.10}$$

This satisfies the boundary conditions and, since it has a nonzero second order derivative, it may be used as an approximate solution to the problem. Note that we can choose other approximations; the only requirement is that the approximation satisfies the boundary conditions and has a nonzero second derivative.

To use this approximation as a solution, we must find a way to evaluate the constant K. Since in general, no value for K will prove to yield an exact solution, the choice of K influences the solution. Thus, the natural approach is to search for the "best" value for K based on some, yet undetermined method. One useful and generic method for the evaluation of K is through the concept of residual. A residual is obtained through substitution of the approximate solution in Eq. (4.5):

$$R(x) = \frac{d^2(Kx(x-d))}{dx^2} + \frac{\rho_0 x(x-d)}{\varepsilon} \neq 0 \tag{4.11}$$

Ideally, we would like to make the residual identically zero throughout the solution domain since this would provide an exact solution to the problem. However, because we chose an approximate solution, which is second order in x, while the exact solution is fourth order in x, no constant can make this solution identical to the exact solution in Eq. (4.8) and therefore, the residual cannot be made identically zero. The next best choice is to make the residual zero for as many values of x as possible. This approach, if properly applied guarantees a minimum error in solution and a convergence towards the exact solution as the number of points at which the residual vanishes increases. This latter consideration is basic to all discrete methods of solution and is the general justification for their application.

4.2.4 Collocation

In our approximate solution, only one parameter (K) can be varied. Thus, for this particular choice, we can only set the residual to zero at one point. We arbitrarily chose this point to be $d/2$ (for lack of a better choice):

$$R(x) = 2K - \frac{\rho_0 d^2}{4\varepsilon} = 0 \tag{4.12}$$

The constant K is therefore:

$$K = \frac{\rho_0 d^2}{8\varepsilon} \tag{4.13}$$

and the approximate solution is

$$V(x) \approx \frac{\rho_0 x^2 d^2}{8\varepsilon} - \frac{\rho_0 x d^3}{8\varepsilon} \tag{4.14}$$

If, for example we chose the collocation point at $x=d/4$, the solution would be

$$V(x) \approx \frac{3\rho_0 x^2 d^2}{32\varepsilon} - \frac{3\rho_0 x d^3}{32\varepsilon} \tag{4.15}$$

These two approximate solutions are plotted together with the exact solution in Figure 4.2 where the quantity $\rho_0/\varepsilon=1$ was used. It is obvious that both approximate solutions are valid, even though they are different.

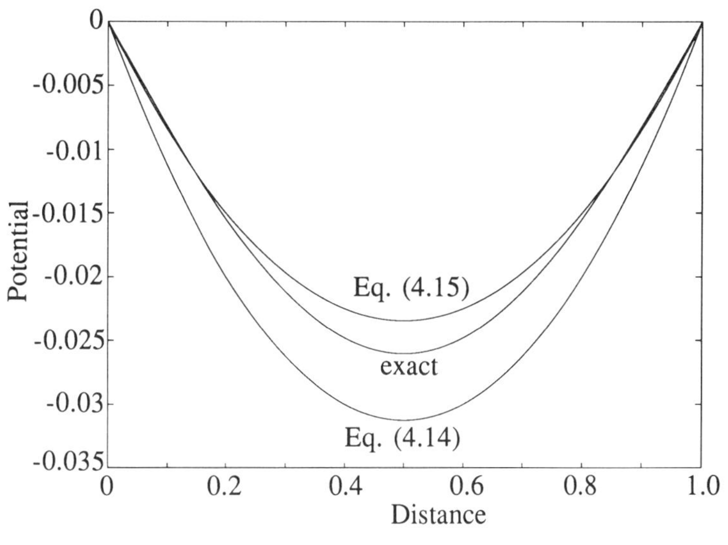

Figure 4.2 Comparison of approximate with exact solutions. Collocation at $x=d/2$ and $x=d/4$ was used.

In our choice of approximate solution in Eq. (4.10) we opted for a polynomial approximation. Of course, the approximation does not have to be a polynomial although in many cases polynomials are convenient as we will see later. For example, a sine function is appropriate for this purpose since it satisfies the two boundary conditions above. The functions

$$V(x) \approx K\sin\frac{\pi x}{2d} \qquad or \qquad V(x) \approx K\sin\frac{n\pi x}{2d} \tag{4.16}$$

also satisfy the boundary conditions of the problem and their second order derivatives do not vanish. When substituting the first of these into the residual (Eq. (4.3)), the solution obtained is

$$V(x) \approx \frac{\rho_0 x^2 d^2}{\pi^2 \varepsilon} - \frac{\rho_0 x d^3}{\pi^2 \varepsilon} \tag{4.17}$$

Figure 4.3 shows this solution together with the exact solution. The solution in Eq. (4.17) is, in fact, better for this problem than the previous solutions in Eqs. (4.14) and (4.15). This however, is by coincidence. One cannot say that sinusoidal approximating functions will be more accurate in general.

We note here that the approximate and exact solutions coincide at the boundaries. If the two solutions match at any other point in the solution domain, this is by coincidence. In general, at any other point in the solution region, the solutions differ.

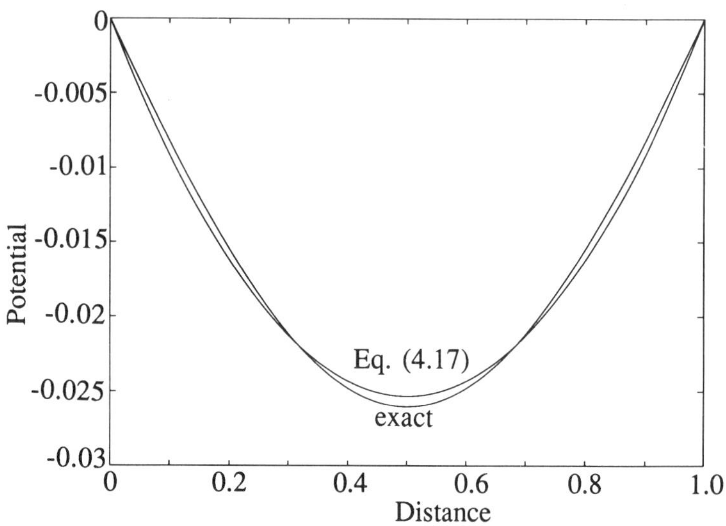

Figure 4.3 Comparison of approximate with exact solutions. A sinusoidal approximation in conjunction with collocation was used for solution.

The solutions in Figures 4.2 and 4.3 are quite close to each other indicating that our approximate solutions are "good". The reason for this is our choice of approximating function, not the method we used to minimize the residual. We just happened to choose a good approximation. To see why this is so, consider another approximating function. If we choose the following approximation:

$$V(x) \approx K\sin\frac{2\pi x}{d} \tag{4.18}$$

the solution we obtain is

$$V(x) \approx \frac{\rho_0 x^2 d^2}{4\pi^2\varepsilon} - \frac{\rho_0 x d^3}{4\pi^2\varepsilon} \tag{4.19}$$

This solution is plotted in Figure 4.4. against the exact solution. We could have anticipated this effect because the approximating function in Eq. (4.18) passes through a zero at the center of the solution domain (i.e. it is not of the same form as the exact solution). Thus, we have now used a "bad" approximation. This is in spite of the fact that this approximation also satisfies the same requirements as the previous solutions.

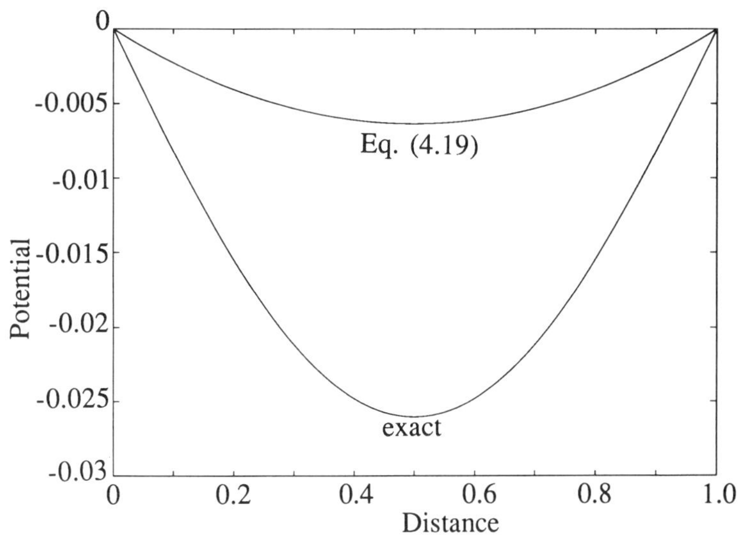

Figure 4.4 Comparison of approximate with exact solutions. The approximation in Eq. (4.18) was used.

4.2.5 Least squares approximation

The collocation method of the previous section guaranteed that the solution will have a minimum error at some point in the solution region. The solution in (4.19) also showed that, depending on our approximation, the error throughout the solution region may be relatively large. Instead of making the residual vanish at a point we may choose to minimize it in an average sense. A simple way of performing this is through a least squares process.

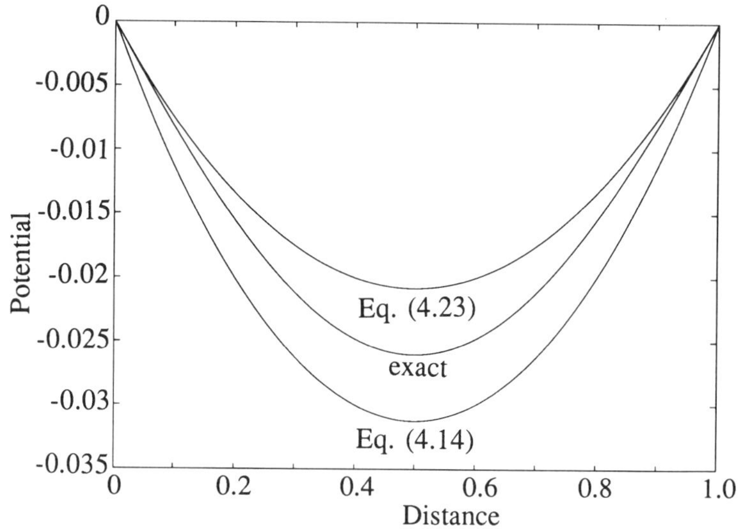

Figure 4.5 Comparison of least squares solution with the exact solution. The collocation method is also shown.

The least squares integral is defined as

$$I = \int_0^d [R(x)]^2 dx \tag{4.20}$$

The purpose here is to minimize this quantity. Ideally, we would like to make I equal to zero. If this were possible, the residual would vanish throughout the solution region. Since in general this is not possible we will try to minimize this value. First, we substitute the approximation in Eq. (4.5). Using the approximation in Eq. (4.10) gives

$$I = 4K^2 d - \frac{4K\rho_0 d^3}{6\varepsilon} + \frac{\rho_0 d^5}{30\varepsilon^2}$$

To minimize this function, we calculate its first derivative with respect to the unknown parameter K

$$\frac{\partial I}{\partial K} = 8Kd - \frac{2\rho_0 d^3}{3\varepsilon} \tag{4.21}$$

Setting this to zero, we obtain a value for K:

$$K = \frac{\rho \alpha d^2}{12\varepsilon} \tag{4.22}$$

The approximate solution is

$$V(x) = \frac{\rho \alpha x^2 d^2}{12\varepsilon} - \frac{\rho \alpha x d^3}{12\varepsilon} \tag{4.23}$$

Figure 4.5 shows this solution together with the exact and the collocation method of solution.

4.2.6 Galerkin's method

The least squares method defined in Eq. (4.20) can be viewed in a more general sense as follows:

$$I = \int_0^d [R(x)]^2 dx = \int_0^d R(x)R(x)dx = \int_0^d W(x)R(x)dx \tag{4.24}$$

where $W(x)$ is a weighting function that we are free to choose. The residual is defined by the choice of the approximate solution but the weighting function is chosen independently. In the least squares method, the weighting function is equal to the residual. In the collocation method, the weighting function is a delta function.

Another choice for the weighting function, one that proves to be very useful in practice is

$$W(x) = \frac{\partial V(x)}{\partial K} \tag{4.25}$$

where $V(x)$ is the approximation. This particular choice for the weighting function leads to the Galerkin procedure. When this is substituted into the integral in Eq. (4.24) we have:

$$I = \int_0^d \frac{\partial V(x)}{\partial K} R(x)dx = 0 \tag{4.26}$$

For the approximation in Eq. (4.10), the weighting function is:

$$W(x) = x^2 - xd \tag{4.27}$$

and together with the residual in Eq. (4.11), the weighted residual method gives

$$\int_0^d W(x)R(x)dx = \int_0^d \left[x^2 - xd\right]\left[2K - \frac{\rho_o x(x-d)}{\varepsilon}\right]dx = 0 \tag{4.28}$$

from which K is found to be (after integration and rearrangement of terms)

$$K = \frac{\rho_o d^2}{10\varepsilon} \tag{4.29}$$

and the approximate solution is

$$V(x) = \frac{\rho_o d^2}{10\varepsilon}\left[x^2 - xd\right] \tag{4.30}$$

The solution in Eq. (4.30) is plotted in Figure 4.6.

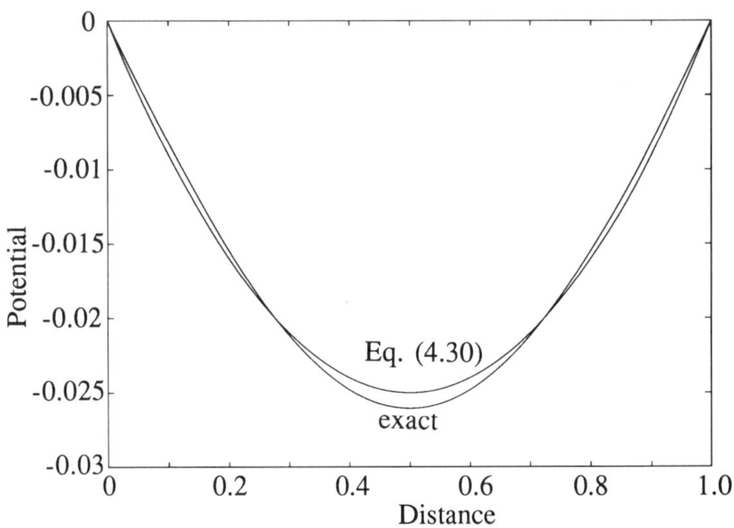

Figure 4.6 Comparison of Galerkin's method solution with the exact solution.

Before continuing, we consider the main points in the solutions obtained so far:

1. All are approximate solutions.

2. All fall under the general category of weighted residual methods.
3. Some seem to be better than others.
4. All approximations must satisfy all boundary conditions.
5. Approximating functions must be second order or higher.
6. Solution is obtained over the whole domain.

4.2.7 The energy functional and variational methods

Before continuing, it is useful to consider here the general form of Galerkin's method, which can be written as:

$$\int_0^d W(x)R(x)dx = \int_0^d \left[\frac{\partial V}{\partial K}\right]\left[\frac{\partial^2 V}{\partial x^2} + \frac{\rho_o x(x-d)}{\varepsilon}\right]dx = 0 \tag{4.31}$$

where the partial derivatives are used to show that this is true in general. First, we separate the two terms in the residual into two integrals:

$$\int_0^d \frac{\partial^2 V}{\partial x^2}\left[\frac{\partial V}{\partial K}\right]dx + \int_0^d \frac{\rho_o x(x-d)}{\varepsilon}\left[\frac{\partial V}{\partial K}\right]dx = 0 \tag{4.32}$$

Integration by parts yields

$$\int_0^d \frac{\partial}{\partial x}\left[\frac{\partial V}{\partial x}\frac{\partial V}{\partial K}\right]dx - \int_0^d \frac{\partial V}{\partial x}\frac{\partial}{\partial K}\left[\frac{\partial V}{\partial x}\right]dx + \int_0^d \frac{\rho_o x(x-d)}{\varepsilon}\left[\frac{\partial V}{\partial K}\right]dx = 0 \tag{4.33}$$

By assuming that

$$\frac{\partial}{\partial x}\left[\frac{\partial V}{\partial K}\right] = \frac{\partial}{\partial K}\left[\frac{\partial V}{\partial x}\right]$$

the first part of the integral is

$$\int_0^d \frac{\partial}{\partial x}\left[\frac{\partial V}{\partial x}\frac{\partial V}{\partial K}\right]dx = \frac{\partial V(d)}{\partial x}\frac{\partial V(d)}{\partial K} - \frac{\partial V(0)}{\partial x}\frac{\partial V(0)}{\partial K} = 0$$

since $V=0$ at $x=0$ and at $x=d$. Using the identity:

$$\frac{\partial V}{\partial x}\frac{\partial}{\partial K}\left[\frac{\partial V}{\partial x}\right]=\frac{1}{2}\frac{\partial}{\partial K}\left[\frac{\partial V}{\partial x}\frac{\partial V}{\partial x}\right]$$

the remaining part of Eq. (4.33) is

$$\int_0^d \left\{\frac{1}{2}\frac{\partial}{\partial K}\left[\frac{\partial V}{\partial x}\right]^2-\frac{\partial}{\partial K}\left[\frac{\rho_0 x(x-d)}{\varepsilon}V\right]\right\}dx=0 \tag{4.34}$$

If the integration is independent of K,

$$\frac{\partial}{\partial K}\int_0^d \left\{\frac{1}{2}\left[\frac{\partial V}{\partial x}\right]^2-\left[\frac{\rho_0 x(x-d)}{\varepsilon}V\right]\right\}dx=0 \tag{4.35}$$

In this relation, the integrand is immediately recognized as the potential energy density in the solution domain and the integral is the total energy in the solution domain. In other words, the solution of the original partial differential equation (4.1) is equivalent to minimizing (actually extremizing) the potential energy in the solution domain. We therefore have yet another method of solving the original partial differential equation. This particular method has certain advantages and, as is shown later, is widely used for solution of electromagnetic fields. We note in particular the following two advantages:

1. The method has a physical interpretation in terms of energy in the system.
2. Only first order derivatives are required for computation. This means that first order approximating functions may be used for the solution. Again, this is an important aspect in finite element formulations.

In the approximation we chose for the solutions above, we are free to adjust one independent parameter. Clearly, we can form a solution that contains more than one parameter. If the approximation has, say, N independent variables that can be calculated, there will be N points at which the collocation method is satisfied while the weighted residual method yields, in general, a more accurate solution. This however, is normally not done in the context of finite element analysis.

For certain problems, those that describe physical phenomena, there exist scalar functions that are stationary when the exact solution is substituted into them. For the problem discussed here, this function is the potential energy of the system. We assume at this point, that the stationary value of the potential energy corresponds to the minimum value of the energy and hence to the exact solution. This then can be used as a measure of our approximation. This method is based on variational principles. It actually involves two steps. One is the functional itself. For the method to be useful we must be able to find a functional which, upon minimization (actually finding a stationary point is a more correct expression but the term minimization is used here because in the case of the energy functional above, the two terms are equivalent). The second step is finding the stationary point which involves finding the variation of the functional and showing that its derivatives correspond to the stationary value of the function.

There is also an additional, necessary step: that of verifying that the stationary value of the functional provides a solution to the original partial differential equation. This is done by evaluating the Euler equation for the functional and is discussed separately.

4.3 THE FINITE ELEMENT METHOD

The approximate solution of the previous section is chosen to be valid throughout the solution domain of the problem, as well as on the boundaries. The function itself, as well as its derivatives are continuous. The problem solved was rather simple and, therefore, it was relatively easy to find a "good" approximation. However, even for this simple problem we are faced with the need to decide on the approximation. Why should we use a second order polynomial? Perhaps a fourth order polynomial is better. Also, do we need to worry about how good the solution is and if so, how do we evaluate the errors involved? It is obvious that the use of a larger number of unknowns in the approximation should improve the solution.

There is however a simpler approach, an approach that is taken in finite element calculations and one that we outline next.

Instead of forming a continuous approximation over the solution domain, we can form piecewise approximations, each valid in part of the solution domain, which together form the solution to the problem. The approximation in each part of the domain is relatively simple, and in most practical applications is of the same order as in other domains. Figure 4.7 shows two possible approximations based on this concept. To perform this type of approximation the continuum is divided into a number of finite size segments, or elements, hence the name finite elements.

In Figure 4.7, the functions chosen to approximate the solution over each element are continuous but their derivatives are not necessarily continuous. Note also that each function can be defined in terms of a finite number of points (i.e. nodal points) and that the size of the individual finite elements (a,b and c or A, B, and C) is immaterial. The number of points necessary depends on the order of approximation required.

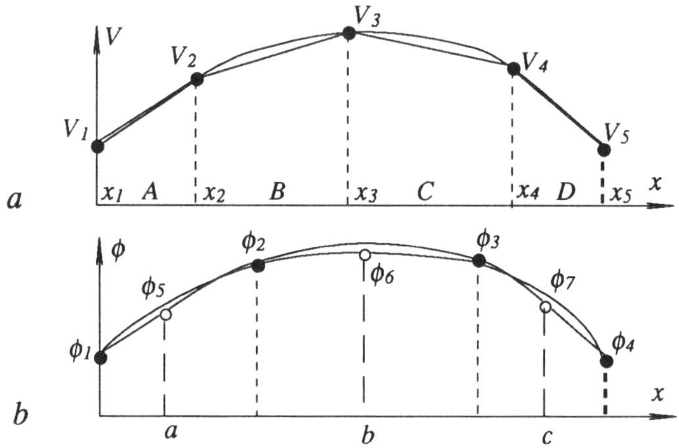

Figure 4.7 Piecewise approximation to a function. a. Linear approximation. b. Parabolic approximation. Element size and number of elements is different in the two approximations.

If the approximation in Figure 4.7a is chosen, the values of the function at points *2*, *3* and *4* must be evaluated while satisfying the boundary conditions at points *1* and *5*. We use the concept of potential energy described above as the basis for calculation. Later on we place all these concepts in a more formal framework but for now we wish to keep the process as simple as possible by not trying to generalize the results. To define the approximation for a one-dimensional finite element, consider Figure 4.8. A linear approximation is

$$V(x) = \alpha + \beta x \tag{4.36}$$

Since this approximation must hold everywhere in the element as well as at the boundaries of the element, the same approximation must hold at points *1* and *2* at which the potentials are V_1 and V_2. These are unknown values to be determined. We can write:

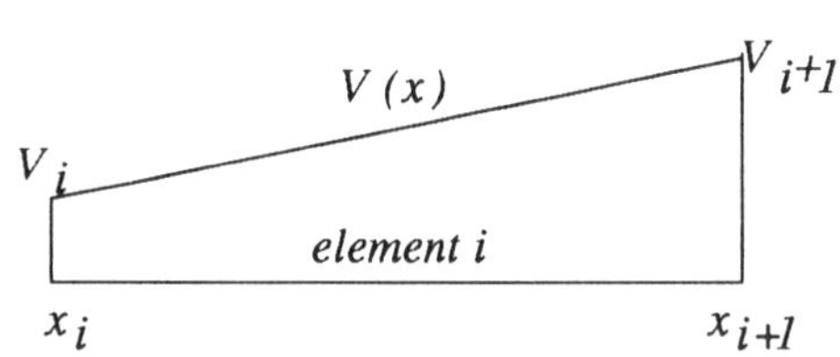

Figure 4.8 Method of defining a one dimensional, linear finite element.

1. The approximation can be written as

$$V_1 = \alpha + \beta x_1$$

$$V_2 = \alpha + \beta x_2$$

From these two equations the value of α and β are

$$\alpha = -\frac{V_2 - V_1}{x_2 - x_1} x_1 + V_1$$

$$\beta = \frac{V_2 - V_1}{x_2 - x_1}$$

The approximation in Eq. (4.36) now is

$$V(x) = V_1 \frac{x_2 - x}{x_2 - x_1} + V_2 \frac{x - x_1}{x_2 - x_1}$$

We note here a number of properties of this approximation, properties we will meet again when discussing more complex finite elements:

$$V(x) = V_1 N_1 + V_2 N_2 = \sum_{i=1}^{2} V_i N_i \tag{4.37}$$

where the functions N_1 and N_2 are geometric functions, independent of potentials. These functions are associated with the nodes of the finite element and are called shape functions or interpolating functions.

2. $N_i=1$ at node i and $N_i=0$ at all other nodes. In this case, $N_1=1$ at node 1, $N_1=0$ at node 2, $N_2=1$ at node 2 and $N_2=0$ at node 1.
3. The sum of the shape functions is equal to 1

$$N_1 + N_2 = \sum_{i=1}^{2} N_i = 1 \tag{4.38}$$

at any point within the finite element.

Although we have only shown these properties to be correct for the element above, they are general and hold for any finite element, regardless of dimensionality and method of defining the finite element.

In addition, we can also calculate the first order derivatives of the approximation which can then be used to calculate the potential energy in the element

$$\frac{\partial V(x)}{\partial x} = \frac{\partial(V_1 N_1)}{\partial x} + \frac{\partial(V_2 N_2)}{\partial x} = V_1\frac{\partial N_1}{\partial x} + V_2\frac{\partial N_2}{\partial x} = \sum_{i=1}^{2} V_i \frac{\partial N_i}{\partial x}$$

where, for the element defined in Figure 4.8 we get

$$\frac{\partial N_1}{\partial x} = \frac{-1}{x_2 - x_1} \qquad \frac{\partial N_2}{\partial x} = \frac{1}{x_2 - x_1} \tag{4.39}$$

and, the derivative of $V(x)$ with respect to x is

$$\frac{\partial V(x)}{\partial x} = V_1 \frac{-1}{x_2 - x_1} + V_2 \frac{1}{x_2 - x_1}$$

Now we can return to the general problem and solve it using the finite element method. Using again Figure 4.7a we view the four elements as being completely general. The approximations are given in Eq. (4.37) and the first derivatives in Eq. (4.39). Before minimizing the energy potential in the system, we calculate the energy in each element separately. Using the definition of the energy functional in Eq. (4.35) for each finite element, gives the following expression for the potential energy in an element

$$PE = \int_{x1}^{x2} \left\{ \frac{1}{2}\left[\frac{\partial V(x)}{\partial x}\right]^2 - \left[\frac{\rho_0 x(x-d)}{\varepsilon} V(x)\right]\right\} dx \tag{4.40}$$

Using the general expressions for $V(x)$ and $\partial V(x)/\partial x$ from Eqs. (4.38) and (4.39) we can write:
For element A, B, C, and D, we get:

$$PE_A = \int_{x1}^{x2} \left\{ \frac{1}{2}\left[V_1\frac{\partial N_1}{\partial x} + V_2\frac{\partial N_2}{\partial x}\right]^2 - \left[\frac{\rho_0 x(x-d)}{\varepsilon}\right]\left[V_1 N_1 + V_2 N_2\right]\right\} dx \tag{4.41}$$

$$PE_B = \int_{x2}^{x3} \left\{ \frac{1}{2}\left[V_1\frac{\partial N_1}{\partial x} + V_2\frac{\partial N_2}{\partial x}\right]^2 - \left[\frac{\rho_0 x(x-d)}{\varepsilon}\right]\left[V_1 N_1 + V_2 N_2\right]\right\} dx \tag{4.42}$$

$$PE_C = \int_{x3}^{x4} \left\{ \frac{1}{2}\left[V_1\frac{\partial N_1}{\partial x} + V_2\frac{\partial N_2}{\partial x}\right]^2 - \left[\frac{\rho_0 x(x-d)}{\varepsilon}\right]\left[V_1 N_1 + V_2 N_2\right] \right\} dx \tag{4.43}$$

$$PE_D = \int_{x4}^{x5} \left\{ \frac{1}{2}\left[V_1\frac{\partial N_1}{\partial x} + V_2\frac{\partial N_2}{\partial x}\right]^2 - \left[\frac{\rho_0 x(x-d)}{\varepsilon}\right]\left[V_1 N_1 + V_2 N_2\right] \right\} dx \tag{4.44}$$

These expressions are identical except for the limits of integration for each element. This property is fundamental to any finite element method: a single element is defined and then the function obtained is integrated over each element separately to obtain the contribution of the element.

Now we have two options. One is to add the potential energy of the independent elements to obtain the total energy in the system and then differentiate the total energy with respect to unknowns V_1 through V_5 to minimize the energy. The second, is to perform the same operation for each element separately (differentiate each one of the expressions above with respect to the unknowns of that element only) and obtain what is often called, an elemental contribution to the minimization process. The first choice is more instructive in the sense that what we do is the same as for a continuum problem. It has the disadvantage that it is cumbersome, especially if the number of elements is large. This is particularly true since the elements are identical except for physical dimensions. We will choose the second option because it is much easier to implement in a computational environment. Substituting the values for $N_1, N_2, \partial N_1/\partial x, \partial N_2/\partial x$ in Eq. (4.41), and integrating, we obtain the following expressions for element A:

$$\begin{aligned} PE_A = {} & \frac{V_1^2}{2L_1} + \frac{V_2^2}{2L_1} - \frac{V_1 V_2}{L_1} - V_1\frac{\rho_0}{\varepsilon}\left[\frac{x_2 L_1^2}{3} - \frac{L_1^3}{4} - \frac{x_2 d L_1}{2} + \frac{d L_1^2}{3}\right] \\ & - V_2\frac{\rho_0}{\varepsilon}\left[\frac{L_1^3}{4} - \frac{x_1 L_1^2}{3} + \frac{x_1 d L_1}{2} - \frac{d L_1^2}{3}\right] \end{aligned} \tag{4.45}$$

where $L_1 = (x_2 - x_1)$ is the length of element A.

Now, we can minimize the energy in element A with respect to the two unknowns in element A: V_1 and V_2. We can write directly:

$$\frac{\partial PE_A}{\partial V_1} = \frac{V_1}{L_1} - \frac{V_2}{L_1} - \frac{\rho_0}{\varepsilon}\left[\frac{x_2 L_1^2}{3} - \frac{L_1^3}{4} - \frac{x_2 d L_1}{2} + \frac{d L_1^2}{3}\right] = 0 \tag{4.46}$$

$$\frac{\partial PE_A}{\partial V_2} = \frac{V_2}{L_1} - \frac{V_1}{L_1} - \frac{\rho_0}{\varepsilon}\left[\frac{L_1^3}{4} - \frac{x_1 L_1^2}{3} + \frac{x_1 d L_1}{2} - \frac{d L_1^2}{3}\right] = 0 \tag{4.47}$$

These two equations can be written as a 2×2 system of equations

$$\begin{bmatrix} \frac{1}{L_1} & -\frac{1}{L_1} \\ -\frac{1}{L_1} & \frac{1}{L_1} \end{bmatrix} \begin{Bmatrix} V_1 \\ V_2 \end{Bmatrix} = \begin{Bmatrix} \frac{\rho_0}{\varepsilon}\left[\frac{x_2 L_1^2}{3} - \frac{L_1^3}{4} - \frac{x_2 d L_1}{2} + \frac{d L_1^2}{3}\right] \\ \frac{\rho_0}{\varepsilon}\left[\frac{L_1^3}{4} - \frac{x_1 L_1^2}{3} + \frac{x_1 d L_1}{2} - \frac{d L_1^2}{3}\right] \end{Bmatrix} = \begin{Bmatrix} R_{A1} \\ R_{A2} \end{Bmatrix} \tag{4.48}$$

A number of observations are in order here: first, the system of equations in Eq. (4.48) is symmetric. Second, the right hand side of the system is due to the charge distribution within element A alone. If there were no charge density in this element, the right hand side would be zero. Third, each equation is associated with a node. The first row is associated with node *1*, the second with node *2*.

We obtain very similar expressions for elements B, C, and D by simply replacing the length of element A by those of elements B, C, and D, and changing x_1 and x_2 by the corresponding values. The result is four systems of equations.

For element B:

$$\begin{bmatrix} \frac{1}{L_2} & -\frac{1}{L_2} \\ -\frac{1}{L_2} & \frac{1}{L_2} \end{bmatrix} \begin{Bmatrix} V_2 \\ V_3 \end{Bmatrix} = \begin{Bmatrix} \frac{\rho_0}{\varepsilon}\left[\frac{x_3 L_2^2}{3} - \frac{L_2^3}{4} - \frac{x_3 d L_2}{2} + \frac{d L_2^2}{3}\right] \\ \frac{\rho_0}{\varepsilon}\left[\frac{L_2^3}{4} - \frac{x_2 L_2^2}{3} + \frac{x_2 d L_2}{2} - \frac{d L_2^2}{3}\right] \end{Bmatrix} = \begin{Bmatrix} R_{B2} \\ R_{B3} \end{Bmatrix} \tag{4.49}$$

For element C:

$$\begin{bmatrix} \frac{1}{L_3} & -\frac{1}{L_3} \\ -\frac{1}{L_3} & \frac{1}{L_3} \end{bmatrix} \begin{Bmatrix} V_3 \\ V_4 \end{Bmatrix} = \begin{Bmatrix} \frac{\rho_0}{\varepsilon}\left[\frac{x_4 L_3^2}{3} - \frac{L_3^3}{4} - \frac{x_4 dL_3}{2} + \frac{dL_3^2}{3}\right] \\ \frac{\rho_0}{\varepsilon}\left[\frac{L_3^3}{4} - \frac{x_3 L_3^2}{3} + \frac{x_3 dL_3}{2} - \frac{dL_3^2}{3}\right] \end{Bmatrix} = \begin{Bmatrix} R_{C3} \\ R_{C4} \end{Bmatrix} \tag{4.50}$$

Finally, for element D we get:

$$\begin{bmatrix} \frac{1}{L_4} & -\frac{1}{L_4} \\ -\frac{1}{L_4} & \frac{1}{L_4} \end{bmatrix} \begin{Bmatrix} V_4 \\ V_5 \end{Bmatrix} = \begin{Bmatrix} \frac{\rho_0}{\varepsilon}\left[\frac{x_5 L_4^2}{3} - \frac{L_4^3}{4} - \frac{x_5 dL_4}{2} + \frac{dL_4^2}{3}\right] \\ \frac{\rho_0}{\varepsilon}\left[\frac{L_4^3}{4} - \frac{x_4 L_4^2}{3} + \frac{x_4 dL_4}{2} - \frac{dL_4^2}{3}\right] \end{Bmatrix} = \begin{Bmatrix} R_{D4} \\ R_{D5} \end{Bmatrix} \tag{4.51}$$

Now, we have the elemental contribution to the minimization process. These are now added together in a system of equations called a "global system of equations". We note that Eq. (4.48) corresponds to nodes *1* and *2*, Eq. (4.49) to nodes *2* and *3*, Eq. (4.50) to nodes *3* and *4*, and Eq. (4.51) to nodes *4* and *5*. Also, to simplify the expressions, the short form notation (for the right hand sides) shown above is used

$$\begin{bmatrix} \frac{1}{L_1} & -\frac{1}{L_1} & 0 & 0 & 0 \\ -\frac{1}{L_1} & \frac{1}{L_1}+\frac{1}{L_2} & -\frac{1}{L_2} & 0 & 0 \\ 0 & -\frac{1}{L_2} & \frac{1}{L_2}+\frac{1}{L_3} & -\frac{1}{L_3} & 0 \\ 0 & 0 & -\frac{1}{L_3} & \frac{1}{L_3}+\frac{1}{L_4} & -\frac{1}{L_4} \\ 0 & 0 & 0 & -\frac{1}{L_4} & \frac{1}{L_4} \end{bmatrix} \begin{Bmatrix} V_1 \\ V_2 \\ V_3 \\ V_4 \\ V_5 \end{Bmatrix} = \begin{Bmatrix} R_{A1} \\ R_{A2}+R_{B2} \\ R_{B3}+R_{C3} \\ R_{C4}+R_{D4} \\ R_{D5} \end{Bmatrix} \tag{4.52}$$

This is the system of equations we need to solve for the unknown values V_1 through V_5. We also note that the system of equations is symmetric, positive definite. In this case it is also a banded system of equations with a small bandwidth (the band around the main diagonal in which the coefficients of the matrix are nonzero).

To solve the system, the boundary conditions of the problem must be applied. In this case, for the problem in Figure 4.1, $V_1=V_5=0$. Thus, to solve the system columns *1* and *5* and rows *1* and *5* are removed and the system is solved for V_2, V_3, and V_4. Although this approach is appropriate here, other methods are often used for larger matrices.

Before solving this system of equations, we observe the following:

1. The size of the system of equations depends on the number of nodes in the system. Each additional element, adds one node to the system (one row and one column).
2. For the finite element used here, the number of nonzero coefficients in the matrix is always equal to or smaller than three in each row/column.
3. The elemental contributions to the matrix are identical in form but differ in values depending on the size of the element.

To solve the system of equations, the actual size of elements must be defined. Refering to Figure 4.7a we use $x_1 = 0.$, $x_2 = 0.4$, $x_3 = 0.5$, $x_4 = 0.75$, and $x_5 = 1.0$. With these, and removing rows/columns *1* and *5* the system is:

$$\begin{bmatrix} 12.5 & -10 & 0 \\ -10 & 14 & -4 \\ 0 & -4 & 8 \end{bmatrix} \begin{Bmatrix} V_2 \\ V_3 \\ V_4 \end{Bmatrix} = \begin{Bmatrix} -0.057583333 \\ -0.042614583 \\ -0.036533329 \end{Bmatrix} \tag{4.53}$$

which, upon solution, yields the following values

$$x_2 = -0.0260 \qquad x_3 = -0.0267 \qquad x_4 = -0.0179 \tag{4.54}$$

This solution is plotted against the exact solution in Figure 4.9. Although not exact, it has one significant advantage over the methods given in section 4.2: it is a simple matter to increase accuracy of the solution by simply using a larger number of finite elements, with a correspondingly larger number of equations to solve. As mentioned before, this is the basic premise of numerical methods in general: as the discretization level increases, the solution approaches the exact solution.

To summarize the process described above we now list the required steps to obtain a solution for a given geometry, partial differential equation and boundary conditions:

1. Division of the solution region into subdomains. The number or size of these domains is not limited in any way, including the choice of a single domain.
2. Approximating or trial functions. Over each subdomain or finite element we use a function to approximate the solution. The function is arbitrary as long as the boundary conditions are satisfied. We will see shortly that some

functions are often used (i.e. polynomials of various orders) for specific reasons.

3. Residual. Regardless of the actual method used to solve the problem, this falls under the generic group of weighted residuals. Finding an appropriate functional is part of the formulation of the problem.
4. Minimization. By setting the variations in the functional, or its first derivatives, with respect to each unknown to zero, we obtain an algebraic equation for each unknown in the solution region. Normally, the system of equations may be quite large, depending on the number of unknowns (independent variables, or degrees of freedom).
5. Solution of the system of equations. While this may seem a trivial operation, it does consume most of the computer resources involved in the solution process. We not only need to perform this step but also to try to do this as efficiently as possible.
6. Post-processing. Although this step is not necessarily associated with the numerical computation, often the variables calculated are not those that are needed for design. For example, if the electric field rather than the electric potential is needed, additional calculations as well as display of data are required. This is done in the post-processing step.

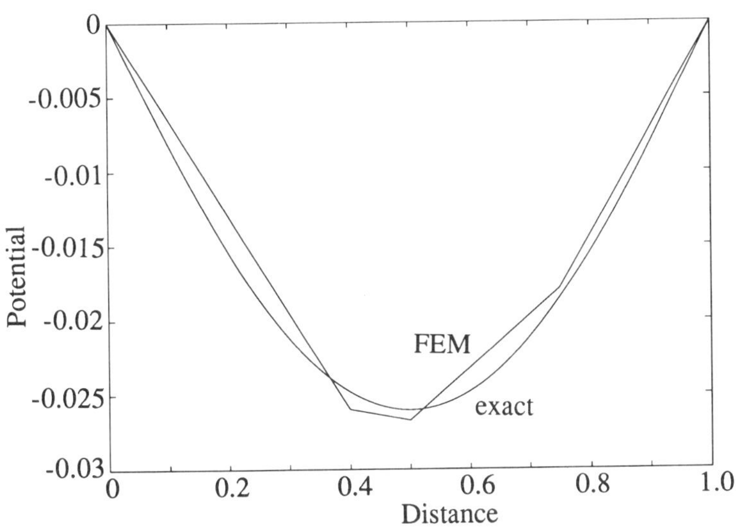

Figure 4.9 Comparison of finite element solution with exact solution. Note how good the solution is in spite of the small number of points used.

4.4 THE FINITE ELEMENT

From the discussion in section 4.3, we can view a finite element as a volume, of defined shape, over which an approximation to a function can be obtained. Up to this point, the shape was not restricted in any way although, in section 4.3, we assumed that the element is one-dimensional. This assumption is not very troublesome since any element can be viewed as a volume element. Thus, a one-dimensional element is a three-dimensional element with two dimensions equal to unity. A two-dimensional element is a volume element with unit depth. Integration, therefore, is always performed over the volume of the element.

Definition of a finite element is the first step in any analysis. We have outlined one general method by which a finite element may be defined in Eqs. (4.36) through (4.39). We now extend the same principle to define other elements. However, this is not the only method of defining an element. It should be noted that in the previous section, the only requirement was that an approximation exists, valid at the nodes of the elements and the approximation must have nonzero first order derivatives if first order derivatives are required for the formulation. Similarly, if higher order derivatives are needed, the approximation must satisfy these as well. These requirements will be formalized shortly but first we wish to define a number of two-dimensional and three-dimensional elements.

Two methods are used. One is the method outlined above which is called a direct method. The second is a method based on mapping of the finite element into a local system of coordinates and is called a parametric method.

To define any element it is necessary to specify the shape functions of the element and its derivatives with respect to each variable. Later, in the context of formulation we will also need to integrate these expressions or products of expressions over the volume (or area) of the element.

In the following sections a number of elements are introduced, all of which are widely used. These include two- and three-dimensional elements, first and second order elements as well as axisymmetric elements. The methods presented can be extended to any other element or any higher order. We start with the direct method and two-dimensional elements as these are easier to visualize.

4.4.1 Triangular elements: direct derivation

A triangular element is shown in Figure 4.10. The triangle has three nodes, therefore it has three shape functions. The approximation within the element is assumed to be linear and of the following form

$$\phi(x,y) = \alpha_1 + \alpha_2 x + \alpha_3 y \tag{4.55}$$

where the coefficients α_1, α_2, and α_3 remain to be determined.

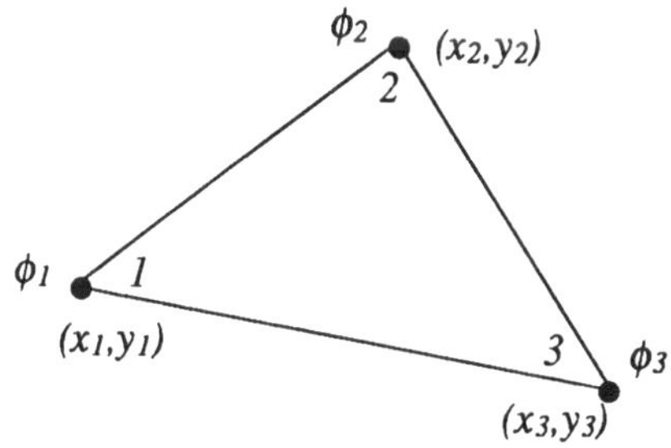

Figure 4.10 Triangular element in the Cartesian system of coordinates.

This approximation is correct everywhere in the element, including at the nodes. To determine the coefficients we use Figure 4.10 and write for the three nodes of the element

$$\phi_1 = \alpha_1 + \alpha_2 x_1 + \alpha_3 y_1$$

$$\phi_2 = \alpha_1 + \alpha_2 x_2 + \alpha_3 y_2$$

$$\phi_3 = \alpha_1 + \alpha_2 x_3 + \alpha_3 y_3$$

Solving these three equations for the three unknowns α_1, α_2, and α_3 gives

$$\alpha_1 = \frac{1}{2\Delta}\left[(x_2 y_3 - x_3 y_2)\phi_1 + (x_3 y_1 - x_1 y_3)\phi_2 + (x_1 y_2 - x_2 y_1)\phi_3\right]$$

$$\alpha_2 = \frac{1}{2\Delta}\left[(y_2 - y_3)\phi_1 + (y_3 - y_1)\phi_2 + (y_1 - y_2)\phi_3\right]$$

$$\alpha_3 = \frac{1}{2\Delta}\left[(x_3 - x_2)\phi_1 + (x_1 - x_3)\phi_2 + (x_2 - x_1)\phi_3\right]$$

where Δ is the area of the triangle and is given by

$$2\Delta = det\begin{bmatrix} 1 & x_1 & y_1 \\ 1 & x_2 & y_2 \\ 1 & x_3 & y_3 \end{bmatrix} = (x_2 y_3 - x_3 y_2) + (x_3 y_1 - x_1 y_3) + (x_1 y_2 - x_2 y_1)$$

These are now substituted into the approximation in Eq. (4.55) to obtain

$$\phi(x,y) = N_1\phi_1 + N_2\phi_2 + N_3\phi_3 \tag{4.56}$$

where the functions N_1, N_2, and N_3 are:

$$N_1 = \frac{1}{2\Delta}\left[(x_2y_3 - x_3y_2) + (y_2 - y_3)x + (x_3 - x_2)y\right] \tag{4.57}$$

$$N_2 = \frac{1}{2\Delta}\left[(x_3y_1 - x_1y_3) + (y_3 - y_1)x + (x_1 - x_3)y\right] \tag{4.58}$$

$$N_3 = \frac{1}{2\Delta}\left[(x_1y_2 - x_2y_1) + (y_1 - y_2)x + (x_2 - x_1)y\right] \tag{4.59}$$

These are the shape functions for the element. The derivatives of the shape functions with respect to x and y are

$$\frac{\partial N_1}{\partial x} = \frac{1}{2\Delta}(y_2 - y_3) \qquad \frac{\partial N_1}{\partial y} = \frac{1}{2\Delta}(x_3 - x_2)$$

$$\frac{\partial N_2}{\partial x} = \frac{1}{2\Delta}(y_3 - y_1) \qquad \frac{\partial N_2}{\partial y} = \frac{1}{2\Delta}(x_1 - x_3)$$

$$\frac{\partial N_3}{\partial x} = \frac{1}{2\Delta}(y_1 - y_2) \qquad \frac{\partial N_3}{\partial y} = \frac{1}{2\Delta}(x_2 - x_1)$$

These provide the derivatives of the approximation as

$$\frac{\partial \phi(x,y)}{\partial x} = \frac{\partial N_1}{\partial x}\phi_1 + \frac{\partial N_2}{\partial x}\phi_2 + \frac{\partial N_3}{\partial x}\phi_3 \tag{4.60}$$

$$\frac{\partial \phi(x,y)}{\partial y} = \frac{\partial N_1}{\partial y}\phi_1 + \frac{\partial N_2}{\partial y}\phi_2 + \frac{\partial N_3}{\partial y}\phi_3 \tag{4.61}$$

4.4.2 Tetrahedral elements: direct derivation

Instead of following the steps used for the definition of the triangular element, we use here a slightly different approach. The shape functions are called natural coordinates since they have a meaning as volume coordinates. However, they are derived in the same manner as for the triangular elements (where the shape functions can be viewed as area coordinates).

Considering a tetrahedron as in Figure 4.11 and assuming a linear variation of the function to be modeled inside the tetrahedron, we can relate the global Cartesian coordinates to a set of shape functions. For the tetrahedral elements these are also called volume coordinates as will be shown shortly. These relations are

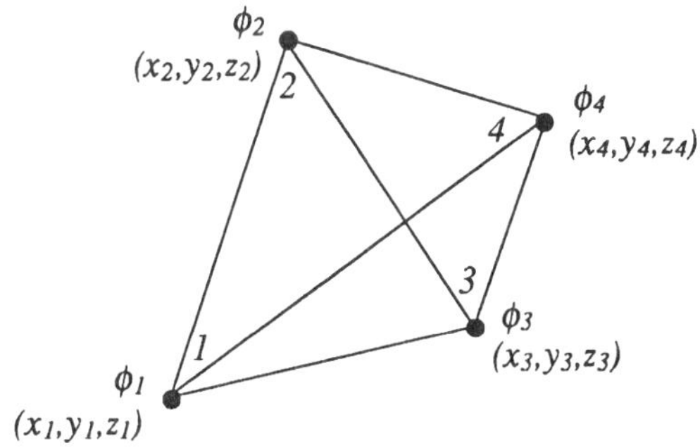

Figure 4.11 A tetrahedral elements and coordinates needed to define its shape functions.

$$x = L_1 x_1 + L_2 x_2 + L_3 x_3 + L_4 x_4$$

$$y = L_1 y_1 + L_2 y_2 + L_3 y_3 + L_4 y_4$$

$$z = L_1 z_1 + L_2 z_2 + L_3 z_3 + L_4 z_4$$

where the shape functions L_1, L_2, L_3, and L_4 satisfy

$$L_1 + L_2 + L_3 + L_4 = 1$$

Inverting the three equations above results in a set of equations for the natural coordinates L_i

$$L_i = \frac{V}{6}(a_i + b_i x + c_i y + d_i z) \qquad i=1,2,3,4 \qquad (4.62)$$

where V is the volume of the tetrahedron and is given by

$$V = \frac{1}{6} det \begin{bmatrix} 1 & x_1 & y_1 & z_1 \\ 1 & x_2 & y_2 & z_2 \\ 1 & x_3 & y_3 & z_3 \\ 1 & x_4 & y_4 & z_4 \end{bmatrix}$$

In a similar way, the coefficients a_i, b_i, c_i and d_i are defined for $i=1$ as:

$$a_1 = det\begin{bmatrix} x_2 & y_2 & z_2 \\ x_3 & y_3 & z_3 \\ x_4 & y_4 & z_4 \end{bmatrix} = x_2(y_3z_4 - y_4z_3) - x_3(y_2z_4 - y_4z_2) + x_4(y_2z_3 - y_3z_2)$$

$$b_1 = -\,det\begin{bmatrix} 1 & y_2 & z_2 \\ 1 & y_3 & z_3 \\ 1 & y_4 & z_4 \end{bmatrix} = -\,(y_3z_4 - y_4z_3) + (y_2z_4 - y_4z_2) - (y_2z_3 - y_3z_2)$$

$$c_1 = -\,det\begin{bmatrix} x_2 & 1 & z_2 \\ x_3 & 1 & z_3 \\ x_4 & 1 & z_4 \end{bmatrix} = (x_3z_4 - x_4z_3) - (x_2z_4 - x_4z_2) + (x_2z_3 - x_3z_2)$$

$$d_1 = -\,det\begin{bmatrix} x_2 & y_2 & 1 \\ x_3 & y_3 & 1 \\ x_4 & y_4 & 1 \end{bmatrix} = -\,(x_3y_4 - x_4y_3) + (x_2y_4 - x_4y_2) - (x_2y_3 - x_3y_2)$$

The other coefficients, for $i=2,3,4$ are obtained through cyclic permutations of the subscripts *1,2,3* and *4*. In expanded form these are:

$$a_2 = -\,x_1(y_3z_4 - y_4z_3) + x_3(y_1z_4 - y_4z_1) - x_4(y_1z_3 - y_3z_1)$$

$$b_2 = (y_3z_4 - y_4z_3) - (y_1z_4 - y_4z_1) - (y_1z_3 - y_3z_1)$$

$$c_2 = (x_1z_4 - x_4z_1) - (x_3z_4 - x_4z_3) - (x_1z_3 - x_3z_1)$$

$$d_2 = (x_3y_4 - x_4y_3) - (x_1y_4 - x_4y_1) + (x_1y_3 - x_3y_1)$$

$$a_3 = x_1(y_2z_4 - y_4z_2) - x_2(y_1z_4 - y_4z_1) + x_4(y_1z_2 - y_2z_1)$$

$$b_3 = (y_1z_4 - y_4z_1) - (y_2z_4 - y_4z_2) - (y_1z_4 - y_4z_1)$$

$$c_3 = (x_2z_4 - x_4z_2) - (x_1z_4 - x_4z_1) + (x_1z_2 - x_2z_1)$$

$$d_3 = (x_1y_4 - x_4y_1) - (x_2y_4 - x_4y_2) - (x_1y_2 - x_2y_1)$$

$$a_4 = -\,x_1(y_2z_3 - y_3z_2) + x_2(y_1z_3 - y_3z_1) - x_3(y_1z_2 - y_2z_1)$$

$$b_4 = (y_2z_3 - y_3z_2) - (y_1z_3 - y_3z_1) + (y_1z_2 - y_2z_1)$$

$$c_4 = (x_1z_3 - x_3z_1) - (x_2z_3 - x_3z_2) - (x_1z_2 - x_2z_1)$$

$$d_4 = (x_2y_3 - x_3y_2) - (x_1y_3 - x_3y_1) - (x_1y_2 - x_2y_1)$$

The coordinates L_i have a physical meaning as volume coordinates. Figure 4.12 describes this concept. Four volumes are defined within the tetrahedron by specifying an internal point p. Then, the four coordinates are defined as the ratio of these volumes to the total volume of the element

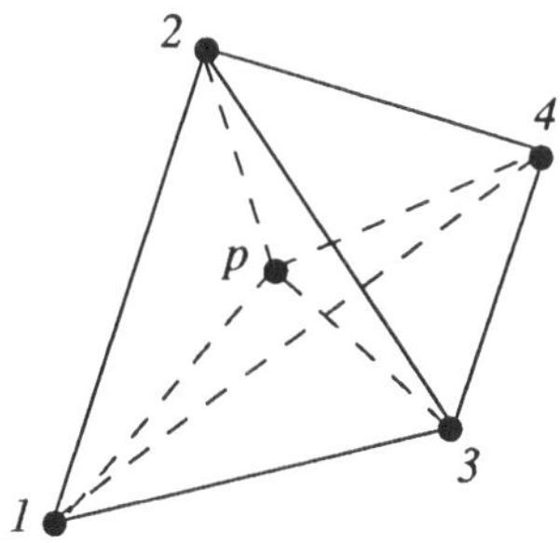

$$L_1 = \frac{V(p,2,3,4)}{V(1,2,3,4)}$$

$$L_2 = \frac{V(p,1,3,4)}{V(1,2,3,4)}$$

$$L_3 = \frac{V(p,1,2,4)}{V(1,2,3,4)}$$

$$L_4 = \frac{V(p,1,2,3)}{V(1,2,3,4)}$$

Figure 4.12 Interpretation of the shape functions as volume coordinates.

$$L_i = \frac{v_i}{V} \qquad i = 1,2,3,4$$

In the relations above, care must be paid to the sign of the equations otherwise negative volumes can arise. To overcome this in a right-handed Cartesian system the nodes of the tetrahedron must be numbered such that node numbers *1*, *2* and *3* are numbered counterclockwise when viewed from node number *4*. Similarly, for any tetrahedron, the first three nodes must be numbered in a counterclockwise direction when viewed from the fourth node.

The appropriate derivative formulae can be found by differentiating Eq. (4.62) with respect to x, y and z.

$$\frac{\partial L_i}{\partial x} = \frac{b_i}{6V} \qquad \frac{\partial L_i}{\partial y} = \frac{c_i}{6V} \qquad \frac{\partial L_i}{\partial z} = \frac{d_i}{6V}$$

These formulae are used in evaluation of the integrals over each element in the formulation process.

Using these definitions, we can now define the approximation of a function throughout the finite element as

$$\phi(x,y,z) = L_1\phi_1 + L_2\phi_2 + L_3\phi_3 + L_4\phi_4 = \sum_{i=1}^{4} L_i\phi_i \qquad (4.63)$$

where ϕ_i is the value of the (unknown) function at node i and L_i is the shape function at the same node. This approximation is again identical in form with that for the triangular element in Eq. (4.56) or, for that matter with that for the one dimensional element in Eq. (4.37).

Similarly, the derivatives of the approximation can be written as

$$\frac{\partial\phi(x,y,z)}{\partial x} = \sum_{i=1}^{4} \frac{\partial L_i}{\partial x}\phi_i = \sum_{i=1}^{4} \frac{b_i}{6V}\phi_i \qquad (4.64)$$

$$\frac{\partial\phi(x,y,z)}{\partial y} = \sum_{i=1}^{4} \frac{\partial L_i}{\partial y}\phi_i = \sum_{i=1}^{4} \frac{c_i}{6V}\phi_i \qquad (4.65)$$

$$\frac{\partial\phi(x,y,z)}{\partial z} = \sum_{i=1}^{4} \frac{\partial L_i}{\partial z}\phi_i = \sum_{i=1}^{4} \frac{d_i}{6V}\phi_i \qquad (4.66)$$

4.4.3 Shape functions for eight-node isoparametric element

The shape functions for the eight-node hexahedral element used here are derived from the so-called 'serendipity family' of elements by requiring that each shape function N_i, be a polynomial, and have a value of 1 at node i and zero at all other nodes of the element. Refering to Figure 4.13, the shape functions are written by multiplying the minimum number of plane equations that contain all the nodes of the element except for the node for which the function is written. For example, for shape function N_5, all nodes except node 5 are on the following planes:

$$u = 1, \qquad v = 1, \qquad w = 0$$

The shape function is therefore

$$N_5 = C(1-u)(1-v)w$$

By requiring that at node 5, $N_5=1$, the constant C is found as $C=1$.

These functions are usually defined in a local system of coordinates where the corners of the elements can take values such as $u=\pm 1$, $v=\pm 1$ and $w=\pm 1$. The coordinate system used here, assumes that all three coordinates vary between 0 and 1 as shown in Figure 4.13.

The differentiation of the shape functions is straightforward. It is done in the local system by differentiating each shape function with respect to u, v and w. The eight shape functions and their derivatives with respect to u, v and w are summarized in Table 4.1.

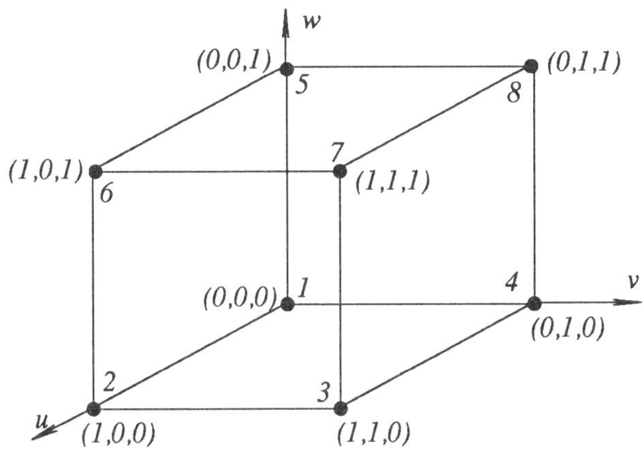

Figure 4.13 Eight-node hexahedral element in the local system of coordinates.

Table 4.1. Shape functions and their derivatives for an eight-node isoparametric element.

Node number	Ni	$\partial N_i/\partial u$
1	$-(1-u)(1-v)(1-w)$	$-(1-v)(1-w)$
2	$u(1-v)(1-w)$	$(1-v)(1-w)$
3	$-uv(1-w)$	$-v(1-w)$
4	$(1-u)(1-v)(1-w)$	$(1-v)(1-w)$
5	$(1-u)(1-v)w$	$(1-v)w$
6	$-u(1-v)w$	$-(1-v)w$
7	uvw	vw
8	$-(1-u)vw$	$-vw$

Node number	$\partial N_i/\partial v$	$\partial N_i/\partial w$
1	$-(1-u)(1-w)$	$-(1-u)(1-v)$
2	$u(1-w)$	$u(1-v)$
3	$-u(1-w)$	$-uv$
4	$(1-u)(1-w)$	$(1-u)(1-v)$
5	$(1-u)w$	$(1-u)(1-v)$
6	$-uw$	$-u(1-v)$
7	uw	uv
8	$-(1-u)w$	$-(1-u)v$

4.4.4 Shape functions for twenty-node isoparametric element

Refering to Figure 4.14 and following the method in the preceding section, the shape functions for the *20* nodes in this element are derived. However, in order to define a corner node, it is necessary to write in addition to the three equations of the planes that do not contain the node, another equation which forms a plane through the three nodes closest to the corner being considered. For mid-side nodes, four planes contain all the rest of the nodes.

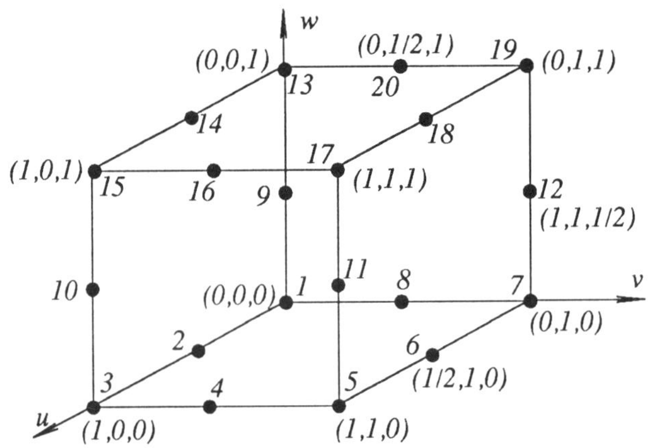

Figure 4.14 Twenty-node hexahedral finite element in a local system of coordinates.

As an example of a corner node consider node number *15* in Figure 4.14. The plane $u=0$, $v=1$ and $w=0$ contain all the nodes of the element except node numbers *10*, *14*, *15*, and *16*. A plane through node numbers *10*, *14*, and *16* is defined as

$$u + v + w = 3/2$$

The shape function N_{15} can be written as

$$N_{15} = Cu(1 - v)w(u + v + w - 3/2)$$

where $C=2$ to ensure that $N_{15}=1$ at the coordinates

$$u = 1, \qquad v = 0, \qquad w = 1$$

For a typical 'mid-side' node such as node number *10* in Figure 4.14, the planes containing the rest of the nodes are

$$u = 0, \qquad v = 1, \qquad w = 0, \qquad \text{and} \qquad w = -1$$

therefore shape function N_{10} is

$$N_{10} = Cu(1 - v)w(1 - w) \tag{4.67}$$

where $C=4$ to ensure that $N_{10}=1$ when the coordinates $u=1$, $v=0$ and $w=1/2$ are substituted in Eq. (4.67).

This procedure is repeated for each node in the element and the shape functions are differentiated with respect to u, v, and w to yield the twenty shape function derivatives which are summarized in Table 4.2 together with the shape functions.

The derivatives in Tables 4.1 and 4.2, are written in the local system of coordinates. To transform them into the global system, we use the Jacobian matrix as shown below. The derivatives are easily obtained in the local system by using the chain rule of differentiation as

$$\frac{\partial N_i}{\partial u} = \frac{\partial N_i}{\partial x}\frac{\partial x}{\partial u} + \frac{\partial N_i}{\partial y}\frac{\partial y}{\partial u} + \frac{\partial N_i}{\partial z}\frac{\partial z}{\partial u}$$

$$\frac{\partial N_i}{\partial v} = \frac{\partial N_i}{\partial x}\frac{\partial x}{\partial v} + \frac{\partial N_i}{\partial y}\frac{\partial y}{\partial v} + \frac{\partial N_i}{\partial z}\frac{\partial z}{\partial v}$$

$$\frac{\partial N_i}{\partial w} = \frac{\partial N_i}{\partial x}\frac{\partial x}{\partial w} + \frac{\partial N_i}{\partial y}\frac{\partial y}{\partial w} + \frac{\partial N_i}{\partial z}\frac{\partial z}{\partial w}$$

Rewriting these equations in matrix form and inverting the system to obtain the derivatives of the shape functions in the global system, we get

$$\begin{bmatrix} \dfrac{\partial N_i}{\partial x} \\[2ex] \dfrac{\partial N_i}{\partial y} \\[2ex] \dfrac{\partial N_i}{\partial z} \end{bmatrix} = [J]^{-1} \begin{bmatrix} \dfrac{\partial N_i}{\partial u} \\[2ex] \dfrac{\partial N_i}{\partial v} \\[2ex] \dfrac{\partial N_i}{\partial w} \end{bmatrix} \tag{4.68}$$

where $[J]$ is the Jacobian matrix.

Table 4.2. Shape functions and derivatives for a 20-node isoparametric element.

Node	N_i	$\partial N_i/\partial u$
1	$-2(1-u)(1-v)(1-w)(u+v+w-1/2)$	$2(1-v)(1-w)(2u+v+w-3/2)$
2	$4u(1-u)(1-v)(1-w)$	$4(1-v)(1-w)(1-2u)$
3	$2u(1-v)(1-w)(u-v-w-1/2)$	$2(1-v)(1-w)(2u-v-w-1/2)$
4	$4uv(1-v)(1-w)$	$4v(1-v)(1-w)$
5	$2uv(1-w)(u+v-w-3/2)$	$2v(1-w)(2u+v-w-3/2)$
6	$4uv(1-u)(1-w)$	$4v(1-w)(1-2u)$
7	$2v(1-u)(1-w)(-u+v-w-1/2)$	$2v(1-w)(2u-v+w-1/2)$
8	$4v(1-u)(1-v)(1-w)$	$-4v(1-v)(1-w)$
9	$4w(1-u)(1-v)(1-w)$	$-4w(1-v)(1-w)$
10	$4uw(1-v)(1-w)$	$4w(1-v)(1-w)$
11	$4uvw(1-w)$	$4vw(1-w)$
12	$4vw(1-u)(1-w)$	$-4vw(1-w)$
13	$2w(1-u)(1-v)(-u-v+w-1/2)$	$2w(1-v)(2u+v-w-1/2)$
14	$4uw(1-u)(1-v)$	$4w(1-v)(1-2u)$
15	$2uw(1-v)(u-v+w-3/2)$	$2w(1-v)(2u-v+w-3/2)$
16	$4uvw(1-v)$	$4vw(1-v)$
17	$2uvw(u+v+w-5/2)$	$2vw(2u+v+w-5/2)$
18	$4uvw(1-u)$	$4vw(1-2u)$
19	$2vw(1-u)(-u+v+w-3/2)$	$2vw(2u-v-w+1/2)$
20	$4vw(1-u)(1-v)$	$-4vw(1-v)$

Node	$\partial N_i/\partial v$	$\partial N_i/\partial w$
1	$2(1-u)(1-w)(u+2v+w-3/2)$	$2(1-u)(1-v)(u+v+2w-3/2)$
2	$-4u(1-u)(1-w)$	$-4w(1-u)(1-v)$
3	$2u(1-w)(-u+2v+w-1/2)$	$2u(1-v)(-u-v+2w-1/2)$
4	$4u(1-w)(1-2v)$	$-4uv(1-v)$
5	$2u(1-w)(u+2v-w-3/2)$	$2uv(-u-v+2w+1/2)$
6	$4u(1-u)(1-w)$	$-4uv(1-u)$
7	$2(1-u)(1-w)(-u+2v-w-1/2)$	$2v(1-u)(u-v+2w-1/2)$
8	$4(1-u)(1-w)(1-2v)$	$-4v(1-u)(1-v)$
9	$-4w(1-u)(1-w)$	$4(1-u)(1-v)(1-2w)$
10	$-4uw(1-w)$	$4u(1-v)(1-2w)$
11	$4uw(1-w)$	$4uv(1-2w)$
12	$4w(1-u)(1-w)$	$4v(1-u)(1-2w)$
13	$2w(1-u)(u+2v-w-1/2)$	$2(1-u)(1-v)(-u-v+2w-1/2)$
14	$-4uw(1-u)$	$4u(1-u)(1-v)$
15	$2uw(-u+2v-w+1/2)$	$2u(1-v)(u-v+2w-3/2)$
16	$4uw(1-2v)$	$4uv(1-v)$
17	$2uw(u+2v+w-5/2)$	$2uv(u+v+2w-5/2)$
18	$4uw(1-u)$	$4uv(1-u)$
19	$2w(1-u)(-u+2v+w-3/2)$	$2v(1-u)(-u+v+2w-3/2)$
20	$4w(1-u)(1-2v)$	$4v(1-u)(1-v)$

The Jacobian is given as

$$[J] = \begin{bmatrix} \dfrac{\partial x}{\partial u} & \dfrac{\partial y}{\partial u} & \dfrac{\partial z}{\partial u} \\ \dfrac{\partial x}{\partial v} & \dfrac{\partial y}{\partial v} & \dfrac{\partial z}{\partial v} \\ \dfrac{\partial x}{\partial w} & \dfrac{\partial y}{\partial w} & \dfrac{\partial z}{\partial w} \end{bmatrix} \qquad (4.69)$$

Similarly we can write

$$dV = dxdy = [J]dudv$$

In formulations of field equations in terms of finite elements, it also becomes necessary to integrate over the area or volume of the finite elements. These integrals are of the form

$$\int_0^{+1}\int_0^{+1}\int_0^{+1} f'(u,v,w)dudvdw \qquad (4.70)$$

or

$$\int_{-1}^{+1}\int_{-1}^{+1}\int_{-1}^{+1} f'(u,v,w)dudvdw \qquad (4.71)$$

depending on the system of coordinates used to define the finite elements. The eight-node element shape functions in Table 4.1. and the 20-node element shape functions in Table 4.2 require Eq. (4.70) for integration. However, the shape functions are often defined in a system of coordinates such that the element nodes are between -1 and $+1$ in which case Eq. (4.71) is required. The reason for listing it here is that this form is normally employed for Gauss-Legendre integration while the integration in Eq. (4.70) is performed by transforming the limits of integration (see chapter 11).

The two methods of defining finite elements demonstrated above for four elements can be used for any other element. Either method can be used for any element. Thus, for example, the idea of local coordinates and mapping can be used with tetrahedral or triangular elements or, the direct method can be used with the three-dimensional brick elements. Similarly, either can be used to define a two-dimensional quadrilateral element. The choice is often a matter of convenience. However, definition by the direct method is only practical for simple finite elements. For complex finite elements the expressions obtained become large and tedious. On the other hand, the use of elements defined in a local system of coordinates in finite element solutions requires numerical integration rather than direct integration.

To see this and point to the differences between different elements we describe next the application of two of these elements to the solution of a simple *3-D* electrostatic problem. At this stage the solution is outlined but not actually solved (this is done in chapter *7*).

4.5 FINITE ELEMENT FORMULATION

4.5.1 The field equation

To show how the finite elements given above can be used, we return now to the electrostatic problem we solved in section 4.2 and look at its general formulation in terms of the finite elements discussed above. However, the electrostatic equation is written here in a more general form in three dimensions as

$$\frac{\partial^2 V}{\partial x^2} + \frac{\partial^2 V}{\partial y^2} + \frac{\partial^2 V}{\partial z^2} = -\frac{\rho(v)}{\varepsilon} \tag{4.72}$$

where the charge density is given as a general function of the volume v. To describe a physical problem we must define a solution domain, including boundary conditions. Now we must decide on a method of formulation of the field equations. Any of the methods in section 4.2 can be used but we choose the energy functional given in Eq. (4.35). This is written as

$$F(V) = \int_V \left\{ \frac{1}{2}\left[\left(\frac{\partial V}{\partial x}\right)^2 + \left(\frac{\partial V}{\partial y}\right)^2 + \left(\frac{\partial V}{\partial z}\right)^2\right] - \left[\frac{\rho(v)}{\varepsilon}V\right]\right\} dv = 0 \tag{4.73}$$

This function must first be shown to have the same solution as Eq. (4.72). To do so we derive the Euler equation for the functional. Writing the functional as

$$F(V) = \int_V f(V, \frac{\partial V}{\partial x}, \frac{\partial V}{\partial y}, \frac{\partial V}{\partial z})\, dv = 0$$

the Euler equation is derived as

$$\frac{\partial}{\partial x}\left(\frac{\partial f(A)}{\partial\left(\frac{\partial V}{\partial x}\right)}\right) + \frac{\partial}{\partial y}\left(\frac{\partial f(V)}{\partial\left(\frac{\partial V}{\partial y}\right)}\right) + \frac{\partial}{\partial z}\left(\frac{\partial f(V)}{\partial\left(\frac{\partial V}{\partial z}\right)}\right) - \frac{\partial f(V)}{\partial V} = 0$$

where $f(v)$ is the integrand of the energy functional in Eq. (4.73). Each of the terms is evaluated separately. Thus, for example, we get for the x component

$$\frac{\partial}{\partial x}\left(\frac{\partial\left[\left\{\frac{1}{2}\left[\left(\frac{\partial V}{\partial x}\right)^2+\left(\frac{\partial V}{\partial y}\right)^2+\left(\frac{\partial V}{\partial z}\right)^2\right]-\left[\frac{\rho(v)}{\varepsilon}V\right]\right\}\right]}{\partial\left(\frac{\partial V}{\partial x}\right)}\right)=$$

$$\frac{\partial}{\partial x}\left(\frac{\partial\left[\left\{\frac{1}{2}\left(\frac{\partial V}{\partial x}\right)^2\right\}\right]}{\partial\left(\frac{\partial V}{\partial x}\right)}\right)=\frac{\partial^2 V}{\partial x^2}$$

Similarly, the second and third terms are $\partial^2 V/\partial y^2$ and $\partial^2 V/\partial z^2$. The last term is

$$\frac{\partial}{\partial V}\left(\left\{\frac{1}{2}\left[\left(\frac{\partial V}{\partial x}\right)^2+\left(\frac{\partial V}{\partial y}\right)^2+\left(\frac{\partial V}{\partial z}\right)^2\right]-\left[\frac{\rho(v)}{\varepsilon}V\right]\right\}\right)=$$

$$\frac{\partial}{\partial V}\left(-\left[\frac{\rho(v)}{\varepsilon}V\right]\right)=-\frac{\rho(v)}{\varepsilon}$$

Summing the four terms we get

$$\frac{\partial^2 V}{\partial x^2}+\frac{\partial^2 V}{\partial y^2}+\frac{\partial^2 V}{\partial z^2}+\frac{\rho(v)}{\varepsilon}=0$$

This is identical to the original equation (Eq. (4.72)). Thus, the energy functional is equivalent to the original equation and therefore the two have the same solution.

The steps involved in deriving the Euler equations are not always explicitly performed but, any time an energy functional is used, it is also implied that the Euler equation has been or can be derived to show that the energy functional is a correct representation of the fields.

4.5.2 Choice of finite elements

Which element should we choose to solve the problem? This question is not always easy to answer and, in some cases, we might need to experiment with a few elements before settling on a particular element. In this case, a *3-D* element must be chosen but we are free to use the tetrahedral or brick elements (or any other properly defined element). First or second order elements can be used. In some instances, we may opt for first order elements since they are more economical in computational resources. If we are concerned with modeling awkward shapes, second order elements may be chosen because of the second order approximation they afford. In this case there is little reason to choose one element over another and therefore, the tetrahedral element is used to demonstrate the use of the direct method. For this purpose, we need the following approximations, given in Eqs. (4.63) through (4.66):

$$\phi(x,y,z) = L_1\phi_1 + L_2\phi_2 + L_3\phi_3 + L_4\phi_4 = \sum_{i=1}^{4} L_i\phi_i \tag{4.74}$$

$$\frac{\partial\phi(x,y,z)}{\partial x} = \sum_{i=1}^{4} \frac{\partial L_i}{\partial x}\phi_i = \sum_{i=1}^{4} \frac{b_i}{6V}\phi_i \tag{4.75}$$

$$\frac{\partial\phi(x,y,z)}{\partial y} = \sum_{i=1}^{4} \frac{\partial L_i}{\partial y}\phi_i = \sum_{i=1}^{4} \frac{c_i}{6V}\phi_i \tag{4.76}$$

$$\frac{\partial\phi(x,y,z)}{\partial z} = \sum_{i=1}^{4} \frac{\partial L_i}{\partial z}\phi_i = \sum_{i=1}^{4} \frac{d_i}{6V}\phi_i \tag{4.77}$$

These approximations are valid in each finite element. The approximations are now substituted into the energy functional in Eq. (4.73). However, this is of little use. Instead, we will rewrite the functional as a function defined over an element and minimize that function for the element.

4.5.3 Functional minimization

The next step is to minimize the functional with respect to each unknown in the solution domain. First, the energy functional in Eq. (4.73) can be written as an energy functional over a single finite element as

$$F(V) = \int_{V_k} \left\{ \frac{1}{2}\left[\left(\frac{\partial V}{\partial x}\right)^2 + \left(\frac{\partial V}{\partial y}\right)^2 + \left(\frac{\partial V}{\partial z}\right)^2\right] - \left[\frac{\rho(v_k)}{\varepsilon_k}V\right]\right\} dv_k \tag{4.78}$$

where k indicates the *kth* element. Also to be noted is that the addition of the index k to ε changes the problem from an isotropic to an anisotropic problem globally, while it is still isotropic on the element level (that is, ε remains constant within each element but may vary from element to element).

To minimize the energy in the element, we calculate the derivatives with respect to the unknown values of the potential at the four nodes of the element and set this to zero. In short form, this is

$$\frac{\partial(F(V))}{\partial V_l} = \frac{\partial}{\partial V_l}\int_{v_k}\left\{\frac{1}{2}\left[\left(\frac{\partial V}{\partial x}\right)^2+\left(\frac{\partial V}{\partial y}\right)^2+\left(\frac{\partial V}{\partial z}\right)^2\right]-\left[\frac{\rho(v_k)}{\varepsilon_k}V\right]\right\}dv_k = 0$$

Interchanging between differentiation and integration gives

$$\int_{v_k}\left\{\frac{1}{2}\left[\frac{\partial}{\partial V_l}\left(\frac{\partial V}{\partial x}\right)^2+\frac{\partial}{\partial V_l}\left(\frac{\partial V}{\partial y}\right)^2+\frac{\partial}{\partial V_l}\left(\frac{\partial V}{\partial z}\right)^2\right]-\frac{\partial}{\partial V_l}\left[\frac{\rho(v_k)}{\varepsilon_k}V\right]\right\}dv_k = 0$$

Using the following identity

$$\frac{\partial V}{\partial x}\frac{\partial}{\partial K}\left[\frac{\partial V}{\partial x}\right] = \frac{1}{2}\frac{\partial}{\partial K}\left[\frac{\partial V}{\partial x}\frac{\partial V}{\partial x}\right]$$

we get

$$\int_{v_k}\left\{\left[\left(\frac{\partial V}{\partial x}\right)\frac{\partial}{\partial V_l}\left(\frac{\partial V}{\partial x}\right)+\left(\frac{\partial V}{\partial y}\right)\frac{\partial}{\partial V_l}\left(\frac{\partial V}{\partial y}\right)+\left(\frac{\partial V}{\partial z}\right)\frac{\partial}{\partial V_l}\left(\frac{\partial V}{\partial z}\right)\right]\right. \tag{4.79}$$
$$\left.-\frac{\partial}{\partial V_l}\left[\frac{\rho(v_k)}{\varepsilon_k}V\right]dv_k\right\} = 0$$

Now, the approximations for the potential V and its derivatives are substituted, and the various terms in the expression evaluated. This is done term by term. We also note that b_i, c_i, and d_i are independent of the coordinates x, y, and z (they only depend on the coordinates of the four nodes of the tetrahedron). Thus the first terms in Eq. (4.79) is

$$\int_{v_k}\left[\left(\frac{\partial V}{\partial x}\right)\frac{\partial}{\partial V_l}\left(\frac{\partial V}{\partial x}\right)\right]dv_k = \int_{v_k}\left[\sum_{i=1}^{4}\frac{b_i}{6v}V_i\frac{\partial}{\partial V_l}\sum_{l=1}^{4}\frac{b_l}{6v}V_l\right]dv_k =$$

$$\int_{v_k}\left[\sum_{i=1}^{4}\frac{b_i}{6v}V_i\sum_{l=1}^{4}\frac{b_l}{6v}\right]dv_k = \frac{1}{36v^2}\int_{v_k}\left[\sum_{i=1}^{4}b_iV_i\sum_{l=1}^{4}b_l\right]dv_k =$$

$$\frac{1}{36v_k}\left[\sum_{i=1}^{4}b_iV_i\sum_{l=1}^{4}b_l\right]$$

While a bit lengthy this expression indicates that the volume integration amounts to simply multiplying the integrand by the volume of the element which has been calculated in section 4.4.2. The remaining three expressions are

$$\int_{v_k}\left[\left(\frac{\partial V}{\partial y}\right)\frac{\partial}{\partial V_l}\left(\frac{\partial V}{\partial y}\right)\right]dv_k = \frac{1}{36v_k}\left[\sum_{i=1}^{4} c_iV_i \sum_{l=1}^{4} c_l\right]$$

$$\int_{v_k}\left[\left(\frac{\partial V}{\partial z}\right)\frac{\partial}{\partial V_l}\left(\frac{\partial V}{\partial z}\right)\right]dv_k = \frac{1}{36v_k}\left[\sum_{i=1}^{4} d_iV_i \sum_{l=1}^{4} d_l\right]$$

and

$$\int_{v_k}\frac{\partial}{\partial V_l}\left[\frac{\rho(v_k)}{\varepsilon_k}V\right]dv_k = \int_{v_k}\frac{\partial}{\partial V_l}\left[\frac{\rho(v_k)}{\varepsilon_k}\sum_{i=1}^{4} L_iV_i\right]dv_k =$$

$$\int_{v_k}\left[\frac{\rho(v_k)}{\varepsilon_k}\sum_{l=1}^{4} L_l\right]dv_k = \frac{v}{4\varepsilon_k}\sum_{l=1}^{4}\rho$$

where, for simplicity, the charge density was assumed to be constant within the element. Evaluating the various terms in the expressions above we can write these as an elemental system of equations as follows

$$[s]\{V\} - \{q\} = 0 \tag{4.80}$$

where the system has four rows/columns and looks as

$$[s] = \frac{1}{36v}\begin{bmatrix} b_1^2+c_1^2+d_1^2 & b_1b_2+c_1c_2+d_1d_2 & b_1b_3+c_1c_3+d_1d_3 & b_1b_4+c_1c_4+d_1d_4 \\ b_2b_1+c_2c_1+d_2d_1 & b_2^2+c_2^2+d_2^2 & b_2b_3+c_2c_3+d_2d_3 & b_2b_4+c_2c_4+d_2d_4 \\ b_3b_1+c_3c_1+d_3d_1 & b_3b_2+c_3c_2+d_3d_2 & b_3^2+c_3^2+d_3^2 & b_3b_4+c_3c_4+d_3d_4 \\ b_4b_1+c_4c_1+d_4d_1 & b_4b_2+c_4c_2+d_4d_2 & b_4b_3+c_4c_3+d_4d_3 & b_4^2+c_4^2+d_4^2 \end{bmatrix} \tag{4.81}$$

$$[q] = -\frac{v}{4\varepsilon}\begin{bmatrix} \rho \\ \rho \\ \rho \\ \rho \end{bmatrix} \tag{4.82}$$

Now we can use these to solve any specific problem of this type simply by defining the solution domain, discretizing it into finite elements, and specifying the correct boundary conditions as discussed above.

Finally, we note that an almost identical expression is obtained for triangular elements for two-dimensional applications. The differences are in the use of surface integration, different values for the coefficients and the number of coefficients in the elemental matrix.

4.5.4 Finite element formulation with isoparametric elements

As an example, we outline here the use of eight-node brick finite elements to formulate the electrostatic problem given above. The steps are similar as we start with the same energy functional. The formulation starts with the energy functional in Eq. (4.79)

$$F(V) = \int_V \left\{ \frac{1}{2}\left[\left(\frac{\partial V}{\partial x}\right)^2 + \left(\frac{\partial V}{\partial y}\right)^2 + \left(\frac{\partial V}{\partial z}\right)^2\right] - \left[\frac{\rho(v)}{\varepsilon}V\right]\right\} dv = 0$$

However, the approximations for $V(x,y,z)$ and $\partial V(x,y,z)/\partial x$, $\partial V(x,y,z)/\partial y$, $\partial V(x,y,z)/\partial z$ are different and are now given by

$$V(x,y,z) = \sum_{j=1}^{8} N_j V_j \tag{4.83}$$

where the shape functions N_i are given in Table 4.1.

The minimization process is also the same and results in

$$\frac{\partial F(V)}{\partial V_i} = 0 \qquad i=1,2,\ldots,8$$

Performing steps identical to those for the tetrahedral element gives

$$\int_{v_k} \left\{ \left[\left(\frac{\partial V}{\partial x}\right)\frac{\partial}{\partial V_l}\left(\frac{\partial V}{\partial x}\right) + \left(\frac{\partial V}{\partial y}\right)\frac{\partial}{\partial V_l}\left(\frac{\partial V}{\partial y}\right) + \left(\frac{\partial V}{\partial z}\right)\frac{\partial}{\partial V_l}\left(\frac{\partial V}{\partial z}\right)\right] - \frac{\partial}{\partial V_l}\left[\frac{\rho(v_k)}{\varepsilon_k}V\right]\right\} dv_k = 0 \tag{4.84}$$

The approximations for the potential V and its derivatives are now substituted into this relation, and the various terms in the expression are integrated. We do this term by term. The first term in Eq. (4.84) is

$$\int_{v_k}\left[\left(\frac{\partial V}{\partial x}\right)\frac{\partial}{\partial V_l}\left(\frac{\partial V}{\partial x}\right)\right]dv_k = \int_{v_k}\left[\sum_{i=1}^{8}\frac{\partial N_i}{\partial x}V_i\frac{\partial N_i}{\partial V_l}\sum_{l=1}^{8}\frac{\partial N_i}{\partial x}V_l\right]dv_k =$$

$$\int_{v_k}\left[\sum_{i=1}^{8}\frac{\partial N_i}{\partial x}V_i\sum_{l=1}^{8}\frac{\partial N_i}{\partial x}\right]dv_k$$

The remaining three expressions are therefore

$$\int_{v_k}\left[\left(\frac{\partial V}{\partial y}\right)\frac{\partial}{\partial V_l}\left(\frac{\partial V}{\partial y}\right)\right]dv_k = \int_{v_k}\left[\sum_{i=1}^{8}\frac{\partial N_i}{\partial y}V_i\sum_{l=1}^{8}\frac{\partial N_i}{\partial y}\right]dv_k$$

$$\int_{v_k}\left[\left(\frac{\partial V}{\partial z}\right)\frac{\partial}{\partial V_l}\left(\frac{\partial V}{\partial z}\right)\right]dv_k = \int_{v_k}\left[\sum_{i=1}^{8}\frac{\partial N_i}{\partial z}V_i\sum_{l=1}^{8}\frac{\partial N_i}{\partial z}\right]dv_k$$

and

$$\int_{v_k}\frac{\partial}{\partial V_l}\left[\frac{\rho(v_k)}{\varepsilon_k}V\right]dv_k = \int_{v_k}\frac{\partial}{\partial V_l}\left[\frac{\rho(v_k)}{\varepsilon_k}\sum_{l=1}^{8}N_lV_l\right]dv_k =$$

$$\int_{v_k}\left[\frac{\rho(v_k)}{\varepsilon_k}\sum_{l=1}^{8}N_l\right]dv_k = \frac{\rho}{\varepsilon}\int_{v_k}\left[\sum_{i=1}^{8}N_l\right]dv_k$$

The individual entries in the elemental matrix are

$$s_{i,j} = \int_{v_k}\left(\frac{\partial N_i}{\partial x}\frac{\partial N_j}{\partial x}+\frac{\partial N_i}{\partial y}\frac{\partial N_j}{\partial y}+\frac{\partial N_i}{\partial z}\frac{\partial N_j}{\partial z}\right)dv_k \tag{4.85}$$

$$q_i = -\frac{\rho}{\varepsilon}\int_{v_k}N_i dv_k \tag{4.86}$$

where the indices i, j, vary from 1 to 8, and the shape functions and their derivatives are given in Table 4.1.

Each coefficient above is numerically integrated over the volume of the element using Gaussian quadrature (see chapter 11) and then summed into a matrix of the form given in Eq. (4.80).

Perhaps the main difference between the two methods of defining the finite elements is in the way the integration is performed. In elements defined in a local system of coordinates we normally need to resort to numerical integration. For elements defined directly, the integration is almost always analytic.

4.5.5 Assembly of elemental matrices

The methods described in the previous two sections result in an elemental system of equations. This system is relatively small (4×4 or 8×8 in the two cases discussed) and is identical in form for the two elements. The next step is assembly of these elemental matrices into a global system of equations before solution can be performed. The assembly is quite simple and consists in essentially creating a large or global system of equations, whose size is equal to the number of unknowns in the finite element discretized domain.

To outline the method for matrix assembly, consider Figure 4.15 where a few elements from a finite element mesh (using *4* node quadrilateral elements since these are easiest to display) are shown with element and node numbering. These numbers are arbitrary and depend on the method of discretization. Figure 4.16 shows the form of the elemental matrices for elements *12* and *18*. The row and column numbers are indicated for each matrix. Note that these correspond to the node number of the corresponding elements. These row and column numbers are the actual location in the global matrix.

The coefficients of the first elemental matrix in Figure 4.16 are entered in the global matrix as follows:

coefficient $s_{1,1}$ is added to coefficient $S_{65,65}$ in the global matrix
coefficient $s_{1,2}$ is added to coefficient $S_{65,64}$ in the global matrix
coefficient $s_{1,3}$ is added to coefficient $S_{65,18}$ in the global matrix
coefficient $s_{1,4}$ is added to coefficient $S_{65,19}$ in the global matrix
coefficient $s_{2,1}$ is added to coefficient $S_{64,65}$ in the global matrix
coefficient $s_{2,2}$ is added to coefficient $S_{64,64}$ in the global matrix
coefficient $s_{2,3}$ is added to coefficient $S_{64,18}$ in the global matrix
coefficient $s_{2,4}$ is added to coefficient $S_{64,19}$ in the global matrix
coefficient $s_{3,1}$ is added to coefficient $S_{18,65}$ in the global matrix
coefficient $s_{3,2}$ is added to coefficient $S_{18,64}$ in the global matrix
coefficient $s_{3,3}$ is added to coefficient $S_{18,18}$ in the global matrix
coefficient $s_{3,4}$ is added to coefficient $S_{18,19}$ in the global matrix
coefficient $s_{4,1}$ is added to coefficient $S_{19,65}$ in the global matrix
coefficient $s_{4,2}$ is added to coefficient $S_{19,64}$ in the global matrix
coefficient $s_{4,3}$ is added to coefficient $S_{19,18}$ in the global matrix
coefficient $s_{4,4}$ is added to coefficient $S_{19,19}$ in the global matrix

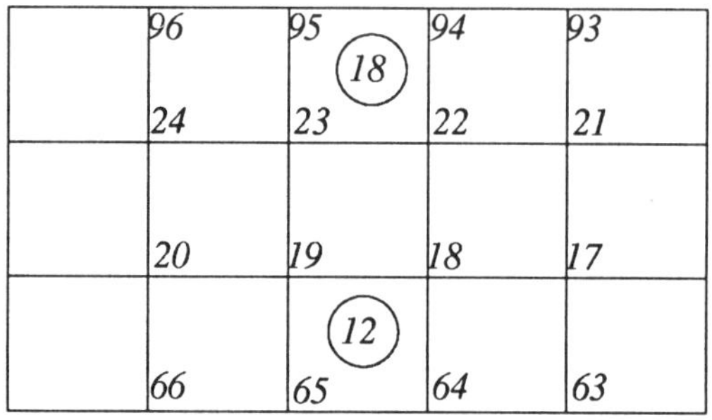

Figure 4.15 A simple finite element mesh used to demonstrate matrix assembly.

Element 12:

	65	64	18	19
65	s_{11}	s_{12}	s_{13}	s_{14}
64	s_{21}	s_{22}	s_{23}	s_{24}
18	s_{31}	s_{32}	s_{33}	s_{34}
19	s_{41}	s_{42}	s_{43}	s_{44}

Element 18:

	23	22	94	95
23	s_{11}	s_{12}	s_{13}	s_{14}
22	s_{21}	s_{22}	s_{23}	s_{24}
94	s_{31}	s_{32}	s_{33}	s_{34}
95	s_{41}	s_{42}	s_{43}	s_{44}

Figure 4.16 Elemental matrices for elements 12 and 18 in Figure 4.15.

This process is repeated for all elements in the solution domain. In terms of the algorithm used for assembly, the normal practice is to generate an elemental matrix and immediately assemble it in the global matrix, before the next elemental matrix is assembled. We also note that an identical process applies to the right hand side vector as well (elemental vectors q_i are assembled in global vector Q). After assembly, a global system of equations is obtained as

$$[S]\{V\} - \{Q\} = 0 \tag{4.87}$$

After applying boundary conditions, this system is solved for the unknown values of V.

4.6 THE FINITE ELEMENT MESH

A necessary step in any finite element analysis is discretization of the solution domain. This discretization defines the finite elements and provides the geometrical data for assembly of elemental matrices. The amount of input data required for finite element analysis, in particular for three-dimensional applications is enormous. A medium sized problem could have thousands of nodes in the region of interest. Each node is specified by two or three coordinates and each element by at least three nodes (in two dimensions) or four nodes (in three dimensions). The number of nodes per element can be as high as *20* for a second order brick element and each node may have from one to four variables (degrees of freedom or *DOF's*) associated with it. The problem of creating, storing and manipulating this data set is rather complicated.

The use of automatic mesh generation to facilitate both the preparation of the data and its checking is necessary if an efficient use of the finite element method is to be made.

An automatic mesh generator should have, as a minimum, the following capabilities:

1. A means of defining the solution domain, including boundaries, material interfaces, sources, and material properties.
2. A systematic way of dividing the solution domain into finite elements, that satisfy the basic criteria for finite element analysis.
3. Checking routines to ensure that elements are correctly defined and connected.

We describe here a very simple mesh generator based on the isoparametric mapping described above. This is done first for an eight-node isoparametric element. The modifications necessary to generate 20-node isoparametric elements and four-node tetrahedral elements are then described briefly followed by modifications needed to obtain triangular and quadrilateral two-dimensional elements.

4.6.1 Eight-node, isoparametric element mesh generator

Isoparametric elements and the isoparametric mapping are defined in section 4.4.4. The main idea behind these elements is to define a standard element in a convenient system of coordinates where operations are rather simple and then use a mapping function to map the space of the real element into the local element. When the mapping function is of the same order as the shape functions of the element the mapping is isoparametric and the elements are isoparametric elements. The most important point in this process is that this transformation maintains the continuity over the elements. Thus, the elements in the solution domain will have the same characteristics as the elements in the local coordinate system. We use the same principle for mesh generation.

For added flexibility in the use of the mesh generator, the region to be discretized is first divided into a number of blocks (or super elements). Each block is a hexahedral volume and a mesh, with the desired number of elements is created within each such block. Since three points are specified as input on each edge of each block, a parabolic fit is created on the boundaries of the solution region. This method, while requiring extra input data for block connectivity and coordinates is very useful in modeling complicated shapes, especially when material boundaries and material discontinuities occur within the discretized region, as is often the case in modeling of electromagnetic *NDT* phenomena. The elements and nodal points are labelled consecutively as the blocks of the mesh are created and then each block is mapped into the solution region. At this point the nodal points of the elements already assembled have coordinates in the (x,y,z) space. A nodal point coordinates array as well as an element array (connectivity array) are created to form part of the output. Boundary values on the outer faces of the blocks (or any other desired plane), if any, are also attached to the corresponding nodes and an array is created for this data. For simplicity, all the nodes on a face of any block, if assigned boundary conditions, are assumed to have the same boundary conditions.

The last part of the mesh generator is a resequencing scheme for bandwidth reduction (or frontwidth reduction if frontal methods are used for solution). This part can reorder the nodes of the mesh specifying new nodal points in such a way as to achieve a smaller bandwidth or frontwidth. A minimum in the bandwidth or frontwidth is normally sought but as will be pointed out later, this is not always possible or desirable. Reduction of bandwidth is only required if the solution method is a direct method such as Gaussian elimination and its sole function is to reduce solution time. Iterative methods can be used without reordering the mesh.

4.6.2 Shape functions

As each block in the mesh is created and divided into elements in a local system of coordinates, it is necessary to derive an appropriate mapping transformation to map the block into the solution domain. These shape functions are given in Table 4.1.

Consider Figure 4.17. Although this figure was used to define the 20-node isoparametric element, we will also use it here as a block or "super element". The block is the building block of the mesh which is composed of any number of connected blocks. The blocks are then subdivided into finite elements. In this block 20 nodes are specified which can define a quadratic function on each edge. In this way a good fitting with distorted boundaries and material interfaces is achieved. The mid-point on each edge is midway between the corner nodes on each edge but in the solution domain this mid-point can be specified anywhere on the edge. This creates an automatic crowding of elements towards the corner to which the mid-point is closer allowing a better distribution of nodes in the solution region, according to the anticipated solution. If the mid-point in the solution region is specified midway between the corner nodes the distribution of nodes on this edge is uniform. The shape functions necessary to define a 20-node block are identical with the 20 shape functions necessary to define a second order isoparametric element as given in Table 4.2.

4.6.3 Mesh assembly

Each block in the mesh and each side within a block are assigned a number. The blocks can be numbered in any convenient way as is the case for side numbering but these should be consistent throughout the mesh. The side numbering used here is indicated in Figure 4.17 and the block numbering in Figure 4.18.

In assembling the mesh, the connectivity data, read as input, is used. This data specifies for each block, which of its sides is connected and to which side of the previously assembled blocks it is connected. This method ensures later a progressive numbering of the nodes in the mesh with a unique number for each. In Figure 4.19 for example, side *No. 5* of block *No. 2* is connected to side *No. 6* of block *No. 1*. However, side *No. 2*, even though connected to side *No. 1* of block *No. 3*, is not specified since block *No. 3* has not yet been assembled. This connection can only be specified for block *No. 3*. For each block, the number of divisions (number of elements) in the x, y, and z directions is also specified as input and the crowding of the elements in each direction is automatically taken care of by the coordinates of the mid-point on each edge of the block. It is important to note that for connected blocks, the number of divisions on both connected sides has to be equal in the two dimensions defining the sides, and the mid-points must coincide. In Figure 4.19 for example, the number of divisions in the y direction of side *No. 6* in block *No. 1* has to be equal to the number of divisions of side *No. 5* of block *No. 2* in the y direction. The same is true for both sides in the z direction. The number of divisions in the x direction is not restricted by this connectivity since the two blocks are connected on a y-z plane. The number of divisions in any direction has no other restrictions.

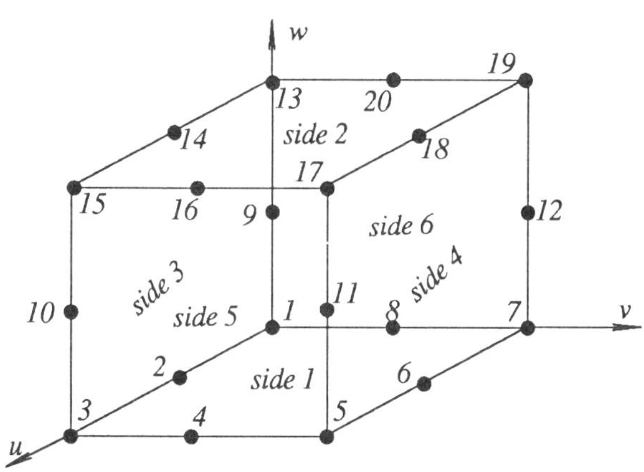

Figure 4.17 Numbering of sides in a block.

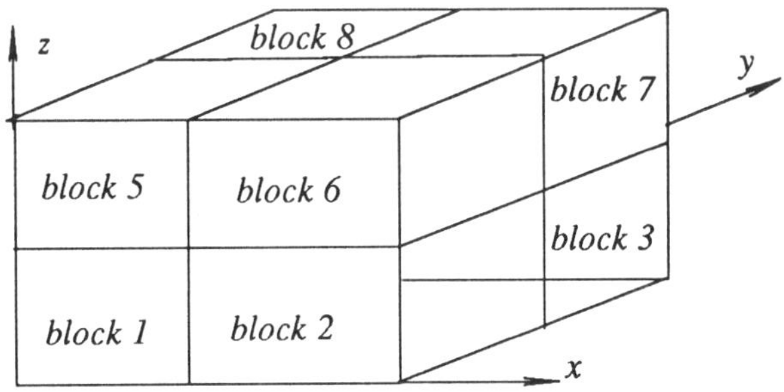

Figure 4.18 *Block numbering and connectivity.*

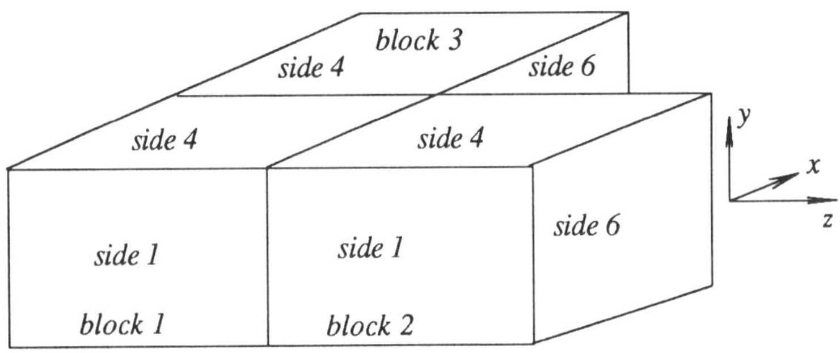

Figure 4.19. Connectivity of blocks.

4.6.4 Nodal point and element numbering

The number of elements and the number of nodal points in the mesh are needed as part of the output. The number of elements is found by multiplying the number of divisions in all three dimensions and summing these values over the number of blocks in the mesh. The number of nodal points is more difficult to calculate because when blocks are connected, the total number of nodal points is not a simple sum of the number of nodal points of the blocks. The calculation is performed as follows: *(1)* the number of nodal points in the current block being assembled is calculated and added to the total number of nodal points already assembled. *(2)* For every two sides connected (as specified in the input connectivity data) the number of nodes on this side is subtracted from the total. *(3)* For every pair of sides (sharing a common edge) the number of nodes on the common edge is added to the total as it was subtracted twice by the previous step. (4) For cases where three adjacent sides are connected (sharing one corner), as is the case when block *No. 8* is assembled the corner node has been added twice by the previous step and therefore is subtracted. Element numbering starts with block *No. 1* and continues to the last block in a simple manner. The labelling proceeds first in the x direction, then in the y direction and then in the z direction of each block. Figure 4.20. shows the element numbering for a two-block mesh.

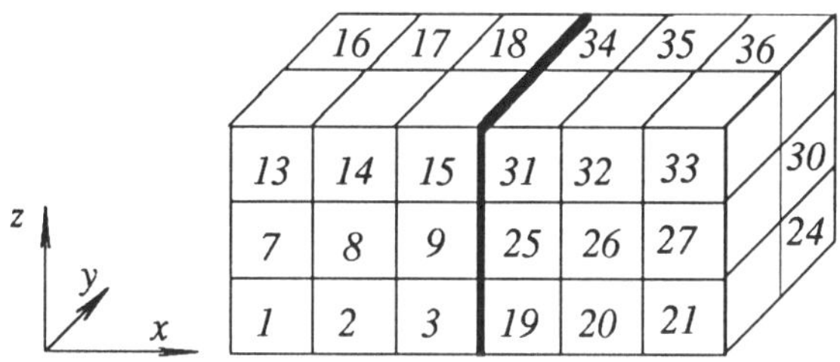

Figure 4.20 Element numbering for a two-block mesh

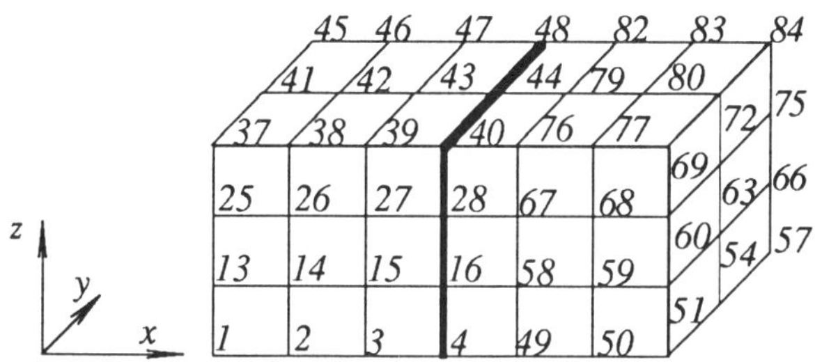

Figure 4.21 Node numbering for a two-block mesh.

Node numbering is more involved since every time a block is connected to an assembled block, the nodes on the connected sides have to be assigned the same numbers and the rest of the nodes numbered consecutively. This is achieved by storing the node numbers of the sides of the assembled blocks in special 'side arrays' and labeling the nodes on the connected sides of the new block with the labels of the sides to which it is connected. As with the element numbering, the node numbering proceeds first in the x direction, then in the y direction and then in the z direction for the remaining nodes of the block. Figure 4.21 shows the node numbering for a two-block mesh.

4.6.5 Reordering of the mesh

The numbering scheme described above is very simple and convenient but it can lead to undesirable results in terms of the bandwidth of the matrix to be solved. The same is true for frontal solution methods in which case the frontwidth of the resulting system can be unnecessarily large. Since the solution time and especially the necessary storage are drastically affected by the bandwidth or the frontwidth, every effort should be made to reduce them when creating the mesh. In three-dimensional solutions this could well make the difference between a solvable and unsolvable problem. Also to be mentioned here is the fact that reordering refers to direct methods only. Iterative methods do not require reordering.

To meet some of the requirements of minimizing the bandwidth for banded solutions and the frontwidth for frontal solutions, a semi-automatic scheme may be adopted. In this approach, the order for the elements is entered as input to the mesh generator and the mesh generator then updates the nodal point and coordinate arrays. This method, while requiring extra data as input to the mesh generator is simple and yields minimum bandwidth for a large variety of meshes. In other cases an absolute minimum may not be achieved although the bandwidth is reduced. It is however dependent on the user's experience and in some cases may require a trial and error approach. The savings in storage and solution time far outweigh the extra time and effort in preparing the mesh.

The reordering of the mesh is optional and should be avoided if the mesh is known to be of minimal bandwidth or frontwidth, such as in the case of a cubic mesh made up from a single block. If the problem solved is small, the reordering of the mesh might be considered unnecessary even though the numbering can be improved. Similarly, adoption of iterative solution techniques should make reordering unnecessary for the majority of practical problems.

As a rule in preparing the ordering of the elements, the elements in the direction with fewer divisions are entered first followed by the elements in the direction with the next larger number of divisions, and so on. This is straight forward with simple rectangular meshes but by trial and error it is possible to reduce the bandwidth or the frontwidth for most practical meshes.

4.6.6 Graphic representation of the mesh

It is of paramount importance to be able to check the mesh generated quickly or unnecessary waste of time and resources may occur when the finite element solution is obtained with an incorrect mesh. Furthermore, if the mesh is only 'slightly out of order', the solution could turn out to be erroneous, without these errors being detected, or detected very late in the solution process. Manual checking for large three-dimensional problems is virtually impossible or otherwise very difficult and lengthy.

A good graphics representation seems to be the only practical way to overcome these difficulties, but, whilst in a two-dimensional mesh a simple x-y line plot is sufficient, in three-dimensional meshes more elaborate techniques are required.

The following graphics operations are considered useful and used for display and checking (for mesh and other data) in any finite element application.

1. Simple line plots. Figure 4.22 shows a mesh with *48* linear elements, and *108* nodes. It is clear that even for small meshes like this very little information can be gained from this representation.
2. Hidden line removal. Figure 4.23 shows the mesh in Figure 4.22 after all hidden lines have been removed. In this way the outline of the mesh can be seen and checked for correctness.
3. Rotation along all three axes. In conjunction with (2) the mesh can now be checked for discrepancies in its geometry. Rotation is either continuous or at a rotation angle, depending on the mesh and display needs in particular cases.

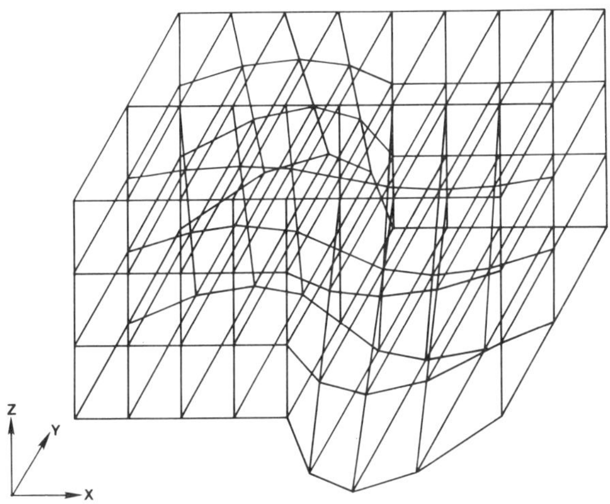

Figure 4.22 Plot of a simple mesh displaying all lines.

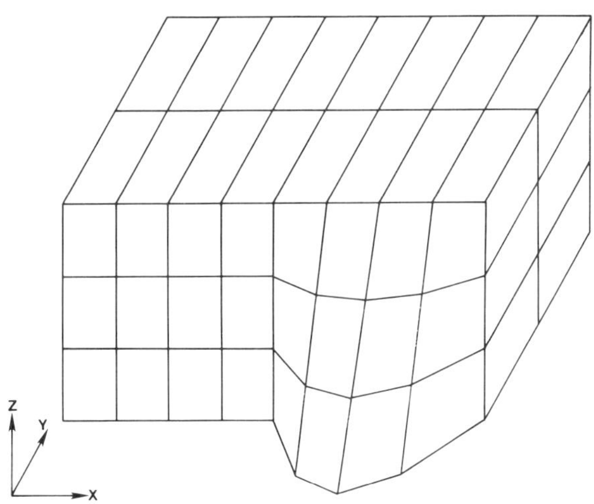

Figure 4.23 Plot of a simple mesh with hidden lines removed.

4. Cut sections through the mesh. This operation is often useful in operating with large, complex meshes where the previous operations are not sufficient to show the details of the mesh. It can be used with rotation and/or hidden line removal to inspect any plane in the mesh, or to display data such as contour lines in a plane through the mesh.
5. Zooming. In practical terms, zooming is the display of part of the data in a preferred section. It can be used in conjunction with any of the previous operations for even greater flexibility in regions of interest within the mesh, such as a small defect or an air gap in an electrical machine.

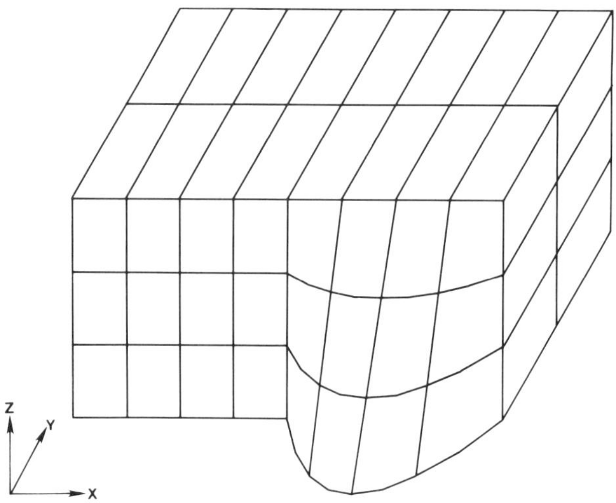

Figure 4.24 Display of the mesh in Figure 4.23 but with second order elements. Note the better fit along the curved boundary.

4.6.7 20-node isoparametric element mesh generator

This mesh generator is identical with the eight-node element mesh generator described in the previous sections in every aspect of the generation. It uses the same input data format, the same shape functions and produces the same output format. It is however a second order element mesh, i.e., uses second order elements, thus allowing a better fit along curved boundaries. Figure 4.24 is a line plot with hidden lines removed, identical with the plot in Figure 4.23 but employing second order elements. The fit along the curved boundary is clearly improved compared to Figure 4.23. This mesh has the same number of elements *(48)* but the number of nodes is *321*.

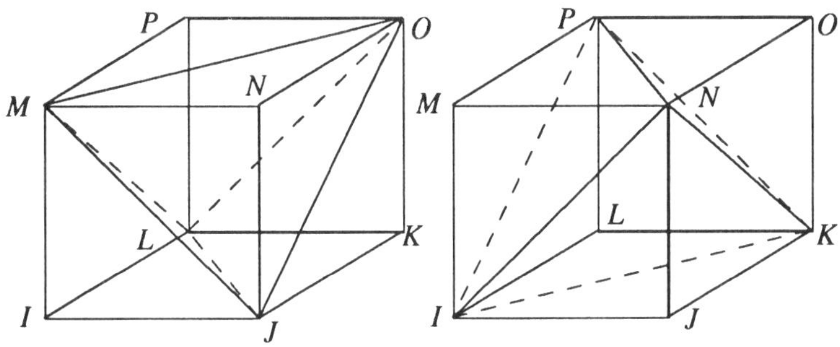

Figure 4.25 Division of neighboring elements to ensure compatibility of tetrahedra (no intersecting edges). Connectivity is shown in Table 4.3.

4.6.8 Tetrahedral element mesh generator.

As pointed out previously, tetrahedral elements are created by first dividing the region of interest into eight-node isoparametric elements. Then, each element (or super element as it is sometimes called) is divided into five tetrahedra. In this way, the problem of generating and displaying tetrahedral meshes is greatly simplified.

When each super element is divided into tetrahedra, it is important to note that no two edges of tetrahedra in adjacent super elements should intersect. This is a requirement that is unique to tetrahedral elements and does not arise when generating meshes with hexahedral elements. Figure 4.25 shows this special arrangement of elements such that edges do not intersect. In generating the mesh, the tetrahedra in adjacent super elements are assigned node numbers according to Table 4.3 (see Figure 4.25), depending on the numbering of the previous element. By alternating between these two possible partitions such that two neighboring super elements are always assigned different numbering sequences, the above requirement is met. The numbering sequences in Table 4.3 also ensure that in each tetrahedron three of the nodes are numbered in a clockwise manner when viewed from the fourth node of the tetrahedron.

Table 4.3 Possible division of neighboring hexahedra to ensure no edges cross each other.

	a	*b*
1	*I, J, L, M*	*I, J, K, N*
2	*J, L, M, O*	*I, K, L, P*
3	*J, M, N, O*	*M, N, I, P*
4	*L, P, M, O*	*O, N, P, K*
5	*J, O, K, L*	*K, N, P, I*

This is necessary for the element formulation, to ensure that the volume of the element is always positive.

After the division of the eight-node element mesh into tetrahedra is completed a new nodal point array (element array) is created to reflect the new elements but all the rest of the data (coordinates, boundary conditions etc.) remain unchanged.

Reordering of the mesh has the same advantages as for the eight- or 20-node element mesh but there appears to be little to be gained by reordering the tetrahedral mesh directly. Rather, the eight-node element mesh is resequenced prior to division into tetrahedra and the element numbers updated to take into account the number of tetrahedra in the mesh. The order of the tetrahedra within each super element is not changed and follows that in Table 4.3.

The display of tetrahedral meshes is more complicated than isoparametric meshes and the fact that edges are not allowed to intersect makes them more difficult to check. Computer graphics techniques are invaluable for these meshes as is exemplified by Figures 4.26 and 4.27. Figure 4.26 is a tetrahedral mesh with *40* tetrahedral elements (only eight super elements) and *27* nodes. Even though very small, this mesh is already difficult to follow. Figure 4.27 is the same mesh with hidden lines removed. Application of other graphics operations such as rotation and cut-sections allow easy checking of all the elements in the mesh.

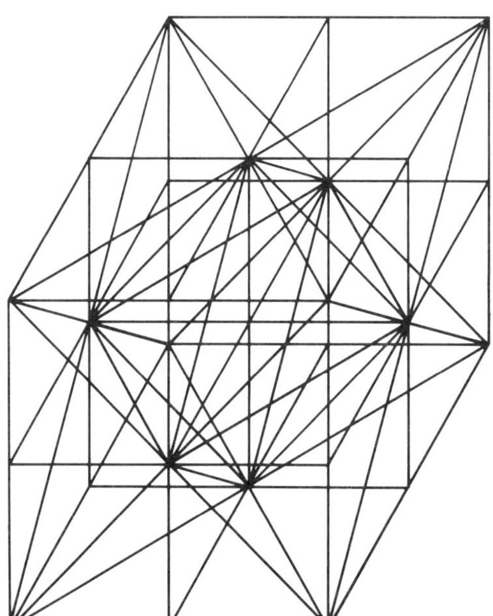

Figure 4.26 A tetrahedral mesh obtained by division of 8 hexahedra.

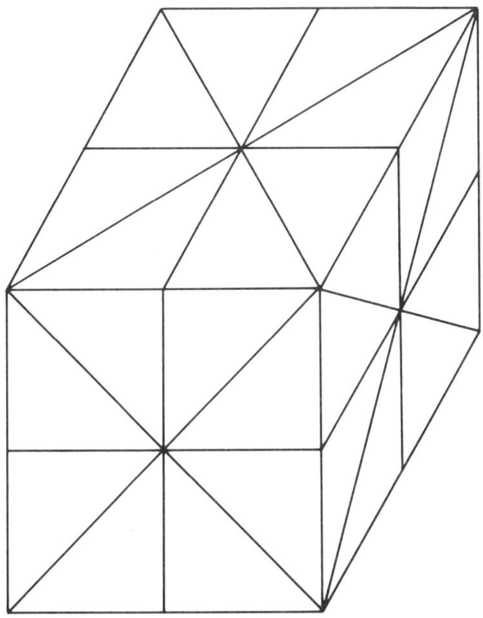

Figure 4.27 The tetrahedral mesh in Figure 4.26 with hidden lines removed.

4.7 TWO-DIMENSIONAL MESH GENERATION

Any three-dimensional mesh generator can be adapted for two-dimensional mesh generation by reducing the dimension of the elements by *1*. This means that elements have unit depth and therefore, the third dimension is unity (an exception to this is discussed in chapter *7* in conjunction with axisymmetric elements). The simplest way to view this is to imagine a three-dimensional element with as many elements as required to describe the cross section of a geometry but with a single element in the third direction. The steps involved are shown in Figure 4.28. Only the nodes on the front surface of this cross section are needed to describe the geometry since the geometry does not vary in the third dimension.

Thus, eight-node, second order, isoparametric, quadrilateral elements are obtained from 20-node elements, four-node elements from eight-node bricks, triangular elements from tetrahedral elements and so on.

The approach used here, namely that of simplifying a *3-D* mesh generator to obtain 2-D meshes is not a common approach but it indicates that *2-D* solutions are always a simplified form of the required *3-D* solution and so are the equivalent meshes.

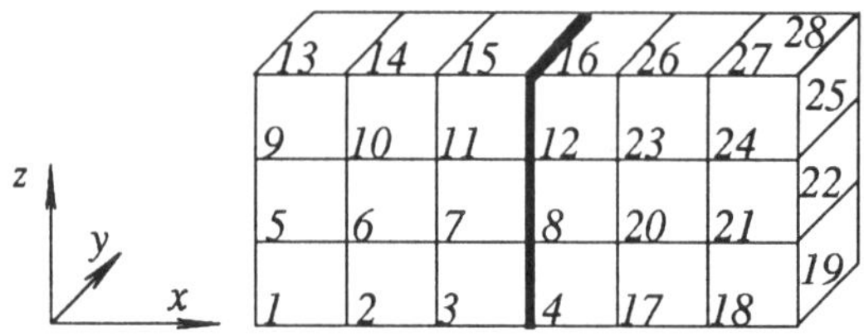

Figure 4.28 Transition from a 3-D to a 2-D mesh. The element is of unity thickness and only the division on the front surface is required. This figure shows only node numbering.

4.8 PRE-PROCESSING SOFTWARE

We note that all steps in the mesh generation process are geometrical in nature and have little to do with finite element analysis in the sense that if we required, say, triangulation over a surface, we would follow identical steps. Thus, if one were to require roofing of a domed building with triangular shingles, one would obtain the same solution as dividing this surface into triangular finite elements. A number of general-purpose packages have been developed to assist in this essential but rather tedious process. Some of these have been specifically designed for finite element analysis, while others have evolved separate from *FEM*. In either case, a package of this type will allow the user to enter the geometry of the problem and then divide it into pre-programmed shapes. The data of these elements then forms the basis of a mesh. Display of the whole geometry, and sections allows relatively easy definition of good meshes. The output data from these programs is not always in the form required by specific *FEM* programs but this can usually be overcome by some simple data translation routines.

The main disadvantage of general-purpose pre-processors is that they are usually much more general than is needed for specific solutions. Thus, a package of this type may have plate or shell elements which are very useful for analysis of shells and plates but not for electromagnetic fields. For this reason, they require considerable learning time, and some proficiency in finite element analysis in continuum mechanics. This aspect is, however, minor compared with the time and effort involved in developing a finite element mesh generator. It should also be remembered that the generation of a mesh for finite element analysis is an auxiliary step which provides no benefit of its own. Also, most available general-purpose packages provide means for displaying and manipulating data, after the finite element solution has been obtained.

For these reasons, it is well worth investing time in using one of these packages. While it is not possible to discuss here the details of available software packages, some are referred to in the bibliography section.

4.9 PROBLEMS

4.1 Consider the problem described in section 4.2.1 (Figure 4.1). Use the following approximation to calculate the potential in the capacitor.

$$V(x) = K_1 sin\frac{x}{2d} + K_2 sin\frac{x}{d}$$

Use least squares and compare with the exact solution given in Eq. (4.8).

4.2 Solve the problem in section 4.3 using the potential energy approach and five finite elements as shown in Figure 4.29. For numerical solution use $d=1$, and $\rho_0/\varepsilon=1$. Repeat the solution with the elements in Figure 4.30. Compare both solutions with the analytic solution. Explain the differences.

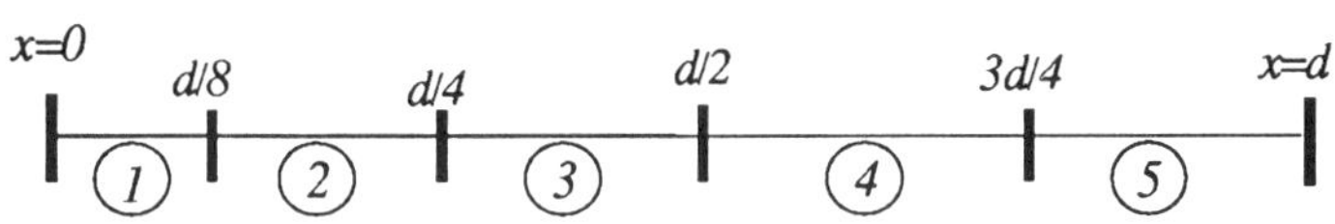

Figure 4.29.

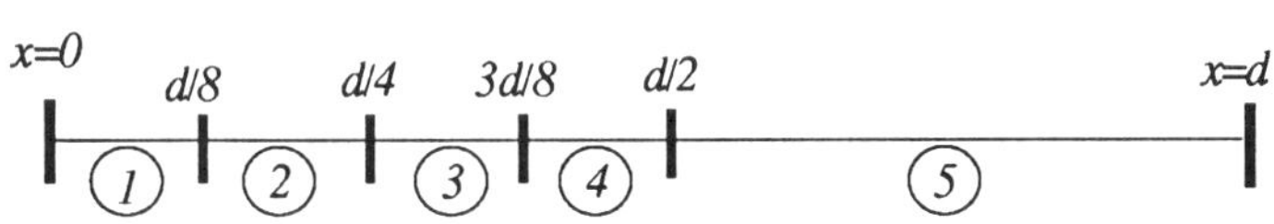

Figure 4.30.

4.3 Write a simple computer program to solve the problem in section 4.3. Use *50* equal elements. Compare the solution with the exact solution and to the solutions in problem 4.2.

4.4 A simple one-dimensional element was defined in section 4.3. Define a second order element by inserting a third node at the center between the end nodes in Figure 4.8. Find the shape functions and their derivatives. What happens as the middle node is moved towards one end or the other?

4.5 Repeat problems 4.2 and 4.3 using the element in problem 4.4.

4.6 Consider the parallel plate capacitor with a dielectric between the plates as shown in Figure 4.31. The dielectric has relative permittivity of 6 and a charge distribution as shown. Calculate the analytic potential distribution in the capacitor. Calculate the electric field intensity inside the capacitor. Set up a finite element solution and solve for the potential. From the potential distribution, calculate the electric field intensity in the capacitor. Sketch the solutions.

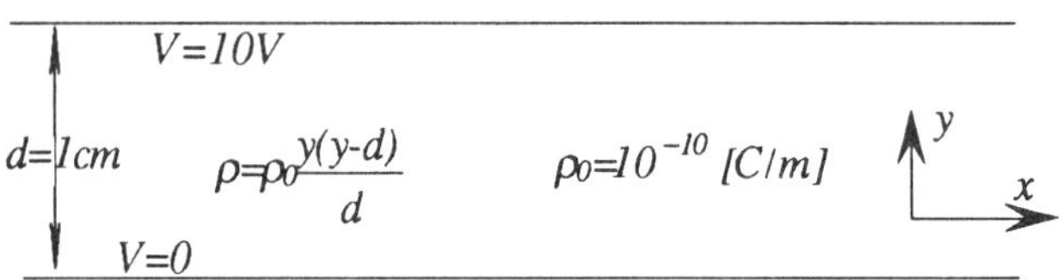

Figure 4.31.

4.7 Figure 4.32 shows a two-dimensional, four-node element (quadrilateral element). Using the direct method of definition, assume an approximation of the form

$$\phi(x,y) = a + bx + cy + dxy$$

and define the shape functions and their derivatives.

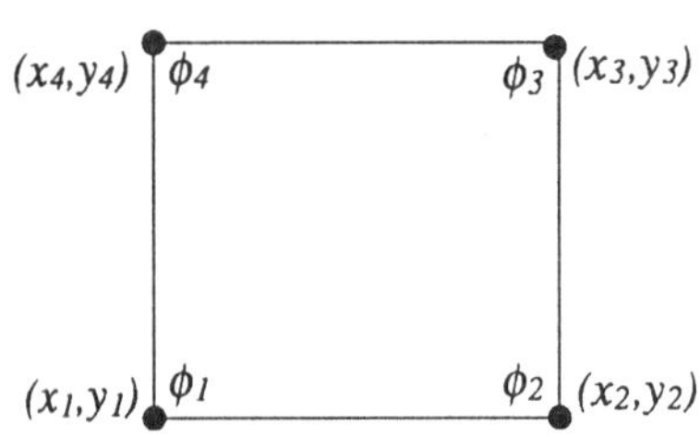

Figure 4.32.

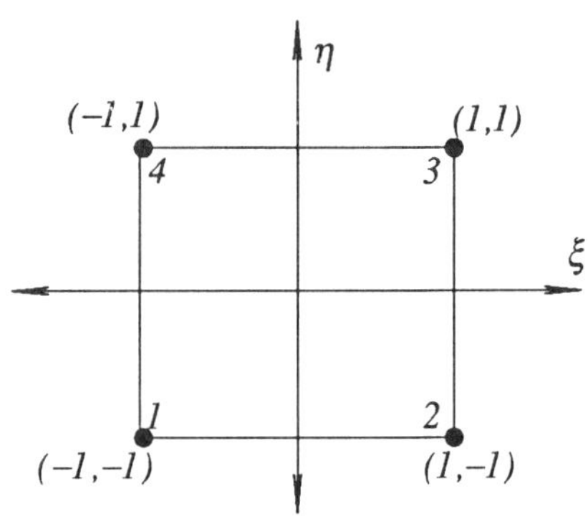

Figure 4.33.

4.8 For the element in Figure 4.32 define a local system of coordinates as in Figure 4.33. Find the shape functions in local coordinates and the shape functions and their derivatives in the global system of coordinates (i.e. in the Cartesian system of coordinates).

4.9 An 8-node element is shown in Figure 4.34. Repeat problem 4.8 for this element. This is a second order element and is often used when compliance with complex boundaries is required or high accuracy is needed. Comment on the elemental matrix required for this element.

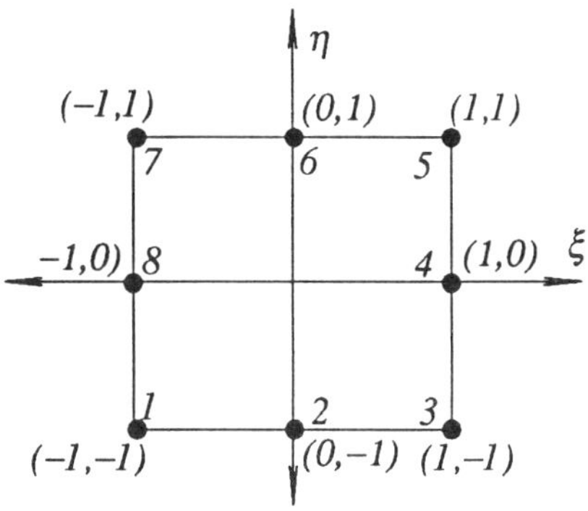

Figure 4.34.

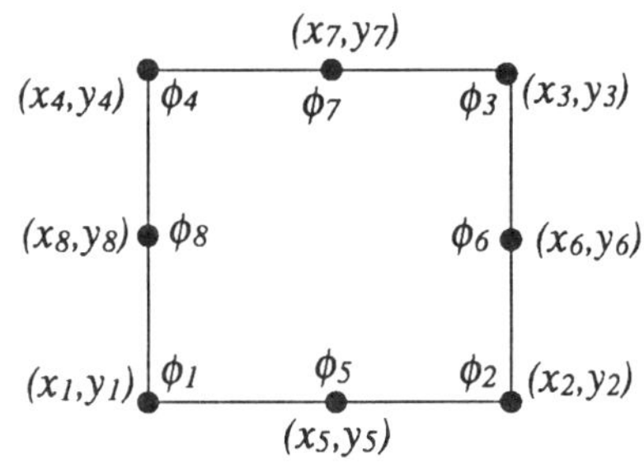

Figure 4.35.

4.10 To see the limitation in defining elements using the direct method consider the 8-node element in Figure 4.35. Use the procedure in section 4.4.1 and the following approximation to define the shape functions and their derivatives

$$\phi(x,y) = a + bx + cy + dxy + ex^2 + fy^2 + gx^2y + hy^2x$$

Comment on the difficulty in defining the shape functions and the prospects of using this method to define *3-D* elements other than the tetrahedral element.

4.11 Define the shape functions and derivatives for the triangular element using the local system of coordinates shown in Figure 4.36. Calculate the Jacobian and show how this element might be used to solve a general *2-D* problem.

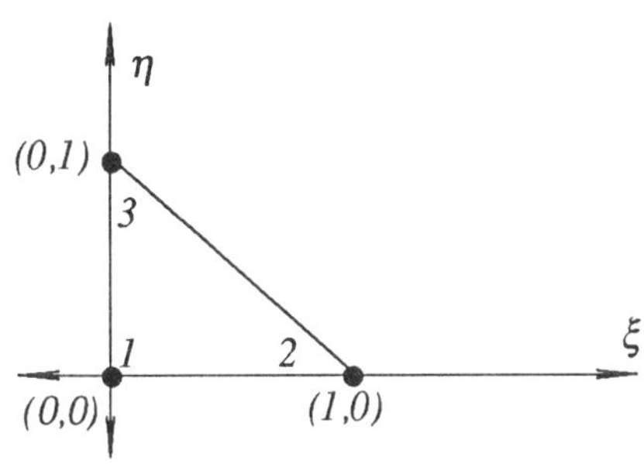

Figure 4.36.

4.12 Repeat problem 4.11 for the tetrahedral element. Use a convenient local system of coordinates. Comment on the integration required to evaluate elemental contributions.

4.13 Solve problem 3.10 using any of the two-dimensional elements defined in this chapter. Use a small number of elements and compute the solution either by hand (lengthy and tedious) or through a computer program (also tedious and lengthy but more fun and more general).

4.14 Solve problem 2.2 using any three-dimensional element defined here (or any other element you wish to define). Compare the solution with the analytical solution obtained in problem 2.2.

4.10 BIBLIOGRAPHY

The finite element method has been described in many texts although the first textbooks, and to date, the majority, are in solid mechanics and structural analysis. However, this is a minor problem since the method can be viewed as a general method for solution of boundary value problems. The general theory is summarized in the two papers by Oden [1,2]. Zienkiewicz's book [4] is considered to be the most complete source on finite elements. Reference [3] is given here because of the simple treatment of the heat equation which readers will find useful in developing solutions for scalar fields. References [5] through [10] deal specifically with electromagnetic field applications.

[1] Oden, J.T. (1969), "A general theory of finite elements I: Topological considerations," *International Journal for Numerical Methods in Engineering*, Vol. 1, No. 2, pp. 205-221.

[2] Oden, J.T. (1969), "A general theory of finite elements II: Applications," *International Journal for Numerical Methods in Engineering*, Vol. 1, No. 3, pp. 247-259.

[3] Huebner, K.H. (1975), *The Finite Element Method for Engineers*, Wiley-Interscience, John Wiley & Sons, New York.

[4] Zienkiewicz, O.C. (1977), *The Finite Element Method in Engineering*, third edition, McGraw-Hill, London.

[5] Steele, C.W. (1987), *Numerical Computation of Electric and Magnetic Fields*, Van Nostrand Reinhold, New York.

[6] Sylvester P.P. and Ferrari, R.P. (1990), *Finite Elements for Electrical Engineers*, Cambridge University Press, Cambridge.

[7] Sabonnadierre, J.C. and Coulomb, J.L. (1989), *Finite Element Methods in CAD*, Springer Verlag, New York.

[8] Hoole, S.R.H. (1989), *Computer-Aided Analysis and Design of Electromagnetic Devices*, Elsevier, New York.

[9] Lowther, D.A. and Sylvester, P.P. (1986), *Computer Aider Design in Magnetics*, Springer Verlag, New York.

[10] Ida, N. and Bastos, J.P.A. (1992), *Electromagnetics and Calculation of Fields*, Springer Verlag, New York.

CHAPTER 5

Elliptic partial differential equations

5.1 INTRODUCTION

As discussed briefly in the introductory chapter, the partial differential equations that we need to solve in conjunction with *NDT* modeling are of three types: elliptic, parabolic and hyperbolic. This classification is based on the form of the equation and affords a simple method of grouping processes using the methods of solution. In this chapter (and also in chapters 9 and 10), the discussion is based on this classification.

We discuss here elliptical processes: processes which lead to static or static-like solutions. Although not necessarily the easiest problems to solve these are discussed first, primarily because they include static and quasi-static applications. The types of problems discussed here are the static electric and magnetic field applications in *NDT*, eddy current methods and microwave applications. Although these are very different types of problems they are treated using similar methods because of the forms of the equations.

5.2 THE GENERAL ELLIPTIC PARTIAL DIFFERENTIAL EQUATION

A general, second order partial differential equation in two variables can be written as

$$a\frac{\partial^2 F(u,v)}{\partial u^2}+b\frac{\partial^2 F(u,v)}{\partial u\,\partial v}+c\frac{\partial^2 F(u,v)}{\partial v^2}=f(u,v) \qquad (5.1)$$

where u,v may represent any two variables such as x,y in space. Also, the first order derivatives as well as constant terms are left out for the moment.

This equation is said to be elliptic in form if the following condition is satisfied:

$$b^2 - 4ac < 0 \tag{5.2}$$

Note here that this is based solely on the form of the equation. The general equation including all terms is

$$a\frac{\partial^2 F(u,v)}{\partial u^2} + b\frac{\partial^2 F(u,v)}{\partial u\,\partial v} + c\frac{\partial^2 F(u,v)}{\partial v^2} + d\frac{\partial F(u,v)}{\partial u} + e\frac{\partial F(u,v)}{\partial v} + gF(u,v) = f(u,v) \tag{5.3}$$

and this is still defined by the property in Eq. (5.2). Thus, the coefficients of the first order derivatives, that of the function itself and the free term are not included in the classification. Also, we note that this classification is quite simple for two variables but does not always hold for three or four variables. Nevertheless, we will use this classification for more than two variables as is shown later. As examples, the following commonly encountered equations are elliptical equations:

Laplace's equation for the electrostatic field:

$$\frac{\partial^2 V(x,y)}{\partial x^2} + \frac{\partial^2 V(x,y)}{\partial y^2} = 0$$

Poisson's equation for the electrostatic field:

$$\frac{\partial^2 V(x,y)}{\partial x^2} + \frac{\partial^2 V(x,y)}{\partial y^2} = -\frac{\rho(x,y)}{\varepsilon}$$

The Helmholtz equation for time-harmonic wave propagation:

$$\frac{\partial^2 E(x,y)}{\partial x^2} + \frac{\partial^2 E(x,y)}{\partial y^2} + k^2 E(x,y) = 0$$

The eddy current equation in two dimensions:

$$\frac{\partial^2 A(x,y)}{\partial x^2} + \frac{\partial^2 A(x,y)}{\partial y^2} - j\omega\sigma A = -J_s$$

These equations have a number of properties in common. First, they are defined by an energy balance relation and are bounded in space. Thus, the solution of any of these equations requires the definition of a solution domain with defined boundaries and boundary conditions. Both the equation and the boundary conditions must be satisfied in the solution process. This is shown schematically in Figure 5.1. where we assume in general that both Dirichlet and Neumann boundary conditions may exist in the solution domain. The domain is closed in all variables (i.e. it is a bounded surface for two-dimensional problems or a bounded volume for three dimensional applications).

There are, however, significant differences in the solution of the general problems above. These are discussed in subsequent chapters but, as an example, the first two equations lead to simple systems of equations although these can be nonlinear. The eddy current equation is solved in complex variables but is always a linear problem. The Helmholtz equation must be solved as a characteristic value (eigenvalue) problem, either in real or complex variables.

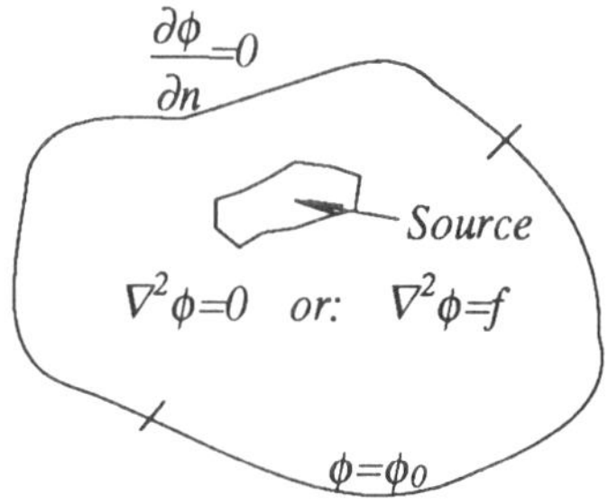

Figure 5.1 A general domain in which Laplace's or Poisson's equation are solved. Both Dirichlet and Neumann boundary conditions may apply.

5.3 CLASSES OF PROBLEMS

As mentioned above, there are three types of applications that fall under the category of elliptic processes. We now discuss these is some detail, with emphasis on general properties of governing equations. In section 5.4 below we relate these equations and their properties with applications to *NDT* and in chapters 6 through 8 use the equations to solve specific *NDT* problems.

5.3.1 Static electric fields

The equations for electrostatic fields are given in chapter 1, in three dimensions. These are either the Laplace equation:

$$\frac{\partial^2 V(x,y,z)}{\partial x^2}+\frac{\partial^2 V(x,y,z)}{\partial y^2}+\frac{\partial^2 V(x,y,z)}{\partial z^2}=0 \tag{5.4}$$

or the Poisson equation:

$$\frac{\partial^2 V(x,y,z)}{\partial x^2}+\frac{\partial^2 V(x,y,z)}{\partial y^2}+\frac{\partial^2 V(x,y,z)}{\partial z^2}=-\frac{\rho(x,y,z)}{\varepsilon} \tag{5.5}$$

In general we treat electrostatic problems as linear, isotropic, and homogeneous but nonlinear, nonhomogeneous, anisotropic applications can also be handled by the same basic equations. The sources of the electrostatic field are, in the first case, potentials on the boundary of the problem. In the second case, we may also have, in addition to boundary potentials, charge densities distributed throughout the solution domain. Figure 5.2a and 5.2b show the general types of solutions possible using these equations.

After formulation into one form or another, the solution is obtained in terms of the potential throughout the solution domain from which fields, energy, and the like can be calculated.

These are static problems as the sources must be static in time.

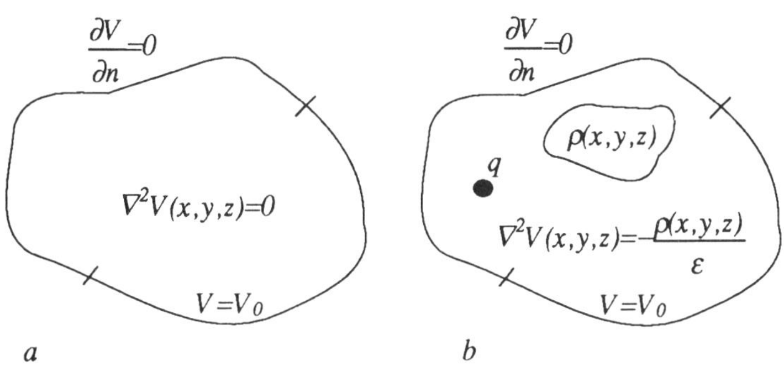

Figure 5.2 Domains, sources and boundary conditions for electrostatic problems. a. Source-free domains (Laplace's equation). b. Charges and/or charge distributions in the solution domain (Poisson's equation).

5.3.2 Static magnetic fields

Static magnetic fields are of two types: one is due to static fields such as fields generated by permanent magnets. The second is due to dc currents which generate static magnetic fields. The two problems are similar but in one the sources may be excluded while the other requires modeling of currents and current densities within the solution domain. As examples, we mention the magnetic scalar potential equation given in Eq. (1.84)

$$\frac{\partial^2 \Phi(x,y,z)}{\partial x^2} + \frac{\partial^2 \Phi(x,y,z)}{\partial y^2} + \frac{\partial^2 \Phi(x,y,z)}{\partial z^2} = 0 \tag{5.6}$$

which is a Laplace-type equation. The solution domain is shown in Figure 5.3a which also shows that the units of Φ are in *Ampere turns*. This equation can be written in terms of a magnetic vector potential but the treatment remains the same.

The second type of magnetostatic problem is shown in Figure 5.3b. This geometry includes current sources. Therefore, in general, it requires a vector function to describe the magnetic field in the solution domain. Two possible forms of describing the problem are given in section 1.13 as

$$\frac{\partial^2 \mathbf{A}(x,y,z)}{\partial x^2} + \frac{\partial^2 \mathbf{A}(x,y,z)}{\partial y^2} + \frac{\partial^2 \mathbf{A}(x,y,z)}{\partial z^2} + \mu \mathbf{J} = 0 \tag{5.7}$$

and

$$\nabla^2 \mathbf{T} = -\mu\sigma \left(\frac{\partial \mathbf{T}}{\partial t} \right) + \mu\sigma \nabla \Omega \tag{5.8}$$

where **A** is the magnetic vector potential, **T** the electric vector potential and **J** the source current density in the solution domain.

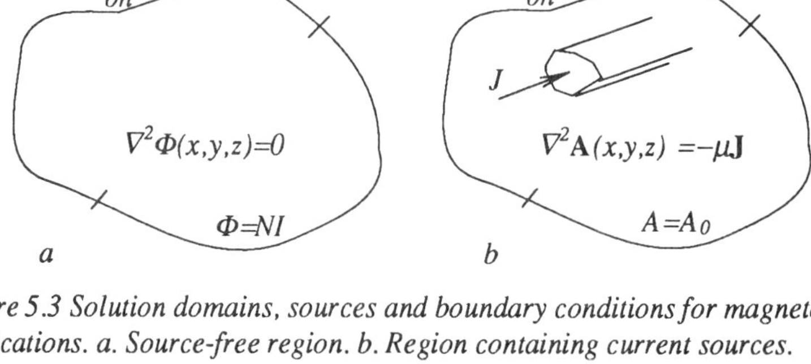

Figure 5.3 Solution domains, sources and boundary conditions for magnetostatic applications. a. Source-free region. b. Region containing current sources.

The linear, anisotropic and nonlinear, isotropic forms are given (for the magnetic vector potential) as

$$\frac{1}{\mu_x}\frac{\partial^2 \mathbf{A}(x,y,z)}{\partial x^2} + \frac{1}{\mu_y}\frac{\partial^2 \mathbf{A}(x,y,z)}{\partial y^2} + \frac{1}{\mu_z}\frac{\partial^2 \mathbf{A}(x,y,z)}{\partial z^2} + \mathbf{J} = 0 \tag{5.9}$$

and

$$\frac{\partial^2 \mathbf{A}(x,y,z)}{\partial x^2} + \frac{\partial^2 \mathbf{A}(x,y,z)}{\partial y^2} + \frac{\partial^2 \mathbf{A}(x,y,z)}{\partial z^2} + \mu(H)\mathbf{J} = 0 \tag{5.10}$$

These are all Poisson-type equations and are treated in similar fashion. The additional complication of nonlinearity, anisotropy and inhomogeneity, are handled in the formulation of the problem and do not affect its classification.

5.3.3 Quasi-static fields: linear eddy currents

Quasi-static fields or, what may be called diffusion fields are defined by sinusoidal excitation and no propagation (displacement currents are neglected). Thus we will deal here with time-harmonic fields. This includes many of the eddy current applications (single frequency, linear applications). The general governing equation is

$$\frac{1}{\mu_x}\frac{\partial^2 \mathbf{A}(x,y,z)}{\partial x^2} + \frac{1}{\mu_y}\frac{\partial^2 \mathbf{A}(x,y,z)}{\partial y^2} + \frac{1}{\mu_z}\frac{\partial^2 \mathbf{A}(x,y,z)}{\partial z^2} + \mathbf{J}_s - j\omega\sigma\mathbf{A} = 0 \tag{5.11}$$

where the electric scalar potential was neglected. We solve this equation with and without the scalar potential but this form is treated first and can be seen as a direct extension of Eq. (5.7) above. The main difference between the two equations is in that the first is solved in real variables, while Eq. (5.11) must be solved in complex variables.

5.3.4 The homogeneous Helmholtz equation: an eigenvalue problem

The Helmholtz equation was treated at some length in chapter 1. Here we merely wish to comment on some of the properties of the equation and its inclusion in this chapter in spite of the fact that it is a source-free wave equation. The type of equation we solve is of the form

$$\nabla^2\mathbf{E} + \omega^2\mu\varepsilon\mathbf{E} = 0 \tag{5.12}$$

where the electric field intensity is used as an example but the same form can be obtained in terms of other variables or potentials as was shown in section 1.15.

Based on the definition in section 5.2 this is an elliptic equation, and we solve it in conjunction with elliptic equations. However, the solution itself is quite different. The equation is a source-free equation and, therefore will be solved as a characteristic value (eigenvalue) problem. The solution is therefore in terms of modes and a solution (eigenvector) exists for each mode. This type of solution is typical of microwave applications, especially in guiding structures such as waveguides and electromagnetic microwave resonators.

5.4 APPLICATIONS TO NDT

Elliptic equations as presented above describe a relatively large subset of electromagnetic fields: static electric and magnetic fields (including nonlinear applications), linear, single frequency eddy current applications (quasi-static fields) and mode analysis in microwave structures under time-harmonic conditions. In terms of applications in *NDT* these include all aspects of testing with static electric and magnetic fields, sinusoidal eddy current applications as well as resonant methods in microwave cavities. Active and residual methods, most eddy current methods, including the remote field methods, and almost all microwave methods of testing are treated in the context of solutions to elliptic equations.

Before introducing methods of formulation and solution, we discuss the equations above in terms of applications in *NDT*. This will allow an easier classification of the types of solution methods applicable to various testing methods.

There are four basic forms of electromagnetic testing that we are concerned with in this chapter. These are the so-called electrostatic methods, magnetostatic methods, eddy current methods, and microwave methods. We describe these next in connection with their governing equations and typical testing geometries that need to be considered.

In the process we also define the various variables that can be used or, in some cases, those that should be used, and the quantities needed for testing purposes. These, to a large extent, define the various formulations and sometimes their solution.

5.4.1 Electrostatic testing

Electrostatic testing relies on the setup of an electric field in various dielectrics and monitoring of the electric field intensity as a measure of dielectric state and condition. Since any variation in dielectric causes changes in the electric field, the method can be used to detect cracks and variations in dielectric materials such as glass, ceramics and the like.

The principle of testing is shown in Figure 5.4 where the dielectric is assumed to be located between the plates of a capacitor. Of course, other methods of generating an electric field can be used. Variations in material properties cause two effects. One is local variations in the dielectric. This can be used for testing by either measuring the field distribution or using a fine dielectric powder to indicate the flaws. The second effect is the change in capacitance due to material changes. The latter is a volume effect and is more indicative of the whole system than of particular defects.

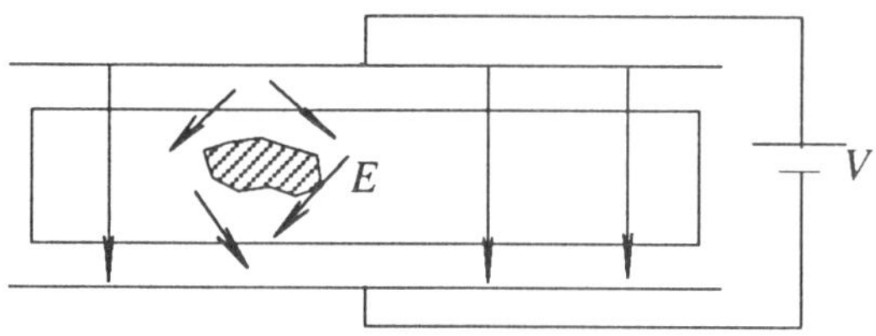

Figure 5.4 A simple electrostatic test geometry. Electric field around discontinuities is disturbed.

In terms of modeling, the calculation of electric field distribution provides all the information needed for characterization of the material. In addition, capacitance and potential distributions can be calculated.

In the generic problem in Figure 5.2a, we assumed that fields are set up by a system of potentials on the boundaries of the test domain. However, it is also possible to set up the fields by a system of charge densities or by a combination of charges, charge densities, and potential distributions as shown in Figure 5.2b. The difference is in the type of equation we need to solve.

The general testing geometry of Figure 5.4 has no charges within the solution domain. It is therefore solvable using Laplace's equation. For general dielectric materials the equation is

$$\frac{\partial}{\partial x}\varepsilon_x\frac{\partial V}{\partial x}+\frac{\partial}{\partial y}\varepsilon_y\frac{\partial V}{\partial y}+\frac{\partial}{\partial z}\varepsilon_z\frac{\partial V}{\partial z}=0 \qquad (5.13)$$

where the equation is written to show that material properties need not be isotropic (although isotropic materials simplify calculations). Boundary conditions for solution are potentials on the boundaries or, occasionally, normal derivatives of the potentials at the boundaries (that is, the electric field is given at the boundary). These are the Dirichlet and Neumann boundary conditions respectively.

If the problem is as in Figure 5.2b, charge densities are also present. The physical problem is now stated by Poisson's equation

$$\frac{\partial}{\partial x}\varepsilon_x\frac{\partial V}{\partial x}+\frac{\partial}{\partial y}\varepsilon_y\frac{\partial V}{\partial y}+\frac{\partial}{\partial z}\varepsilon_z\frac{\partial V}{\partial z}=-\rho \tag{5.14}$$

Boundary conditions for this problem are as in the previous case since charge densities are either in the solution domain or outside in which case they cause a boundary potential which must be specified.

If, in addition to charge densities there also exist point charges, these must be replaced by an equivalent charge density since the equation above is a differential equation.

In some cases, we use an integral form of the field equations. In this case, point charges are more readily taken into account than charge densities. This however, is discussed separately in chapter 8.

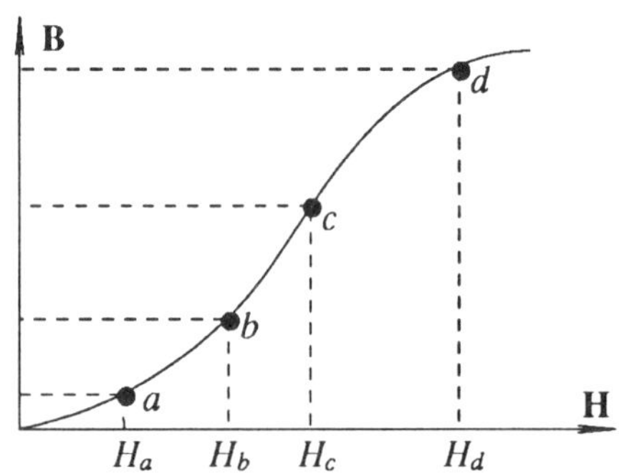

Figure 5.5 Points on the initial magnetization curve as used for active leakage testing.

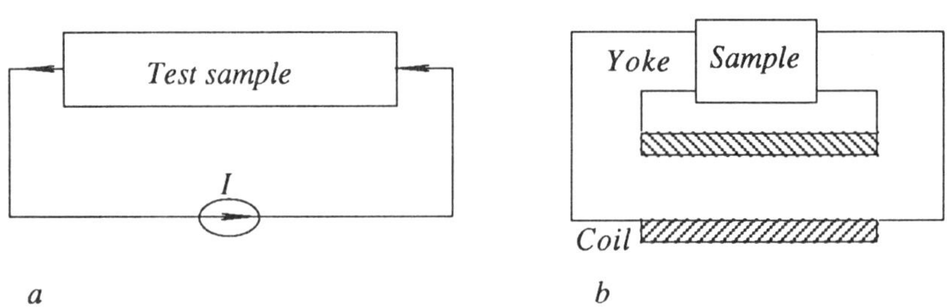

Figure 5.6 Two methods of generating a field in a sample. a. Passing a current through the sample or its vicinity. b. By a yoke in which a field is set up by a coil or permanent magnet.

5.4.2 Magnetostatic testing

Testing in the magnetostatic field includes two basic methods: residual and active leakage fields. The two methods are distinguished by the method of excitation. In active leakage fields a field is set up by external sources while in residual methods, internal magnetization is the source. From a modeling point of view the methods are quite different as they operate in different quadrants of the magnetization curve.

(a) Active leakage fields

Active leakage fields are generated by a dc field in a space in which the test material is located. Active leakage fields are created around defects or inhomogeneities in the material. These fields can be detected by a suitable flux sensitive device such as a Hall element or moving coil. The active leakage methods of *NDT* are therefore characterized by their operation on the initial magnetization curve at points such as *a*,*b*,*c* or *d* in Figure 5.5. The field itself can be generated either by a dc current as shown schematically in Figure 5.6a or, alternatively, by a magnetic yoke as shown in Figure 5.6b. In either case, the magnetic fields in the sample are affected by flaws and variations in the magnetic permeability. However, the two methods are rather different in their modeling. In the first case, the current must be modeled while in the second, the field is applied externally and there are no currents in the sample.

The fact that currents must be modeled indicates that a vector function must be used either to model the whole domain or, alternatively, a vector function must be used in current regions and a scalar function in regions void of currents. Thus, the magnetic vector potential or the electric vector potential can be used for this purpose. The equation for magnetostatic applications of the type shown in Figure 5.6a in terms of the magnetic vector potential is

$$\frac{\partial}{\partial x}\frac{1}{\mu_x}\frac{\partial \mathbf{A}}{\partial x}+\frac{\partial}{\partial y}\frac{1}{\mu_y}\frac{\partial \mathbf{A}}{\partial y}+\frac{\partial}{\partial z}\frac{1}{\mu_z}\frac{\partial \mathbf{A}}{\partial z}=-\mathbf{J} \qquad (5.15)$$

where the Coulomb gage is used as discussed in section 1.13. Boundary conditions are normally given as Dirichlet boundary conditions in terms of **A** although Neumann boundary conditions are often used on symmetry boundaries.

A second possible equation is that for the electric vector potential which is defined as

$$\nabla\times\mathbf{T}=\mathbf{J} \qquad (5.16)$$

and

$$\mathbf{H}=\mathbf{T}-\nabla\Omega \qquad (5.17)$$

For magnetostatic applications it is also possible to use the two scalar potentials discussed in section 1.14. The reduced scalar potential Ψ is used in current regions and the total scalar potential Φ in current-free regions and the two potentials are coupled at the interfaces between the two regions. In a current region of constant permeability, the reduced scalar potential is used:

$$-\nabla\mu\nabla\Psi + \nabla\mu\mathbf{H}_c = 0 \tag{5.18}$$

and in current-free regions containing magnetic materials the total scalar potential is employed:

$$-\nabla\mu\nabla\Phi + \nabla\mu\mathbf{H}_m = 0 \tag{5.19}$$

The coupling between Φ and Ψ is imposed by requiring continuity of the normal components of **B** and the tangential components of **H** at the interfaces of the regions. This representation is shown schematically in Figure 5.7 where the total scalar potential is used in current regions and the reduced scalar potential in a region without current densities. This method is most attractive in terms of economy in solution and is often employed for solution of linear and nonlinear magnetostatic fields such as those encountered in active and residual leakage field applications.

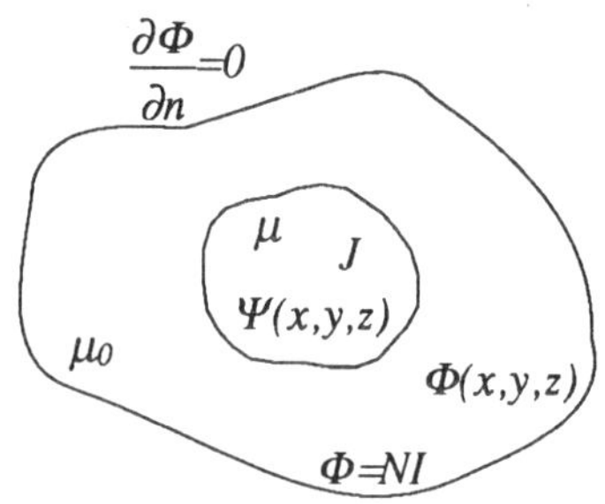

Figure 5.7 The use of the reduced and total scalar potential. The reduced scalar potential is used in current regions and the total scalar potential elsewhere.

If a yoke is used as the driving source for leakage fields, the problem is simplified considerably. The problem is shown in Figure 5.8 which also indicates the boundary conditions. In this case, it is assumed that Neumann boundary conditions can be used on all boundaries which are not in contact with the yoke. The simplest approach here is to use the magnetic scalar potential

$$\frac{\partial}{\partial x}\mu(H)_x\frac{\partial\Phi}{\partial x} + \frac{\partial}{\partial y}\mu(H)_y\frac{\partial\Phi}{\partial y} + \frac{\partial}{\partial z}\mu(H)_z\frac{\partial\Phi}{\partial z} = 0 \tag{5.20}$$

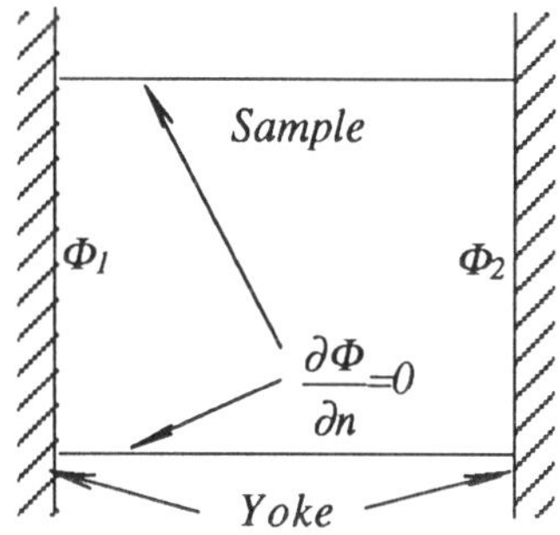

Figure 5.8 Solution domain for an externally generated field such as by a yoke.

where the nonlinear nature of the magnetic permeability is explicitly shown. Boundary conditions are given in terms of Φ as indicated in Figure 5.8. This is a particularly attractive approach because of its simplicity and the economy associated with the use of a single scalar function. The magnetic vector potential can be used as well by simply setting the current density in Eq. (5.15) to zero. However, this is more complicated since boundary conditions are given in terms of **H** and are required in terms of **A**. In addition, the use of a vector will require the solution of three variables at each point in the solution domain. In some cases, however, this may be acceptable especially since the magnetic vector potential is more general and can solve both types of problems.

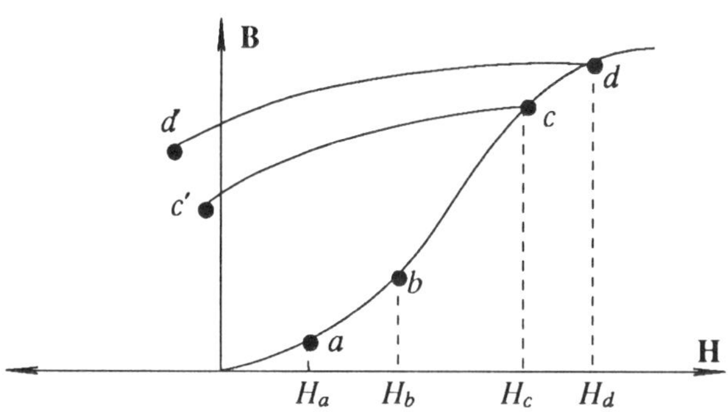

Figure 5.9 Residual field in ferromagnetic material is generated by relaxation of fields into the second quadrant.

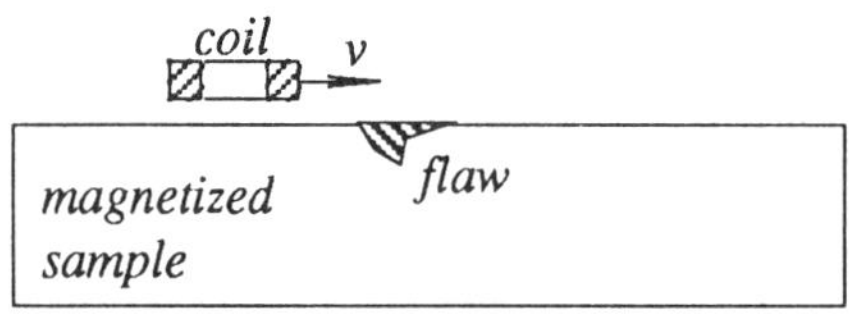

Figure 5.10 Method of measuring dc *leakage fields in the vicinity of flaws.*

(b) Residual leakage fields

If the driving current in an active leakage test is discontinued, the working points relax to points in the second quadrant (for example, c to c', d to d' in Figure 5.9), maintaining a residual leakage field around the defect. This field is not different than the field of a permanent magnet which also operates in the second quadrant.

The process of generating residual leakage fields also indicates the calculation method: we start with an active leakage field (normally a rather high field) and decrease the field, or the current in the sample, to zero. By doing so, the magnetization curve relaxes to a point in the second quadrant.

Thus, the equations are the same but a nonlinear process and the magnetization curve must be used. Normally a Newton Raphson technique is used for linearization of the nonlinear equation. The Newton Raphson technique is discussed in chapter 11 and applied in chapter 7 for the solution of nonlinear fields.

Leakage fields can also be generated by movement of a sample in a static field. However, modeling of motion in the magnetic field presents rather specialized aspects and these are not discussed here (see section 2.6.1).

Test parameters that need to be calculated are normally the magnetic flux density, induced voltages in moving coils, and voltages in Hall elements. These all require detailed knowledge of the magnetic flux density above the sample. Flux density is calculated either from the vector potentials using their definitions in Eq. (1.85) or from the magnetic scalar potential using the definition in Eq. (1.62). Induced voltages in moving coils rely on the fact that a changing flux is generated in a moving coil based on Faraday's law

$$emf = -N\frac{d\Phi}{\partial t}$$

where the minus sign indicates an induced *emf* (electromotive force, or voltage), N the number of turns in the coil and Φ is the total flux through the coil. The rate of change of the flux is directly proportional to the velocity at which the coil moves and is zero for a stationary coil in a dc field.

This method is shown schematically in Figure 5.10. The induced voltage is an integrated function since it requires the coil to have a finite cross sectional area (i.e. $d\Phi = \mathbf{B}.\mathbf{ds}$ where s is the cross sectional area over which the integration takes place and **B** is the flux density). For accurate measurement and interpretation the velocity of movement of the coil must be constant. Of course, the alternative way is to move the specimen through or by the coil. The use of coils for detection of leakage fields is particularly good for larger defects. For smaller defects, the coils must be small and in close proximity to the material. Computationally, this means that correct prediction of the induced voltage of moving coil sensors relies on accurate calculation of the magnetic flux density **B**.

More commonly, leakage field is measured by use of a Hall element. Since the Hall voltage is directly proportional to flux density, the same computation requirements apply here. The Hall voltage is given as

$$V_h = kIB$$

where k is a constant of the Hall element that includes the Hall coefficient and thickness of the element (i.e. $k=C/t$ where C=Hall coefficient, t=thickness of element). This constant is temperature dependent. I is the current through the element and B is the component of the flux density normal to the surface of the Hall element.

For a constant driving current, the Hall voltage is directly proportional to the component of the magnetic flux density normal to the surface of the element. Because the Hall coefficient is very small, the Hall voltage is also very small (about 30 $\mu V/Tesla$). The Hall element output is either a dc signal for a dc field or an ac signal for an ac field. It is also possible to produce an ac signal proportional to the dc field by using an ac bias current.

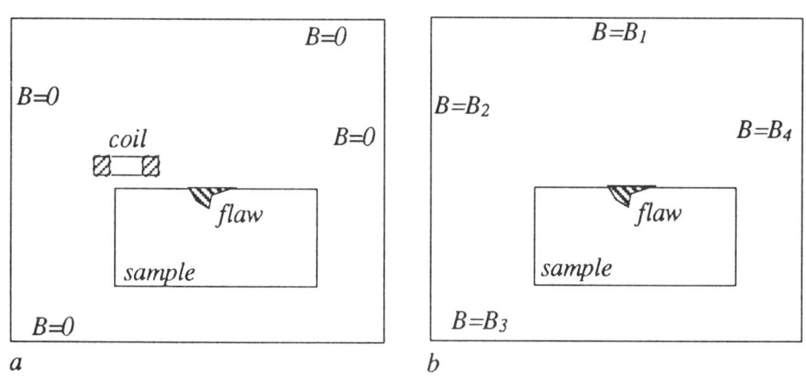

Figure 5.11 Eddy current testing. a. Source of eddy currents is in the solution domain. b. Eddy currents in solution domain are generated by external fields.

5.4.3 Sinusoidal eddy current testing

In the third method of testing, the specimen is excited with alternating currents inducing eddy currents in conducting materials. The interaction of these secondary currents with defects and inhomogeneities causes a change in the exciting field which can be detected by measuring the change in the impedance of the exciting coil or by a search coil. As the source moves over the defect, changes in its impedance occur which carry information on the shape, size and location of the defect. These can be deduced from the complex voltage signal obtained on a suitable display such as an oscilloscope screen. The excitation levels for eddy current testing are usually low and therefore the operating point is around the origin of the B/H curve (origin in Figure 5.5).

Similarly, one may define ac leakage methods as those that employ a normal hysteresis curve rather than the initial magnetization curve. The operating point now is on the major loop.

The basic test geometry considered here is shown in Figure 5.11. It consists of a source, driven by a sinusoidal current and both conducting and nonconducting materials. In Figure 5.11a the source is in the solution domain while in Figure 5.11b, it is outside the solution domain. For simplicity, the source is shown as a coil but this need not be the case. Induced currents can be generated by an external field (shown as boundary conditions in Figure 5.11b). Eddy currents are induced in conducting materials by the time variation of the magnetic field intensity.

As with active leakage fields, modeling requires a vector function to describe the sources. However, a vector function is only required in conducting materials where currents exists (both source and eddy currents). There are a number of vector functions and combinations of vector functions that can be used for solution of this type of problem. The simplest, is an extension of the magnetostatic equation given in Eq. (5.15) which includes Coulomb's gage:

$$\frac{\partial}{\partial x}\frac{1}{\mu_x}\frac{\partial \mathbf{A}}{\partial x}+\frac{\partial}{\partial y}\frac{1}{\mu_y}\frac{\partial \mathbf{A}}{\partial y}+\frac{\partial}{\partial z}\frac{1}{\mu_z}\frac{\partial \mathbf{A}}{\partial z}=j\omega\sigma\mathbf{A}-\mathbf{J} \tag{5.21}$$

This form is often used, particularly in two dimensions and in axisymmetric geometries. While it can be used in three dimensions, it is often necessary to use a different gage condition because of the need to include the electric scalar potential in the equation. An example is the so-called constrained eddy current equation in which the electric scalar potential is included in the equation, combined with the Coulomb gage but a second equation is added as a constraint on the eddy currents in the solution domain

$$\frac{\partial}{\partial x}\frac{1}{\mu_x}\frac{\partial \mathbf{A}}{\partial x}+\frac{\partial}{\partial y}\frac{1}{\mu_y}\frac{\partial \mathbf{A}}{\partial y}+\frac{\partial}{\partial z}\frac{1}{\mu_z}\frac{\partial \mathbf{A}}{\partial z}=j\omega\sigma\mathbf{A}+\sigma\nabla V-\mathbf{J} \tag{5.22}$$

$$\nabla\cdot(j\omega\sigma\mathbf{A}+\sigma\nabla V)=0 \tag{5.23}$$

These two equations must be solved together and, while the result in Eq. (5.21) is in terms of the three components of **A**, the solution of Eqs. (5.22) and (5.23) results in the three components of **A** and one component of V. The solution then produces both potentials.

In eddy current applications in which the conductivity is constant, a modified magnetic vector potential can be used so that there is no need to calculate the scalar potential directly (and therefore reduce the number of variables needed for solution).

$$\frac{\partial}{\partial x}\frac{1}{\mu_x}\frac{\partial \mathbf{A}^*}{\partial x}+\frac{\partial}{\partial y}\frac{1}{\mu_y}\frac{\partial \mathbf{A}^*}{\partial y}+\frac{\partial}{\partial z}\frac{1}{\mu_z}\frac{\partial \mathbf{A}^*}{\partial z}=j\omega\sigma\mathbf{A}^*-\mathbf{J} \qquad (5.24)$$

where the modified magnetic vector potential $\mathbf{A}^*$ is given as

$$\mathbf{A}^*=\mathbf{A}+\int_t \nabla V dt \qquad \textit{and} \qquad \nabla\cdot\sigma\mathbf{A}^*=0$$

and **A** is the magnetic vector potential as defined in Eq. (1.63) and as used in Eq. (5.21).

Another vector that is often used to calculate fields in eddy current applications is the electric vector potential. In this case, the solution is in terms of the electric vector potential and the magnetic scalar potential in current regions as

$$\frac{\partial^2\mathbf{T}}{\partial x^2}+\frac{\partial^2\mathbf{T}}{\partial y^2}+\frac{\partial^2\mathbf{T}}{\partial z^2}=-j\omega\mu\sigma\mathbf{T}+\mu\sigma\nabla\Omega \qquad (5.25)$$

and in terms of the magnetic scalar potential in regions of no current as

$$\frac{\partial}{\partial x}\frac{\partial\Omega}{\partial x}+\frac{\partial}{\partial y}\frac{\partial\Omega}{\partial y}+\frac{\partial}{\partial z}\frac{\partial\Omega}{\partial z}=0 \qquad (5.26)$$

where constant conductivity and permeability are assumed.

The eddy current equations as used above, have been obtained by neglecting displacement currents in the field equations. However, neglection of displacement currents is a rather arbitrary decision based on the fact that eddy currents occur in highly conducting materials and, therefore, eddy currents are dominant in these applications. This, however is not always true. In low loss dielectrics at high frequencies, displacement currents are not necessarily negligible. Thus, testing of composite materials at higher frequencies (in the microwave range) may require both displacement and conduction currents for proper modeling. Displacement currents can be taken into account by starting with the general, time-harmonic wave equation

$$\frac{\partial^2 \mathbf{A}}{\partial x^2} + \frac{\partial^2 \mathbf{A}}{\partial y^2} + \frac{\partial^2 \mathbf{A}}{\partial z^2} = j\omega\mu\sigma\mathbf{A} - \omega^2\mu\varepsilon\mathbf{A} - \mathbf{J} \tag{5.27}$$

where this is written as a linear equation. The second term on the right hand side is the displacement current density term. This particular form is equivalent to Eq. (5.21) although the form in Eq. (5.27) assumes isotropy of materials while Eq. (5.21) allows for anisotropy. The same can be done with the constrained eddy current equation in Eqs. (5.22) and (5.23). Now the system is

$$\frac{\partial^2 \mathbf{A}}{\partial x^2} + \frac{\partial^2 \mathbf{A}}{\partial y^2} + \frac{\partial^2 \mathbf{A}}{\partial z^2} = j\omega\mu\sigma\mathbf{A} - \omega^2\mu\varepsilon\mathbf{A} + \sigma\mu\nabla V - \mathbf{J} \tag{5.28}$$

$$\nabla.(j\omega\mu\sigma\mathbf{A} - \omega^2\mu\varepsilon\mathbf{A} + \sigma\mu\nabla V) = 0 \tag{5.29}$$

The addition of displacement currents allows extension of eddy current calculations into the microwave domain. This is possible even if the eddy currents themselves are negligible. Thus, this method can be used to calculate geometries with lossy, and lossless dielectrics in microwave waveguides, electromagnetic cavity resonators, as well as scattering, without the need for modal solutions.

It was mentioned at the beginning of this section that the basic requirement for eddy current calculations is the use of a vector function to describe current sources. This means that we can also use the field variables themselves including **E**, **H**, **B** or **D**. The equations can be written in either variable from the relations in chapter 1 and then solved. As an example, we can write the equations in terms of the magnetic field intensity using Eq. (1.61) as

$$\nabla^2\mathbf{H} + \omega^2\mu\varepsilon\mathbf{H} = \nabla\times\mathbf{J} \tag{5.30}$$

Using the fact that $\mathbf{J}=\mathbf{J}_s+\mathbf{J}_e$ where $\mathbf{J}_s$ is the source current density and $\mathbf{J}_e$ the eddy current density and that $\mathbf{J}_e=\sigma\mathbf{E}$ we get

$$\nabla^2\mathbf{H} + \omega^2\mu\varepsilon\mathbf{H} = \nabla\times\mathbf{J}_s + \nabla\times\mathbf{J}_e = \nabla\times\mathbf{J}_s + \nabla\times\sigma\mathbf{E}$$

Substituting $j\omega\mu\mathbf{H}$ for $\nabla\times\mathbf{E}$ (From Eq. (1.94)) we get

$$\nabla^2\mathbf{H} + \omega^2\mu\varepsilon\mathbf{H} = \nabla\times\mathbf{J}_s - j\omega\mu\sigma\mathbf{H} \tag{5.31}$$

This approach is, however, much more complicated than, say, the use of the magnetic vector potential.

In terms of computed quantities in eddy current testing, we are often more interested in calculating secondary parameters such as inductance, impedance, and induced voltages, rather than primary parameters such as electric and magnetic field intensities. Normally, we calculate the primary parameters first, from which other parameters are calculated based on the definitions given in chapter 1. Thus, for example, inductance is calculated from energy considerations while induced voltage is evaluated from the electric field intensity.

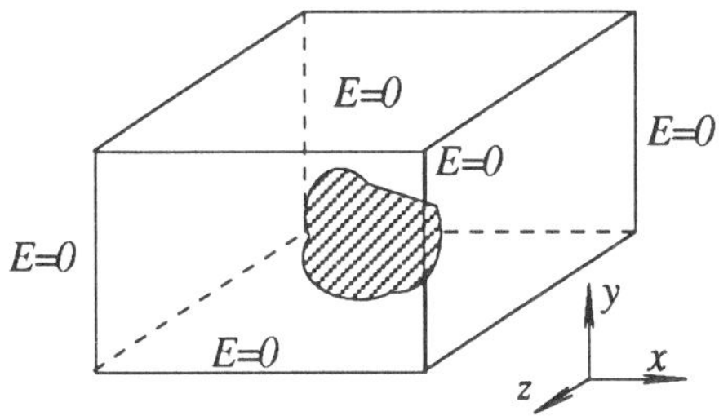

Figure 5.12 Testing in a microwave cavity. Boundary conditions shown are for TE modes (see chapter 1).

5.4.4 Testing in waveguides and resonant cavities

Testing in waveguides and cavity resonators can take two basic forms; the first assumes that the test sample is included within the waveguide or cavity resonator. In a waveguide the main parameter is the cutoff frequency of the waveguide and the shift in cutoff frequency due to insertion of the material. Also, losses, the reflection coefficient and transmission coefficient can be calculated provided that the modes can be correctly identified. However, insertion of samples inside waveguides is not practical. On the other hand, testing of materials in cavity resonators is common and the method is shown schematically in Figure 5.12. Due to insertion of the sample, the resonant frequency as well as the quality factor (Q-factor) of the cavity changes and these properties can be related with the condition of the material.

A third possibility of testing in the microwave region is a scattering method as shown in Figure 5.13. In this case, the material is illuminated by a wave and the scattered field is calculated. From a computational point of view, the main difference between the methods is in the application of the boundary conditions and methods of solution. Waveguide applications usually require the solution of the homogeneous (source free) Helmholtz equation in two dimensions while the cavity resonator requires solution in three dimensions. The scattering problem is handled by solving the general wave equation in an open domain.

Thus, the following equation is solved for waveguides

$$\frac{\partial^2 \mathbf{E}}{\partial x^2} + \frac{\partial^2 \mathbf{E}}{\partial y^2} + k^2 \mathbf{E} = 0 \tag{5.32}$$

where k^2 is the cutoff frequency of the waveguide. This is normally given as $k=\omega^2\mu\varepsilon$ and represents the eigenvalues of the system. A solution for **E** is obtained for any eigenvalue of the system as discussed in section 1.20.

For cavity resonators, the same equation in three dimensions applies:

$$\frac{\partial^2 \mathbf{E}}{\partial x^2}+\frac{\partial^2 \mathbf{E}}{\partial y^2}+\frac{\partial^2 \mathbf{E}}{\partial z^2}+k^2\mathbf{E}=0 \tag{5.33}$$

For open domain problems and scattering applications we can solve the general wave equation. One general form is given in Eq. (5.31) as

$$\frac{\partial^2 \mathbf{H}}{\partial x^2}+\frac{\partial^2 \mathbf{H}}{\partial y^2}+\frac{\partial^2 \mathbf{H}}{\partial z^2}+\omega^2\mu\varepsilon\mathbf{H}=\nabla\times\mathbf{J}_s-j\omega\mu\sigma\mathbf{H} \tag{5.34}$$

This is the inhomogeneous wave equation and reduces to the Helmholtz equation for domains void of current sources (including eddy currents). In this case, the only difference between the cavity resonator and the scattering problem is in the application of boundary conditions.

As mentioned before, the main parameters we need to calculate are the resonant frequency of the system (eigenvalues) and, especially in open problems, the fields in space.

The equations above are written for the electric or magnetic field intensities. Essentially identical forms can be obtained for other field variables as discussed in section 1.20.

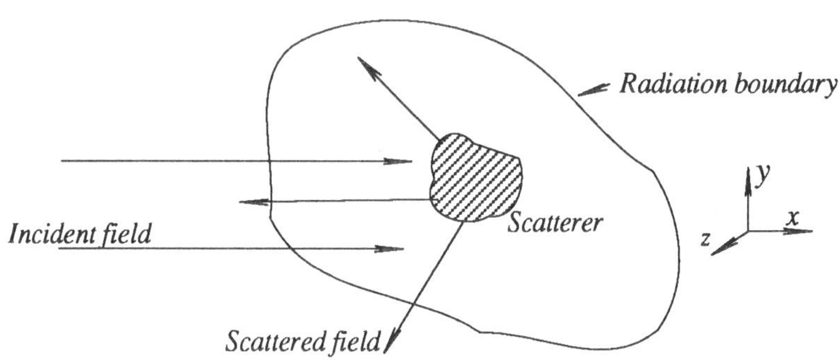

Figure 5.13 General scattering test. The source is outside the solution domain and a radiation boundary condition is imposed on the artificial boundary.

5.5 2-D, AXISYMMETRIC, AND 3-D APPLICATIONS: DIFFERENCES AND SIMILARITIES

We complete this chapter with a discussion on dimensionality. In general, the solution domain is a three-dimensional geometry and a full three-dimensional solution is required. This means that sources, fields and geometry are three-dimensional. Fields and sources may also be time dependent. In some cases, a correct solution to a particular problem can be obtained by a simplified method using a two-dimensional problem either in Cartesian coordinates or in cylindrical coordinates. The latter is often referred to as an axisymmetric geometry. In still other cases, the two-dimensional solution is not a correct solution in the sense that the approximations needed modify the problem but the solution may still be a good approximation to the correct solution.

It should also be noted at this point that a particular application may be two-dimensional in geometry but may still require three dimensional treatment because the sources or the fields are three-dimensional.

Since the differences between two-dimensional and three-dimensional solutions are very significant both in computational effort and in formulations, it is extremely important to solve the problem under the most efficient conditions; solution of a two-dimensional problem in *3-D* is wasteful, lengthy and less accurate than solution in *2-D*. On the other hand, only a correct solution or a good approximation to the correct solution has value. Solution of a *3-D* problem in two dimensions is only meaningful if the solution obtained is a good representation of the physical problem.

5.5.1 Two-dimensional applications

Two-dimensional applications are defined as those in which the fields do not vary in one dimension. As an example, the current carrying conductor, shown in Figure 5.14 is a two-dimensional problem since the magnetic flux density only varies in the x and y directions. The field is constant in the direction perpendicular to the plane of the figure.

On the other hand, the same current carrying conductor, in the presence of a cube of ferromagnetic material cannot be considered a two dimensional problem since now the fields are disturbed by the *3-D* geometry. Thus, in terms of the components of the magnetic flux density, only two components are allowed in a two dimensional geometry.

From Figure 5.14 it is also obvious that in terms of currents, a *2-D* geometry is defined by a single component of the current, perpendicular to the plane of the problem. If we use the magnetic vector potential for solution, the magnetic vector potential has a single component in the direction perpendicular to the solution plane. This can be best understood from the fact that the magnetic vector potential defines a flux density through the curl relation

$$\mathbf{B} = \nabla\times\mathbf{A} = \hat{\mathbf{x}}\left(\frac{\partial A_y}{\partial z} - \frac{\partial A_z}{\partial y}\right) + \hat{\mathbf{y}}\left(\frac{\partial A_z}{\partial x} - \frac{\partial A_x}{\partial z}\right) + \hat{\mathbf{z}}\left(\frac{\partial A_x}{\partial y} - \frac{\partial A_y}{\partial x}\right)$$

In Figure 5.14 the direction of **A** is in the y direction, therefore, only that component of **A** exists in this relation

$$\mathbf{B} = \nabla\times(\hat{\mathbf{y}}A_y) = \hat{\mathbf{x}}\left(\frac{\partial A_y}{\partial z}\right) - \hat{\mathbf{z}}\left(\frac{\partial A_y}{\partial x}\right)$$

The two components of **B** are in the x and z directions and perpendicular to **A** as required.

This also explains why the magnetic vector potential is a good choice for solution in two dimensions; a single component of the vector is required, making the solution identical to a scalar solution. In general, with three-dimensional applications, all three components are required.

Although a *2-D* problem is defined in the cross section of an infinite geometry, for purposes of calculation we often need to define such values as energy density and total energy which are only properly defined in a finite volume. Thus, we often refer to a surface as being a volume of unit depth (that is *1m* deep). The solution domain for a *2-D* application is therefore a plane, although a unit depth is assumed. Therefore, whenever we calculate the surface of a cross section, the corresponding volume (that is surface multiplied by unit depth) is assumed. The solution domain for Figure 5.14 is shown in Figure 5.15. The fact that unit depth is assumed, allows calculation in two dimensions but also computation of volumetric parameters. These quantities, such as energy, are calculated per unit depth rather than total values.

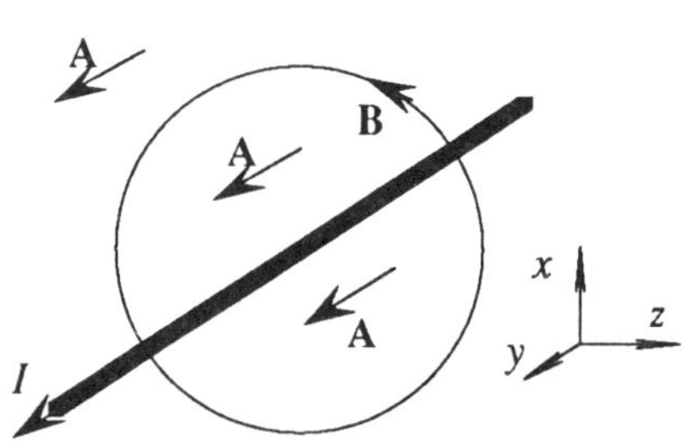

Figure 5.14 A two-dimensional problem. Conductor is assumed to be infinitely long and the magnetic field is in the plane perpendicular to the current.

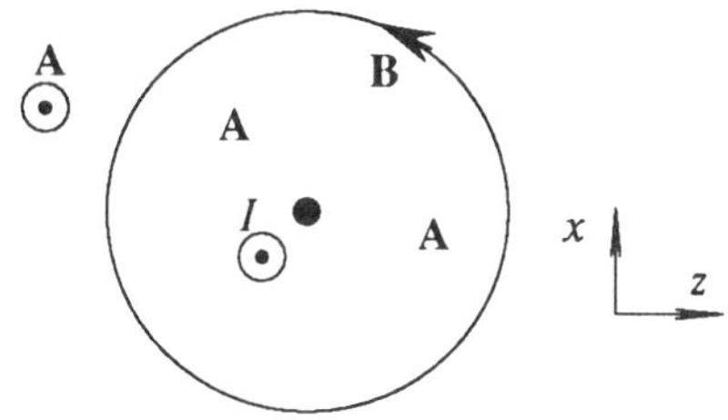

Figure 5.15 Solution domain and relations between current and magnetic flux density for the two dimensional problem in Figure 5.14.

5.5.2 Three-dimensional applications

Three-dimensional applications are the most obvious since all variables used so far are assumed to vary in all three directions. As a simple example, consider the simple loop in free space shown in Figure 5.16. Since the current density has two components at any point in the coil (x and y components in this orientation), the magnetic flux density must have the following three components in Cartesian coordinates

$$\mathbf{B} = \nabla\times(\hat{\mathbf{x}}A_x + \hat{\mathbf{y}}A_y) = \hat{\mathbf{x}}\left(\frac{\partial A_y}{\partial z}\right) - \hat{\mathbf{y}}\left(\frac{\partial A_x}{\partial z}\right) + \hat{\mathbf{z}}\left(\frac{\partial A_x}{\partial y} - \frac{\partial A_y}{\partial x}\right)$$

and this is clearly a *3-D* application.

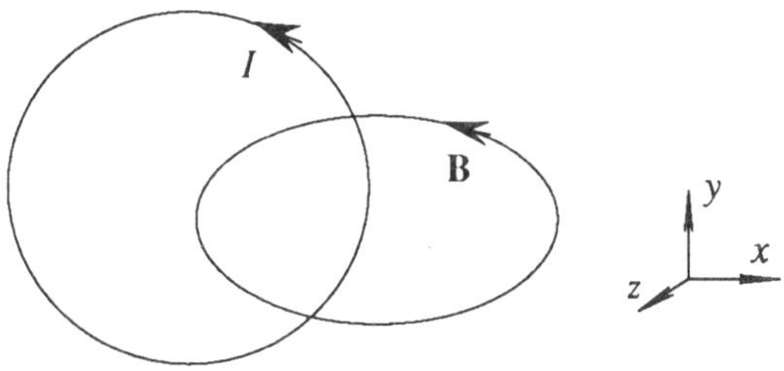

Figure 5.16 A current loop in Cartesian coordinates modeled as a three dimensional problem.

5.5.3 Axisymmetric applications

Consider again the geometry in Figure 5.16. While in the Cartesian coordinate system the current density has two components, in cylindrical coordinates the current density can be viewed as being directed in the ϕ direction as shown in

Figure 5.17. The magnetic flux density has two components and is always perpendicular to the current density. Thus, **B** is in the r-z plane.

If we define the solution domain in the r-z plane, then the problem is similar to that of Figure 5.14 since the current density is perpendicular to this plane. However, there are some very significant differences.

Figure 5.14 describes a conductor of infinite length of which we choose a volume of unit depth. The surface in which the fields are calculated is the cross section of the conductor and the space around it (this surface extends to infinity, but we normally limit this surface to a finite size). The surface in Figure 5.17 contains the cross section of the coil which is finite in circumference, and the area extending from the axis of the coil to infinity (again, we will limit this area artificially). Now the depth of the geometry cannot be taken as unit depth but, rather, as the circumference of the coil (average circumference). The domain we are treating is shown in Figure 5.18. In this geometry, the volume of the conductor is taken as $2\pi r_0 S$ where S is the cross sectional area of the conductor. Thus, calculation of volumetric quantities such as energy, provide total quantities rather than quantities per unit depth.

Although the two geometries are different, we also note that as the radius of the coil tends to infinity, the two geometries become the same, provided that a unit depth is taken rather than the circumference.

Axisymmetric geometries are extremely useful in that they simplify a *3-D* problem and allow its calculation in the cross section. However, this must be used carefully since problems that may appear axisymmetric, are not. Two examples are shown in Figure 5.19 where a coil, enclosing a conducting bar is shown. In Figure 5.19a, both coil and bar are circular but they are offset. If a cross section is taken at two different angles (two of these are shown), they produce different geometries. In Figure 5.19b the coil is axisymmetric but the bar is not and, again, the problem must be solved in three dimensions.

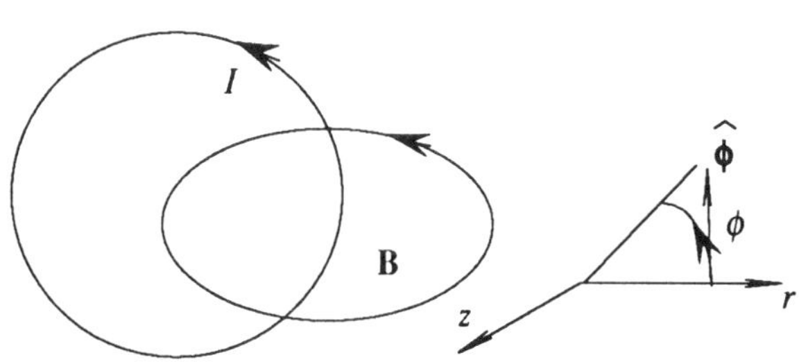

Figure 5.17 The geometry in Figure 5.16 redefined in the cylindrical coordinate system.

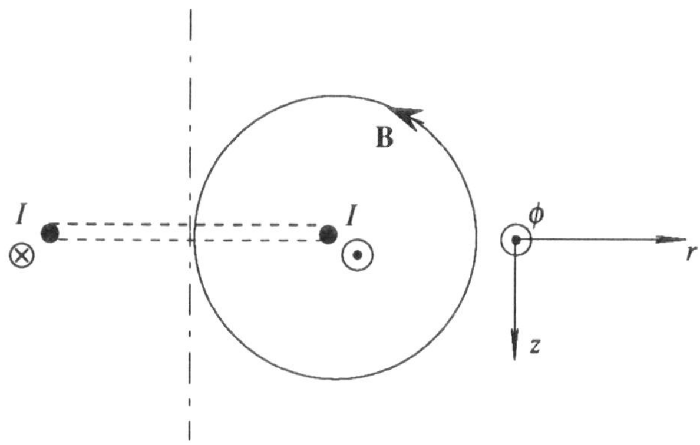

Figure 5.18 Definition of an axisymmetric geometry. The geometry only contains the area to the right or left of the symmetry line.

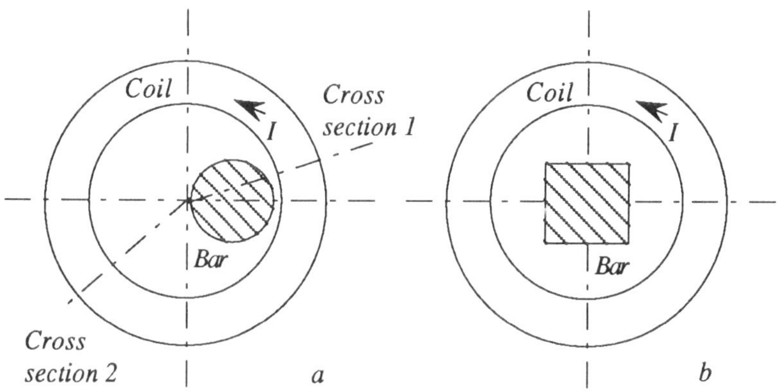

Figure 5.19 Two geometries which cannot be treated as axisymmetric even though the source (coil), by itself, is axisymmetric.

5.6 BIBLIOGRAPHY

Classification of partial differential equations as elliptic, parabolic and hyperbolic can be found in most texts on partial differential equations. Reference [1,2] discuss these in the context of computation of general equations while [3] discusses classification of the electromagnetic field equations. Field equations in terms of various potential functions as described in this chapter can be found in many references. Some of these are [4] through [12].

[1] James, M.L. Smith, G.M. and Wolford, J.C. (1993), *Applied Numerical Methods for Digital Computation*, 4th edition, Harper-Collins.

[2] Courant, R. and Hilbert, D. (1961), *Methods of Computational Physics*, Vol. 2, Wiley Interscience, New York.

[3] Sadiku, M.N.O. (1992), *Numerical Techniques in Electromagnetics*, CRC Press, Boca Raton.

[4] Ida, N. and Bastos, J.P.A. (1992), *Electromagnetics and Calculation of Fields*, Springer Verlag, New York.

[5] Steele, C.W. (1987), *Numerical Computation of Electric and Magnetic Fields*, Van Nostrand Reinhold, New York.

[6] Sylvester, P.P. and Ferrari, R.P. (1990), *Finite Elements for Electrical Engineers*, Cambridge University Press, Cambridge.

[7] Hoole, S.R.H. (1989), *Computer-Aided Analysis and Design of Electromagnetic Devices*, Elsevier, New York.

[8] Konrad, A. (1985) "A direct three-dimensional finite element method for the solution of magnetic fields in cavities", *IEEE Transactions on Magnetics*, Vol. MAG-21, No. 6, pp. 2276-2279.

[9] Ida, N. (1991), "Microwave NDT", in *Electrosoft*, special issue on NDT Vol. 2, No. 2/3, pp. 215-237.

[10] Wang, J.S. and N. Ida, N. (1991), "Eigenvalue analysis in electromagnetic cavities using divergence free finite elements", in *IEEE Transactions on Magnetics*, Vol. 27, No. 5, pp. 3978-3981.

[11] Wang, J.S. and N. Ida, N. (1991), "A numerical model for characterization of specimens in a microwave cavity", in *Review of Progress in Quantitative Nondestructive Evaluation*, D. O. Thompson and D. E. Chimenti, Eds., Vol. 10A, Plenum Press, New York, pp. 567-574.

[12] Pissanetzky, S. (1982), "Solution of three-dimensional, anisotropic, nonlinear problems of magnetostatics using two scalar potentials, finite and infinite multipolar elements and automatic mesh generation", *IEEE Transactions on Magnetics*, Vol. MAG-18, No. 2, pp. 346-350.

CHAPTER 6

Finite difference solution of elliptic processes

6.1 INTRODUCTION

After discussing the finite difference method in chapter *3* and the general properties of elliptic processes in chapter *5* we now have the tools to formulate and solve realistic problems in two and three dimensions using the finite difference technique. In the process, a number of properties are presented which, unlike the general properties given in chapter 5 are best discussed as examples. The applications chosen are those that emphasize the various aspects of modeling. While it is not possible to give all the different variations of formulations and methods, most of these are at least indicated and those that are not are covered in problems at the end of the chapter. A number of applications are solved completely, others are formulated but the results are not given for two reasons. First, giving results for all problems would require too much space but, more important, once a problem is formulated, it is a simple matter to obtain a solution.

6.2 ELLIPTIC PROCESSES: APPLICATIONS IN 2-D AND 3-D ELECTROSTATICS

Solution of electrostatic problems is normally accomplished through the use of the electric scalar potential (voltage). While other functions can be used, there is no incentive to doing so since formulation with the electric potential is the most efficient method both in two and three dimensions. Also, for electrostatic methods, the extension to three dimensions is merely a matter of adding a third partial derivative. Thus, we describe here a number of two-dimensional solutions and then indicate their extensions to three dimensions since two-dimensional solutions are easier to visualize. The steps involved in solution are general and apply to the formulation of any problem. These methods are general and can be used for applications such as nondestructive testing with the potential drop method.

6.2.1 The problem

We consider here a simple electrostatic problem with two extensions:

1. Figure 6.1a shows the basic geometry considered. It consists of a conducting box, *2 m* high and *1 m* wide. The box is infinitely long in the third dimension, thus making this a two-dimensional problem. The figure shows the cross-section of the actual geometry. Three sides of the box are connected to zero potential while the top side is at a potential of *10 V*. These potentials are the boundary conditions of the problem. Note that the top side is assumed to be isolated from the other sides. The material inside is free space with permittivity ε_0. This problem was discussed briefly in chapter 3.
2. An extension to this application is shown in Figure 6.1b. It shows the same box with the same boundary conditions but with an area of charge density at the center. The charge density is distributed within a material of different permittivity than the rest of the box. Charge density is given as 1.0×10^{-07} $[C/m^2]$. The relative permittivity of this material is equal to *80*.
3. A second extension is shown in Figure 6.1c. The box in Figure 6.1a is shown again, with the same boundary conditions but now a line of charge q is located at the center of the box. The charge is assumed to be 5.0×10^{-11} $[C/m]$.

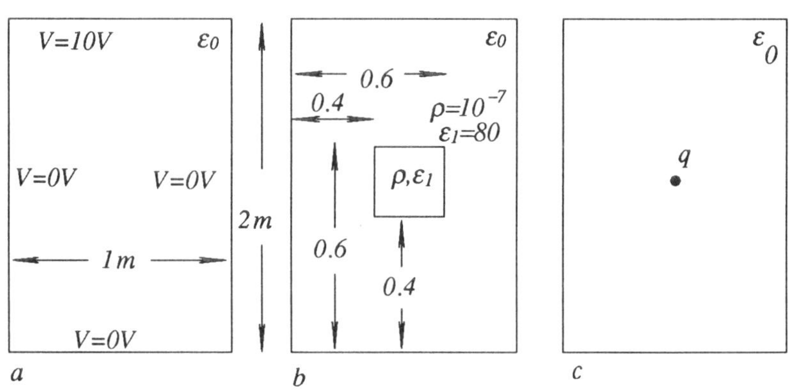

Figure 6.1 Basic geometries for a two-dimensional electrostatic problem.

6.2.2 Formulation

The problem in Figure 6.1a is solved by Laplace's equation in two dimensions:

$$\frac{\partial^2 V}{\partial x^2} + \frac{\partial^2 V}{\partial y^2} = 0 \tag{6.1}$$

The finite difference scheme for this operator is obtained from Eq. (3.20) as

$$V_{i+1,j} + V_{i-1,j} + V_{i,j+1} + V_{i,j-1} - 4V_{i,j} = 0 \tag{6.2}$$

where a finite difference formula is used on a uniform grid: $\Delta x = \Delta y$. We note here that this approximation is independent of material permittivity because the right hand side of Eq. (6.1) is zero.

In terms of the formulation, we must also decide if an explicit or implicit solution is to be used and also define the finite difference grid. In general we opt for explicit solution as it is almost always more efficient and much easier to obtain than an implicit solution because a system of equations is not created (see chapter *3* for details).

A grid for the problem is defined over the solution domain with the restriction $\Delta x = \Delta y$. In this particular case, any grid spacing produces a correct solution. However, since the potentials change very rapidly at the upper two corners, we may wish to use a relatively small grid spacing to allow an accurate solution, especially if the solution around the corners is important. If we choose a grid spacing of $\Delta x = \Delta y = 0.01 m$ the grid contains 201×101 points or *20,301* nodes. Out of these there are *600* boundary nodes which do not need to be evaluated. Thus, the mesh contains *19,701* nodes at which the approximation above must be applied and the potentials evaluated.

Note that if we chose an implicit solution, a system of *19,701* equations in *19,701* unknowns would need to be solved. In an explicit solution the potentials at these nodes must be evaluated iteratively until convergence is obtained.

6.2.3 Boundary and symmetry conditions

The problem is defined in terms of boundary potentials and therefore we use these as Dirichlet boundary conditions. The simplest method is to solve the problem as given: the solution domain is *2 m* by *1 m* with the given boundary conditions.

An alternative approach, one that should be adopted whenever possible, is to exploit the inherent symmetry in the problem and minimize the solution domain. In this case we observe that the geometry has a two-way symmetry while the potentials have a one-way symmetry. Thus, we can reduce the problem to half its original size by dividing it vertically with a symmetry line as shown in Figure 6.2a. This can be done since the potentials at two points, at equal distances about the symmetry line, are equal.

While this reduces the solution domain, it also "upsets" the specification of boundary conditions. The outer boundary conditions remain the same, but boundary conditions on the symmetry line are not defined. We observe, however, that because this is a symmetry line we can write

$$\frac{\partial V}{\partial n} = 0 \tag{6.3}$$

or, in other words, a Neumann boundary condition can be used on this boundary. This, can be done in many cases but, in the approximation in Eq. (6.2) we can only use potential values rather than their first derivatives. To overcome this we add an additional layer of nodes to the left of the symmetry line as shown in Figure 6.2b. Now, since potentials at any node with index $i+1$ and $i-1$ are the same, the approximation of potentials on the symmetry line is given as

$$2V_{i-1,j} + V_{i,j+1} + V_{i,j-1} - 4V_{i,j} = 0 \tag{6.4}$$

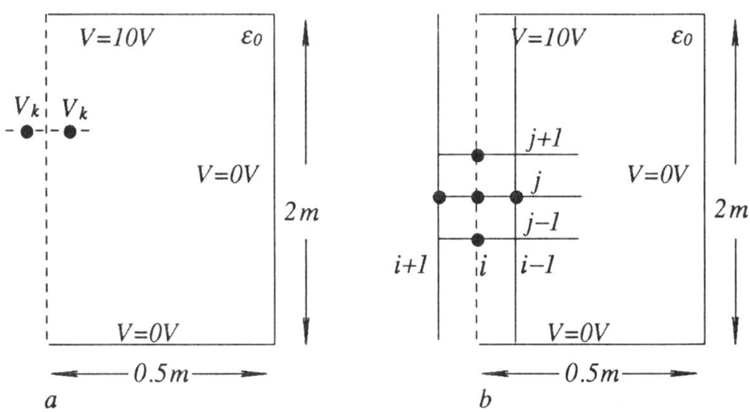

Figure 6.2 Use of symmetry to reduce size of solution domain. a. Solution domain with symmetry line showing equal potentials on both sides of symmetry line. b. Addition of a virtual layer to account for symmetry.

From this, it is obvious that the additional layer is a virtual layer and does not actually exist. All that needs to be done is to modify the approximation to that in Eq. (6.4) for potentials on the symmetry line. It should be noted however, that in this case, the finite difference approximation is applied on all internal nodes and on nodes on the symmetry line. Thus, using the same mesh spacing as above, the finite difference mesh has a total of 201×51 nodes ($10{,}251$

nodes) out of which *451* nodes are known boundary conditions and are not evaluated. The remaining *9,800* nodes must be evaluated. As expected, use of symmetry reduces the size of the problem without affecting results.

6.2.4 Solution

The solution is straightforward and in fact, is identical to that given in sections 3.4.1 and 3.4.2. There are two decisions to be made at this stage: One is the choice between explicit and implicit solution. If the solution is explicit (as is the case here) then an error criterion must also be used to stop the solution. Error criteria were discussed in section 3.4.2. In this case we choose the following:

$$\frac{1}{N}\sum_{i=1}^{N}\left|V_i^n - V_i^{n-1}\right| \le 10^{-5} \qquad (6.5)$$

This choice indicates that the change in solution between subsequent iterations is smaller than 10^{-5} (an average tolerance per node). This value is arbitrary and must be given by the user. It should not be too small as that results in a large number of iterations but should not be too large since then solution accuracy will suffer. Experimentation with this value may indicate an appropriate range of values that can be used. While the number of iterations will depend on many parameters, including geometry, material properties, grid spacing, and tolerance, it is not possible to predict beforehand how many iterations will be required. For the problem above, some *200* iterations were needed.

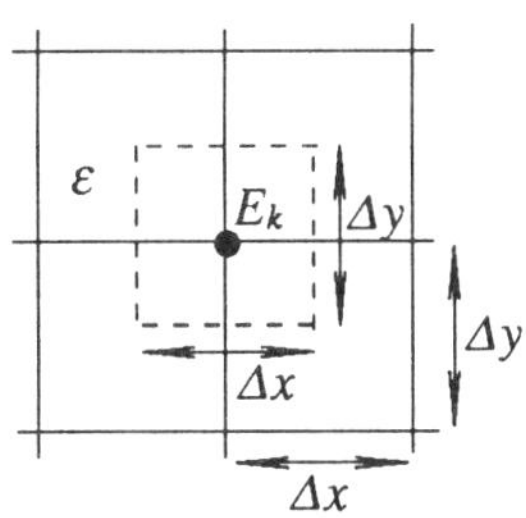

Figure 6.3 The use of domain of a node to calculate volumetric quantities.

6.2.5 Presentation of results

The solution above is obtained in terms of the electric scalar potential (V). If this is what we need, then all that is necessary is to display this potential distribution in a suitable form. This can take the form of a contour plot as shown in Figure 3.15 or a *3-D* plot (magnitude above a plane) as shown in Figure 3.16. Both plots also indicate the electric field intensity as the slope between lines. The electric field intensity can be calculated using the definitions:

$$E_x = \frac{\partial V}{\partial x} \qquad \text{and} \qquad E_y = \frac{\partial V}{\partial y} \tag{6.6}$$

and these are most easily evaluated using the central difference scheme in Eq. (3.7). The electric field intensity may be displayed similarly or other parameters such as maximum field intensity, total field intensity, etc. may be calculated.

If electrostatic energy in the system (or in any part of the system) is required, it is evaluated from the energy density in a volume which is given as

$$w_e = \frac{1}{2}\varepsilon E^2 \tag{6.7}$$

and the total energy in a volume v is given by

$$W = \int_v w_e dv = \frac{1}{2}\int_v \varepsilon E^2 dv \tag{6.8}$$

Since in this case, the electric field intensity can only be calculated as point values, we use an approximation based on the idea of a domain of the node discussed in section 3.6 and shown schematically in Figure 6.3. We assume that the electric field is constant within this domain and calculate the energy associated with this node as

$$w_{node} = \frac{1}{2}\frac{\varepsilon_0 E_{node}^2}{\Delta x \Delta y} \tag{6.9}$$

Note however, that this is energy per unit depth of the structure as it should be for a two-dimensional geometry. The total energy per unit depth is now

$$W = \sum_{i=1}^{N} \frac{1}{2}\frac{\varepsilon_0 E_i^2}{\Delta x \Delta y} \tag{6.10}$$

This calculation can be very accurate, especially if the number of nodes N is relatively large.

Other parameters such as capacitance can also be calculated from the potential or the electric field intensity based on the definition of the respective parameter. Finally, any of these parameters can be related to an *NDT* parameter such as resistance or reactance of an eddy current probe.

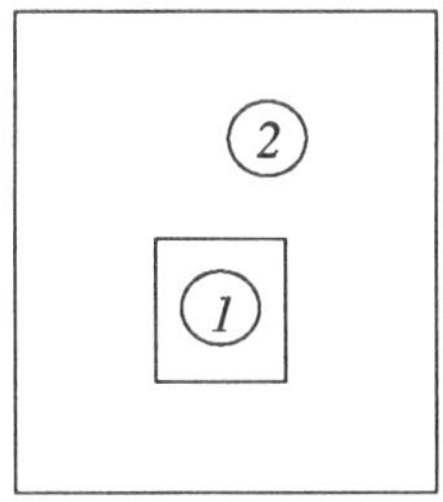

Figure 6.4 Division of geometry in Figure 6.1b into two domains. Domain (1) is described by Poisson's equation while domain (2) by Laplace's equation.

6.2.6 Variations in formulation

Computation of potentials in Figure 6.1b is similar except that now we must consider two domains as shown in Figure 6.4. In domain *(1)*, there are charge densities while in domain *(2)* there are no charges. Thus, in domain *(1)* the approximation is that required for Poisson's equation and is given as

$$\frac{V_{i+1,j} + V_{i-1,j} + V_{i,j+1} + V_{i,j-1} - 4V_{i,j}}{h^2} = -\frac{\rho}{\varepsilon_r \varepsilon_0} \tag{6.11}$$

or, substituting the values given in Figure 6.1b

$$V_{i+1,j} + V_{i-1,j} + V_{i,j+1} + V_{i,j-1} - 4V_{i,j} = -45\pi h^2 \tag{6.12}$$

where $h=\Delta x=\Delta y$, $\varepsilon_0=(1/36\pi)\times 10^{-9}$, $\varepsilon_r=80$, and $\rho=10^{-7}$ $[C/m^2]$ are used.

In domain *(2)* we use the expression in Eq. (6.2) since there are no charges in this domain and Laplace's equation applies directly.

Problem setup as well as the solution follow identical processes to those described above. In this case, the solution domain can also be cut by a symmetry line as for the previous problem but now, the expressions on the symmetry line depend on where the point is. If it is in free space, Eq. (6.4) must be used. If the point is in the charge region and on the symmetry line the following modification of Eq. (6.4) should be used

$$2V_{i-1,j} + V_{i,j+1} + V_{i,j-1} - 4V_{i,j} = -45\pi h^2 \tag{6.13}$$

The various approximations necessary are shown in Figure 6.5.

After solution, using the same grid and boundary conditions as in the previous problem, the results are displayed as in the previous solution. These are shown in Figures 6.6 and 6.7 as contour and *3-D* plots respectively. Note that the charge density has significantly altered the solution, especially around the charge distribution.

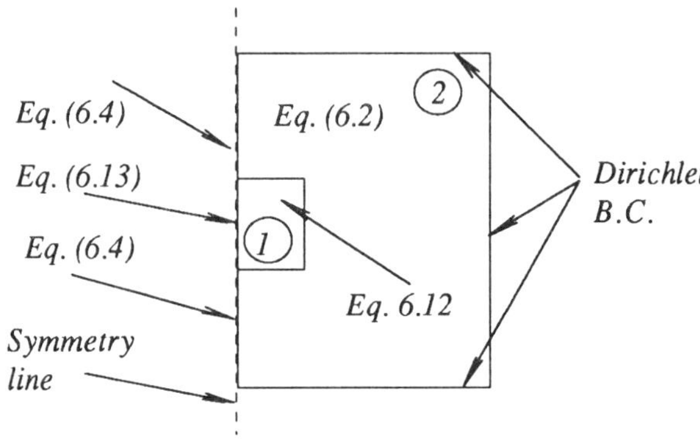

Figure 6.5 Solution domain and required finite difference approximations.

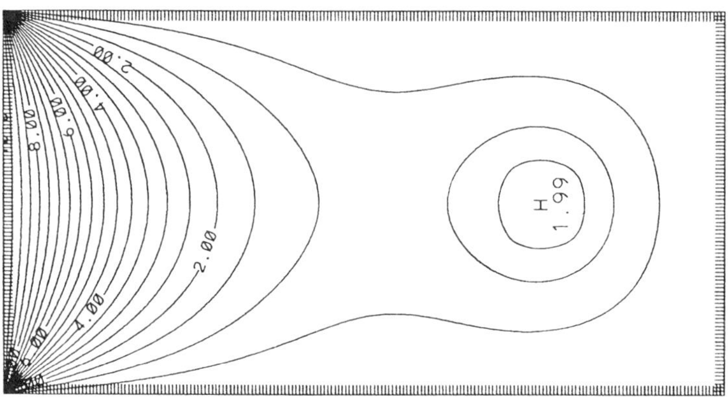

Figure 6.6 Contour plot of potential in Figure 6.1b.

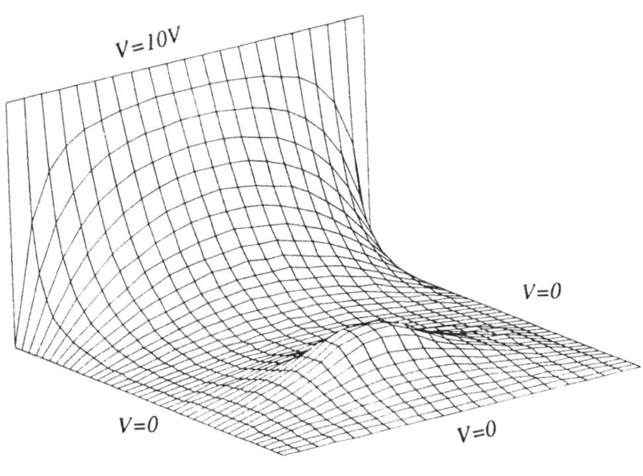

Figure 6.7 3-D plot of potential in Figure 6.1b.

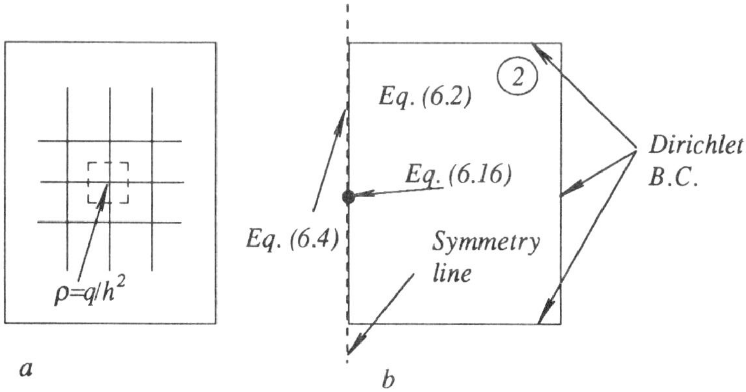

Figure 6.8 a. Representation of point charge as a charge density. b. Finite difference approximations at various points in the grid. A symmetry line through the point charge is used.

For the problem in Figure 6.1c, we must modify the charge to that of a charge density so that we may use the same approximation as in the previous applications. This is necessary since Poisson's equation contains charge densities but not point charges. For lack of a better method, we distribute the point charge over a single cell of the grid. The equivalent charge density is therefore $\rho=q/(\Delta x\Delta y)$ and this charge density is assumed to exist only at a point as shown schematically in Figure 6.8a.

Now the approximation everywhere in the solution domain, except at the point charge is given by Eq. (6.2). At the point charge, the approximation is given by

$$\frac{V_{i+1,j}+V_{i-1,j}+V_{i,j+1}+V_{i,j-1}-4V_{i,j}}{h^2}=-\frac{\rho}{\varepsilon_0}=-\frac{q}{h^2\varepsilon_0} \qquad (6.14)$$

or, for the values given in Figure 6.1c we get

$$V_{i+1,j}+V_{i-1,j}+V_{i,j+1}+V_{i,j-1}-4V_{i,j}=-\ 1.13097\times10^{-5} \qquad (6.15)$$

where $h=0.01\ m$ is used.

The problem can also be solved using symmetry in which case Eq. (6.2) is used everywhere except on the symmetry line. Eq. (6.4) is used everywhere on the symmetry line except for the charge point (which now lies on the symmetry line) and the following is used at the symmetry point

$$2V_{i-1,j}+V_{i,j+1}+V_{i,j-1}-4V_{i,j}=-\ 1.13097\times10^{-5} \qquad (6.16)$$

The various approximations necessary are shown schematically in Figure 6.8b.

The solution to this variation of the problem is shown in Figure 6.9 in terms of contour lines and in Figure 6.10 in terms of magnitudes over a plane.

We note here that the *3-D* extension to these calculations is very simple. All that is necessary is to add a partial derivative term to Eq. (6.1), include this term in the approximations and solve in exactly the same fashion. The only difficulty is in the display of data which must be done in cross sections, on surfaces or on lines through the geometry. We do not solve *3-D* problems using the finite difference method because the extension from *2-D* to *3-D* is rather straight forward.

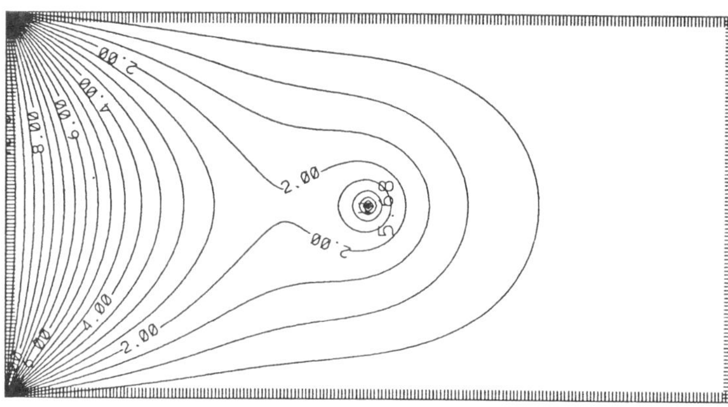

Figure 6.9 Contour plot of solution of problem 6.1c.

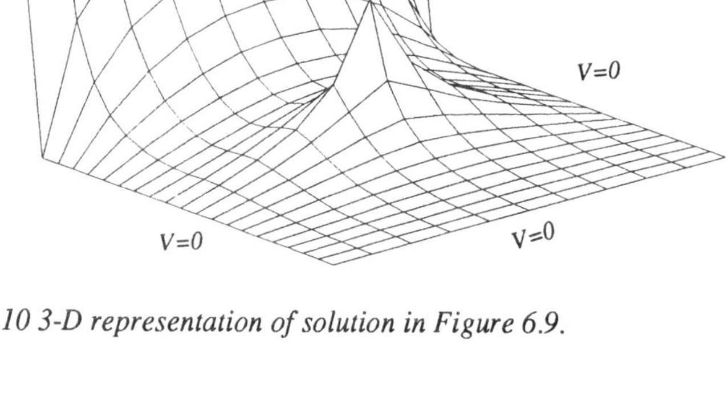

Figure 6.10 3-D representation of solution in Figure 6.9.

6.3 MAGNETOSTATIC APPLICATIONS

As a first example for two dimensional solutions of a magnetostatic problem we consider the geometry shown in Figure 6.11. A current carrying busbar is located above a conducting material. The lower conductor is assumed to be thick but any thickness can be modeled. Flaws and inclusions in the conductor can be included but, under two dimensional assumptions these are infinitely long in the direction perpendicular to the plane of the figure.

We choose to use the magnetic vector potential for this purpose. Since it is a *2-D* problem, the magnetic vector potential has a single component in the *z* direction (perpendicular to the plane of the figure).

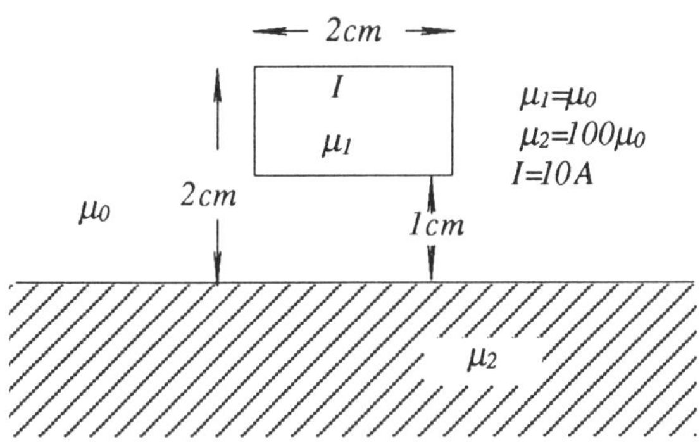

Figure 6.11 A magnetostatic problem.

6.3.1 The governing equation

To define the governing equations, we identify three domains or materials as shown in Figure 6.12. In material *(1)*, we have a current density $J=I/S$ $[A/m^2]$ where S is the cross-sectional area of the busbar ($2\ cm^2$ in this case). Thus, the field equation in this domain is

$$\frac{\partial^2 A}{\partial x^2} + \frac{\partial^2 A}{\partial y^2} = -\mu_1 J \qquad \textit{in material 1} \qquad (6.17)$$

In materials *(2)* and *(3)*, there are no current densities and we can write

$$\frac{\partial^2 A}{\partial x^2} + \frac{\partial^2 A}{\partial y^2} = 0 \qquad \textit{in materials 2 and 3} \tag{6.18}$$

In addition, since the normal component of the magnetic flux density is continuous across interfaces with different magnetic permeabilities, we must introduce an interface condition between materials *(2)* and *(3)*. There is no need for an interface condition between materials *(1)* and *(2)*. The interface condition is based on the fact that the normal derivative of the magnetic vector potential must be continuous across the interface (that is: the normal component of the magnetic flux density must be continuous). The interface condition is (Figure 6.13):

$$\frac{A_{y2}}{\mu_2} = \frac{A_{y3}}{\mu_3} \qquad \textit{interface 2 and 3 (i.e.} \; \frac{\partial A}{\partial y} \; \textit{is continuous)} \tag{6.19}$$

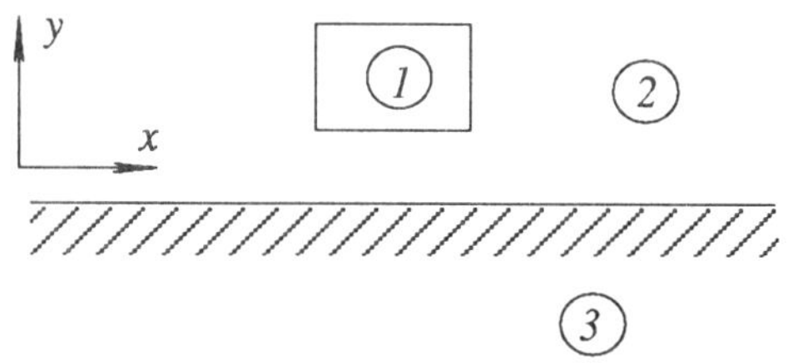

Figure 6.12 Various domains of the geometry in Figure 6.11

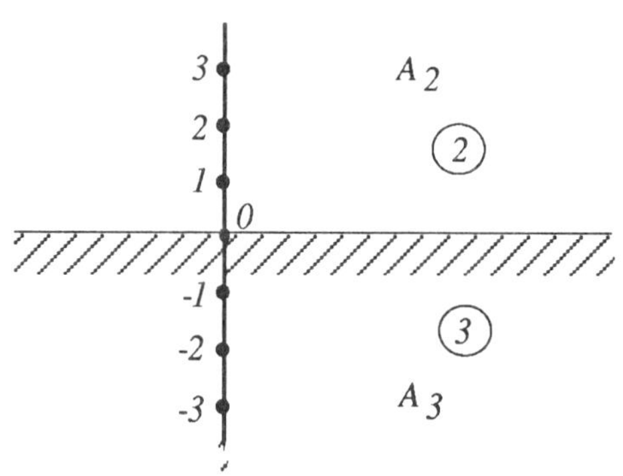

Figure 6.13 Modeling of the interface between materials with different permeabilities.

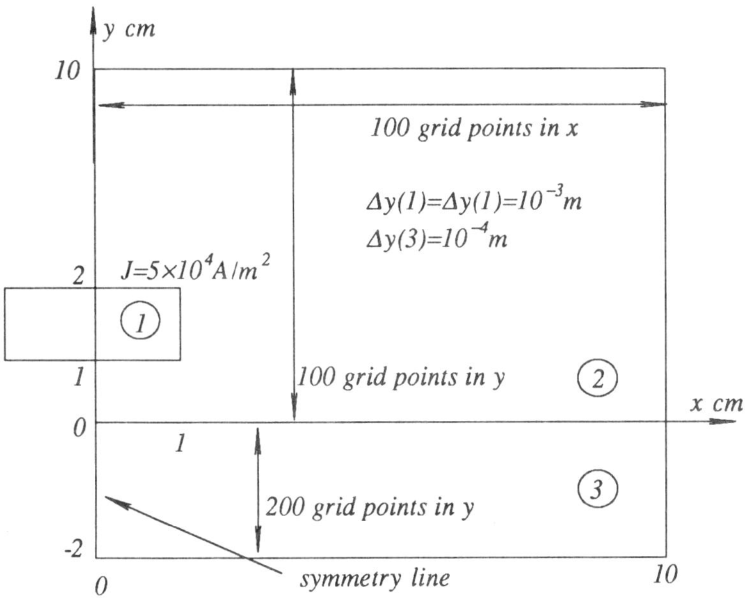

Figure 6.14 Finite difference solution domain for Figure 6.11. The left boundary is a symmetry line.

6.3.2 The finite difference grid

The computational domain is discretized as shown in Figure 6.14. This grid is unique in that the grid spacing in the y direction in the lower conductor is 10 times smaller than in the rest of the mesh. In materials *(1)* and *(2)* we use $\Delta x = \Delta y = 0.001 m$ while in material *(3)* $\Delta x = 0.001 m$ and $\Delta y = 0.0001 m$. As with previous solutions it is advantageous to use symmetry along a vertical line that bisects the busbar. The left boundary is a symmetry boundary and should be treated as a Neumann boundary condition. The other three boundaries are Dirichlet boundary conditions. In this case, for lack of any other choice, we assume that the boundaries are far enough (that is, the magnitude of the potential is low enough) so that the effect of the boundary condition is negligible. We assume that the magnetic vector potential is zero on these boundaries.

The grid contains a total of 301×101 nodes or *30,401* nodes out of which *501* nodes are given as Dirichlet boundary conditions and *299* nodes are on the symmetry line (the nodes of the symmetry line that also lie on the upper and lower boundary are specified as Dirichlet boundary nodes). Thus the total number of unknown values of the magnetic vector potential is *29,900*, and only an explicit solution is feasible.

6.3.3 Approximation

The finite difference approximation in material *(1)* is

$$A_{i+1,j} + A_{i-1,j} + A_{i,j+1} + A_{i,j-1} - 4A_{i,j} = -\mu_1 h^2 J_{i,j} \tag{6.20}$$

where $h=\Delta x=\Delta y$. In material *(2)* the approximation is

$$A_{i+1,j} + A_{i-1,j} + A_{i,j+1} + A_{i,j-1} - 4A_{i,j} = 0 \tag{6.21}$$

However, in material *(3)* we chose a different Δy because we expect a much higher gradient in the solution. A smaller Δy provides a more accurate solution. We choose $\Delta y=\Delta x/10$. Thus, the approximation in this domain is

$$\frac{A_{i+1,j} - 2A_{i,j} + A_{i-1,j}}{h^2} + \frac{10(A_{i,j+1} - 2A_{i,j} + A_{i,j-1})}{h^2} = 0$$

or, in terms of V_{ij} and using $\Delta x=h$ and $\Delta y=h/10$ we get

$$A_{i,j} = \frac{A_{i+1,j} + A_{i-1,j} + 10(A_{i,j+1} + A_{i,j-1})}{12} \tag{6.22}$$

The interface condition is written (using Figure 6.13) as

$$\begin{aligned} &\frac{1}{\mu^{(2)}2\Delta y^{(2)}}\left(-A_2^{(2)} + 4A_1^{(2)} - 3A_0^{(2)}\right) \\ &= \frac{1}{\mu^{(3)}2\Delta y^{(3)}}\left(3A_0^{(3)} - 4A_{-1}^{(3)} + A_{-2}^{(3)}\right) \end{aligned} \tag{6.23}$$

where superscripts indicate domain numbers as given in Figure 6.13. The following forward difference formula is used in material *(2)*

$$\frac{1}{\mu^{(2)}2\Delta y^{(2)}}\left(-A_2^{(2)} + 4A_1^{(2)} - 3A_0^{(2)}\right)$$

and the following backward difference formula in material *(3)*

$$\frac{1}{\mu^{(3)}2\Delta y^{(3)}}\left(3A_0^{(3)} - 4A_{-1}^{(3)} + A_{-2}^{(3)}\right)$$

Equating the two gives the relation in Eq. (6.23). On the interface we can write:

$$A_0^{(2)} = A_0^{(3)} = A_0 \tag{6.24}$$

where A_0 is the value of the magnetic vector potential at the interface and this relation indicates that the value is the same in either material.

Rewriting Eq. (6.23) for the magnetic vector potential at the interface we get

$$A_0 = \frac{1}{3\left(1 + \frac{\Delta y^{(3)}}{\Delta y^{(2)}}\frac{\mu^{(3)}}{\mu^{(2)}}\right)}\left[4A_{-1}^{(3)} - A_{-2}^{(3)} + \frac{\Delta y^{(3)}}{\Delta y^{(2)}}\frac{\mu^{(3)}}{\mu^{(2)}}\left(A_1^{(2)} - A_2^{(2)}\right)\right] \quad (6.25)$$

where superscripts indicate domain number. On the symmetry line we write:

Outside the current region:

$$2A_{i-1\,j} + A_{i\,j+1} + A_{i\,j-1} - 4A_{i\,j} = 0 \quad (6.26)$$

For nodes in the current region:

$$2A_{i-1\,j} + A_{i\,j+1} + A_{i\,j-1} - 4A_{i\,j} = -h^2\mu_l J \quad (6.27)$$

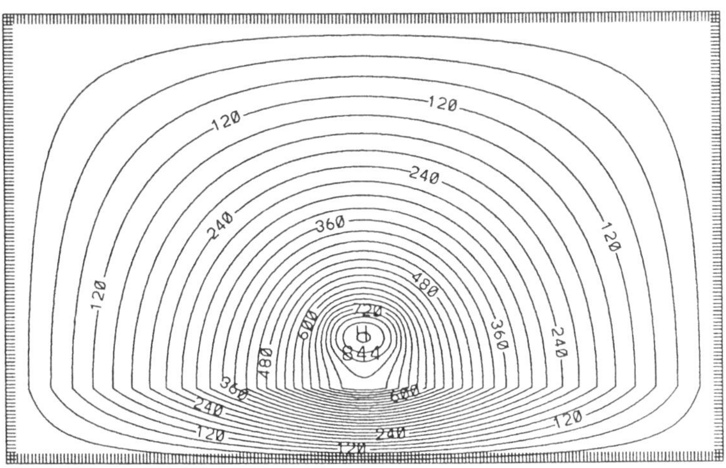

Figure 6.15. Contour lines of the magnetic field for geometry in Figure 6.11. Note the effect of the high permeability in the lower conductor.

6.3.4 Solution and results

Once the approximations have been defined we apply (Eq. (6.20) at all interior nodes in material *(1)*, Eq. (6.21) at all interior nodes in material *(2)*, Eq. (6.22) at all interior nodes in material *(3)*. Eq. (6.26) is applied at nodes on the symmetry line outside current regions and Eq. (6.27) at nodes on the symmetry line in current regions. On the interface between materials *(2)* and *(3)*, Eq. (6.25) is used. Using the same tolerance criterion as in Eq. (6.5) we obtain a solution as shown in Figure 6.15. This is a contour plot of the magnetic vector potential. The effects of permeability are clearly shown by the higher density of lines (higher gradient of the magnetic vector potential or, alternatively, higher flux density) as we would expect.

6.4 EDDY CURRENT APPLICATIONS

To see how an eddy current problem is defined and solved, we consider the problem in Figure 6.11 again but now the current in the busbar is an ac current at *1,000 Hz*. Also, because this is an eddy current application, material conductivities must be included in the calculation. The geometry and parameters are shown in Figure 6.16. Although this is the same geometry as in Figure 6.11, it is given here again because of the different parameters needed for computation. Unlike the geometry in Figure 6.11, conductivity of materials is very important as is the skin depth in conductors.

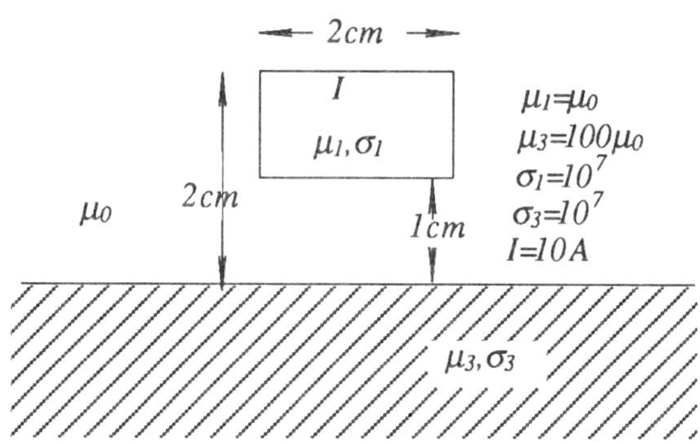

Figure 6.16 A two-dimensional eddy current geometry.

6.4.1 Governing equations

The parameters needed to define the governing equations are

$$A(t) = Ae^{j\omega t}, \qquad \omega = 2\pi\times 1000 \ rad/sec, \qquad J = \frac{10^5}{2} \ A/m^2 \tag{6.28}$$

Where A is now a phasor and J is the magnitude of the current density in the bar. The governing equations are:

$$\frac{\partial^2 A}{\partial x^2} + \frac{\partial^2 A}{\partial y^2} = -\mu_1 J + j\omega\mu_1\sigma_1 A \qquad \text{in material (1)} \tag{6.29}$$

$$\frac{\partial^2 A}{\partial x^2} + \frac{\partial^2 A}{\partial y^2} = 0 \qquad \text{in material (2)} \tag{6.30}$$

$$\frac{\partial^2 A}{\partial x^2} + \frac{\partial^2 A}{\partial y^2} = j\omega\mu_3\sigma_3 A \qquad \text{in material (3)} \tag{6.31}$$

Thus, in material *(1)* there are both source current and eddy currents, in material *(2)* there are no currents and in material *(3)* there are only eddy currents. If we substitute the values for frequency and current density, the equations are:

$$j8\pi^2 \times 10^3 A = \nabla^2 A + 2\pi\times 10^{-6} \qquad \text{in material 1}$$

$$\nabla^2 A = 0 \qquad \text{in material 2}$$

$$j8\pi^2 \times 10^5 A = \nabla^2 A \qquad \text{in material 3}$$

These can now be approximated as for the magnetostatic case except that all variables are complex instead of real. However, here, and in some other cases, it is also possible to scale the problem. The purpose of scaling is to simplify the solution and, in some cases, to avoid loss of accuracy by avoiding manipulation of very large or very small values. In this particular application we use scaling to show how this can be done. We scale both the magnetic vector potential and grid spacing by defining scaled values as

$$A^* = A\frac{10^{10}}{2\pi} \qquad \text{and} \qquad x = 10^{-2}x^*$$

The scaled equations are:

$$j8\pi^2 \times 10^3 \frac{2\pi}{10^{10}} A^* = \frac{2\pi}{10^{10}\times 10^{-4}} \nabla^2 A^* + 2\pi\times 10^{-6} \qquad \text{in material 1}$$

$$\frac{2\pi}{10^{10}\times 10^{-4}} \nabla^2 A^* = 0 \qquad \text{in material 2}$$

$$j8\pi^2 \times 10^5 \frac{2\pi}{10^{10}} A^* = \frac{2\pi}{10^{10}\times 10^{-4}} \nabla^2 A^* \qquad \text{in material 3}$$

Or, after simplification:

$$j8\pi^2 \times 10^{-1} A^* = \nabla^2 A^* + 1 \qquad \text{in material 1}$$

$$\nabla^2 A^* = 0 \qquad \text{in material 2}$$

$$j8\pi^2 \times 10^1 A^* = \nabla^2 A^* \qquad \text{in material 3}$$

Next we must define the interface condition between material *(2)* and *(3)*. The interface condition is identical to that given in Eq. (6.19). Note that in the relations that follow, explicit scaling is not required: that is, the magnetic vector potential in the following three relations can be taken either as scaled or as the actual potential. For this reason, the scaled notation is dropped.

Between material *(1)* and *(2)* there is no need for an interface condition because both materials have the same permeability. The interface condition between material *(2)* and *(3)* is:

$$\frac{A_y^{(2)}}{\mu^{(2)}} = \frac{A_y^{(3)}}{\mu^{(3)}} \quad \text{at interface (2) and (3) (i.e.} \frac{\partial A}{\partial y} \text{ is continuous)}$$

After discretization, the interface condition is

$$\frac{1}{\mu^{(2)} 2\Delta y^{(2)}} \left(-A_2^{(2)} + 4A_1^{(2)} - 3A_0^{(2)} \right) = \frac{1}{\mu^{(3)} 2\Delta y^{(3)}} \left(3A_0^{(3)} - 4A_{-1}^{(3)} + A_{-2}^{(3)} \right)$$

$$A_0^{(2)} = A_0^{(3)} = A_0$$

That is:

$$A_0 = \frac{1}{3\left(1 + \dfrac{\Delta y^{(3)}}{\Delta y^{(2)}} \dfrac{\mu^{(3)}}{\mu^{(2)}}\right)} \left[4A_{-1}^{(3)} - A_{-2}^{(3)} + \frac{\Delta y^{(3)}}{\Delta y^{(2)}} \frac{\mu^{(3)}}{\mu^{(2)}} \left(A_1^{(2)} - A_2^{(2)} \right) \right]$$

The computational domain is discretized as shown in Figure 6.14.

The solution is the same as for the magnetostatic case above but in complex variables. Also, because of the normalization: $\Delta y^* = 10^{-1}$ in domain *(2)*, and $\Delta y^* = 10^{-2}$ in domain *(3)*. Without normalization $\Delta y = 10^{-3}$ in material *(2)* and $\Delta y = 10^{-4}$ in material *(3)*. Since the solution is in complex variables, the current density is written as $J = (5 \times 10^4 + j0)$, or, as is more practical, the term $\mu_1 J$ is written as $\mu_1 J = (2\pi \times 10^{-2} + j0)$.

All other aspects of approximation, grid and solution are exactly the same as for the magnetostatic case, The only difference is that the solution must now proceed in complex variables.

Figures 6.17 and 6.18 show the real and imaginary parts of the magnetic vector potential as examples of the type of results we may need to display. In particular, note that the penetration of the magnetic field in the lower conductor is very low. This, of course, is to be expected since this is a high conductivity, high permeability material in which the skin depth is rather small (skin depth is *0.5 mm*, as calculated using Eq. (1.144)).

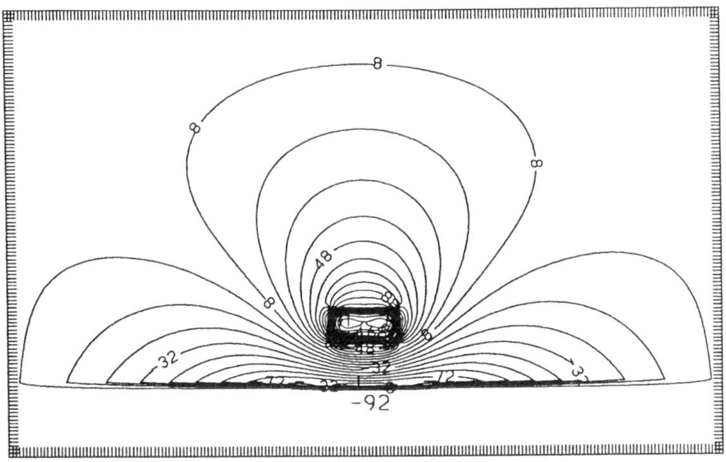

Figure 6.17 Real part of the magnetic vector potential for the eddy current problem in Figure 6.16.

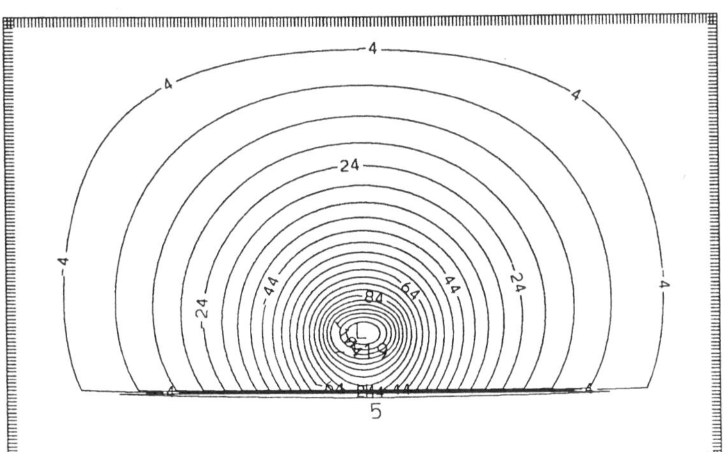

Figure 6.18 Imaginary part of the magnetic vector potential for the eddy current problem in Figure 6.16.

6.4.2 Other calculations and extensions

It is also possible to calculate other variables as mentioned in chapter 1 but these are considered secondary variables (variables that have not been calculated directly). Magnetic flux density, field intensity, losses, eddy currents, etc. can all be calculated. These secondary results are discussed in more detail in chapter 7 in conjunction with finite element analysis of eddy current coils but we mention here that the magnetic flux density is calculated based on its definition $\mathbf{B}=\nabla\times\mathbf{A}$ and, in two dimensions, this reduces to:

$$\mathbf{B} = \nabla\times(\hat{\mathbf{y}}A_y) = \hat{\mathbf{x}}\left(\frac{\partial A_y}{\partial z}\right) - \hat{\mathbf{z}}\left(\frac{\partial A_y}{\partial x}\right)$$

As expected, the partial derivatives are calculated using central differences at most interior nodes and backward or forward difference formulae at nodes next to boundaries. Refering to Figure 6.19, the magnetic flux density at an interior node i, j, is calculated as

$$\mathbf{B} = \hat{\mathbf{x}}\frac{A_{i+1,j} - A_{i-1,j}}{2h} + \hat{\mathbf{y}}\frac{A_{i,j+1} - A_{i,j-1}}{2h}$$

while at nodes near boundaries, the backward and forward difference formulae in Eqs (3.17) and (3.18) may be used. As an example, at node j,n, the finite difference for the x component must be a backward difference formula while the y component is calculated as a central difference formula. The result at any node next to the right boundary except for the upper and lower corner nodes is:

$$\mathbf{B} = \hat{\mathbf{x}}\frac{A_{j,n} - A_{j-1,n}}{h} + \hat{\mathbf{y}}\frac{A_{j,n+1} - A_{j,n-1}}{2h}$$

Similar expressions apply at other locations as indicated in Figure 6.19.
Eddy current density at a node i is calculated as

$$J_{ei} = j\omega\sigma_i A_i$$

The total eddy current through the cross section of a conductor is calculated using the method of section 3.6. Eddy current at a node is calculated by multiplying the current density above by $\Delta x\Delta y$. Eddy currents through a given section are calculated using the nodal values and summing over all nodes in the cross section in which we wish to calculate the current

$$I = \sum_{i=1}^{n} J_{ei}\Delta x\,\Delta y = \sum_{i=1}^{n} j\omega\sigma_i A_i\Delta x\,\Delta y$$

As with the magnetostatic formulation, the extension of this formulation to three dimensions and to axisymmetric geometries is rather simple. It is also possible to modify the formulation for anisotropic materials by simply leaving permeabilities in the respective terms of the Laplacian as in Eq. (5.21). Thus, if permeability in material *(3)* is anisotropic ($\mu_x \neq \mu_y$), equations (6.29) and (6.30) remain the same but Eq. (6.31) is modified as

$$\frac{1}{\mu_{3x}}\frac{\partial^2 A}{\partial x^2} + \frac{1}{\mu_{3y}}\frac{\partial^2 A}{\partial y^2} = j\omega\sigma_3 A \qquad \text{in material (3)}$$

The remainder of the solution is the same although the interface condition in Eq. (6.23) must also take into account the permeability of both materials as described above.

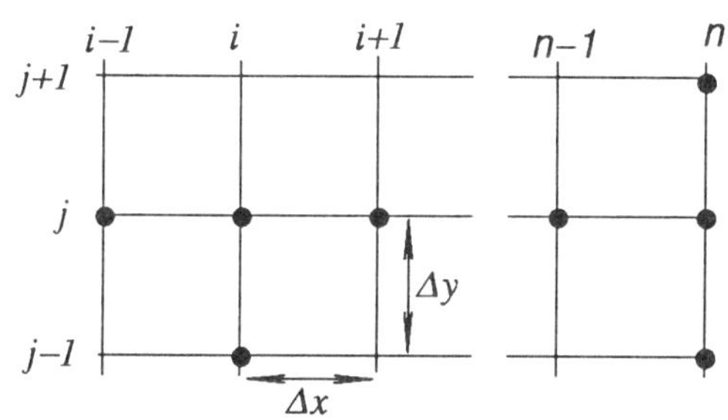

Figure 6.19 Use of the magnetic vector potential as calculated by the finite difference method to calculate the magnetic flux density at point i, j.

6.5 TIME-HARMONIC WAVE PROPAGATION

As an introduction to this type of solution we consider a simple, empty, rectangular, electromagnetic cavity resonator, *1m×1m×1m* in size. Modifications to this simple problem are then discussed. The geometry is shown in Figure 6.20a with boundary conditions shown for *TM* modes. Extensions to this simple geometry are shown in Figures 6.20b and 6.20c. Figure 6.20b shows the same cavity resonator filled with a low permittivity dielectric. This is typical of cavities filled with gas or vapors as the cavities are sometimes used for monitoring of gases and curing processes. Figure 6.20c shows a test sample material inside the cavity. This is typical in testing lossless and lossy dielectrics in microwave cavities.

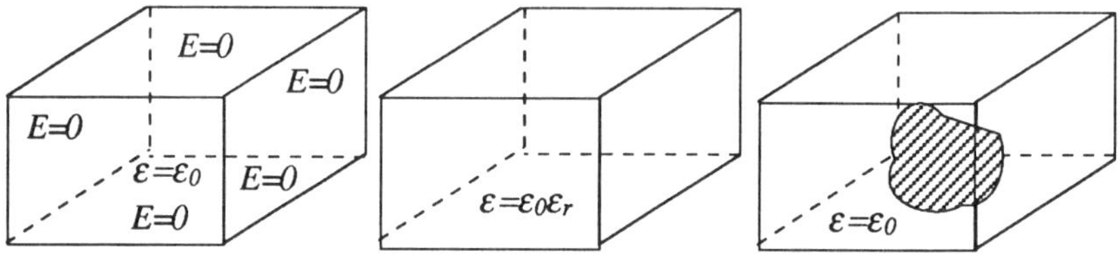

Figure 6.20 A cavity resonator and boundary conditions for TM modes. a. Empty cavity. b. Cavity filled with a dielectric. c. Test sample inside a cavity.

6.5.1 Formulation

Although there are two modes possible, the solution of the two is similar; the main difference is in boundary conditions. For *TM* modes, the equation to solve is

$$\frac{\partial^2 E_z}{\partial x^2} + \frac{\partial^2 E_z}{\partial y^2} + k_c^2 E_z = 0\ , \qquad k_c^2 = \gamma^2 + k^2 \tag{6.32}$$

with Dirichlet boundary conditions $(E_z=0)$ on the conducting surfaces of the cavity as shown in Figure 6.20a. Similarly, for *TE* modes, the equation to solve is

$$\frac{\partial^2 H_z}{\partial x^2} + \frac{\partial^2 H_z}{\partial y^2} + (\gamma^2 + k^2)H_z = 0 \tag{6.33}$$

with Neumann boundary conditions $(\partial H_z/\partial x=0,\ \partial H_z/\partial y=0)$. We concentrate on *TM* modes. In both equations *E* and *H* are phasors.

First, we note that since γ is the propagation constant, at cutoff, $\gamma^2=0$. Thus, we can write for *TM* modes at cutoff

$$k_c^2 = \omega^2\mu\varepsilon = \omega^2\mu_0\varepsilon_0\mu_r\varepsilon_r = k_0^2\mu_r\varepsilon_r \tag{6.34}$$

where ε_r and μ_r are the relative permeability and permittivity in the cavity respectively. Thus, Eq. (6.32) becomes

$$\frac{1}{\mu_r}\left(\frac{\partial^2 E_z}{\partial x^2} + \frac{\partial^2 E_z}{\partial y^2}\right) + k_0^2\varepsilon_r E_z = 0 \tag{6.35}$$

In this relation k_0 is the free space wavenumber and once calculated can be directly interpreted as the cutoff frequency of a particular mode of propagation.

We further note that ε_r is in general the complex permittivity, depending on the material.

The solution to either the *TM* or the *TE* application leads to an eigenvalue problem. This means that we can only find a solution for distinct, known values of k_c which are the eigenvalues of the problem. The eigenvalues then give the resonant frequencies of the system from Eq. (6.34). Thus, we normally wish to know the eigenvalues as well as the eigenvectors (field distribution in the cavity resonator) since these can then be used to calculate such parameters as the *Q*-factor and energy in the cavity (see section 1.20.3).

6.5.2 Discretization

A finite difference grid is superposed over the solution domain. This simply consists of dividing the volume into equally spaced domains. In this case, the grid has $5\times5\times5$ internal points (total of $6\times6\times6$ points, including boundary points) with a grid spacing of $0.2\ m$. The grid has the form shown in Figure 3.5 although the third dimension here is spatial not time.

Now we are in a position to define an approximation to Eq. (6.35) over the grid.

$$E_{i-1,j,p} + E_{i+1,j,p} + E_{i,j+1,p} + E_{i,j-1,p} + E_{i,j,p-1} + E_{i,j,p+1} - 6E_{i,j,p} + h^2k_c^2E_{i,j,p} = 0 \tag{6.36}$$

Or, in a more conventional form

$$E_{i-1,j,p} + E_{i+1,j,p} + E_{i,j+1,p} + E_{i,j-1,p} + E_{i,j,p-1} + E_{i,j,p+1} - \left(6 - h^2k_c^2\right)E_{i,j,p} = 0 \tag{6.37}$$

Boundary conditions are handled exactly as for any other Dirichlet boundary given above.

6.5.3 Solution

We note from Eq. (6.37) that this is an eigenvalue problem where the eigenvalues are

$$\lambda = h^2k_c^2$$

To solve the problem (that is: to find the eigenvalues of the system or, at least some of them, and the corresponding eigenvectors), we can proceed using one of two basic methods. The first assumes that a system of equations is formed as an implicit process and the eigenvalues of the matrix are evaluated. The second is an iterative process whereby the eigenvalues are assumed and the iteration modifies the eigenvalues until an acceptable solution is obtained. The latter is equivalent to an explicit process. We discuss the matrix method first.

By applying Eq. (6.37) at all nodes of the mesh (and introducing the boundary conditions) we get a matrix system of the form

$$[A - \lambda I]\{E_z\} = 0 \tag{6.38}$$

This is a standard eigenvalue system and can be solved for the eigenvalues λ (resonant frequencies) and for the eigenvector E_z. Thus, a solution means that both the resonant frequency and field distribution of the corresponding mode is known. In this case, the modes are easily identifiable but in other, more complex geometries, modes are much more difficult to identify.

The solution to the matrix eigenvalue problem in Eq. (6.38) is obtained using any appropriate method of solving the eigenvalue problem. Some methods for this purpose are given in chapter 11. Computer packages for solution of eigenvalue problems also exist and these are also discussed in chapter 11. Most eigenvalue methods also provide the eigenvectors of the system.

The first three resonant frequencies of the cavity resonator in Figure 6.20a are given in Table 6.1. Note that these are within about 5% of the exact resonant frequency, calculated using Eq. (1.167).

Table 6.1 Numerical calculation of resonant frequencies for the first three modes in empty cubic cavity (1×1×1m). Analytic values were found from Eq. (1.167).

Mode	Resonant frequency, numerical	Resonant frequency, analytic
TM_{110}	*199.6 mHz*	*212.13 mHz*
TM_{111}	*248.88 mHz*	*259.8 mHz*
TM_{210}	*325.3 mHz*	*335.4 mHz*

An alternative method is to use an iterative solution. The iterative process has two advantages from a computational point of view:

1. The process can be explicit, avoiding the generation of the matrix eigensystem in Eq. (6.38). This is particularly important when the system is very large or when the system is complex (as is the case with lossy dielectrics and composite materials).
2. Accuracy can be improved, especially for higher order modes which tend to be less accurate in matrix eigenvalue solutions.

To proceed with an iterative solution, we must have an initial estimate of the eigenvalues. This is a critical issue in this type of calculation since, the better our estimate, the faster the solution. An estimate can be obtained either from analytical solution of a simplified problem or by calculating the eigenvalues on a much smaller grid by use of the matrix eigenvalue method given above.

In the first case, we may be able to use approximations to the actual problem. For example, if a test sample is located in the cavity, we could use the eigenvalues of the empty cavity (from an analytical solution) as a first approximation and use these to calculate the actual values for the loaded cavity. Similarly, methods like the power method may be used to estimate the largest or lowest eigenvalues (dominant mode corresponds to lowest eigenvalue) from which successive eigenvalues may also be estimated with relatively little effort.

The second method is more systematic and requires only the solution of a small eigensystem. For example, a grid $5\times5\times5$ may be used which results in a 125×125 system. This is small enough to solve on any computer in a very short time. After an initial estimate of the eigenvalues is obtained, the iteration may commence. The process is as follows:

1. Start with an estimate for the eigenvalue. Assume initial eigenvalue is λ_0.
2. Assume an initial solution for the eigenvector (that is, for the electric field intensity at all nodes of the finite difference grid. A choice might be $E=1.0$ at all nodes, unless a better estimate can be obtained.
3. Iterate through the approximation in Eq. (6.37). The potential at a node (i, j,p) is calculated as

$$E_{i,j,p} = \frac{E_{i-1,j,p} + E_{i+1,j,p} + E_{i,j+1,p} + E_{i,j-1,p} + E_{i,j,p-1} + E_{i,j,p+1}}{6 - \lambda_0} \qquad (6.39)$$

   This iteration is repeated until the solution has converged to within a specified tolerance. Although this is the standard iteration in an explicit process, in this case the tolerance should not be too strict since the solution is supposed to converge to an "incorrect" solution because the eigenvalue is only an approximation of the correct eigenvalue.
4. Update the eigenvalues using the Rayleigh quotient and current solution:

$$k^2 \le -\frac{\displaystyle\int E\,\nabla^2 E\,dS}{\displaystyle\int E^2 dS} \qquad (6.40)$$

5. Check if the updated value of k^2 is within a given tolerance of the previous value of k^2. If so, this is the correct eigenvalue (within the tolerance specified). If not, repeat steps 3 through 5 until convergence to an eigenvalue occurs. When it does, the field values at the nodes of the grid are the eigenvector corresponding to the eigenvalues.
6. Repeat steps 1 through 5 for any other eigenvalues that may be required.

This process is particularly useful for very large matrices since we are free to choose the tolerance in solution as small as we wish (as long as it is larger than the roundoff error of the computer). Thus, we obtain results which can be more accurate than using the matrix eigensystem. However, the two tolerances used here, one for calculation of field values and the other for the quotient must be chosen judiciously. They should not be too small to avoid long calculations but should not be too large to avoid inaccuracies. A possible initial value is a field tolerance of perhaps *0.1-1%* and an eigenvalue tolerance of about *1%*. If necessary, these can be reduced.

6.5.4 Dielectric filled cavity

If the cavity is filled with a dielectric, the resonant frequency of the cavity is lower than that for the empty cavity. This situation is shown in Figure 6.20b. For a uniformly filled cavity, the solution above applies directly since the only difference is in the values for k. The results for the cavity in Figure 6.20b are given in Table 6.2 for a relative permittivity $\varepsilon_r=1.1$. Note that the results (obtained by the iterative solution with a *1%* tolerance for both field and eigenvalues using the same mesh as in the previous calculation), are within *3%* of the exact solution. A lower tolerance can improve these results.

Table 6.2 Numerical calculation of resonant frequencies for the first three modes in a loaded cubic cavity ($\varepsilon_r=1.1$). Analytic values were found using Eq. (1.167).

Mode	Resonant frequency, numerical	Resonant frequency, analytic
TM_{110}	*199.8 mHz*	*202.25 mHz*
TM_{111}	*242.2 mHz*	*247.7 mHz*
TM_{210}	*315.6 mHz*	*319.8 mHz*

6.5.5 Lossy dielectric sample in a cavity

The problem shown in Figure 6.20c is much more difficult to solve. An example of this type of application is shown in Figure 6.21a where a small cubic sample of lossy dielectric is shown inside the cavity. In this case the dielectric constant is a complex variable (Eq. (1.45)):

$$\varepsilon = \varepsilon' + j\varepsilon''$$

This implies losses in the cavity as well as internal reflections. More important, in addition to boundary conditions, interface conditions between materials must also be specified. Again, as in section 6.3.1, interface conditions are based on continuity of tangential components of the electric field intensity. For *TE* modes, solution is in terms of the magnetic field intensity, in which case continuity of the tangential components of the magnetic field intensity is enforced. The solution here is for *TM* modes, therefore tangential components of **E** or **H** must be enforced on the interfaces between the lossy dielectric and free space inside the cavity. First, we note that on the surfaces parallel to the *xz* and *yz* planes, the electric field is parallel to the surface. Therefore, the electric field intensity is continuous and need not be enforced on these surfaces. However, the sample used in Figure 6.21a is somewhat artificial because it has parallel surfaces. For an arbitrary sample, there may be no surface parallel to the z direction and special treatment of interfaces may be needed.

To calculate the required interface conditions on surfaces of the dielectric which are parallel to the *xy* plane we use Figure 6.21b. On surface *(1)*, the electric field intensity is normal to the surface. Following the convention of vectors (that is, that a vector is positive if it points out of a volume), the electric field intensity on this surface is taken as positive. On this interface we can write:

$$\varepsilon_0 E_{1z} = \varepsilon E_{2z} \qquad (6.41)$$

This cannot be easily implemented in finite differences but, from Maxwell's first equation we note the following

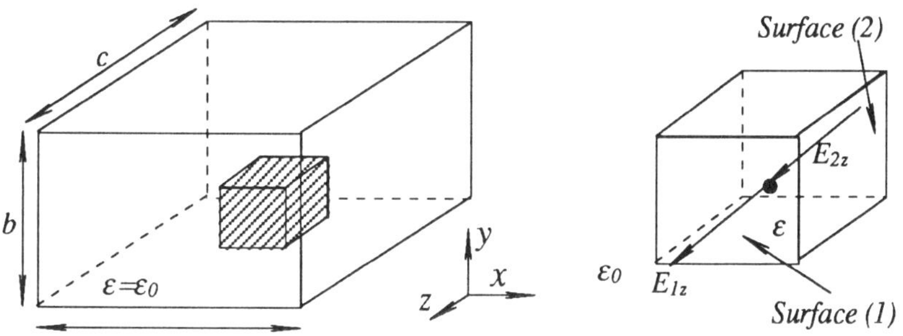

Figure 6.21 Dielectric sample inside resonant cavity. a. Geometry. b. Derivation of interface conditions between dielectric and free space in cavity.

$$\varepsilon_0 \nabla\times E_{1z} = \varepsilon_0\left(\hat{\mathbf{x}}\frac{\partial E_{1z}}{\partial y} - \hat{\mathbf{y}}\frac{\partial E_{1z}}{\partial x}\right) = -j\omega\mu_1\varepsilon_0(H_{1x} + H_{1y}) \tag{6.42}$$

Thus, enforcing this alternative condition, enforces continuity of the tangential components of **H** on this interface. This is completely analogous to the condition in Eq. (6.19) where the tangential components of **H** were enforced instead of continuity of **A**. Thus, the condition needed is

$$\varepsilon_0\left(\hat{\mathbf{x}}\frac{\partial E_{1z}}{\partial y} - \hat{\mathbf{y}}\frac{\partial E_{1z}}{\partial x}\right) = \varepsilon\left(\hat{\mathbf{x}}\frac{\partial E_{2z}}{\partial y} - \hat{\mathbf{y}}\frac{\partial E_{2z}}{\partial x}\right) \tag{6.43}$$

Similarly, on surface *(2)* in Figure 6.21b, the electric field intensity is negative and we have

$$\varepsilon_0\left(-\hat{\mathbf{x}}\frac{\partial E_{1z}}{\partial y} + \hat{\mathbf{y}}\frac{\partial E_{1z}}{\partial x}\right) = \varepsilon\left(-\hat{\mathbf{x}}\frac{\partial E_{2z}}{\partial y} + \hat{\mathbf{y}}\frac{\partial E_{2z}}{\partial x}\right) \tag{6.44}$$

These conditions are implemented using finite differences on surfaces *(1)* and *(2)*. The implementation of these relations is rather simple and requires first order derivatives. However, either forward or backward difference formulae are used rather than central differences. This was chosen for simplicity of implementation since in explicit methods the central point is evaluated and central difference approximations to the first order derivative do not contain the central point. Figure 6.22 shows the combination of difference formulae used for x and y components of the magnetic field intensity in Eqs. (6.43) and (6.44). The central point in the grid on the surface is implemented using either forward or backward differences.

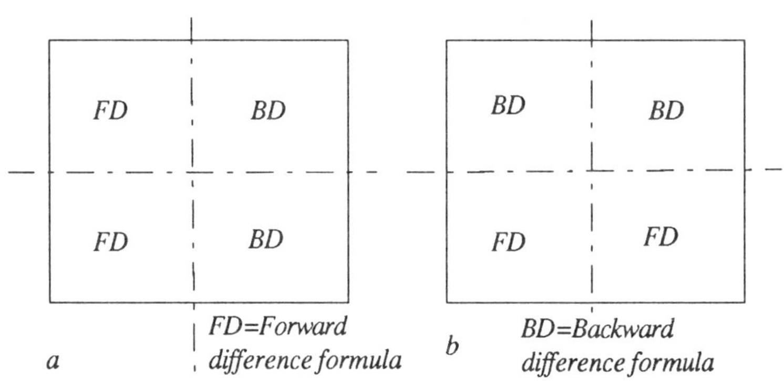

Figure 6.22 Implementation of interface conditions on surface (1) and (2) in Figure 6.21b. a. Implementation of derivatives with respect to x. b. Implementation of derivatives with respect to y.

Discretization into finite differences is identical to that for the geometry in Figure 6.21a but this geometry is discretized into $11\times11\times11$ points. This discretization is rather coarse as finite difference grids go but even so, it requires *1,331* nodes out of which *1,000* are internal nodes. Note that the third index (*p*) is treated the same as *i* or *j* for the two-dimensional case.

At all internal nodes of the sample, we apply Eq. (6.37) as

$$E_{i-1,j,p} + E_{i+1,j,p} + E_{i,j+1,p} + E_{i,j-1,p} + E_{i,j,p-1} + E_{i,j,p+1} - \left(6 - h^2 k_c^2\right) E_{i,j,p} = 0 \tag{6.47}$$

where

$$k_c^2 = k_0^2 \mu_r \varepsilon_r = \omega_0^2 \mu_0 \varepsilon_0 \mu_r (\varepsilon'_r + j\varepsilon''_r)$$

At nodes inside the cavity but not in the sample, the following approximation is used

$$E_{i-1,j,p} + E_{i+1,j,p} + E_{i,j+1,p} + E_{i,j-1,p} + E_{i,j,p-1} + E_{i,j,p+1} - \left(6 - h^2 k_0^2\right) E_{i,j,p} = 0 \tag{6.48}$$

For solution, we must rewrite these in terms of eigenvalues. These are:

$$\lambda_0 = h^2 k_0^2$$

and, therefore, Eqs. (6.47) and (6.48) are written as

$$E_{i-1,j,p} + E_{i+1,j,p} + E_{i,j+1,p} + E_{i,j-1,p} + E_{i,j,p-1} + E_{i,j,p+1} - (6 - \lambda_0)\varepsilon_r E_{i,j,p} = 0 \tag{6.49}$$

$$E_{i-1,j,p} + E_{i+1,j,p} + E_{i,j+1,p} + E_{i,j-1,p} + E_{i,j,p-1} + E_{i,j,p+1} - (6 - \lambda_0) E_{i,j,p} = 0 \tag{6.50}$$

These, together with the interface conditions above are assembled into an eigenvalue matrix system as in Eq. (6.38) and the system is solved for the eigenvalues λ_0. These are then the free-space wavenumbers of the various modes from which the resonant frequency is calculated as

$$\omega_0^2 = \frac{\lambda_0}{\mu_0 \varepsilon_0}$$

Note that although the calculation is in terms of the free-space wavenumber, it represents the sample material through calculation of the eigenvalues λ_0 which are different than those for the empty cavity. Thus, a sample inside the cavity causes a shift in the resonant frequency.

For the sample shown in Figure 6.21a, which is $10cm \times 10cm \times 10cm$, with a relative permittivity of 2, the resonant frequency for mode TM_{111} as calculated is $258.2\ MHz$ while in the empty cavity the exact resonant frequency for the same mode is $259.8\ MHz$. The sample is at the center of the cavity. The shift in resonant frequency of $1.5\ MHz$ is quite large and easily measurable.

6.6 NONLINEAR APPLICATIONS

6.6.1 Nonlinear equations for static fields

A number of applications in NDT are classified as nonlinear in nature. By nonlinear is meant that the quantity of interest, normally the magnetic flux density, is nonlinear. This type of field is discussed at some length in section 1.7. Nonlinearity of magnetic materials is defined through the hysteresis curve and dictates special treatment in solving for electromagnetic fields.

In terms of solution, however, nonlinearity influences only material properties not the solution itself, that is, the nonlinear method of solution of field problems is a method that allows an iterative definition of material properties such as permeability while the solution for fields is the same as for linear properties. This aspect of modeling is discussed in detail in chapter 11 where the Newton Raphson method is presented. Here we outline the method in general terms since the finite difference method is not modified due to nonlinearity.

6.6.2 The magnetization curve

Before a nonlinear solution can commence, the nonlinear relation must be known. This means that for a given magnetic field intensity, the magnetic flux density can be obtained. This can be in the form of a magnetization curve such as the curve in Figure 5.9, which shows the type of curve one would use for nonlinear calculations for active leakage fields. Other options for magnetization data can be in the form of tables or functional relationships. Representation of magnetization data is a difficult problem in electromagnetics because of the complex relationship between field intensity and flux density in ferromagnetic material.

6.6.3 Methods of solution

As mentioned above, the main point in nonlinear solution is the nonlinear material properties definition which is separate from the finite difference calculation. Using the Newton Raphson method we can write for any given point in the finite difference grid

$$x_{n+1} = x_n - \left[\frac{\partial f(x_n)}{\partial x}\right]^{-1} f(x_n) \tag{6.51}$$

where n indicates the iteration number in the nonlinear process (as opposed to the iteration number for the explicit process in the finite difference solution if explicit solution is used). Performing this for each node in the solution we obtain a new value for x. If a matrix solution (implicit solution) is used, the Newton-Raphson method is written as

$$\{x\}_{n+1} = \{x\}_n - \left[\frac{\partial [A]_n\{x\}_n}{\partial x}\right]^{-1} [[A]_n\{x\}_n - \{Q\}] \tag{6.52}$$

where we have assumed that $\{Q\}$ does not depend on the solution vector $\{x\}$.

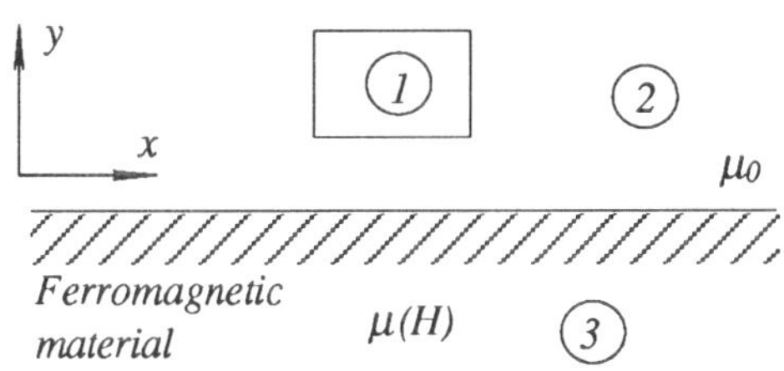

Figure 6.23. A general nonlinear problem. Domain (1) contains current sources.

6.6.4 Implementation

The relations in Eqs. (6.51) and (6.52) must be modified slightly for solution since in using the magnetization curve, we actually need to update the values of permeability so that the coefficients of the matrix [A] can be created. If an iterative process is used, the approximation at a given node requires knowledge of permeability before the new solution is obtained.

To put these ideas in more concrete terms we consider the problem in section 6.3 again but now as a nonlinear problem and assume that the nonlinear behavior of the materials is known.

As was done in Figure 6.12. we define three domains. These are shown again in Figure 6.23 but now, material *(3)* is nonlinear. In material *(1)*, we have a current density $J=I/S\ [A/m^2]$ where S is the cross-sectional area of the busbar (in this case, $2\ cm^2$) and the equation is identical to Eq. (6.17):

$$\frac{\partial^2 A}{\partial x^2}+\frac{\partial^2 A}{\partial y^2}=-\mu_1 J \qquad \text{in material 1} \tag{6.53}$$

This material is linear (that is, permeability μ_1 is constant and independent of the magnitude of field in the material). Therefore, treatment of this equation is identical to that in section 6.3.

In material *(2)* there are no current densities and we can write

$$\frac{\partial^2 A}{\partial x^2}+\frac{\partial^2 A}{\partial y^2}=0 \qquad \text{in material 2} \tag{6.54}$$

and again, nonlinearity does not enter into consideration.

However, in material (3) we must write

$$\frac{1}{\mu_{2x}}\frac{\partial^2 A}{\partial x^2}+\frac{1}{\mu_{2y}}\frac{\partial^2 A}{\partial y^2}=0 \qquad \text{in material 3} \tag{6.55}$$

where the dependency of permeability on the magnetic vector potential (and therefore on **B**) is evident.

The finite difference approximations to Eqs. (6.53) through (6.55) are:
In material *(1)*:

$$A_{i+1,j}+A_{i-1,j}+A_{i,j+1}+A_{i,j-1}-4A_{i,j}=-\mu_1 h^2 J_{i,j} \tag{6.56}$$

where $h=\Delta x=\Delta y$.
In material *(2)*:

$$A_{i+1,j}+A_{i-1,j}+A_{i,j+1}+A_{i,j-1}-4A_{i,j}=0 \tag{6.57}$$

In material *(3)*:

$$\frac{A_{i+1,j}-2A_{i,j}+A_{i-1,j}}{\mu_{3x}h^2}+\frac{(A_{i,j+1}-2A_{i,j}+A_{i,j-1})}{\mu_{3y}h^2}=0 \tag{6.58}$$

The last equation requires values of μ_{3x} and μ_{3y} before it can be applied. Unlike the approximation in Eq. (6.22) no scaling was used here to keep the discussion simple. Both scaling and different values of h may be used as was done in section 6.3. The interface condition is written (using Figure 6.13) as

$$\frac{1}{\mu_y^{(2)} 2\Delta y^{(2)}}\left(-A_2^{(2)}+4A_1^{(2)}-3A_0^{(2)}\right) = \frac{1}{\mu_y^{(3)} 2\Delta y^{(3)}}\left(3A_0^{(3)}-4A_{-1}^{(3)}+A_{-2}^{(3)}\right) \tag{6.59}$$

which is identical to the condition in Eq. (6.23) except for the use of different permeabilities. Note also that only the y components of permeability are used for the interface condition.

The dependency in Eq. (6.55) indicates the problem with nonlinear calculations; the magnetic vector potential can only be calculated after permeability is known, while on the other hand, permeability is obtained from magnetization data (such as the hysteresis curve) for which the magnetic field must be known. A solution to this dilemma is in the following process:

1. Assume a solution for the magnetic vector potential at each node in the solution domain. This initial solution which we denote as A_n is the starting point of the finite difference solution.
2. From A_n, calculate B_{xn} and B_{yn} by using the derivatives of A as in section 6.4.2. This gives the two components of the magnetic flux density B_x and B_y for the current iteration.
3. Locate the magnitude of B_x and B_y on the magnetization curve in Figure 5.9 (that is, on the curve that represents the magnetization of the material used). From B_{xn} and the corresponding point H_{xn}, the permeability μ_{3xn} is now calculated. Similarly, the value of μ_{3yn} is evaluated. These values of permeability are used for calculation in Eqs. (6.58) and (6.59)
4. Perform the finite difference solution either as an explicit or implicit method using the approximations in Eqs. (6.56) through (6.59) for the linear case but with permeabilities as calculated in step 3. A new solution for the magnetic vector potential is now obtained which we will denote as A_{n+1}. Note that if an explicit solution is used, this step includes as many iterations as required.
5. Apply the Newton Raphson method to obtain an updated value of the solution using Eq. (6.51) or (6.52).
6. Repeat steps 2 through 5 until the solution for the magnetic vector potential does not change or, more appropriately, until the change in solution is below a certain threshold as discussed in chapter 3 and 11.

From the steps above it is clear that the finite difference iteration is not affected by the nonlinear iteration and the two iterations are, in this sense, independent.

6.7 PROBLEMS

6.1 Following the steps in sections 6.2.1 through 6.2.4, calculate the voltage inside the conducting box shown in Figure 6.24.

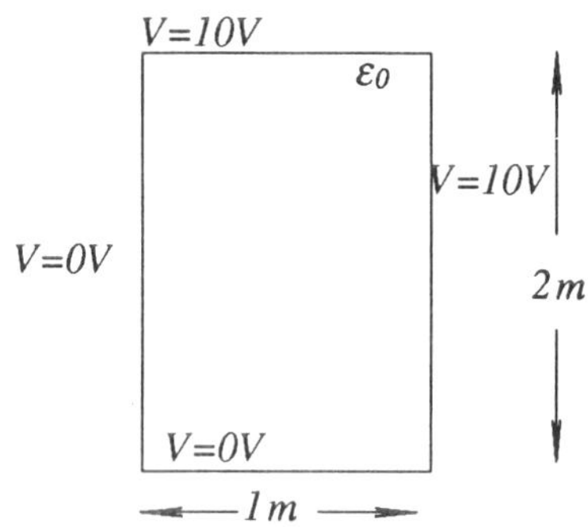

Figure 6.24.

6.2 For the previous problem, calculate the electric field intensity everywhere and the total energy stored in the box (per unit length). From the electric field intensity, calculate charge density on all conducting surfaces.

6.3 Define symmetry conditions and energy relations in *3-D* geometries by extending the relations in Eqs. (6.6) and (6.10). Note that there are three types of nodes on an interface plane: (1) nodes in the symmetry plane, (2) nodes on an edge between two symmetry planes, and (3) nodes at intersections between edges formed by intersecting planes. Treat each of these separately.

6.4 To extend the methods in this chapter consider the three-dimensional problem in Figure 6.25. All boundaries are at zero potential except the top surface which is at *10 V*. Use three symmetry planes to divide the problem into *4* equal sections (using two planes, one parallel to the *yz* plane, one parallel to the *yx* plane). Solve for potential. Sketch the solution and compare with the exact solution in problem 2.2.

6.5 A circular coil is placed over a conducting plane as shown in Figure 6.26. For the dimensions given, calculate the magnetic field everywhere. Assume that on the external boundaries the magnetic field is zero (that is, locate these boundaries far enough from the source). Use the axisymmetric form of equations to find a solution. Assuming there are *100* turns in the coil, calculate the inductance of the coil.

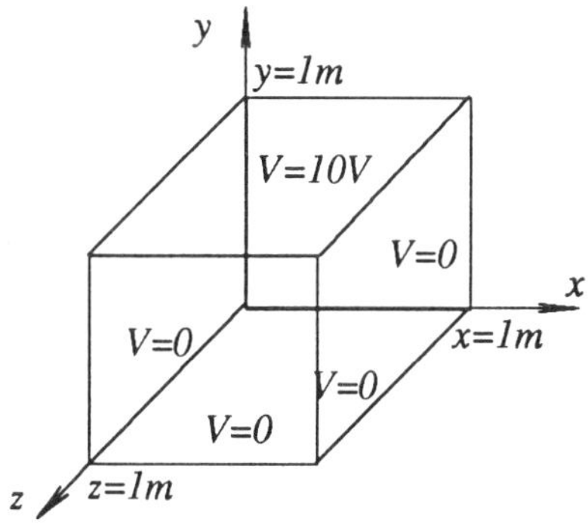

Figure 6.25.

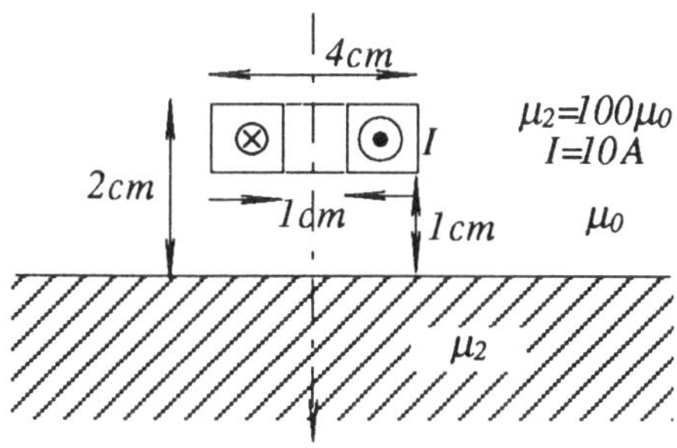

Figure 6.26.

6.6 Reconsider Problem 6.5 but assume now a flat bottomed hole, $1cm$ deep and $1cm$ in diameter is drilled in the flat conductor, exactly below the center of the coil (Figure 6.27). Calculate the inductance of the coil. Comment on the possible use of this method for defect characterization.

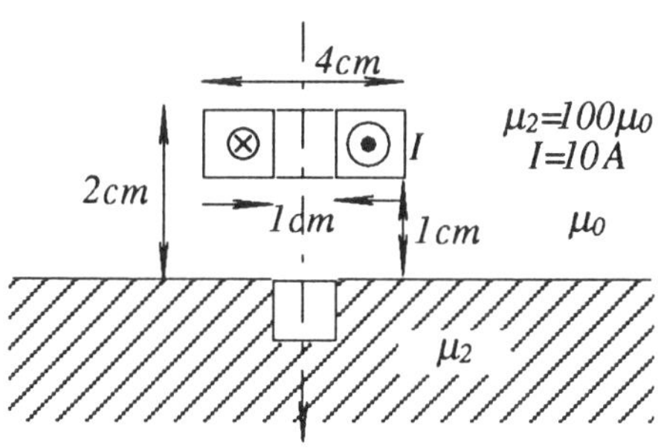

Figure 6.27.

6.7 A variation of problem 6.5. is to use a square coil with the same dimensions as given in Figure 6.26. Now the problem must be solved as a three-dimensional problem. Calculate the magnetic field and inductance of the coil. Use symmetry if possible, to reduce the size of the solution domain.

6.8 Figure 6.28 shows an eddy current coil over a conducting slab. For the dimensions shown, calculate the magnetic field everywhere. Use a frequency of *100 kHz* and neglect induced currents in the coil (that is, assume it is perfectly conducting). Use the same grid as in problem 6.5.

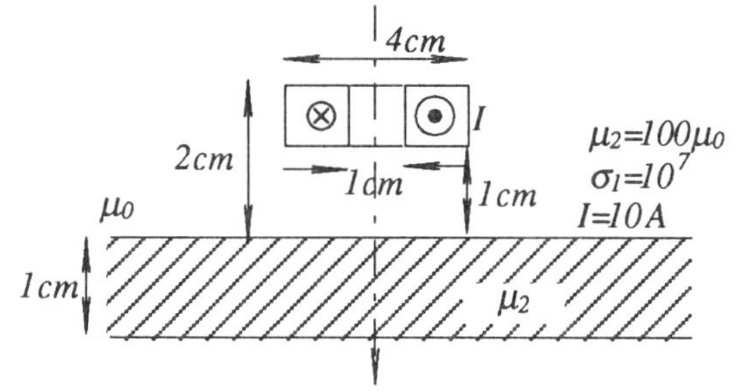

Figure 6.28.

6.9 Repeat problem 6.8 but with a through hole through the flat conductor just below the center of the coil. Calculate the change in impedance of the coil due to a *1cm* hole. Repeat the calculation with smaller diameter and larger diameter holes.

6.10 In a test, the coil in Figure 6.8 is moved from left to right over a long slot, *1cm* deep and *1cm* wide. Assuming the velocity is low (that is, the coil is moved to a location, fields are calculated for the stationary coil and then the coil is moved to a new location) calculate the impedance of the coil due to the slot. Move the coil one grid-cell at a time. Figure 6.29 shows how the coil is moving. Note that this must be solved as a *3-D* problem.

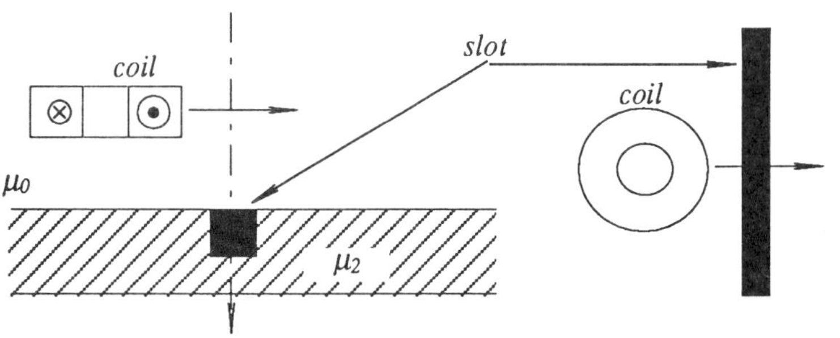

Figure 6.29.

6.11 Calculate the cutoff frequencies of a rectangular waveguide, *1 cm* wide, and *2 cm* high as shown in Figure 6.30. The waveguide is filled with a gas with relative permittivity ε_r=*2.0*. Compare the numerical calculation with the exact solution in Eq. (1.155).

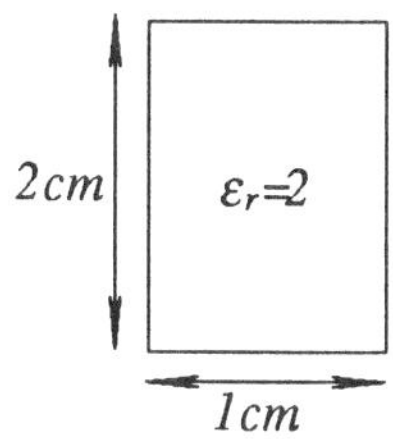

Figure 6.30.

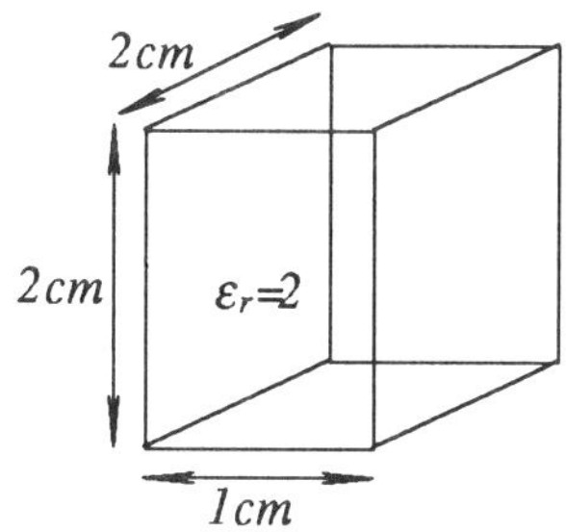

Figure 6.31.

6.12 Calculate the resonant frequencies of a cavity resonator with dimensions as in Figure 6.31. Compare with exact solution in Eq. (1.166).

6.13 Consider again problem 6.5 but the ferromagnetic material is nonlinear and the relation between magnetic field intensity and magnetic flux density is $B=H^2$. Calculate the flux density everywhere in the solution domain and the inductance of the coil. Compare with the solution in problem 6.5.

6.14 Repeat problem 6.13 but increase the current by a factor of *100*. Compare with results in problem 6.13.

6.8 BIBLIOGRAPHY

General references on finite difference methods are given in chapter *3*. The references given here deal with applications. A number of applications are given in [1-8]. Specific applications in time-harmonic wave propagation are given in [9-12] and nonlinear methods in [13-15]. The question of modeling the hysteresis curve for purposes of numerical calculations is only mentioned briefly in the context of nonlinear calculations. References [16-22] expand on this and give additional information.

[1] Binns, K.J. and Lawrenson, P.J. (1975), *Analysis and Computation of Electric and Magnetic Field Problems*, Pergamon Press, 2nd edition.

[2] Muller, W. (1982), "A new iteration method for solving stationary eddy current problems using the method of finite differences", *IEEE Transactions on Magnetics*, Vol. MAG-18, No. 2, pp. 588-593.

[3] Trutt, F.C. Erdelyi, E.A. and Jackson, R.F. (1963), "The non-linear potential equation and its numerical solution for highly saturated

electrical machines", *IEEE Transactions on Aerospace*, Vol. 1, pp. 430-440.

[4] Southwell, R.V. (1940), *Relaxation Methods in Engineering Science*, Oxford University Press, Oxford.

[5] Motz, H. and Worthy, W.D. (1945), "Calculation of the magnetic field in dynamoelectric machines using Southwell's relaxation method", *Journal IEE*, Vol. 92, Pt. 2, pp. 522-528.

[6] Stoll, R.L. (1970), "Solution of linear steady state eddy current problems by complex successive over-relaxation", *IEE Proceedings*, Pt. A, Vol. 117, pp. 1317-1323.

[7] Ahmed, S.V. and Erdelyi, E.A. (1966), "Nonlinear theory of salient pole machines", *IEEE Transactions on Power Apparatus and Systems*, Vol. PAS-85, pp. 61-70.

[8] Mamak, R.S. and Laithwaite, E.R. (1960), "Numerical evaluation of inductance and AC resistance", *IEE Proceedings*, pp. 252-258.

[9] Davies, J.B. (1966), "Numerical solution of uniform hollow waveguides with boundaries of arbitrary shapes", *Proceedings of the IEEE*, Vol. 113, No. 2, pp. 277-284.

[10] Beaubien, M.J. and Wexler, A. (1968), "An accurate finite-difference method for higher order waveguide modes", *IEEE Transactions on Microwave Theory and Techniques*, Vol. MTT-16, No. 12, pp. 1007-1017.

[11] Bierwith, K. Schultz, N. and Arndt, F. (1986), "Finite-diffference analysis of rectangular dielectric waveguide structures", *IEEE Transactions on Microwave Theory and Techniques*, Vol. MTT-34, No. 11, pp. 1104-1114.

[12] Davies, J.B. (1972), "Review of methods for numerical solution of the hollow waveguide problem", *IEE Proceedings*, Vol. 119, No. 1, pp. 33-37.

[13] Erdelyi, E.A. and Fuchs, E.F. (1970), "Nonlinear magnetic field analysis of DC machines, Part I", *IEEE Transactions on Power Apparatus and Systems*, Vol. PAS-89, No. 7, pp. 1546-1583.

[14] Erdelyi, E.A. and Fuchs, E.F. (1970), "Nonlinear magnetic field analysis of DC machines, Part II", Application of the improvement method", *IEEE Transactions on Power Apparatus and Systems*, Vol. PAS-89, No. 7, pp. 1546-1583.

[15] Erdelyi, E.A. and Fuchs, E.F. (1970), D.H. Binkley, "Nonlinear magnetic field analysis of D.C. machines, Part III :equipotential plots drawn by computer", *IEEE Transactions on Power Apparatus and Systems*, Vol. PAS-89, No. 7, pp. 1565-1583.

[16] Mayergoyz, I. (1991), *Mathematical Models of Hysteresis*, Springer Verlag, New York.

[17] Asher, G.M. Williams, J.T. Walters, C.R. Joyce, H. and Paul, R.J.A. (1982), "Modeling of hysteresis in thin superconducting screens for

mixed-mu suspension systems", *IEEE Transactions on Magnetics*, Vol. MAG-18, No. 2, pp. 540-545.

[18] Savini, A. (1982), "Modeling hysteresis loops for finite element magnetic field calculations", *IEEE Transactions on Magnetics*, Vol. MAG-18, No. 2, pp. 552-557.

[19] Burais, N. and Grellet, G. (1982), "Numerical modelling of iron losses in ferromagnetic steel plates", *IEEE Transactions on Magnetics*, Vol. MAG-18, No. 2, pp. 558-562.

[20] Penman, J. and Kamar, A.M.A. (1982), "Linearization of saturable field problems, including eddy currents", *IEEE Transactions on Magnetics*, Vol. MAG-18, No. 2, pp. 563-566.

[21] Denegri, G.B. Giardinio, P. Molfino, P and Viviani, A. (1982), "A computer approach to the nonlinear hysteresis problem of electromagnetic field computation in hard superconductors for loss evaluation", *IEEE Transactions on Magnetics*, Vol. MAG-18, No. 2, pp. 567-572.

[22] Udpa, S.S. and Lord, W. (1985), "A Fourier descriptor model of hysteresis loop phenomena", *IEEE Transactions on Magnetics*, Vol. MAG-21, No. 6, pp. 2370-2373.

CHAPTER 7

Finite element formulation

7.1 INTRODUCTION

In the previous chapter, we discussed the use of finite differences for the solution of elliptic problems. The main point perhaps, is that the equations are approximated directly in terms of finite difference approximations of the partial derivatives. The finite element method, as discussed in chapter 4 is substantially different in that an approximation of the function itself (the unknowns that need to be solved for) is defined within the finite element while the solution is obtained by substituting the approximation in an appropriate statement of the physical problem followed by minimization (actually extremization) of that function with respect to the unknowns.

All steps in the solution are more complex: the discretization step requires the definition of volumes, surfaces and nodes: in the approximation stage we must first find an appropriate function such as a weighted residual or energy function; the solution, because minimization is with respect to all unknowns, must be done implicitly. Thus, a matrix system must be solved.

Nevertheless, finite element methods have some very important advantages over finite difference methods. These include:

1. The approximation is valid throughout the volume of the element, including nodes. Therefore, both nodal and distributed material properties and sources can be handled.
2. Discretization of complex material boundaries and interfaces is easily accomplished.
3. Interface conditions are not normally required since these are natural or implicit conditions in finite element formulations. For this reason, anisotropic material properties can be easily accommodated.

A number of two-dimensional, axisymmetric and three-dimensional formulations are presented in this chapter. Some are presented in detail as representative of classes of problems, others are only formulated but not solved. As in the previous chapter this is done in the interest of brevity. The general approach is to show three-dimensional formulations since their reduction to 2-D applications is rather simple.

7.2 CHOICE OF FORMULATIONS AND FINITE ELEMENTS (2-D AND 3-D)

The first step in any solution is formulation of the problem. By formulation is meant the setup of the field equations, boundary conditions, and any other condition within the solution domain that must be considered, in a form amenable to finite element representation. We must decide on the field variable or potential function to be used and consider the alternatives. This is not a minor decision; much of the subsequent effort in formulation and computation as well as results are directly affected by this decision.

Since the number of variables we can use is large (see chapter 1), we often have alternatives, some better than others. The final decision must take into account both current needs and any foreseeable future needs. Time investment in finite element solutions is heavy and we must ensure that any time invested in developing a formulation is not wasted when the formulation is extended to other, similar problems. Thus, for example, if magnetostatic problems without current sources are our main interest at the moment, we may consider a scalar potential formulation. However, if we expect also to apply these methods to applications with currents, or to eddy current problems, a vector potential formulation might be a better choice. The only general rule is that vector potentials or vector field variables must be used in current regions while scalar or vector functions may be used in regions void of currents. This also implies that more than one function (scalar or vector) may be used in the same solution, within the same region in the solution domain or in neighboring domains, provided that the solutions in terms of the various functions are matched at interfaces.

Dimensionality of the problem is also of prime importance as it affects the computational effort. If two-dimensional solutions are sufficient, or an axisymmetric solution is a fair representation of the problem, there may not be a need for three-dimensional solutions. As a rule, three-dimensional solutions are considerably more complex, less accurate, and much more expensive than two-dimensional or axisymmetric solutions.

The field equations are written in alternative forms in terms of both vector and scalar potentials in chapter 1. Here we pursue the formulation of some of the alternative field equations in terms of two rather general methods: the variational method (using what is often called the "energy related functional") and Galerkin's method. As with the various field variables, the choice is not always obvious. However, as a rule, we use the variational method whenever an energy functional (or energy related functional) exists while Galerkin's method must be used when such a functional does not exist or cannot be found. If a functional exists, both formulations are equivalent (see section (4.2)).

The choice of finite elements is somewhat less obvious than that of field variables or method of formulation. It would seem at first glance that any finite element can be used as long as the continuity and completeness conditions are satisfied. While this is in general true, some finite elements perform better than others, some are more accurate, or may result in smaller or more sparse matrices. First and second order elements are often used but higher order elements may be employed. The method of defining the finite element affects both the integration process (analytical or numerical integration) accuracy, and to a certain extent, computer resources. First order elements provide a linear approximation in terms of the primary variables (that calculated in the finite element solution) but only constant approximation in its derivatives. If this is not acceptable, higher order elements, or a different variable must be used.

Within a single group of elements, various finite elements offer some advantages. For example, tetrahedral and hexahedral elements are both linear elements. However, tetrahedral elements have only four nodes, are integrated analytically, and are very convenient for describing complex shapes. Hexahedral elements, on the other hand, have eight nodes, are integrated numerically and are most convenient for curvilinear objects. They are much more accurate than tetrahedral elements and easier to visualize. Similarly, for the same number of nodes, a tetrahedral mesh contains five times the number of elements compared to hexahedral elements. Therefore, the choice between the two is not trivial.

As a rule, we choose the simplest element that can be used for the class of application we intend to solve. First order (linear) triangular elements are most often used for two-dimensional and axisymmetric applications, followed by quadrilateral elements. In three dimensions, hexahedral elements are more commonly used than tetrahedral elements but tetrahedral elements are also common. Higher order elements are only used when necessary, such as with very complex geometries where the interpolation properties of the higher order elements are important.

Next we present two formulations: one a variational and the second based on Galerkin's method for the general eddy current problem. While these are certainly not the only formulations possible, they are rather general and cover most types of elliptic equations. Additional modifications required in particular applications are given in examples for specific applications

7.3 FORMULATION USING AN ENERGY FUNCTIONAL: VARIATIONAL APPROACH

7.3.1 The equations

Consider the eddy current problem (such as in Eq. (5.11)) where we have written the field equations in terms of the magnetic vector potential. This was done by substituting the definition of the magnetic vector potential for **B** in the field equations. The following steps are given in detail in chapter 1 but are repeated here for completeness. The magnetic vector potential is defined as $\mathbf{B}=\nabla\times\mathbf{A}$. Substituting this in Eq. (1.1) we get

$$\nabla\times\mathbf{E} = -\nabla\times\left(\frac{\partial\mathbf{A}}{\partial t}\right) \tag{7.1}$$

From this, we note that the time dependent electric field can be written as

$$\mathbf{E} = -\frac{\partial\mathbf{A}}{\partial t}$$

However, since this is defined through the curl relation, the gradient of a scalar potential can be added as follows

$$\nabla\times\mathbf{E} = -\nabla\times\left(\frac{\partial\mathbf{A}}{\partial t} + \nabla V\right)$$

where V is the electric scalar potential and, although V as defined here is in general time dependent it also means that in the static case the electric field is due only to the gradient of V as is well known from electrostatics (see chapter 1). Thus, the general electric field intensity is given as

$$\mathbf{E} = -\frac{\partial\mathbf{A}}{\partial t} - \nabla V \tag{7.2}$$

Since we are interested here in time-harmonic fields, we replace $\partial/\partial t$ by $j\omega$ and get

$$\nabla\times\mathbf{E} = -\nabla\times(j\omega\mathbf{A} + \nabla V) \tag{7.3}$$

$$\mathbf{E} = -j\omega\mathbf{A} - \nabla V \tag{7.4}$$

Substituting the above into Maxwell's second equation (Eq. (1.2)) gives

$$\nabla\times\left(\frac{1}{\mu}\nabla\times\mathbf{A}\right) = \mathbf{J}_s + \mathbf{J}_e + j\omega\varepsilon\left(-\nabla V - j\omega\mathbf{A}\right) \tag{7.5}$$

where the current density is separated into an eddy current and a source term. The eddy current term is now written in terms of the electric field $\mathbf{J}_e=\sigma\mathbf{E}$ and the equation expanded using the identity in Eq. (1.71). We also assume that μ is linear. This gives:

$$\nabla\left(\nabla\cdot\mathbf{A}\right) - \nabla^2\mathbf{A} = \mu\mathbf{J}_s + \mu\sigma\left(-\nabla V - j\omega\mathbf{A}\right) + \mu\varepsilon j\omega\left(-\nabla V - j\omega\mathbf{A}\right)$$

or

$$-\nabla\left(\nabla\cdot\mathbf{A}\right) + \nabla^2\mathbf{A} = -\mu\mathbf{J}_s + \mu\sigma\nabla V + j\omega\mu\sigma\mathbf{A} + j\omega\mu\varepsilon\nabla V - \omega^2\mu\varepsilon\mathbf{A} \tag{7.6}$$

This is the general, time-harmonic field equation. For eddy current applications, the propagation term (last term on the right hand side) is neglected. Similarly, the Lorentz gage

$$\nabla\cdot\mathbf{A} = -j\omega\mu\varepsilon\nabla V$$

is used to define the divergence of $\mathbf{A}$ in Eq. (7.6).

Also, the term $\mu\sigma\nabla V$ is assumed to be negligible (although we take this term into account in the following section). Substitution of these relations in Eq. (7.6) gives

$$\nabla^2\mathbf{A} = -\mu\mathbf{J}_s + j\omega\mu\sigma\mathbf{A} \tag{7.7}$$

Eq. (7.7) is often referred to as the eddy current equation. In expanded form this is:

$$\frac{\partial^2\mathbf{A}}{\partial x^2} + \frac{\partial^2\mathbf{A}}{\partial y^2} + \frac{\partial^2\mathbf{A}}{\partial z^2} = -\mu\mathbf{J}_s + j\omega\mu\sigma\mathbf{A} \tag{7.8}$$

This is a vector equation in complex variables. At this point we are interested in the formulation of the equation in a suitable form for finite element solution.

7.3.2 The formulation

The approach to formulation is to assume that a functional, equivalent to Eq. (7.8) exists and then show that this assumption is correct. We formulate the general three-dimensional eddy current equation in terms of the magnetic vector potential. A general energy statement for electromagnetic field applications can be written as

$$F = \int_v (w_e + w_m + w_d - w_i)dv \tag{7.9}$$

where w_e is the stored electric energy density, w_m the stored magnetic energy density, w_d the dissipated energy density and w_i the input energy density. In terms of the magnetic flux density and the magnetic vector potential we can write

$$F(\mathbf{A}) = \int_v \left(\int_B \mathbf{H}\cdot d\mathbf{B} + \int_A j\omega\sigma \mathbf{A}\cdot d\mathbf{A} - \int_A \mathbf{J}\cdot d\mathbf{A} \right) dv = \int_v f dv \tag{7.10}$$

where

$$f = \int \nu_x B_x dB_x + \int \nu_y B_y dB_y + \int \nu_z B_z dB_z + \int j\omega\sigma A_x dA_x + \int j\omega\sigma A_y dA_y$$
$$+ \int j\omega\sigma A_z dA_z - \int J_x dA_x - \int J_y dA_y - \int J_z dA_z$$

and $\nu = 1/\mu$. Performing the integration we can write in short form

$$f = \frac{1}{2}(\mathbf{B}\cdot\mathbf{H}) + \frac{j\omega\sigma}{2}(\mathbf{A}\cdot\mathbf{A}) - \mathbf{J}\cdot\mathbf{A} \tag{7.11}$$

carrying out the dot products in Eq. (7.11) gives

$$\mathbf{B}\cdot\mathbf{H} = \nu_x B_x^2 + \nu_y B_y^2 + \nu_z B_z^2 \tag{7.12}$$

$$\mathbf{J}\cdot\mathbf{A} = J_x A_x + J_y A_y + J_z A_z \tag{7.13}$$

$$\mathbf{A}\cdot\mathbf{A} = A_x^2 + A_y^2 + A_z^2 \tag{7.14}$$

Also, from the relation between **B** and **A**, in Eq. (1.63) the components of flux density are

$$B_x = \frac{\partial A_z}{\partial y} - \frac{\partial A_y}{\partial z} \tag{7.15}$$

$$B_y = \frac{\partial A_x}{\partial z} - \frac{\partial A_z}{\partial x} \tag{7.16}$$

$$B_z = \frac{\partial A_y}{\partial x} - \frac{\partial A_x}{\partial y} \tag{7.17}$$

Using the notation $A_{jk}=\partial A_j/\partial_k$ for convenience, substituting Eq. (7.12) through (7.14) into Eq. (7.10), and using Eq. (7.15) through Eq. (7.17), the energy functional becomes

$$F(A) = \int_v \frac{1}{2}\Big[\nu_x(A_{zy} - A_{yz})^2 + \nu_y(A_{xz} - A_{zx})^2 + \nu_z(A_{yx} - A_{xy})^2 + \frac{j\omega\sigma}{2}\left(A_x^2 + A_y^2 + A_z^2\right) - (J_xA_x + J_yA_y + J_zA_z)\Big] dv \tag{7.18}$$

We assume in the formulation above that the energy relations in Eq. (7.9) apply here. However, because the variables used are complex variables, this functional should be properly called an energy-related functional rather than an energy functional. However, in the static case, the variables are real and the terms of the functional are actual energy terms.

To verify that this functional is equivalent to the governing partial differential equation the integrand *(f)* is substituted into the appropriate Euler equation. The Euler equation is derived by noting that in standard variational form the functional is of the type:

$$F(A) = \int_v f(x,y,z,A_x,A_y,A_z,A_{xx},A_{yy},A_{zz},A_{xy},A_{xz},A_{yx},A_{yz},A_{zx},A_{zy})dxdydz \tag{7.19}$$

where the notation given above for partial derivatives is used. Thus, for example, $A_{xy}=\partial A_x/\partial y$.

To find a stationary point of this functional the first variation must be equal to zero:

$$\begin{aligned}\delta F(A) = \int_v \Bigg(& \frac{\partial f}{\partial A_x}\delta A_x + \frac{\partial f}{\partial A_y}\delta A_y + \frac{\partial f}{\partial A_z}\delta A_z + \\ & + \frac{\partial f}{\partial A_{xx}}\delta A_{xx} + \frac{\partial f}{\partial A_{xy}}\delta A_{xy} + \frac{\partial f}{\partial A_{xz}}\delta A_{xz} + \\ & + \frac{\partial f}{\partial A_{yx}}\delta A_{yx} + \frac{\partial f}{\partial A_{yy}}\delta A_{yy} + \frac{\partial f}{\partial A_{yz}}\delta A_{yz} + \\ & + \frac{\partial f}{\partial A_{zx}}\delta A_{zx} + \frac{\partial f}{\partial A_{zy}}\delta A_{zy} + \frac{\partial f}{\partial A_{zz}}\delta A_{zz}\Bigg) dxdydz = 0\end{aligned} \tag{7.20}$$

Using the identity

$$\frac{\partial f}{\partial A_{jk}}\,\delta A_{jk} = \frac{\partial f}{\partial A_{jk}}\frac{\partial}{\partial k}(\delta A_j)$$

$$\begin{aligned}
\delta F(\mathbf{A}) &= \int_v \left(\frac{\partial f}{\partial A_x}\,\delta A_x + \frac{\partial f}{\partial A_y}\,\delta A_y + \frac{\partial f}{\partial A_z}\,\delta A_z\right)dxdydz \\
&+\int_v \left(\frac{\partial f}{\partial A_{xx}}\frac{\partial}{\partial x}(\delta A_x) + \frac{\partial f}{\partial A_{xy}}\frac{\partial}{\partial y}(\delta A_x) + \frac{\partial f}{\partial A_{xz}}\frac{\partial}{\partial z}(\delta A_x)\right)dxdydz \\
&+\int_v \left(\frac{\partial f}{\partial A_{yx}}\frac{\partial}{\partial x}(\delta A_y) + \frac{\partial f}{\partial A_{yy}}\frac{\partial}{\partial y}(\delta A_y) + \frac{\partial f}{\partial A_{yz}}\frac{\partial}{\partial z}(\delta A_y)\right)dxdydz \\
&+\int_v \left(\frac{\partial f}{\partial A_{zx}}\frac{\partial}{\partial x}(\delta A_z) + \frac{\partial f}{\partial A_{zy}}\frac{\partial}{\partial y}(\delta A_z) + \frac{\partial f}{\partial A_{zz}}\frac{\partial}{\partial z}(\delta A_z)\right)dxdydz
\end{aligned} \tag{7.21}$$

Applying the formula for integration by parts to the last three integrals gives

$$\begin{aligned}
\delta F(\mathbf{A}) &= \int_v \left(\frac{\partial f}{\partial A_x}\,\delta A_x + \frac{\partial f}{\partial A_y}\,\delta A_y + \frac{\partial f}{\partial A_z}\,\delta A_z\right)dxdydz \\
&+\int_v \left(\frac{\partial}{\partial x}\left(\frac{\partial f}{\partial A_{xx}}\,\delta A_x\right) + \frac{\partial}{\partial y}\left(\frac{\partial f}{\partial A_{xy}}\,\delta A_x\right) + \frac{\partial}{\partial z}\left(\frac{\partial f}{\partial A_{xz}}\,\delta A_x\right)\right)dxdydz \\
&-\int_v \left(\frac{\partial}{\partial x}\left(\frac{\partial f}{\partial A_{xx}}\right)\delta A_x + \frac{\partial}{\partial y}\left(\frac{\partial f}{\partial A_{xy}}\right)\delta A_x + \frac{\partial}{\partial z}\left(\frac{\partial f}{\partial A_{xz}}\right)\delta A_x\right)dxdydz \\
&+\int_v \left(\frac{\partial}{\partial x}\left(\frac{\partial f}{\partial A_{yx}}\,\delta A_y\right) + \frac{\partial}{\partial y}\left(\frac{\partial f}{\partial A_{yy}}\,\delta A_y\right) + \frac{\partial}{\partial z}\left(\frac{\partial f}{\partial A_{yz}}\,\delta A_y\right)\right)dxdydz \\
&-\int_v \left(\frac{\partial}{\partial x}\left(\frac{\partial f}{\partial A_{yx}}\right)\delta A_y + \frac{\partial}{\partial y}\left(\frac{\partial f}{\partial A_{yy}}\right)\delta A_y + \frac{\partial}{\partial z}\left(\frac{\partial f}{\partial A_{yz}}\right)\delta A_y\right)dxdydz \\
&+\int_v \left(\frac{\partial}{\partial x}\left(\frac{\partial f}{\partial A_{zx}}\,\delta A_z\right) + \frac{\partial}{\partial y}\left(\frac{\partial f}{\partial A_{zy}}\,\delta A_z\right) + \frac{\partial}{\partial z}\left(\frac{\partial f}{\partial A_{zz}}\,\delta A_z\right)\right)dxdydz \\
&-\int_v \left(\frac{\partial}{\partial x}\left(\frac{\partial f}{\partial A_{zx}}\right)\delta A_z + \frac{\partial}{\partial y}\left(\frac{\partial f}{\partial A_{zy}}\right)\delta A_z + \frac{\partial}{\partial z}\left(\frac{\partial f}{\partial A_{zz}}\right)\delta A_z\right)dxdydz
\end{aligned} \tag{7.22}$$

The second, fourth and sixth volume integrals can be evaluated using Green's theorem:

$$\int_v \nabla \cdot \mathbf{Q} dv = \int_S \mathbf{Q} \cdot \mathbf{ds}$$

and written as surface integrals as:

$$\delta F(\mathbf{A}) = \int_s \left(\hat{\mathbf{x}} \frac{\partial f}{\partial A_{xx}} \delta A_x + \hat{\mathbf{y}} \frac{\partial f}{\partial A_{xy}} \delta A_x + \hat{\mathbf{z}} \frac{\partial f}{\partial A_{xz}} \delta A_x \right) . \mathbf{ds}$$
$$+ \int_s \left(\hat{\mathbf{x}} \frac{\partial f}{\partial A_{yx}} \delta A_y + \hat{\mathbf{y}} \frac{\partial f}{\partial A_{yy}} \delta A_y + \hat{\mathbf{z}} \frac{\partial f}{\partial A_{yz}} \delta A_y \right) . \mathbf{ds}$$
$$+ \int_s \left(\hat{\mathbf{x}} \frac{\partial f}{\partial A_{zx}} \delta A_z + \hat{\mathbf{y}} \frac{\partial f}{\partial A_{zy}} \delta A_z + \hat{\mathbf{z}} \frac{\partial f}{\partial A_{zz}} \delta A_z \right) . \mathbf{ds} \tag{7.23}$$
$$+ \int_v \left(\frac{\partial f}{\partial A_x} \delta A_x - \frac{\partial}{\partial x}\left(\frac{\partial f}{\partial A_{xx}}\right) \delta A_x - \frac{\partial}{\partial y}\left(\frac{\partial f}{\partial A_{xy}}\right) \delta A_x - \frac{\partial}{\partial z}\left(\frac{\partial f}{\partial A_{xz}}\right) \delta A_x \right) dxdydz$$
$$+ \int_v \left(\frac{\partial f}{\partial A_y} \delta A_y - \frac{\partial}{\partial x}\left(\frac{\partial f}{\partial A_{yx}}\right) \delta A_y - \frac{\partial}{\partial y}\left(\frac{\partial f}{\partial A_{yy}}\right) \delta A_y - \frac{\partial}{\partial z}\left(\frac{\partial f}{\partial A_{yz}}\right) \delta A_y \right) dxdydz$$
$$+ \int_v \left(\frac{\partial f}{\partial A_z} \delta A_z - \frac{\partial}{\partial x}\left(\frac{\partial f}{\partial A_{zx}}\right) \delta A_z - \frac{\partial}{\partial y}\left(\frac{\partial f}{\partial A_{zy}}\right) \delta A_z - \frac{\partial}{\partial z}\left(\frac{\partial f}{\partial A_{zz}}\right) \delta A_z \right) dxdydz = 0$$

Since the variations δA_x, δA_y, and δA_z are arbitrary throughout the volume the last three integrals must vanish independently. The three Euler equations are therefore

$$\frac{\partial f}{\partial A_x} - \frac{\partial}{\partial x}\left(\frac{\partial f}{\partial A_{xx}}\right) - \frac{\partial}{\partial y}\left(\frac{\partial f}{\partial A_{xy}}\right) - \frac{\partial}{\partial z}\left(\frac{\partial f}{\partial A_{xz}}\right) = 0 \tag{7.24}$$

$$\frac{\partial f}{\partial A_y} - \frac{\partial}{\partial x}\left(\frac{\partial f}{\partial A_{yx}}\right) - \frac{\partial}{\partial y}\left(\frac{\partial f}{\partial A_{yy}}\right) - \frac{\partial}{\partial z}\left(\frac{\partial f}{\partial A_{yz}}\right) = 0 \tag{7.25}$$

$$\frac{\partial f}{\partial A_z} - \frac{\partial}{\partial x}\left(\frac{\partial f}{\partial A_{zx}}\right) - \frac{\partial}{\partial y}\left(\frac{\partial f}{\partial A_{zy}}\right) - \frac{\partial}{\partial z}\left(\frac{\partial f}{\partial A_{zz}}\right) = 0 \tag{7.26}$$

Similarly, the first three integrals must vanish independently. These determine the natural boundary conditions on the surface S:

$$\left(\hat{\mathbf{x}} \frac{\partial f}{\partial A_{xx}} + \hat{\mathbf{y}} \frac{\partial f}{\partial A_{xy}} + \hat{\mathbf{z}} \frac{\partial f}{\partial A_{xz}} \right) \cdot \mathbf{ds} = 0 \tag{7.27}$$

$$\left(\hat{\mathbf{x}}\frac{\partial f}{\partial A_{yx}}+\hat{\mathbf{y}}\frac{\partial f}{\partial A_{yy}}+\hat{\mathbf{z}}\frac{\partial f}{\partial A_{yz}}\right)\cdot \mathbf{ds}=0 \tag{7.28}$$

$$\left(\hat{\mathbf{x}}\frac{\partial f}{\partial A_{zx}}+\hat{\mathbf{y}}\frac{\partial f}{\partial A_{zy}}+\hat{\mathbf{z}}\frac{\partial f}{\partial A_{zz}}\right)\cdot \mathbf{ds}=0 \tag{7.29}$$

These last three relations are particularly important since they define the natural boundary conditions in the formulation. Thus, Neumann-type boundary conditions are satisfied implicitly in a variational formulation and need not be specified in a finite element formulation. Only Dirichlet boundary conditions are applied.

Now the terms of the Euler equations in Eqs. (7.24) through (7.26) are evaluated explicitly using Eq. (7.18). The various terms are:

$$\frac{\partial f}{\partial A_x}=-J_x+j\omega\sigma A_x \qquad \frac{\partial f}{\partial A_y}=-J_y+j\omega\sigma A_y \qquad \frac{\partial f}{\partial A_z}=-J_z+j\omega\sigma A_z \tag{7.30}$$

$$\frac{\partial f}{\partial A_{xx}}=0 \qquad \frac{\partial f}{\partial A_{yy}}=0 \qquad \frac{\partial f}{\partial A_{yz}}=0 \tag{7.31}$$

$$\frac{\partial f}{\partial A_{xy}}=-\nu_z B_z \qquad \frac{\partial f}{\partial A_{xz}}=\nu_y B_y \qquad \frac{\partial f}{\partial A_{yz}}=\nu_z B_z \tag{7.32}$$

$$\frac{\partial f}{\partial A_{yz}}=-\nu_x B_x \qquad \frac{\partial f}{\partial A_{zx}}=-\nu_y B_y \qquad \frac{\partial f}{\partial A_{zy}}=-\nu_z B_z \tag{7.33}$$

Substituting the appropriate terms into Eq. (7.24), this equation becomes

$$J_x-j\omega\sigma A_x-\frac{\partial}{\partial y}(-\nu_z B_z)-\frac{\partial}{\partial z}(\nu_y B_y)=0$$

or, reordering and writing **B** in terms of **A**

$$\frac{\partial}{\partial y}(\nu_z(A_{yx}-A_{xy}))-\frac{\partial}{\partial z}(\nu_y(A_{xz}-A_{zx}))=J_x-j\omega\sigma A_x \tag{7.34}$$

Similarly, by substituting the appropriate terms into Eq. (7.25) and (7.26) we get

$$\frac{\partial}{\partial z}(\nu_x(A_{zy}-A_{yz}))-\frac{\partial}{\partial x}(\nu_z(A_{yx}-A_{xy}))=J_y-j\omega\sigma A_y \tag{7.35}$$

$$\frac{\partial}{\partial x}(\nu_y(A_{xz}-A_{zx}))-\frac{\partial}{\partial y}(\nu_x(A_{zy}-A_{yz}))=J_z-j\omega\sigma A_z \tag{7.36}$$

and, since Equations (7.34), (7.35) and (7.36) are identical with the original partial differential equations in Eqs. (7.7), finding a stationary point of the functional implicitly satisfies the field equation in the volume considered.

Turning back to Equations (7.27) through (7.29) the natural boundary conditions are determined. Along Dirichlet portions of the boundary s, the variations δA_x, δA_y and δA_z in Eq. (7.23) are by definition zero. The surface integrals in Eq. (7.23) are therefore automatically satisfied. Equations (7.27), (7.28) and (7.29) define the Neumann-type boundary conditions. Without loss of generality one can take the normal to the surface in the x direction, i.e. $\mathbf{ds}=\hat{\mathbf{x}}|ds|$. As mentioned above, the variations δA_x, δA_y and δA_z are arbitrary and therefore the three integrands in Eq. (7.23) must vanish independently. For this condition the following must be satisfied

$$\frac{\partial f}{\partial A_{xx}} = 0 \qquad \textit{from Eq. (7.27)} \tag{7.37}$$

$$\frac{\partial f}{\partial A_{yx}} = 0 \qquad \textit{from Eq. (7.28)} \tag{7.38}$$

$$\frac{\partial f}{\partial A_{zx}} = 0 \qquad \textit{from Eq. (7.29)} \tag{7.39}$$

Equation (7.37) is identically zero from the relations in Eq. (7.31). From Eq. (7.33) it follows that Eq. (7.38) must satisfy the following:

$$\frac{\partial f}{\partial A_{yx}} = \nu_z B_z = 0$$

Similarly from Eq. (7.33), equation (7.39) becomes

$$\frac{\partial f}{\partial A_{zx}} = -\,\nu_x B_y = 0$$

Thus to satisfy these equations B_y and B_z must be zero along the boundary s. The natural (Neumann-type) boundary conditions are therefore

$$B_{tangential} = 0 \tag{7.40}$$

The formulation, while lengthy, is rather simple. It consists of finding a functional and showing that this functional is a correct representation of the field equations. The fact that Neumann boundary conditions are natural in this formulation is particularly useful since it implies that symmetry boundary conditions do not need to be explicitly entered in the solution process.

The same formulation can be used for magnetostatic applications by removing the eddy current terms.

There is however, one point that we have not addressed: can we always find a functional for a given problem and if not, what should be done in cases for which a functional cannot be found? The answer to this question is rather simple. While many applications do have an equivalent energy functional and therefore can be formulated using the method above, some do not. These can be formulated using Galerkin's method. We demonstrate Galerkin's method in section 7.4 for a rather simple eddy current problem which does not have an energy functional.

7.3.3 The finite element

The results of the previous section have nothing to do with finite elements or, for that matter, with any discretization of the continuum. The discretization is introduced now and by doing so, the variational method is combined with the finite element method to provide a finite element solution to the problem.

At least in principle, we can choose any finite element. The formulation above is three-dimensional, therefore, we must choose a three-dimensional element. Since only first order derivatives are necessary, first order elements are sufficient. For example, the tetrahedral element in section 4.4.2 or the hexahedral element in section 4.4.3 can be used here. We first assume a generic finite element without reference to a particular finite element shape and then discuss the formulation in terms of the eight-node hexahedral element.

For the variational expressions in the previous section to be equivalent to the original equations, a stationary point of the functional must be found. This is simply done by finding the derivatives of the functional with respect to the unknowns and setting these to zero throughout the solution domain

$$\frac{\partial F(\mathbf{A})}{\partial A_{ki}} = 0, \qquad i=1,2,3.....,N; \qquad k=x,y,z \tag{7.41}$$

where N is the total number of nodes in the solution region. Instead of performing this process over the entire region it is convenient to do it element by element and then sum the contribution of individual elements to obtain $3 \times N$ simultaneous linear algebraic equations in $3 \times N$ unknown components of the magnetic vector potential in the entire solution region. In this way a repeatable process is performed on each element, one which is well adapted for automatic assembly on a computer. The functional is differentiated with respect to each unknown A_x, A_y, and A_z at each node of the element and equating each individual differential to zero yields a set of $3 \times M$ simultaneous equations in $3 \times M$ unknowns. Although the term 'equations' is used here these are not necessarily the equations to be solved since an equation is formed by one or more elements and in general the 'equations' formed in any particular element are not complete. In a linear hexahedral element there are eight nodes, thus the element contributes to *24* equations. The size of the elemental system of equations is *24×24*. In a second order element there are *20* nodes (*60* unknowns) and therefore the size of the elemental matrix is *60×60* but the process is identical in every other respect.

Applying Eq. (7.41), the derivatives of the functional are of the form

$$\begin{aligned}\frac{\partial F(A)}{\partial A_{ki}} = \int_v \Bigg[& \frac{\partial A_z}{\partial y}\frac{\partial}{\partial A_{ki}}\frac{\partial A_z}{\partial y} + \frac{\partial A_y}{\partial z}\frac{\partial}{\partial A_{ki}}\frac{\partial A_y}{\partial z} + \frac{\partial A_x}{\partial z}\frac{\partial}{\partial A_{ki}}\frac{\partial A_x}{\partial z} \\ & + \frac{\partial A_z}{\partial x}\frac{\partial}{\partial A_{ki}}\frac{\partial A_z}{\partial x} + \frac{\partial A_x}{\partial y}\frac{\partial}{\partial A_{ki}}\frac{\partial A_x}{\partial y} + \frac{\partial A_y}{\partial x}\frac{\partial}{\partial A_{ki}}\frac{\partial A_y}{\partial x} \\ & - \frac{\partial}{\partial A_{ki}}\left(\frac{\partial A_z}{\partial y}.\frac{\partial A_y}{\partial z}\right) - \frac{\partial}{\partial A_{ki}}\left(\frac{\partial A_x}{\partial z}.\frac{\partial A_z}{\partial x}\right) - \frac{\partial}{\partial A_{ki}}\left(\frac{\partial A_y}{\partial x}.\frac{\partial A_x}{\partial y}\right) \\ & + j\omega\sigma A_x \frac{\partial A_x}{\partial A_{ki}} + j\omega\sigma A_y \frac{\partial A_y}{\partial A_{ki}} + j\omega\sigma A_z \frac{\partial A_z}{\partial A_{ki}} \\ & - J_x \frac{\partial A_x}{\partial A_{ki}} - J_y \frac{\partial A_y}{\partial A_{ki}} - J_z \frac{\partial A_z}{\partial A_{ki}} \Bigg] dxdydz = 0\end{aligned}$$

where $i=1,2,3......M$, $k=x,y,z$. For the linear hexahedral element, $M=8$, for the second order hexahedral element, $M=20$, and for the linear tetrahedral element, $M=4$.

Calculating the derivative above and substituting the approximation of the magnetic vector potential in terms of the shape functions in Table 4.1 (see Eq. (4.4)), the following sub-matrix results from each node of an element

$$S_{3i-2,3j-2} = \int_{v_i} \left(\frac{\partial N_i}{\partial y} v_z \frac{\partial N_j}{\partial y} + \frac{\partial N_i}{\partial z} v_y \frac{\partial N_j}{\partial z} \right) dxdydz \tag{7.42}$$

$$S_{3i-1,3j-1} = \int_{v_i} \left(\frac{\partial N_i}{\partial z} v_x \frac{\partial N_j}{\partial z} + \frac{\partial N_i}{\partial x} v_z \frac{\partial N_j}{\partial x} \right) dxdydz \tag{7.43}$$

$$S_{3i,3j} = \int_{v_i} \left(\frac{\partial N_i}{\partial y} v_x \frac{\partial N_j}{\partial y} + \frac{\partial N_i}{\partial x} v_y \frac{\partial N_j}{\partial x} \right) dxdydz \tag{7.44}$$

$$S_{3i-1,3j-2} = -\int_{v_i} \frac{\partial N_i}{\partial x} v_z \frac{\partial N_j}{\partial y} dxdydz \tag{7.45}$$

$$S_{3i,3j-1} = -\int_{v_i} \frac{\partial N_i}{\partial y} \nu_x \frac{\partial N_j}{\partial z} dxdydz \qquad (7.46)$$

$$S_{3i,3j-2} = -\int_{v_i} \frac{\partial N_i}{\partial x} \nu_y \frac{\partial N_j}{\partial z} dxdydz \qquad (7.47)$$

$$R_{3i-2,3j-2} = R_{3i-1,3j-1} = R_{3i,3j} = \int_{v_i} N_i N_j \, dxdydz \qquad (7.48)$$

$$Q_{3i-2} = \int_{v_i} J_x N_i \, dxdydz \qquad (7.49)$$

$$Q_{3i-1} = \int_{v_i} J_y N_i \, dxdydz \qquad (7.50)$$

$$Q_{3i} = \int_{v_i} J_z N_i \, dxdydz \qquad (7.51)$$

where the indices i, j, vary from 1 to M and all coefficients are symmetric about the diagonal.

In this formulation, reluctivities are allowed to vary spatially while the conductivity is assumed to be single valued in each element. Conductivity can also have spatial variation by introducing its components in the imaginary coefficients instead of multiplying the imaginary part of the matrix but, in practice, this is almost never necessary.

Each coefficient (Eqs. (7.42) through (7.51)) is either numerically integrated using Gaussian quadrature (for hexahedral elements, or for any other elements defined as a parametric element, see chapter 11) or, if possible, analytically integrated and summed into a complex elemental matrix of the form

$$\{[S] + j\omega\sigma[R]\}\{A\} = \{Q\} \qquad (7.52)$$

where $[S]$ is the real part of the 24×24 elemental matrix, $[R]$ is the imaginary part and $\{A\}$ and $\{Q\}$ are the 24×1 unknown and source vectors respectively.

The elemental matrices are summed over the total number of elements to provide a global system of $3\times N$ simultaneous linear algebraic equations in $3\times N$ unknown components of the magnetic vector potential, where N is the total number of nodes in the solution region. Assembly of the elemental matrices into the global matrix is similar to that shown in section 4.5.5 except that here each node represents a 3×3 space in the global matrix.

The elemental and global matrices above are written in complex notation for the general eddy current problem. For magnetostatic applications only the real part of the matrices is assembled and the vectors {A} and {Q} become real. In this case the matrix equation becomes

$$[S]\{A\} = \{Q\} \tag{7.53}$$

The system of equations in Eq. (7.52) or Eq. (7.53) can now be solved by any standard solution technique such as Gauss elimination, or the conjugate gradient method.

This completes the finite element formulation but the following observations are in order before we discuss Galerkin's method:

1. The energy functional is almost always "guessed" and then proven to be correct by use of Euler's equation. The functional is either a statement of energy in the system or related to energy as shown above.
2. Neumann boundary conditions are natural and need not be specified.
3. The stationary point of the functional is found by setting the first derivative of the functional with respect to each unknown to zero.
4. The system of equations wc obtain is symmetric.

7.4 FORMULATION USING GALERKIN'S METHOD

Formulation based on Galerkin's procedure is rather straight-forward and requires substitution of the finite element approximation in the field equation to obtain a residual. A set of weighting functions is defined as the derivative of the approximating function with respect to the unknown variables and the resulting weighted residual integral is set to zero. This method is introduced in section 4.2. We repeat here its basic elements:

1. Define an approximation $A(x,y,z)=\Sigma f(x,y,z)A_i$
2. Find the residual by substituting the approximation into the field equation.
3. A set of weighting functions is now defined:

$$W(x,y,z) = \frac{\partial f(x,y,z)}{\partial K} \tag{7.54}$$

where K represents the unknowns in the solution domain.

The following integral is defined:

$$\int_v W(x,y,z)R(x,y,z)dv = 0 \tag{7.55}$$

This is a general approach using weighted residuals. The unique aspect of Galerkin's method is the choice of the weighting functions as the derivatives with respect to the unknowns of the approximating functions.

7.4.1 The field equations

To see how this formulation is implemented, the eddy current equation in the previous example is used again but, we also assume that the source current density moves relative to the sample. This is a realistic situation in eddy current testing and in testing with the remote field effect method where coils move at relatively high velocities.

In considering the inclusion of velocity terms in the eddy current equations, we choose a slightly different method of formulation. Unlike the previous case, we assume here that the electric scalar potential cannot be neglected. Inclusion of the scalar potential is, in general, nonzero and in this case, it is important to include it in the formulation to minimize the sources of error in solution. Thus, the general electric field intensity can be written as

$$\mathbf{E} = \mathbf{u}\times\nabla\times\mathbf{A} - j\omega\mathbf{A} - \nabla V \tag{7.56}$$

where $\mathbf{u}$ is the velocity vector. The first term in Eq. (7.56) is the electric field due to velocity, the second is the eddy current term while the third is due to the gradient of the electric scalar potential. Therefore, Maxwell's second equation can be written as

$$\nabla\times\frac{1}{\mu}\nabla\times\mathbf{A} = J_s + \sigma\mathbf{u}\times\nabla\times\mathbf{A} - j\omega\sigma\mathbf{A} - \sigma\nabla V \tag{7.57}$$

where the displacement current term is neglected and the magnetic field intensity $\mathbf{H}$ is written in terms of the magnetic vector potential $\mathbf{A}$ using Eq. (1.63). By assuming Coulomb's gage ($\nabla\cdot\mathbf{A}=0$), the equation is modified as

$$\nabla\times\frac{1}{\mu}\nabla\times\mathbf{A} - \nabla\frac{1}{\mu}\nabla\cdot\mathbf{A} = J_s + \sigma\mathbf{u}\times\nabla\times\mathbf{A} - j\omega\sigma\mathbf{A} - \sigma\nabla V \tag{7.58}$$

While this modification may seem arbitrary, inspection of Eq. (7.58) shows that by doing so, the divergence term in Eq. (7.57) disappears and the left hand side of Eq. (7.58) is identical to the Laplacian. Eq. (7.58) is now our governing equation but, since both the magnetic vector potential and the electric scalar potential are present, we must find a second equation to be able to solve for the two variables. The equation we choose is $\nabla \cdot \mathbf{J} = 0$. However, since the source current density is constant, it is sufficient to write

$$\nabla \cdot (\sigma \mathbf{u} \times \nabla \times \mathbf{A} - j\omega\sigma\mathbf{A} - \sigma\nabla V) = 0 \tag{7.59}$$

We view this equation as a constraint equation since it constrains the eddy currents through the divergence relation. Thus, the governing equations are Eqs. (7.58) and (7.59).

7.4.2 Approximating and weighting functions

To continue we must define the approximation over a finite element. The general approximation for a finite element is

$$A = \sum_{i=1}^{M} N_i A_i \tag{7.60}$$

for the vector potential and

$$V = \sum_{i=1}^{M} N_i V_i \tag{7.61}$$

for the scalar potential. Since the vector potential has three components, each approximated by an equation as in Eq. (7.60), for economy in derivation we use the following short form notation:

$$W = N \qquad \textit{and} \qquad \overline{W} = \overline{N}$$

where W are the weighting functions and are chosen to be equal to the shape functions N. The bar indicates the following form

$$\overline{W} = \overline{N} = \begin{bmatrix} N & 0 & 0 \\ 0 & N & 0 \\ 0 & 0 & N \end{bmatrix}$$

This notation indicates that a shape function is associated with a node. The same shape function is used for each degree of freedom at the same node.

Now we substitute these into Eqs. (7.58) and (7.59) and integrate over the volume of the element

$$\int_v \left\{ \nabla\times\overline{\mathbf{W}}\cdot\frac{1}{\mu}\nabla\times\mathbf{A} - \nabla\cdot\overline{\mathbf{W}}\frac{1}{\mu}\nabla\cdot A - \overline{\mathbf{W}}\cdot\sigma(\mathbf{u}\times\nabla\times\mathbf{A}) + j\omega\sigma\overline{\mathbf{W}}\cdot\mathbf{A} + \sigma\overline{\mathbf{W}}\cdot\nabla V \right\} dv$$
$$= \int_v \overline{\mathbf{W}}\cdot\mathbf{J}dv \tag{7.62}$$

$$\int_v \{ \nabla W\cdot(\sigma\mathbf{u}\times\nabla\times\mathbf{A}) - j\omega\sigma\nabla W\cdot\mathbf{A} - \sigma\nabla W\cdot\nabla V \} dv = 0 \tag{7.63}$$

or in matrix form

$$\begin{bmatrix} a_{11} & a_{12} \\ a_{21} & a_{22} \end{bmatrix} \begin{Bmatrix} \mathbf{A} \\ V \end{Bmatrix} = \begin{Bmatrix} Q \\ 0 \end{Bmatrix} \tag{7.64}$$

where $\mathbf{A}$ is a vector (three components). The terms of the matrix are found by replacing $\mathbf{A}$ and V by the approximations in Eqs. (7.60) and (7.61) as:

$$a_{11} = \int_v \left\{ (\nabla\times\overline{\mathbf{W}})^T\cdot\frac{1}{\mu}\nabla\times\mathrm{N} - (\nabla\cdot\overline{\mathbf{W}})^T\frac{1}{\mu}\nabla\cdot\overline{\mathrm{N}} - \overline{\mathbf{W}}^T\cdot\sigma(\mathbf{u}\times\nabla\times\overline{\mathrm{N}}) \right\} dv$$
$$+ \int_v \left\{ j\omega\sigma\overline{\mathbf{W}}^T\cdot\overline{\mathrm{N}} \right\} dv \tag{7.65}$$

$$a_{12} = \int_v \{ \sigma\overline{\mathbf{W}}^T\cdot\nabla\mathrm{N} \} dv \tag{7.66}$$

$$a_{21} = \int_v \{ (\nabla W)^T\cdot(\sigma\mathbf{u}\times\nabla\times\overline{\mathrm{N}}) \} dv + \int \{ j\omega\sigma(\nabla\mathrm{W})^T\cdot\overline{\mathrm{N}} \} dv \tag{7.67}$$

$$a_{22} = \int_v \sigma(\nabla W)^T\cdot\nabla N dv \tag{7.68}$$

We note however that a_{11} is a 3×3 matrix, a_{12} is a 1×3 row vector and a_{21} is a 3×1 column vector. Similarly, $\{Q\}$ is a 3×1 column vector of (in this case) currents. Also, we can write:

$$\mathbf{u}\times\nabla\times\mathbf{A} = \hat{\mathbf{x}}\left[u_y\left(\frac{\partial A_y}{\partial x}-\frac{\partial A_x}{\partial y}\right)-u_z\left(\frac{\partial A_x}{\partial z}-\frac{\partial A_z}{\partial x}\right)\right]+$$
$$+\hat{\mathbf{y}}\left[u_z\left(\frac{\partial A_z}{\partial y}-\frac{\partial A_y}{\partial z}\right)-u_x\left(\frac{\partial A_y}{\partial x}-\frac{\partial A_x}{\partial y}\right)\right]+$$
$$+\hat{\mathbf{z}}\left[u_x\left(\frac{\partial A_x}{\partial z}-\frac{\partial A_z}{\partial x}\right)-u_y\left(\frac{\partial A_z}{\partial y}-\frac{\partial A_y}{\partial z}\right)\right]$$

Evaluating now the various terms in the matrix $[a]$ in Eq. (7.64) we write first

$$(\nabla\times\overline{\mathrm{W}})^T\frac{1}{\mu}\nabla\times\overline{\mathrm{N}} - (\nabla\cdot\overline{\mathrm{W}})^T\frac{1}{\mu}\nabla\cdot\overline{\mathrm{N}} - \overline{\mathrm{W}}^T\cdot\sigma(\mathbf{u}\times\nabla\times\overline{\mathrm{N}}) = \begin{bmatrix} b_{11} & b_{12} & b_{13} \\ b_{21} & b_{22} & b_{23} \\ b_{31} & b_{32} & b_{33} \end{bmatrix}$$

where the coefficients are:

$$b_{11} = \frac{\partial W^T}{\partial x}\frac{1}{\mu_x}\frac{\partial N}{\partial x} + \frac{\partial W^T}{\partial y}\frac{1}{\mu_z}\frac{\partial N}{\partial y} + \frac{\partial W^T}{\partial z}\frac{1}{\mu_y}\frac{\partial N}{\partial z} + \sigma_x u_y W^T\frac{\partial N}{\partial y} + \sigma_x u_z W^T\frac{\partial N}{\partial z}$$

$$b_{22} = \frac{\partial W^T}{\partial x}\frac{1}{\mu_z}\frac{\partial N}{\partial x} + \frac{\partial W^T}{\partial y}\frac{1}{\mu_y}\frac{\partial N}{\partial y} + \frac{\partial W^T}{\partial z}\frac{1}{\mu_x}\frac{\partial N}{\partial z} + \sigma_y u_x W^T\frac{\partial N}{\partial x} + \sigma_y u_z W^T\frac{\partial N}{\partial z}$$

$$b_{33} = \frac{\partial W^T}{\partial x}\frac{1}{\mu_y}\frac{\partial N}{\partial x} + \frac{\partial W^T}{\partial y}\frac{1}{\mu_x}\frac{\partial N}{\partial y} + \frac{\partial W^T}{\partial z}\frac{1}{\mu_z}\frac{\partial N}{\partial z} + \sigma_z u_y W^T\frac{\partial N}{\partial y} + \sigma_z u_x W^T\frac{\partial N}{\partial x}$$

$$b_{12} = \frac{\partial W^T}{\partial x}\frac{1}{\mu_y}\frac{\partial N}{\partial y} - \frac{\partial W^T}{\partial y}\frac{1}{\mu_z}\frac{\partial N}{\partial x} - \sigma_x u_y W^T\frac{\partial N}{\partial x}$$

$$b_{13} = \frac{\partial W^T}{\partial x}\frac{1}{\mu_z}\frac{\partial N}{\partial z} - \frac{\partial W^T}{\partial z}\frac{1}{\mu_y}\frac{\partial N}{\partial x} - \sigma_x u_z W^T\frac{\partial N}{\partial x}$$

$$b_{21} = \frac{\partial W^T}{\partial y}\frac{1}{\mu_x}\frac{\partial N}{\partial x} - \frac{\partial W^T}{\partial x}\frac{1}{\mu_z}\frac{\partial N}{\partial y} - \sigma_y u_x W^T\frac{\partial N}{\partial y}$$

$$b_{23} = \frac{\partial W^T}{\partial y}\frac{1}{\mu_z}\frac{\partial N}{\partial z} - \frac{\partial W^T}{\partial z}\frac{1}{\mu_x}\frac{\partial N}{\partial y} - \sigma_y u_z W^T\frac{\partial N}{\partial y}$$

$$b_{31} = \frac{\partial W^T}{\partial z}\frac{1}{\mu_x}\frac{\partial N}{\partial x} - \frac{\partial W^T}{\partial x}\frac{1}{\mu_y}\frac{\partial N}{\partial z} - \sigma_z u_x W^T\frac{\partial N}{\partial z}$$

$$b_{32} = \frac{\partial W^T}{\partial z}\frac{1}{\mu_y}\frac{\partial N}{\partial y} - \frac{\partial W^T}{\partial y}\frac{1}{\mu_x}\frac{\partial N}{\partial z} - \sigma_z u_y W^T \frac{\partial N}{\partial z}$$

$$j\omega\sigma \overline{W}^T \cdot \overline{\mathrm{N}} = j\omega\sigma \begin{bmatrix} \overline{W}^T N & 0 & 0 \\ 0 & \overline{W}^T N & 0 \\ 0 & 0 & \overline{W}^T N \end{bmatrix} = j\omega\sigma[c]$$

$$\sigma \overline{W}^T \cdot \nabla N = \sigma \begin{bmatrix} \overline{W}^T \frac{\partial N}{\partial x} \\ \overline{W}^T \frac{\partial N}{\partial y} \\ \overline{W}^T \frac{\partial N}{\partial z} \end{bmatrix} = [d]$$

$$j\omega\sigma(\nabla \overline{W}^T) \cdot \overline{\mathrm{N}} = j\omega\sigma \begin{bmatrix} \frac{\partial \overline{W}^T}{\partial x} & \frac{\partial \overline{W}^T}{\partial y} & \frac{\partial \overline{W}^T}{\partial z} \end{bmatrix} = [e]$$

$$\sigma(\nabla \overline{W}^T) \cdot \nabla N = \sigma \begin{bmatrix} \frac{\partial \overline{W}^T}{\partial x}\frac{\partial N}{\partial x} & \frac{\partial \overline{W}^T}{\partial y}\frac{\partial N}{\partial y} & \frac{\partial \overline{W}^T}{\partial z}\frac{\partial N}{\partial z} \end{bmatrix} = [f]$$

$$-(\nabla \overline{W}^T) \cdot \sigma (\mathbf{u} \times \nabla \times \overline{\mathrm{N}})$$

$$= \begin{bmatrix} \sigma_x u_y \frac{\partial \overline{W}^T}{\partial x}\frac{\partial N}{\partial y} + \sigma_x u_z \frac{\partial \overline{W}^T}{\partial x}\frac{\partial N}{\partial z} - \sigma_y u_x \frac{\partial \overline{W}^T}{\partial y}\frac{\partial N}{\partial y} - \sigma_z u_x \frac{\partial \overline{W}^T}{\partial z}\frac{\partial N}{\partial z} \\ \sigma_y u_x \frac{\partial \overline{W}^T}{\partial y}\frac{\partial N}{\partial x} + \sigma_y u_z \frac{\partial \overline{W}^T}{\partial y}\frac{\partial N}{\partial z} - \sigma_x u_y \frac{\partial \overline{W}^T}{\partial x}\frac{\partial N}{\partial x} - \sigma_z u_y \frac{\partial \overline{W}^T}{\partial z}\frac{\partial N}{\partial z} \\ \sigma_z u_y \frac{\partial \overline{W}^T}{\partial z}\frac{\partial N}{\partial y} + \sigma_z u_x \frac{\partial \overline{W}^T}{\partial z}\frac{\partial N}{\partial x} - \sigma_x u_z \frac{\partial \overline{W}^T}{\partial x}\frac{\partial N}{\partial x} - \sigma_y u_z \frac{\partial \overline{W}^T}{\partial y}\frac{\partial N}{\partial y} \end{bmatrix}^T = [g]$$

Now we can write the coefficients of the matrix $[a]$ as

$$[a_{11}] = \int_{v_i} [b]dv + \int_{v_i} [c]dv \tag{7.69}$$

$$[a_{12}] = \int_{v_i} [c]dv \tag{7.70}$$

$$[a_{21}] = \int_{v_i} [g]dv + \int_{v_i} [e]dv \tag{7.71}$$

and

$$[a_{22}] = \int_{v_i} [f]dv \tag{7.72}$$

The vector $\{Q\}$ is identical to that given in Eqs (7.49) through (7.51).

The coefficients of $[a]$ are similar to those of the matrix in Eq. (7.52) except for the obvious differences in that velocity coefficients are added and the matrix $[a]$ is nonsymmetric. If we were to set velocities to zero we would obtain exactly the result in Eq. (7.52). The system to solve is therefore that of Eq. (7.64]. This system of equations is nonsymmetric but otherwise is similar to other finite element matrices, including the normally banded form associated with finite element analysis. The velocity terms make this system a nonsymmetric system of equations. Without velocity, the system becomes symmetric again. The terms of the system are numerically or analytically integrated (similar to the previous section) for each element and introduced in the global system of equations in Eq. (7.64).

The system in Eq. (7.64) can also be solved by standard methods although with considerable more effort and computer resources. Some methods of solution such as the conjugate gradient methods cannot be used directly but, as a rule, the system above can be solved relatively easily.

In summary, Galerkin's method can be used to formulate any equation but it is more often used when variational methods cannot be applied. One example to this type of equation is the eddy current equation with velocity effects. This particular equation is typical of the types of equations that cannot be formulated using variational methods.

7.5 EXAMPLES: STATIC APPLICATIONS

7.5.1 Electrostatic formulations

The finite element formulation for electrostatic applications follows the same principles as in section 7.3 or 7.4. Normally, the variational approach is adopted since the functional is clearly a statement of potential energy in the system.

Two examples of classes of problems are given here: one is a problem with charges as sources and is typical of solution of Poisson's equations in terms of the electric scalar potential. The second is a problem defined in terms of boundary potentials and is typical of solution of Laplace's equation. In both cases the electric scalar potential is used but in the second application, the current density in the solution domain is required. The latter is also used in the following section to calculate the magnetic flux density due to a defect, using the magnetostatic formulation.

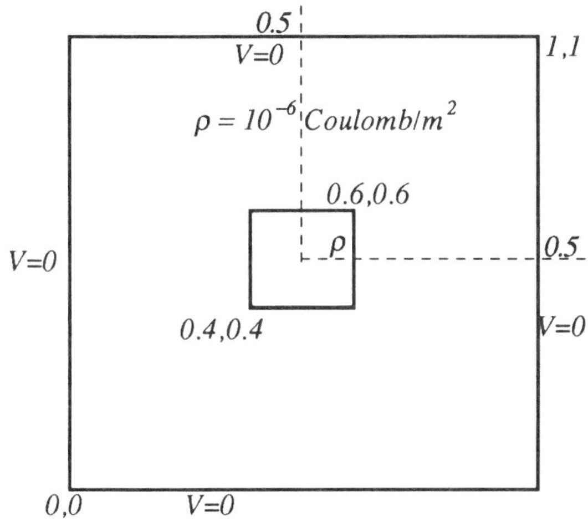

Figure 7.1 An electrostatic geometry. The geometry modeled is shown by the dashed liens (one quarter of the geometry).

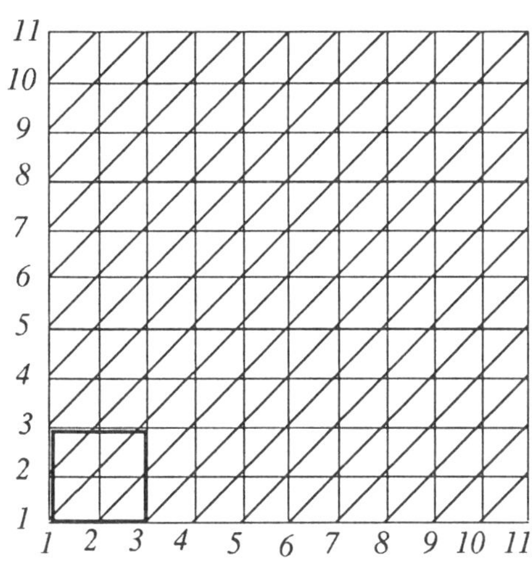

Figure 7.2 A uniform finite element mesh over the geometry in Figure 7.1. Numbers indicate point numbers on each side.

7.5.2 Poisson's equation: the electric scalar potential

The problem considered here is shown in Figure 7.1. Boundary conditions are also shown in the figure but, because of symmetry in geometry, the solution domain adopted is that shown in Figure 7.2. Thus, only a quarter of the geometry is shown. The left and bottom boundaries have Neumann boundary conditions which are not enforced. Only the Dirichlet boundary conditions on the right and top boundaries are enforced. A charge density is distributed on a square area at the center as shown. The electric potential distribution is required.

The equation to solve is:

$$\nabla^2 V = \frac{\partial^2 V}{\partial x^2} + \frac{\partial^2 V}{\partial y^2} = -\frac{\rho}{\varepsilon} \tag{7.73}$$

This is formulated using the variational approach of section 7.3. An energy functional can be written as

$$F(V) = \int_v \left[\frac{1}{2} \left\{ \left(\varepsilon \frac{\partial V}{\partial x} \right)^2 + \varepsilon \left(\frac{\partial V}{\partial y} \right)^2 \right\} - \rho V \right] dxdy = 0 \tag{7.74}$$

A finite element must now be chosen for which, an elemental matrix is defined as

$$[s]_e \{V\} = \{q\}_e \tag{7.75}$$

where the terms of the matrix $[s]$ and $[q]$ are

$$s_{ij} = \int_{v_i} \left(\frac{\partial N_i}{\partial x} \varepsilon \frac{\partial N_j}{\partial x} + \frac{\partial N_i}{\partial y} \varepsilon \frac{\partial N_j}{\partial y} \right) dxdy\,, \qquad i, j = 1,2,\ldots.M$$

$$q_i = -\int_{v_i} \rho N_i dxdy, \qquad i, j = 1,2,\ldots.M$$

and where the integration is over the volume of the element (element thickness is equal to unity), M is the number of nodes of the finite element and N_i are the shape functions for the element chosen.

To solve this particular geometry we choose two-dimensional triangular elements. The shape functions for the triangular elements are given as

$$N_1 = \frac{1}{2\Delta} \left[(x_2 y_3 - x_3 y_2) + (y_2 - y_3)x + (x_3 - x_2)y \right] \tag{7.76}$$

$$N_2 = \frac{1}{2\Delta}\left[(x_3y_1 - x_1y_3) + (y_3 - y_1)x + (x_1 - x_3)y\right] \tag{7.77}$$

$$N_3 = \frac{1}{2\Delta}\left[(x_1y_2 - x_2y_1) + (y_1 - y_2)x + (x_2 - x_1)y\right] \tag{7.78}$$

and the derivatives of the shape functions are:

$$\frac{\partial N_1}{\partial x} = \frac{y_2 - y_3}{2\Delta} \qquad \frac{\partial N_1}{\partial y} = \frac{x_3 - x_2}{2\Delta} \tag{7.79}$$

$$\frac{\partial N_2}{\partial x} = \frac{y_3 - y_1}{2\Delta} \qquad \frac{\partial N_2}{\partial y} = \frac{x_1 - x_3}{2\Delta} \tag{7.80}$$

$$\frac{\partial N_3}{\partial x} = \frac{y_1 - y_2}{2\Delta} \qquad \frac{\partial N_2}{\partial y} = \frac{x_2 - x_1}{2\Delta} \tag{7.81}$$

where Δ is the area of the triangular element. These are derived in detail in section 4.4.1 but we list them again to show how integration is carried out. Substituting the terms for the partial derivatives in the general term of the matrix $[s]$, we get

$$[s] = \int_{V_i} \begin{bmatrix} \frac{\partial N_1}{\partial x}\varepsilon\frac{\partial N_1}{\partial x} + \frac{\partial N_1}{\partial y}\varepsilon\frac{\partial N_1}{\partial y} & \frac{\partial N_1}{\partial x}\varepsilon\frac{\partial N_2}{\partial x} + \frac{\partial N_1}{\partial y}\varepsilon\frac{\partial N_2}{\partial y} & \frac{\partial N_1}{\partial x}\varepsilon\frac{\partial N_3}{\partial x} + \frac{\partial N_1}{\partial y}\varepsilon\frac{\partial N_3}{\partial y} \\ \frac{\partial N_2}{\partial x}\varepsilon\frac{\partial N_1}{\partial x} + \frac{\partial N_2}{\partial y}\varepsilon\frac{\partial N_1}{\partial y} & \frac{\partial N_2}{\partial x}\varepsilon\frac{\partial N_2}{\partial x} + \frac{\partial N_2}{\partial y}\varepsilon\frac{\partial N_2}{\partial y} & \frac{\partial N_2}{\partial x}\varepsilon\frac{\partial N_3}{\partial x} + \frac{\partial N_2}{\partial y}\varepsilon\frac{\partial N_3}{\partial y} \\ \frac{\partial N_3}{\partial x}\varepsilon\frac{\partial N_1}{\partial x} + \frac{\partial N_3}{\partial y}\varepsilon\frac{\partial N_1}{\partial y} & \frac{\partial N_3}{\partial x}\varepsilon\frac{\partial N_2}{\partial x} + \frac{\partial N_3}{\partial y}\varepsilon\frac{\partial N_2}{\partial y} & \frac{\partial N_3}{\partial x}\varepsilon\frac{\partial N_3}{\partial x} + \frac{\partial N_3}{\partial y}\varepsilon\frac{\partial N_3}{\partial y} \end{bmatrix} dv$$

$$q_i = \int_{v_i} \begin{bmatrix} \rho N_1 \\ \rho N_2 \\ \rho N_3 \end{bmatrix} dv$$

With the values above, multiplying both sides of the equation by 2Δ and, since ε is constant throughout the solution domain, dividing both sides by ε, we get

$$[s] = \int_{v_i} \begin{bmatrix} (y_3-y_2)(y_3-y_2)+(x_3-x_2)(x_3-x_2) & (y_3-y_2)(y_3-y_1)+(x_3-x_2)(x_1-x_3) & (y_3-y_2)(y_1-y_2)+(x_3-x_2)(x_2-x_1) \\ (y_3-y_1)(y_3-y_2)+(x_1-x_3)(x_3-x_2) & (y_3-y_1)(y_3-y_1)+(x_1-x_3)(x_1-x_3) & (y_3-y_1)(y_1-y_2)+(x_1-x_3)(x_2-x_1) \\ (y_1-y_2)(y_3-y_2)+(x_2-x_1)(x_3-x_2) & (y_1-y_2)(y_3-y_1)+(x_2-x_1)(x_1-x_3) & (y_1-y_2)(y_1-y_2)+(x_2-x_1)(x_2-x_1) \end{bmatrix} dv$$

$$q_i = \int_{v_i} \begin{bmatrix} \dfrac{\rho}{\varepsilon} 2\Delta\left[(x_2y_3 - x_3y_2) + (y_2 - y_3)x + (x_3 - x_2)y\right] \\ \dfrac{\rho}{\varepsilon} 2\Delta\left[(x_3y_1 - x_1y_3) + (y_3 - y_1)x + (x_1 - x_3)y\right] \\ \dfrac{\rho}{\varepsilon} 2\Delta\left[(x_1y_2 - x_2y_1) + (y_1 - y_2)x + (x_2 - x_1)y\right] \end{bmatrix} dv$$

The integration of these terms is straightforward since, for example, none of the terms in $[s]$ depends on x or y. Therefore, integration over the element simply implies multiplying the terms of $[s]$ by the volume of the element or, in this case, by 2Δ. After this is performed, the global system of equations is assembled. The elemental system of equations after integration has the following form

$$[s] = \frac{1}{4\Delta} \begin{bmatrix} (y_3-y_2)(y_3-y_2)+(x_3-x_2)(x_3-x_2) & (y_3-y_2)(y_3-y_1)+(x_3-x_2)(x_1-x_3) & (y_3-y_2)(y_1-y_2)+(x_3-x_2)(x_2-x_1) \\ (y_3-y_1)(y_3-y_2)+(x_1-x_3)(x_3-x_2) & (y_3-y_1)(y_3-y_1)+(x_1-x_3)(x_1-x_3) & (y_3-y_1)(y_1-y_2)+(x_1-x_3)(x_2-x_1) \\ (y_1-y_2)(y_3-y_2)+(x_2-x_1)(x_3-x_2) & (y_1-y_2)(y_3-y_1)+(x_2-x_1)(x_1-x_3) & (y_1-y_2)(y_1-y_2)+(x_2-x_1)(x_2-x_1) \end{bmatrix} dv \tag{7.82}$$

$$[q]_i = \frac{\rho\Delta}{3\varepsilon}\begin{bmatrix} 1 \\ 1 \\ 1 \end{bmatrix} \tag{7.83}$$

where, in effect, the total charge in an element ($\rho\Delta$) is distributed equally among the three nodes of the element. This approach avoids the need for integration over the element. Thus, because of the method of defining the finite element, the integration is rather simple.

Boundary conditions (Dirichlet only) are set as in the general method in section (4.3). The system is solved using the *ICCG* method (chapter 11).

The electric potential distribution is shown in Figure 7.3. From these results, the electric field may be calculated by simply calculating the gradient of the potential. This can be performed in each element using the general approximation for a single triangular element (nodes 1,2,3)

$$E_x = -\frac{\partial V}{\partial x} = -\sum_{i=1}^{3}\frac{\partial N_i V_i}{\partial x} = -\frac{y_2 - y_3}{2\Delta}V_1 - \frac{y_3 - y_1}{2\Delta}V_2 - \frac{y_1 - y_2}{2\Delta}V_3 \tag{7.84}$$

and

$$E_y = -\frac{\partial V}{\partial y} = -\sum_{i=1}^{3}\frac{\partial N_i V_i}{\partial y} = -\frac{x_3 - x_2}{2\Delta}V_1 - \frac{x_1 - x_3}{2\Delta}V_2 - \frac{x_2 - x_1}{2\Delta}V_3 \tag{7.85}$$

The electric field is, therefore, constant in each element because the approximation for potentials is linear. This process can now be repeated for all elements in the solution domain to obtain an electric field distribution. Quantities such as energy in the solution domain are calculated similarly.

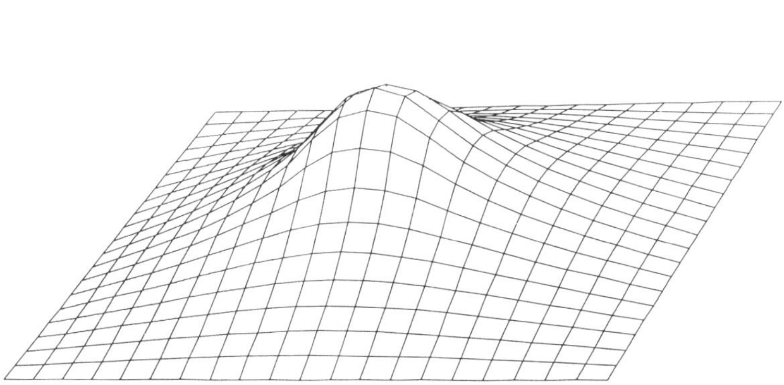

Figure 7.3 Potential distribution in the geometry in Figure 7.1.

Finally, we note that in the finite element formulation of this problem both charge densities and total charge can be easily incorporated. In Eq. (7.83) the charge density is integrated over the element and the equivalent charge is located at the corresponding nodes of the mesh. Instead, the charge at the node can be specified directly without the need for integration over the finite element. This is in contrast with the finite difference method where only charge density could be specified.

We also note that while the calculation above is performed in an isotropic material, anisotropic materials can also be included. All that is necessary is to use ε_x and ε_y in the corresponding terms of Eq. (7.74) and retain these permittivities throughout the derivation rather than dividing both sides of the equation by ε as is done in Eqs. (7.82) and (7.83).

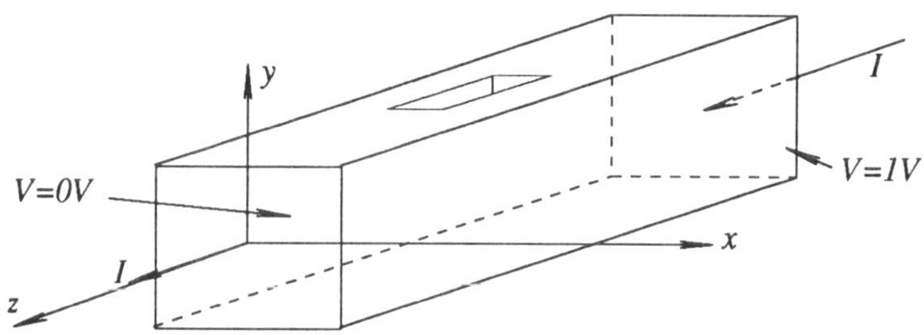

Figure 7.4 A steel bar with slot as used for calculation of current density distribution and active leakage fields.

7.5.3 Laplace's equation: calculation of current densities

The solution of Laplace's equation is essentially the same as in the previous case. However, we show here how this can be done in three dimensions in a conductor, for the purpose of calculating the potential distribution around a defect and the current density in the conductor. This result is then used to calculate the active magnetic leakage field for the same defect.

The geometry considered is shown in Figure 7.4. It consists of a steel bar with a slot at the center of the top surface. The slot is *12.5 mm* long, *6 mm* wide and *6 mm* deep. This is hardly a "defect" in the sense that it is too large to be of any practical interest but, because it is large, the effects are more pronounced, and therefore suitable for demonstration of the method. The bar is *31.75×31.75 mm* in cross section. A potential difference of *1 V* is applied on the bar as shown. The two ends of the bar are therefore Dirichlet boundary conditions while the other sides, including the defect have unknown values of the potential on the surface. However, since the electric field is always normal to a conducting surface, these surfaces can be considered to be Neumann boundary conditions. In addition, the problem can be reduced significantly by considering the symmetry through the bar as shown in Figure 7.5a. In this particular case, it is also possible to define a symmetry line as shown in Figure 7.5b. The reason for this is that the electric field distribution or the potential, will not change in the section considered (one quarter of the bar) due to this change. However, the problem is actually modified in the process: instead of a single slot, the effect is that of two slots. Thus, while the electric field and potential in the section considered are correct, they are not correct in the lower part of the bar. We consider the two way symmetry which is the correct calculation for two slots on opposite surfaces of the bar.

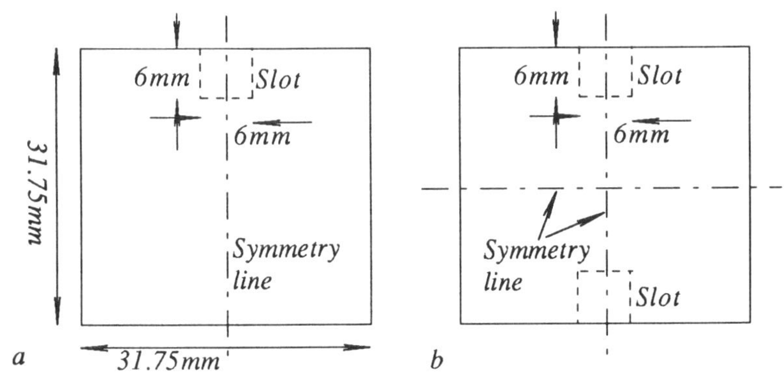

Figure 7.5. Symmetry boundaries in the geometry in Figure 7.4. a. Vertical line through center of slot. b. Additional horizontal line and resulting geometry.

Formulation of this problem is similar to the previous formulation except that now there are no sources. The equation to solve is Laplace's equation:

$$\nabla^2 V = \frac{\partial^2 V}{\partial x^2} + \frac{\partial^2 V}{\partial y^2} + \frac{\partial^2 V}{\partial z^2} = 0 \qquad (7.86)$$

The energy functional for this equation is

$$F(V) = \frac{1}{2}\int_v \left[\left(\frac{\partial V}{\partial x}\right)^2 + \left(\frac{\partial V}{\partial y}\right)^2 + \left(\frac{\partial V}{\partial z}\right)^2\right] dxdydz = 0 \qquad (7.87)$$

where the material is assumed to be isotropic.

The rest of the formulation is similar to that in the previous section except that now we use a three dimensional element (eight-node hexahedron). The general term of the elemental system is

$$s_{i,j} = \int_{v_i} \left(\frac{\partial N_i}{\partial x}\frac{\partial N_j}{\partial x} + \frac{\partial N_i}{\partial y}\frac{\partial N_j}{\partial y} + \frac{\partial N_i}{\partial z}\frac{\partial N_j}{\partial z} \right) dxdydz, \qquad i,j = 1,2,....8$$

and the elemental system is

$$[s]_e \{V\} = \{0\}$$

Integration of these terms is done numerically since the element is isoparametric (see section 4.4.3 for definition of the element and chapter 11 for numerical integration). The elemental systems are assembled into a global system, boundary conditions are applied to the global system and the resulting matrix equation is solved.

A suitable mesh for the solution is shown in Figure 7.6. Note that only the bar is modeled, using eight-node hexahedral elements. A total of *360* elements and *539* nodes are used. This leads to a rather small matrix with *539* rows/columns and a semi-bandwidth of *26*. Solution to this system of equations provides the electric potential everywhere inside the bar and on its surface.

The potential distribution in a cross-section of the bar passing through the center of the slot is shown in Figure 7.7. This clearly shows that the potential is uniform away from the slot but bends around the corners of the slot. Current density is perpendicular to potential lines and "bends" around the corners of the slot, creating the three-dimensional nature of the leakage fields around the defect.

To calculate the current density, we must first calculate the electric field intensity. In this case, the current may be in any direction, therefore all three components of the electric field intensity are calculated. These are:

$$E_x = -\frac{\partial V}{\partial x} = -\sum_{i=1}^{8} \frac{\partial N_i}{\partial x} V_i$$

$$E_y = -\frac{\partial V}{\partial y} = -\sum_{i=1}^{8} \frac{\partial N_i}{\partial y} V_i$$

$$E_z = -\frac{\partial V}{\partial z} = -\sum_{i=1}^{8} \frac{\partial N_i}{\partial z} V_i$$

and, the current density is calculated from the relation $\mathbf{J}=\sigma\mathbf{E}$. Although the conductivity can be anisotropic, we assume that it is isotropic (this is true for most conductors) and write

$$J_x = -\sigma \sum_{i=1}^{8} \frac{\partial N_i}{\partial x} V_i$$

$$J_y = -\sigma \sum_{i=1}^{8} \frac{\partial N_i}{\partial y} V_i$$

$$J_z = -\sigma \sum_{i=1}^{8} \frac{\partial N_i}{\partial z} V_i$$

We note here however, that these current densities are only correct for the potential difference applied on the ends of the bar. If this potential is not known, the current density is not calculable. A simple method to overcome this is given in the following section.

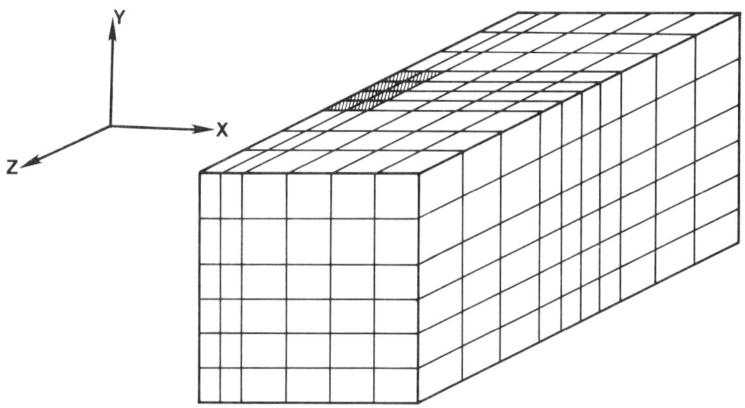

Figure 7.6 Finite element mesh of the bar and slot as used for calculation of potential distribution in the bar.

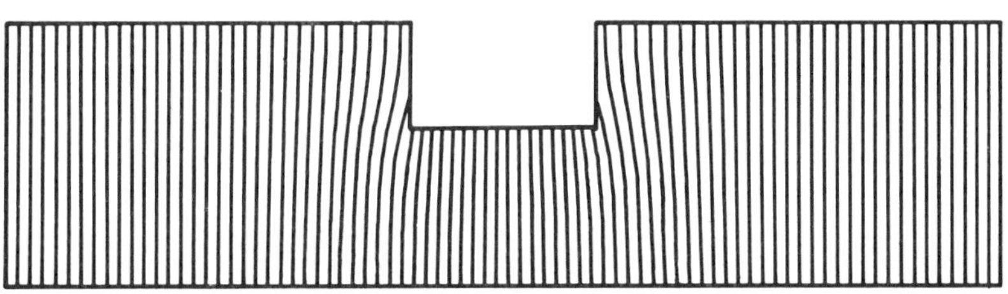

Figure 7.7 Potential distribution in a longitudinal cross-section through the slot. Current is perpendicular to potential lines (not shown).

To gain some insight into the current density distribution consider Table 7.1. It lists the minimum, maximum and nominal normalized current density. Nominal current density is the current density at the Dirichlet boundary (where the current density is uniform and directed entirely in the z direction. This current density is taken as $1A/m^2$ (normalized). Current densities everywhere are normalized with respect to the current density in the uniform region. The table shows that the x and y components are significant and the z component in the narrow region is over 25% higher than in the uniform region.

Table 7.1. Minimum, maximum, and nominal normalized current densities in the bar of Figure 7.4.

	Minimum	Maximum	Nominal
J_x	−0.0322 A	0.241A	0
J_y	−0.178A	0.241A	0
J_z	0.739A	1.258A	1.0 A

7.5.4 Magnetostatic formulations

The formulation of magnetostatic applications can be divided into two basic methods. The first is governed by a linear equation and the second by a nonlinear equation. The linear formulation is presented here while the nonlinear formulation is presented separately in section 7.9.3.

The formulation here is based on the magnetic vector potential. The governing equation is therefore

$$\frac{\partial^2 \mathbf{A}}{\partial x^2} + \frac{\partial^2 \mathbf{A}}{\partial y^2} + \frac{\partial^2 \mathbf{A}}{\partial z^2} = -\mu \mathbf{J}_s \tag{7.88}$$

This equation is identical to that used in section 7.3 except that the eddy current term does not exist. We note that the only material property is the magnetic permeability which, in this case is assumed to be isotropic and linear. The anisotropic, linear case can be easily obtained as

$$\frac{1}{\mu_x}\frac{\partial^2 \mathbf{A}}{\partial x^2} + \frac{1}{\mu_y}\frac{\partial^2 \mathbf{A}}{\partial y^2} + \frac{1}{\mu_z}\frac{\partial^2 \mathbf{A}}{\partial z^2} = -\mathbf{J}_s \tag{7.89}$$

Thus, formulation of Eq. (7.88) or (7.89) is identical to that in section 7.3. The system of equations is therefore

$$[S]\{A\} = \{Q\}$$

where the terms of $[S]$ and $[Q]$ are the assembly of elemental matrices as given in Eqs. (7.42) through (7.47) and (7.49) through (7.51) respectively.

The formulation in section 7.4 can be used just as well, but now both the magnetic vector potential and the electric scalar potential are used leading to four variables per node. Again, all that is necessary is to remove the eddy current term in Eqs. (7.57), (7.58), and (7.59).

As an example of the use of this formulation we consider modeling the magnetic flux density over the steel bar in Figure 7.4. The magnetic flux density is often the measured quantity in an active leakage field test. Thus, the ability to model the field profile accurately is extremely important.

To excite an active field, the bar in Figure 7.4 is used again and a current of *125 A* is passed through the bar. The bar has a cross section of $0.00108\ m^2$. As a result, the current density in the uniform region of the bar (close to one of the ends) is $124{,}000\ A/m^2$ and is directed in the z direction.

To calculate the magnetic vector potential, the current density everywhere in the solution domain must be known before the solution can commence.

The solution is performed in two steps:

1. First, the current density distribution everywhere in the bar is calculated as in section 7.5.3. This provides a normalized current density distribution in the bar. The normalized current density distribution is now scaled by the correct current density in the bar ($124{,}000\ A/m^2$). As an example, the maximum current density is $155{,}992\ A/m^2$, as can be seen from Table 7.1.
2. Using the current density calculated in the previous step, one of the magnetostatic formulations is applied.

Calculation of the current density distribution is discussed in the previous section and is not repeated here. We note however that the same type of elements are used for calculation of the current density as for the magnetic vector potential. A suitable mesh for solution is shown in Figure 7.8. The following should be noted with regard to the finite element mesh:

1. The mesh in the bar is identical to the mesh in Figure 7.6.
2. Only one quarter of the bar is modeled.
3. Boundary conditions for calculation of the magnetic vector potential are as shown in Figure 7.9. This assumes that boundaries parallel to the two symmetry planes are far enough away not to influence the solution significantly. Note in particular that the boundary condition for A_z on the ends of the geometry are calculated (see below).
4. The mesh is nonuniform with larger elements at higher distances from the bar, where the flux density is lower and therefore the errors involved in using large elements is small. The small element layer above the bar is used to

allow calculation of the flux density just above the bar (at *0.25 mm* over the bar).

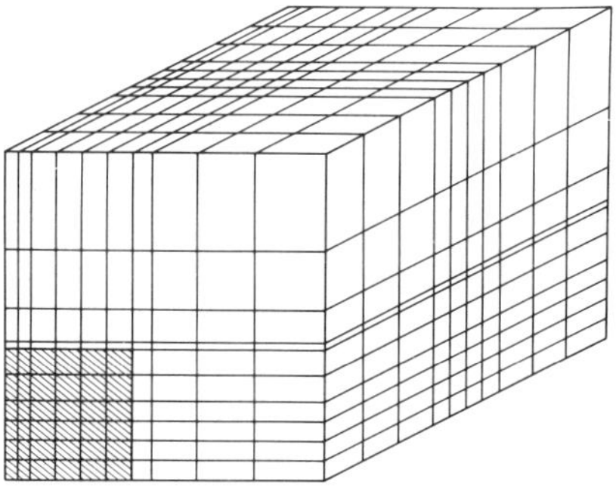

Figure 7.8 3-D finite element mesh for leakage field calculation. The bar is shaded at the bottom left corner of the mesh. Note mesh grading.

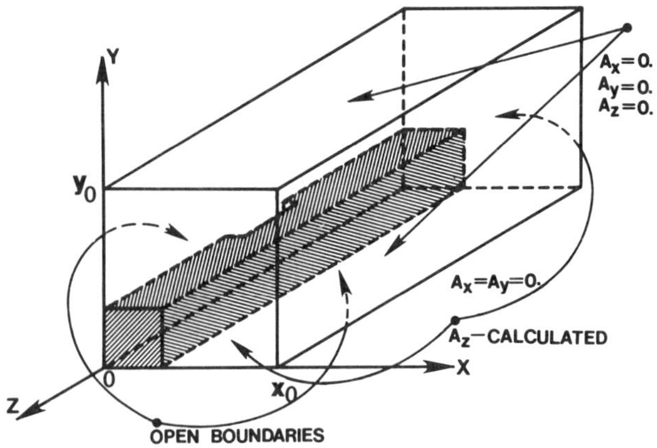

Figure 7.9 Boundary conditions for calculation of the magnetic vector potential.

The mesh in Figure 7.8 contains *1,000* elements with *1,331* nodes. The total number of variables to be calculated is *3,993* (using the formulation in section 7.3). This produces a system of *3,993* equations with a semi-bandwidth of *181*. Solution can be performed using Gaussian elimination but it is much more efficient to solve a system of this size using the conjugate gradient method.

Before the system of equations can be solved, boundary conditions must be incorporated into the system of equations. Boundary conditions are known everywhere except on the ends of the bar and adjoining space as shown in Figure 7.9. Here there are two possibilities in specifying boundary conditions. The first, and the simplest, is to leave them unspecified. This means that the end surfaces assume Neumann boundary conditions. The justification for this choice is the fact that current density is uniform in the bar, producing a flux density which has only transverse components in the plane of the boundary. The magnetic vector potential is not known but its normal derivative is zero anywhere on the two end boundaries. However, these boundary conditions can also be calculated from a two-dimensional calculation in the cross-section of the bar. The latter approach is preferred here because the mesh is relatively coarse and specification of accurate values for boundary conditions can improve accuracy of the overall solution. Since the current density is uniform on the end boundaries, we can assume a two-dimensional cross section with current density perpendicular to the boundary and calculate the magnetic vector potential in this cross section using only the outer boundary conditions which can be assumed to be zero.

This approach follows similar steps as for the previous calculation. As an example, we use triangular elements as for the electrostatic solution in section 7.5.2. The governing equation for the two-dimensional magnetostatic solution is

$$\frac{1}{\mu_x}\frac{\partial^2 A_z}{\partial x^2} + \frac{1}{\mu_y}\frac{\partial^2 A_z}{\partial y^2} = -J_s \tag{7.90}$$

Since the form of this equation is identical to that in Eq. (7.88) (except for the obvious change in dimensionality of the problem) we can also write an energy functional for the *2-D* problem as

$$F(A) = \int_v \left[\frac{1}{2}\nu_x\left(\frac{\partial A_z}{\partial y}\right)^2 + \nu_y\left(\frac{\partial A_z}{\partial x}\right)^2 - J_z A_z\right] dv \tag{7.91}$$

which is obtained from Eq. (7.18) by setting: $A_x = A_y = 0$ and $\sigma = 0$. The functional is written to allow anisotropic materials, although materials must be homogeneous in each finite element.

To formulate the *2-D* problem the process in sections 7.3 or 7.4 can be followed. A much simpler approach is to note that this functional is similar to the functional for Poisson's equation in Eq. (7.74) and, since only one component of the magnetic vector potential is necessary, Eq. (7.90) is a scalar equation. Therefore, we can use the results in Eqs. (7.74) through (7.83) directly by simply replacing V by A_z, ε by $1/\mu$, and ρ by J_z. For constant permeability the matrix equation can be written directly as

$$[S]\{A_z\} = \{Q\}$$

where the entries in the elemental matrix are

$$s_{i,j} = \int_{v_i} \left(\frac{\partial N_i}{\partial x} \frac{1}{\mu_x} \frac{\partial N_j}{\partial x} + \frac{\partial N_i}{\partial y} \frac{1}{\mu_y} \frac{\partial N_j}{\partial y} \right) dxdy\,, \qquad i,j = 1,2,3$$

$$q_i = \int_{v_i} J_z N_i dxdy, \qquad i,j = 1,2,3$$

Since we chose to use triangular elements, the elemental matrices in Eqs (7.82) and (7.83) with the modifications mentioned above may be used:

$$[s] = \frac{1}{4\Delta} \begin{bmatrix} (y_3-y_2)\frac{1}{\mu_x}(y_3-y_2)+(x_3-x_2)\frac{1}{\mu_y}(x_3-x_2) & (y_3-y_2)\frac{1}{\mu_x}(y_3-y_1)+(x_3-x_2)\frac{1}{\mu_y}(x_1-x_3) & (y_3-y_2)\frac{1}{\mu_x}(y_1-y_2)+(x_3-x_2)\frac{1}{\mu_y}(x_2-x_1) \\ (y_3-y_1)\frac{1}{\mu_x}(y_3-y_2)+(x_1-x_3)\frac{1}{\mu_y}(x_3-x_2) & (y_3-y_1)\frac{1}{\mu_x}(y_3-y_1)+(x_1-x_3)\frac{1}{\mu_y}(x_1-x_3) & (y_3-y_1)\frac{1}{\mu_x}(y_1-y_2)+(x_1-x_3)\frac{1}{\mu_y}(x_2-x_1) \\ (y_1-y_2)\frac{1}{\mu_x}(y_3-y_2)+(x_2-x_1)\frac{1}{\mu_y}(x_3-x_2) & (y_1-y_2)\frac{1}{\mu_x}(y_3-y_1)+(x_2-x_1)\frac{1}{\mu_y}(x_1-x_3) & (y_1-y_2)\frac{1}{\mu_x}(y_1-y_2)+(x_2-x_1)\frac{1}{\mu_y}(x_2-x_1) \end{bmatrix} dv \tag{7.93}$$

$$\{q\} = \frac{J_z \Delta}{3} \begin{Bmatrix} 1 \\ 1 \\ 1 \end{Bmatrix} \tag{7.94}$$

A finite element mesh is next defined over the solution domain as shown in Figure 7.10. This particular mesh is designed to exactly overlap the *3-D* mesh in any cross section. Doing so simplifies specification of boundary conditions because the nodes of the triangular-element mesh are identical to the nodes of the *3-D* mesh in cross section. However, there is no reason why the mesh cannot be denser and, in fact, in most cases it should be. Using this mesh, the magnetic vector potential is calculated in a fashion identical to that in section 7.5.2.

Now that boundary conditions have been calculated, the solution to the *3-D* magnetostatic problem can commence. Solution is rather straightforward and, again, the incomplete Choleski conjugate gradient (*ICCG*) method is used for this purpose.

The type of result that may be of interest here is the magnetic flux density at different locations over the bar. Figure 7.11 shows the magnetic flux density (magnitude) at a number of locations across the bar as shown in the inset, and a lift-off of *0.25mm.* The flux density at corners is highest, as expected. It is also possible to calculate the flux density at different lift-offs as shown in Figure

7.12. The plots in this figure are for the field above the corner of the slot at lift-offs of *0.25mm, 5mm,* and *10mm.*

When modeling lift-off with the formulation given here it is important to design the mesh such that the location of measurement coincides with the center of elements. The reason for this is that the element we use is linear and therefore, the magnetic flux density inside any element is constant. The components of the magnetic flux density are

$$B_x = \sum_{i=1}^{8} \frac{\partial N_i}{\partial y} A_{zi} - \sum_{i=1}^{8} \frac{\partial N_i}{\partial z} A_{yi} \tag{7.95}$$

$$B_y = \sum_{i=1}^{8} \frac{\partial N_i}{\partial z} A_{xi} - \sum_{i=1}^{8} \frac{\partial N_i}{\partial x} A_{zi} \tag{7.96}$$

$$B_z = \sum_{i=1}^{8} \frac{\partial N_i}{\partial x} A_{yi} - \sum_{i=1}^{8} \frac{\partial N_i}{\partial z} A_{xi} \tag{7.97}$$

It is also possible to calculate (or measure) the flux density as a scan and produce a continuous profile such as the one shown in Figure 7.13. Generation of data for a display of this sort requires a relatively dense mesh and interpolation of data points to produce smooth curves. However, it allows visualization of the field over the entire surface of the bar.

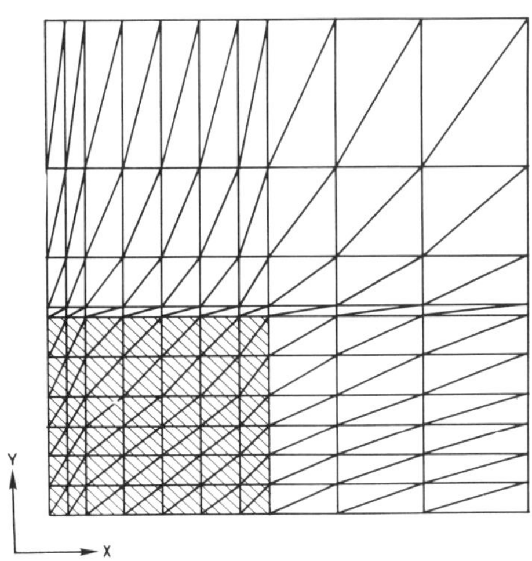

Figure 7.10 Two-dimensional mesh through bar as used to calculate boundary conditions in the uniform region of the bar.

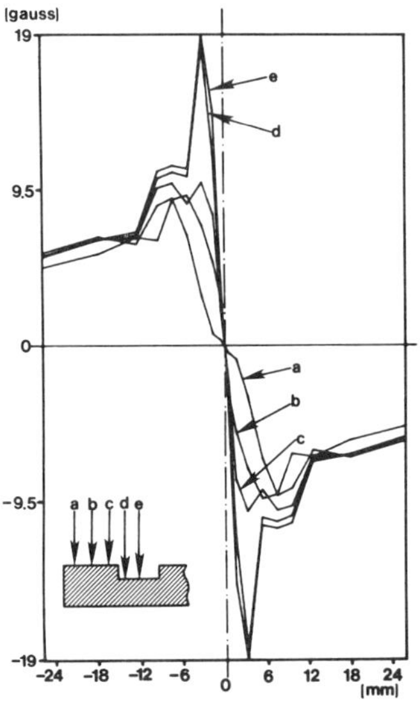

Figure 7.11 Leakage field profile (magnetic flux density) at a number of locations above the slot, at a lift-off of 0.25 mm above surface of the bar.

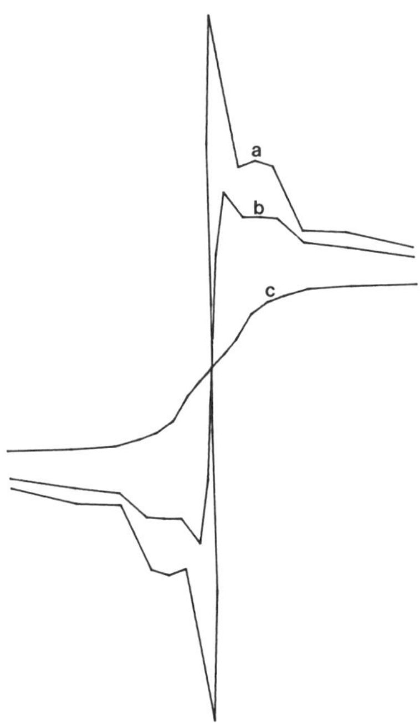

Figure 7.12 Leakage field profiles at edge of slot and different lift-offs. a. at 0.25 mm, b. at 5 mm and c. at 10 mm lift-off.

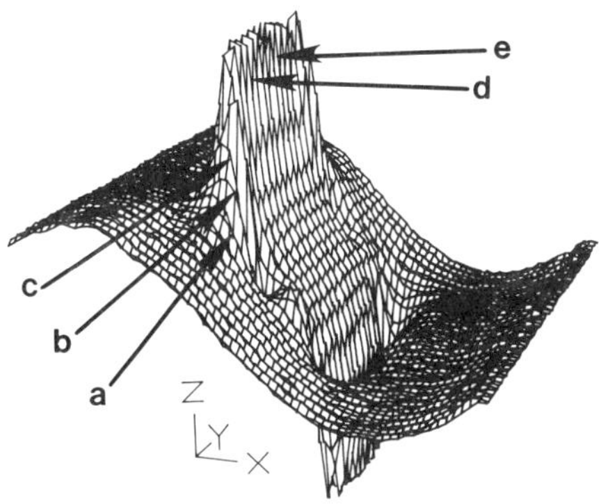

Figure 7.13 A 3-D field profile above the slot obtained from the 3-D solution. Marked curves correspond to those marked in Figure 7.11.

7.6 EXAMPLES: EDDY CURRENT APPLICATIONS

Eddy current applications form a very significant part of testing of materials and therefore of modeling. This is the main reason for choosing the general *3-D* eddy current problem in developing the two methods of formulation in sections 7.3 and 7.4.

While eddy current formulations follow the same pattern as magnetostatic formulations, the quantities of interest are often very different. Calculation of impedance (resistance and reactance) and induced (eddy) currents are often of more interest than magnetic flux densities. Phase of fields are almost always calculated and, unlike magnetostatic calculations, eddy current calculations are often performed in low intensity field environments.

In this section we demonstrate the use of the magnetic vector potential to calculation of impedances in axisymmetric geometries as used for testing of tubular materials, two-dimensional applications, the remote eddy current fields and to three-dimensional applications in a range of frequencies. As in previous sections, only modifications to formulations are given, and elements are chosen to demonstrate their formulation rather than to indicate which element should be used.

7.6.1 Two-dimensional applications

As an example of a *2-D* application consider the geometry in Figure 7.14a. It consists of a thick conductor between two flat, thick sheets. A current density of

1000 A/m² at *100 Hz* is present in the central conductor. The conducting sheets are highly conductive and ferromagnetic. The magnetic flux density is required. The geometry in Figure 7.14b is used for calculation. Two symmetry lines were used here to reduce the problem to a quarter of the geometry in Figure 7.14a.

This geometry is typical of *2-D* applications and we use triangular elements for the formulation.
The *2-D* eddy current equation is

$$\frac{\partial^2 A_z}{\partial x^2} + \frac{\partial^2 A_z}{\partial y^2} = -\mu J_s + j\omega\mu\sigma A_z \tag{7.98}$$

The finite element approximation for three-node triangular elements is

$$A = \sum_{i=1}^{3} A_i N_i$$

where A_i are the nodal vector potentials. The shape functions for the element are given in Eqs. (7.76) through (7.78) and their derivatives in Eqs. (7.79) through (7.81).

The integration to be performed (see the general energy functional in Eq. (7.18)) is

$$\int_v \frac{1}{2}\left(\left[\frac{\partial A}{\partial x}\right]^2 + \left[\frac{\partial A}{\partial y}\right]^2 + j\omega\sigma\mu_0\mu_r A^2 - \mu_r J A\right) dv = 0 \tag{7.99}$$

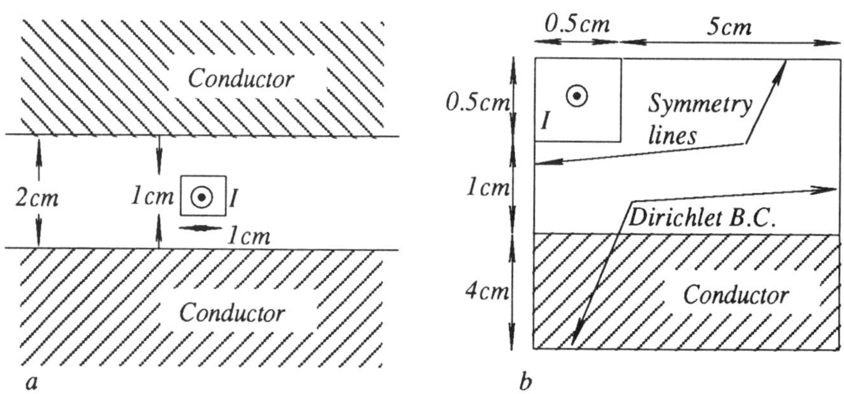

Figure 7.14. A two-dimensional geometry. a. Geometry. b. Use of symmetry lines and part of geometry used for computation.

The approximation for A is substituted in Eq. (7.99) and, after integration over element volume, we obtain the following matrix equation for an element:

$$\begin{bmatrix} S_{1,1} & S_{1,2} & S_{1,3} \\ S_{2,1} & S_{2,2} & S_{2,3} \\ S_{3,1} & S_{3,2} & S_{3,3} \end{bmatrix} \begin{bmatrix} A_1 \\ A_2 \\ A_3 \end{bmatrix} = \mu_0 \mu_r J \begin{bmatrix} d_1 \\ d_2 \\ d_3 \end{bmatrix} \tag{7.100}$$

where the individual coefficients are

$$S_{i,j} = \int_{v_i} \left[\frac{\partial N_i}{\partial x}\frac{\partial N_j}{\partial x} + \frac{\partial N_i}{\partial y}\frac{\partial N_j}{\partial y} + j\omega\sigma\mu_0\mu_r N_i N_j \right] dv \tag{7.101}$$

$$d_i = \int_{v_i} N_i dv \tag{7.102}$$

S_{ij} and d_i are evaluated for each element using the shape functions and their derivatives as given above. The various integrals are evaluated analytically as:

$$\int_{v_i} \left(\frac{\partial N_1}{\partial x}\right)^2 dv = \frac{(y_2 - y_3)^2}{(2\Delta)^2}\Delta = \frac{(y_2 - y_3)^2}{4\Delta}$$

$$\int_{v_i} \left(\frac{\partial N_1}{\partial y}\right)^2 dv = \frac{(x_3 - x_2)^2}{(2\Delta)^2}\Delta = \frac{(x_3 - x_2)^2}{4\Delta}$$

$$\int_{v_i} \frac{\partial N_1}{\partial x}\frac{\partial N_2}{\partial x} dv = \frac{(y_2 - y_3)(y_3 - y_1)}{4\Delta}$$

$$\int_{v_i} \frac{\partial N_1}{\partial x}\frac{\partial N_3}{\partial x} dv = \frac{(y_2 - y_3)(y_1 - y_2)}{4\Delta}$$

$$\int_{v_i} \frac{\partial N_2}{\partial x}\frac{\partial N_3}{\partial x} dv = \frac{(y_3 - y_1)(y_1 - y_2)}{4\Delta}$$

$$\int_{v_i} \frac{\partial N_1}{\partial y}\frac{\partial N_2}{\partial y} dv = \frac{(x_3 - x_2)(x_1 - x_3)}{4\Delta}$$

$$\int_{v_i} \frac{\partial N_1}{\partial y}\frac{\partial N_3}{\partial y}dv = \frac{(x_3 - x_2)(x_2 - x_1)}{4\Delta}$$

$$\int_{v_i} \frac{\partial N_2}{\partial y}\frac{\partial N_3}{\partial y}dv = \frac{(x_1 - x_3)(x_2 - x_1)}{4\Delta}$$

$$\int_{v_i} \left(\frac{\partial N_2}{\partial x}\right)^2 dv = \frac{(y_3 - y_1)^2}{4\Delta}$$

$$\int_{v_i} \left(\frac{\partial N_2}{\partial y}\right)^2 dv = \frac{(x_1 - x_3)^2}{4\Delta}$$

$$\int_{v_i} \left(\frac{\partial N_3}{\partial x}\right)^2 dv = \frac{(y_1 - y_2)^2}{4\Delta}$$

$$\int_{v_i} \left(\frac{\partial N_3}{\partial y}\right)^2 dv = \frac{(x_2 - x_1)^2}{4\Delta}$$

$$\int_{v_i} [j\omega\sigma\mu_0\mu_r N_i N_j]dv = j\frac{\omega\sigma\mu_0\mu_r\Delta}{12} \qquad \text{for } i\neq j, \qquad i,j=1,2,3$$

$$\int_{v_i} [j\omega\sigma\mu_0\mu_r N_i N_j]dv = j\frac{\omega\sigma\mu_0\mu_r\Delta}{6} \qquad \text{for } i=j, \qquad i,j=1,2,3$$

$$\int_{v_i} N_i dv = \frac{\Delta}{3} \qquad i=1,2,3$$

The elemental matrices are now entered in the global matrix as shown in section 4.5.5. This produces a system of equations with N unknown values of the magnetic vector potential.

A finite element mesh must be defined over the geometry. A very simple mesh, which can even be entered by hand is shown in Figure 7.15. The important aspects of the mesh are specification of material properties, and correct discretization of geometries. This mesh contains a total of *50* elements and *36* nodes. Since there are only two elements in the central conductor and very few in the flat sheet, this mesh produces some errors in calculations but the errors can be easily reduced by increasing the discretization level. The mesh in Figure 7.16 contains *200* triangular elements and *121* nodes. This mesh is a more refined mesh and shows how a mesh may be refined to improve a solution.

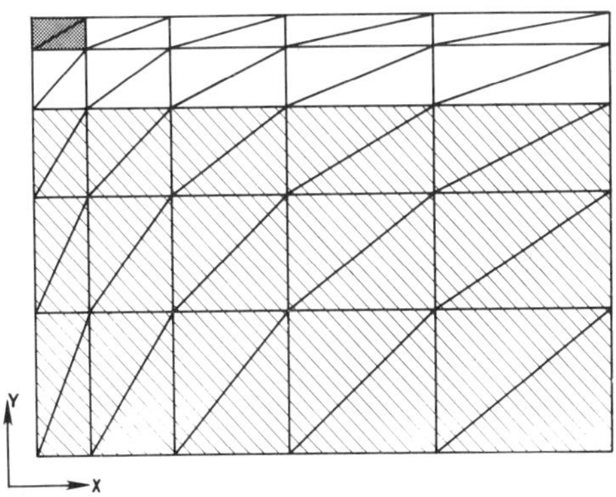

Figure 7.15 Triangular mesh over the geometry. Current carrying bar and thick sheet are shown shaded.

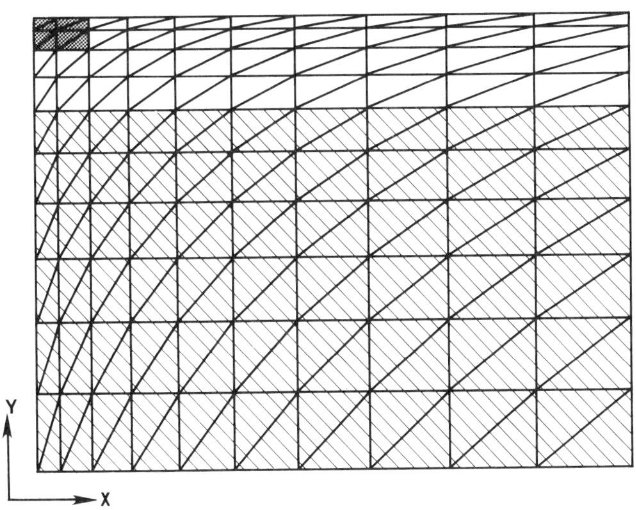

Figure 7.16 A refined mesh of the geometry in Figure 7.14b.

Boundary conditions are now imposed. The simplest method is to apply zero boundary conditions on the outer contour of the boundary, assuming that the flux density at the boundary is negligible.

Solution of the system of equations provides the magnetic vector potential everywhere in the solution domain in complex variables. Results of this calculation are given in Figures 7.17 and 7.18. Figure 7.17 shows the absolute value of the magnetic vector potential (magnetic flux density contour lines) clearly showing the skin effect. Figure 7.18 shows lines of constant eddy current density. Verification of either result can only be done experimentally.

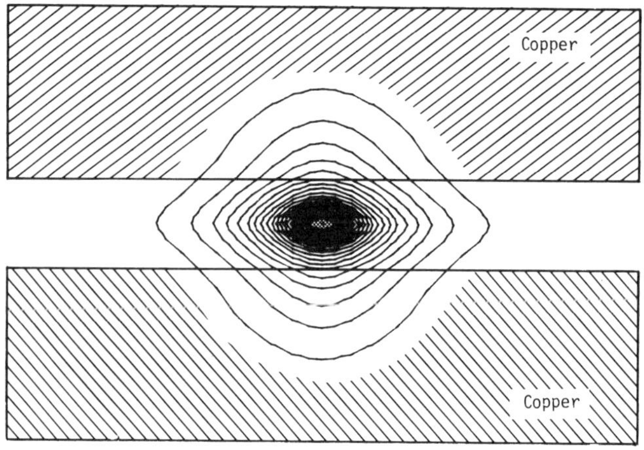

Figure 7.17 Contours of constant magnetic vector potentials.

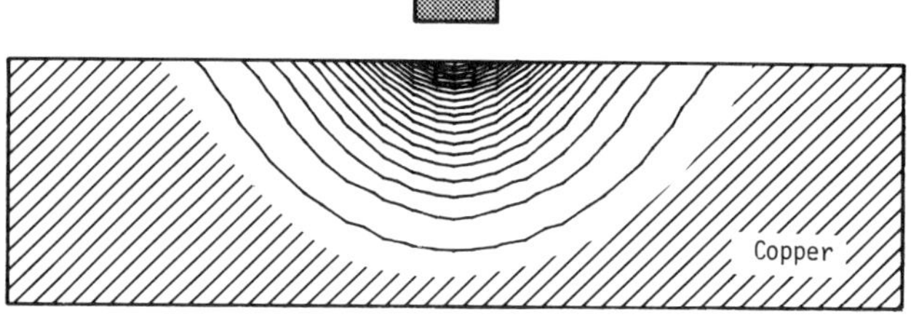

Figure 7.18 Contours of constant eddy current density in the thick sheets.

Flux density may be calculated using Eq. (7.95) for the x component and Eq. (7.96) for the y component. For the three-node triangular element these are:

$$B_x = \sum_{i=1}^{3} \frac{\partial N_i}{\partial y} A_{zi} \tag{7.103}$$

$$B_y = -\sum_{i=1}^{3} \frac{\partial N_i}{\partial x} A_{zi} \tag{7.104}$$

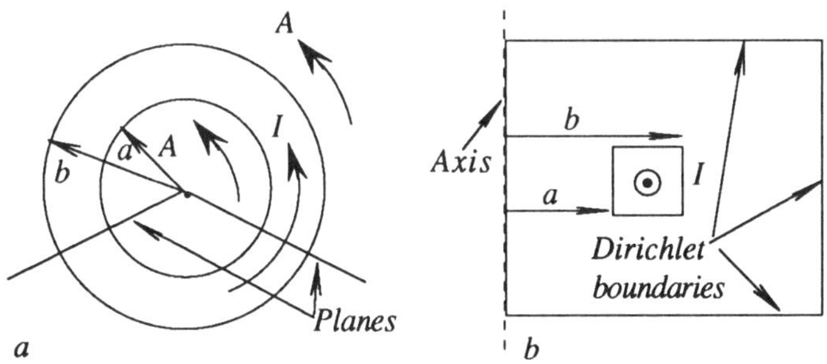

Figure 7.19 Basis of the axisymmetric formulation: all variables remain the same in any radial cross-section of the geometry (indicated by the two planes).

7.7 EXAMPLES: AXISYMMETRIC APPLICATIONS

A relatively large number of applications involve cylindrical objects and coils which, in many cases, can be analyzed in cylindrical coordinates. Examples are the remote field effect in steel tubes and eddy current testing of tubing. The main advantage of axisymmetric formulations is in the fact that applications which are three dimensional can be analyzed by *2-D* methods. Therefore, the solution is much more economical and more accurate.

As was pointed out at the beginning of this chapter, the idea of analysis of axisymmetric geometries is to superpose the cylindrical coordinate system over the Cartesian system and consider the special case in which there is no variation in the primary variable in the ϕ direction. This is shown schematically in Figure 7.19. The method is based on the fact that the field of a circular loop is circular; that is, if the current is in the ϕ direction, the flux density on any plane perpendicular to the current has only components in that plane. Since the magnetic flux density is calculated as the curl of the magnetic vector potential, the magnetic vector potential must be directed in the ϕ direction (perpendicular to the flux density) and must be constant along any circle parallel to the current loop. Therefore, the magnetic vector potential can be calculated in any plane perpendicular to the current.

However, we must modify the *2-D* method to account for the curvature of the geometry. This is done by modifying the definition of the finite element as given in section 4.4.

First, the field equations are written in cylindrical coordinates. The general eddy current equation in axisymmetric geometries is

$$\frac{\partial^2 A_\phi}{\partial r^2} + \frac{1}{r}\frac{\partial A_\phi}{\partial r} + \frac{\partial^2 A_\phi}{\partial z^2} - \frac{A_\phi}{r^2} = -\mu J_s + j\omega\mu\sigma A_\phi \qquad (7.105)$$

This equation is quite similar to the two-dimensional equation in Eq. (7.99) in that only variables r and z are present while the magnetic vector potential has only a ϕ direction in space. Quantities derived from the primary function (in this case the magnetic vector potential) are also modified. As an example, in evaluating the magnetic flux density from the magnetic vector potential we use the relation $\mathbf{B}=\nabla\times\mathbf{A}$. In cylindrical coordinates, this expression is

$$\nabla\times\mathbf{A} = \hat{\mathbf{r}}\left(\frac{1}{r}\frac{\partial A_z}{\partial \phi} - \frac{\partial A_\phi}{\partial z}\right) + \hat{\boldsymbol{\phi}}\left(\frac{\partial A_r}{\partial z} - \frac{\partial A_z}{\partial r}\right) + \hat{\mathbf{z}}\left(\frac{1}{r}\frac{\partial (rA_\phi)}{\partial r} - \frac{1}{r}\frac{\partial A_r}{\partial \phi}\right)$$

however, since $A_r = A_z = 0$, we get

$$\nabla\times\mathbf{A} = \hat{\mathbf{r}}\left(-\frac{\partial A_\phi}{\partial z}\right) + \hat{\mathbf{z}}\left(\frac{1}{r}\frac{\partial (rA_\phi)}{\partial r}\right)$$

Therefore, the magnetic flux density has only two components, one in the r direction, the second in the z direction. These are:

$$B_r = -\frac{\partial A}{\partial z} \qquad (7.106)$$

$$B_z = \frac{1}{r}\frac{\partial (rA)}{\partial r} = \frac{A}{r} + \frac{\partial A}{\partial r} \qquad (7.107)$$

and $\mathbf{A}$ is perpendicular to the plane r-z as required. The same general modifications apply to other quantities such as the gradient of scalar potential functions.

Second, we note that unlike a two-dimensional element which is considered to be of unit depth, the differential volume of an axisymmetric element is written as $dv = r dr d\phi dz$, and this is dependent on r. Thus, we consider the actual element volume rather than use the unit depth used in the two-dimensional case.

Figure 7.20 shows the shape of an axisymmetric element in the cylindrical coordinate system (r,ϕ,z).

In summary, the formulation in axisymmetric geometries requires two basic modifications. First, the volume integration over an element must be modified and second, the coefficients of the elemental matrices must be updated to take into account these modifications.

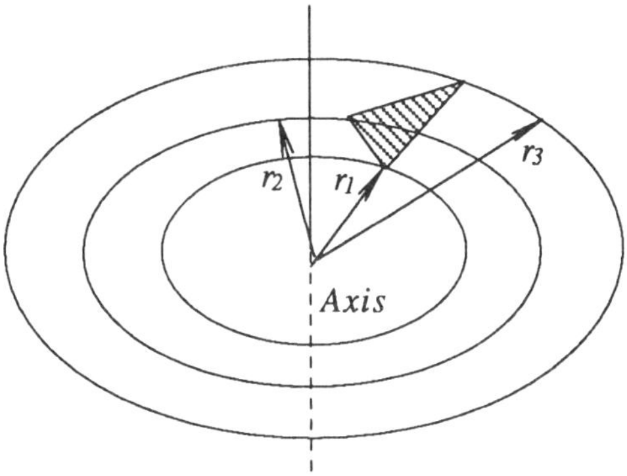

Figure 7.20 Definition of an axisymmetric element. Volume of element is finite and depends on radius r.

7.7.1 Formulation

We use the variational approach for the formulation, starting with the following energy-related functional

$$F(A) = \int_v \left(\frac{1}{2}\left[\nu_z \left(\frac{\partial A_\phi}{\partial z}\right)^2 + \nu_r \left(\frac{\partial A_\phi}{\partial r} + \frac{A}{r}\right)^2 - J_\phi A_\phi + j\frac{\omega\sigma\partial A_\phi}{2} \right) rdrdz \quad (7.108)$$

The formal approach requires that we first derive the Euler equations to show that this functional is in fact equivalent to Eq. (7.105). Assuming that this is correct, we still need to take the derivatives with respect to the unknowns and set these to zero to obtain the discretized form of the system and integrate each term separately. The approach here is somewhat different in that we use the derivation in the previous section and observe that Eq. (7.99) and (7.108) are of similar form provided we make the following equivalence:

1. Replace z with y.
2. Replace r with x.
3. Replace $rdrdz$ by $r_c dxdy$, and r in A/r by r_c, where r_c is some properly defined constant value.

4. Modify the second term in the integral to take into account the term A/r_C.

At first glance, these modifications seem to be rather extensive but, in fact they are not. The first two are merely notations that are designed to bring the two equations to the same basic forms.

The third modification is also rather straightforward. Since the integration is dependent on the distance from the axis of symmetry, it is different than in the two-dimensional case. This is shown in Figure 7.21a where the values for r at the three nodes of the element are shown together with a general value r in the element. Therefore, evaluation of the integrals for any node or for any point is properly defined except for the situation in Figure 7.21b where one of the nodes is on the axis of symmetry.

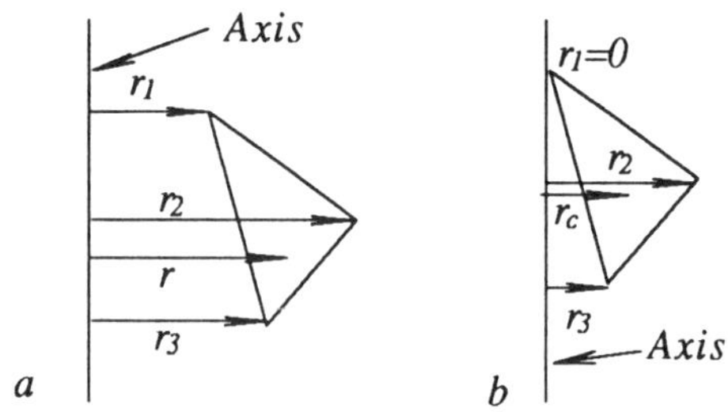

Figure 7.21 Radial distances in an axisymmetric element. a. An element away from symmetry line. b. One node is on symmetry line; r_C is centroidal distance.

To avoid this difficulty and to allow the axisymmetric evaluation to use the terms already defined for the *2-D* application, we use an average value for the distance r. One possible choice is

$$r_c = \frac{r_1 + r_2 + r_3}{3}$$

There are other methods of approximating r_C but this method is adequate for most applications. r_C is called the centroidal value of r.

With these modifications, the energy-related functional is

$$F(A) = \int_v \frac{1}{2}\left[v_y\left(\frac{\partial A_z}{\partial y}\right)^2 + v_x\left(\frac{\partial A_z}{\partial x} + \frac{A_z}{r_0}\right)^2 - J_z A_z + j\,\frac{\omega\sigma\partial A_z}{2}\right] r_0 dxdy \qquad (7.109)$$

The fourth modification is performed in the process of evaluating the coefficients of the elemental matrices. Taking the derivative with respect to each unknown and setting it to zero we get the following elemental matrix system

$$\{[S]_e + j\omega\sigma[R]_e\}\{A\}_e = \{Q\}_e \tag{7.110}$$

Performing direct integration on the terms of Eq. (7.109), we obtain the coefficients of the elemental matrices as

$$[s] = \frac{r_c}{4\Delta}\begin{bmatrix}
\left(y_3-y_2+\frac{2\Delta}{3r_c}\right)\frac{1}{\mu_x}\left(y_3-y_2+\frac{2\Delta}{3r_c}\right)+(x_3-x_2)\frac{1}{\mu_y}(x_3-x_2) & \left(y_3-y_2+\frac{2\Delta}{3r_c}\right)\frac{1}{\mu_x}\left(y_3-y_1+\frac{2\Delta}{3r_c}\right)+(x_3-x_2)\frac{1}{\mu_y}(x_1-x_3) & \left(y_3-y_2+\frac{2\Delta}{3r_c}\right)\frac{1}{\mu_x}\left(y_1-y_2+\frac{2\Delta}{3r_c}\right)+(x_3-x_2)\frac{1}{\mu_y}(x_2-x_1) \\
\left(y_3-y_1+\frac{2\Delta}{3r_c}\right)\frac{1}{\mu_x}\left(y_3-y_2+\frac{2\Delta}{3r_c}\right)+(x_1-x_3)\frac{1}{\mu_y}(x_3-x_2) & \left(y_3-y_1+\frac{2\Delta}{3r_c}\right)\frac{1}{\mu_x}\left(y_3-y_1+\frac{2\Delta}{3r_c}\right)+(x_1-x_3)\frac{1}{\mu_y}(x_1-x_3) & \left(y_3-y_1+\frac{2\Delta}{3r_c}\right)\frac{1}{\mu_x}\left(y_1-y_2+\frac{2\Delta}{3r_c}\right)+(x_1-x_3)\frac{1}{\mu_y}(x_2-x_1) \\
\left(y_1-y_2+\frac{2\Delta}{3r_c}\right)\frac{1}{\mu_x}\left(y_3-y_2+\frac{2\Delta}{3r_c}\right)+(x_2-x_1)\frac{1}{\mu_y}(x_3-x_2) & \left(y_1-y_2+\frac{2\Delta}{3r_c}\right)\frac{1}{\mu_x}\left(y_3-y_1+\frac{2\Delta}{3r_c}\right)+(x_2-x_1)\frac{1}{\mu_y}(x_1-x_3) & \left(y_1-y_2+\frac{2\Delta}{3r_c}\right)\frac{1}{\mu_x}\left(y_1-y_2+\frac{2\Delta}{3r_c}\right)+(x_2-x_1)\frac{1}{\mu_y}(x_2-x_1)
\end{bmatrix} dxdy \tag{7.111}$$

$$[R]_e = j\,\frac{\omega\sigma\Delta r_c}{12}\begin{bmatrix} 2 & 1 & 1 \\ 1 & 2 & 1 \\ 1 & 1 & 2 \end{bmatrix} \tag{7.112}$$

$$\{q\}_e = \frac{J_z\Delta r_c}{3}\begin{Bmatrix} 1 \\ 1 \\ 1 \end{Bmatrix} \tag{7.113}$$

The last two matrices are identical to those obtained for the two-dimensional eddy current formulation except for the the multiplication by the centroidal value of r for the element. The modifications in $[s]$ are also straightforward and only include the term $2\Delta/r_C$ for each term due to the additional term A/r in the functional.

The main advantage of this formulation is that the two-dimensional elements, as defined previously, can be used directly. However, the matrices would be somewhat different if instead of using triangular elements, we used isoparametric elements. Because in two dimensions, there is little advantage in using other elements, or for that matter, in using functions other than the magnetic vector potential, we do not present the formulation with other elements, or Galerkin's method here. The bibliography section points to other formulations that use different elements as well as other methods of formulation.

As examples of axisymmetric formulations, we consider next two applications: calculation of probe impedance in eddy current testing of tubular products and induced voltage (as well as fields) in coils, using the remote field effect method. The latter is particularly difficult to model because of sensitivity to location of boundary conditions, very low field intensities, and large extent of the solution domain. However, from a formulation point of view, the two methods are identical.

7.7.2 Modeling of absolute and differential eddy current probes

The typical geometries considered here are shown in Figures 7.22a and 7.22b. In Figure 7.22a a single coil inside a tube is shown. The tube contains an axisymmetric slot on the outer surface of the tube. In this type of application, the coil is moved past the defect and the impedance of the coil is monitored as it moves past the defect. The change in probe impedance is the quantity of interest since it indicates the existence of defects. A typical application as used for testing of nonferromagnetic tubes at *100 kHz* is shown. Before designing a mesh for computation, we must consider a number of aspects of the geometry.

First, the dimensions of the solution domain must be considered. Since the coil must move from some distance away from the defect to some distance past the defect, the solution domain must be large enough to accommodate this movement. Normally at least one coil diameter on either side of the defect is required for accurate solution (depending on coil diameter: larger coils require larger distances). Also, the field of the coil extends some additional distance away from the coil in the radial direction and boundaries must be located far enough apart not to interfere with this field since we assume zero boundary conditions on the outer boundaries of the geometry. A distance of between *1* to *3* coil diameters or more is required to minimize this effect. While it is not possible to decide on this dimension in a clear-cut fashion, a typical choice is shown in Figure 7.23 where the dimensions of the mesh are also shown. In this case, we could have used a much larger domain since in axisymmetric geometries, large meshes can be accommodated. However, to show the whole mesh and also to identify errors in solution, a rather small mesh, with relatively small dimensions is chosen here. Normally, the dimensions of the geometry are chosen based on experience or some simple analytic calculation. Perhaps the best method is to solve the problem with one set of boundaries and then move the boundaries some distance away and see if the solution changes significantly. If it does, new boundaries may be chosen until a satisfactory solution is obtained. The mesh shown in Figure 7.23 uses four-node quadrilateral, isoparametric elements but triangular elements can also be used. The mesh contains *600* elements and *671* nodes.

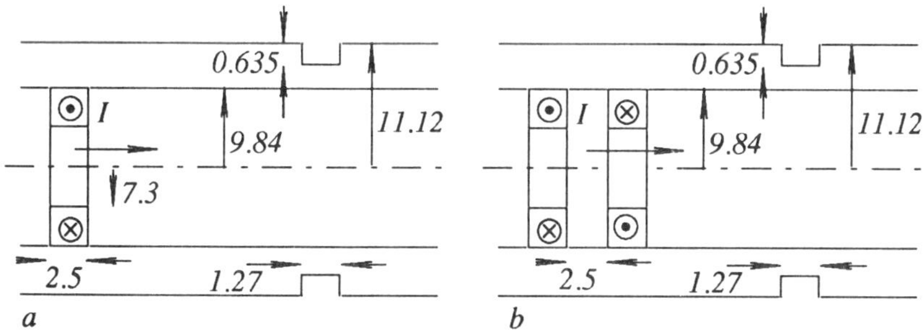

Figure 7.22 Axisymmetric geometries. a. Axisymmetric slot tested with absolute probe. b. Axisymmetric slot tested with differential probe.

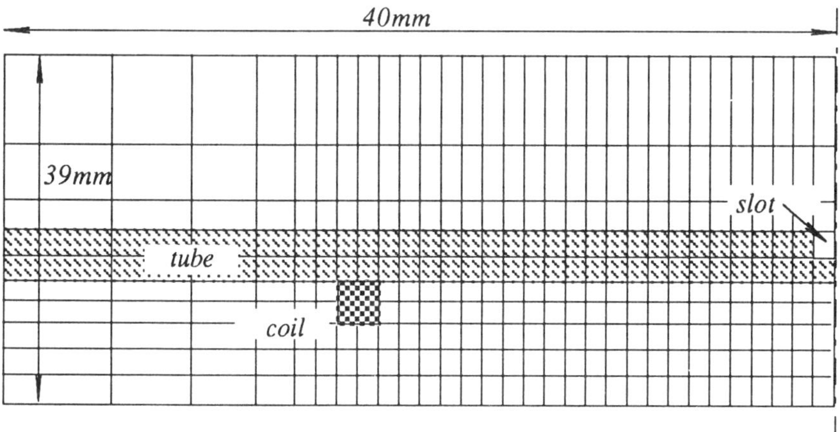

Figure 7.23 A mesh for the problem in Figure 7.22a. Only half the mesh is shown. Note the uniform region in which the probe moves.

Second, the domain in which the coil moves must be divided such that the coil does not change dimensions as it moves. This normally means that the solution domain in the region of coil movement is uniformly divided in the direction of movement. Also, a minimum number of elements are required to model the coil and this must be specified. We must note here that coil movement here means that the coil is moved to a new location and the fields and impedance calculated for this location before moving the coil to a new location. Thus the speed of the coil is not considered in the calculation.

Third, the element density in various sections as well as element size must be defined. In conducting and ferromagnetic materials, this is done by considering the skin depth. While there are no specific rules for discretization, *1* to *2* elements per skin depth are required in areas of large gradients in fields, with lower element densities in areas of low gradients. Again, this is a matter of experimentation but the basic rules are to use small, dense elements in regions of expected high gradients in fields and fewer, larger elements in areas of low gradients.

The mesh used here is not necessarily optimal for the problem but it is a simple mesh and, because the boundaries are rather far from the source, produces good results. To be noted in particular is the graded structure of the mesh with larger elements in areas of low fields. The coil is initially located at one end of the uniform region and for each subsequent calculation it is moved one element layer to the right to produce the next impedance point.

While the formulation of the problem is identical to that given above, the following modifications must be considered.

As the coil moves from one position to the next, the problem must be solved anew. It would appear that for each probe position, a new matrix must be assembled and the system solved again. This however is not the case.

Since at each probe position, all materials, and material locations (including defects) do not change position, the only part of the matrix system that must be modified is the right hand side of the system in Eq. (7.110). This means that for the first probe position, the left hand side as well as the right hand side are eliminated (say by a Gaussian elimination procedure). For each subsequent probe position, only the right hand side matrix is assembled and modified using identical steps as for the first probe position. Subsequent probe positions are treated in an identical manner. In this fashion, solution of each subsequent probe position after the first is rather fast and is essentially a backsubstitution process (see chapter 11).

This approach can also be used to modify other parameters such as frequency or current density, provided these are not part of the left hand side matrix, allowing parametric study of the problem. However, if any material property is modified, it affects the left hand side and the matrix must be reassembled and solved again for each variation in material properties.

We also note that the method described above cannot be used with iterative methods of solution unless all modifying factors at all iterations are stored so that the modification process for the right hand side can be recreated. This is normally not practical and the system is solved again, regardless of modifications.

After solution for each probe position, the magnetic vector potential at all nodes is obtained. From this, a number of important quantities are evaluated in addition to the magnetic flux density. These include coil impedances, energy, and in fact, any other quantity including forces. As an example, coil impedance is calculated from the magnetic vector potential directly. To do so we start with the general formula for calculation of impedance of a length of wire carrying a current. This is given as

$$z = \frac{j\omega}{I}\int_l \mathbf{A}\cdot d\mathbf{l} \tag{7.114}$$

where the current (rms) in the wire is I. If the wire is straight and the magnetic vector potential in the same direction as the wire, the impedance per unit length of the wire is

$$z = \frac{j\omega \mathbf{A}}{I}$$

If instead of a single conductor there are n conductors per square meter (turn density), each carrying a current I, the total impedance per unit length of a bundle of these wires that fills an element of cross sectional area Δ is

$$z = \frac{j\omega n \Delta A_c}{I}$$

where the magnetic vector potential within the element is assumed to be constant and equal to an average value A_c. If the coil occupies k elements, the impedance per unit length is

$$z = \frac{j\omega n}{I} \sum_{i=1}^{k} \Delta_i A_{ci}$$

Normally, the turn density is not given but, if the total current in the coil is known, the turn density can be written as $n=J/I$. Using this, the expression is

$$z = \frac{j\omega J_s}{I_s^2} \sum_{i=1}^{k} \Delta_i A_{ci} \tag{7.115}$$

where the index s is added to indicate a source.

This however gives the impedance for straight conductors. For curved conductors such as a coil, the impedance is found by multiplying by $2\pi r$ where r is the radial distance to the centroid of the element as given above. The impedance of a coil that occupies k elements in cross-section is

$$z = \frac{j\omega 2\pi J_s}{I_s^2} \sum_{i=1}^{k} \Delta_i A_{ci} r_{ci} \tag{7.116}$$

Using this method, the impedance of the coil at each probe position is calculated. This is then plotted as shown in Figure 7.24 to produce an impedance plot which is characteristic of the defect. Note that in this case, the impedance plot is rather coarse and not entirely symmetric. This is because of the very coarse mesh we chose. However, as the number of elements increases and the boundaries of the mesh are taken further away from the source, the solution improves (as will be shown in the next example).

We must note here that the expression in Eq. (7.116) is only the ac impedance: that is, the resistance of the coil (due to resistance of the wires) is not included in this calculation. If necessary, it can be added to the real part of the expression in Eq. (7.116) but, since it does not change with probe position, it is usually not required in calculation or in evaluation of the defect. Thus, the values of reactance and resistance in Figure 7.24 are only the ac values.

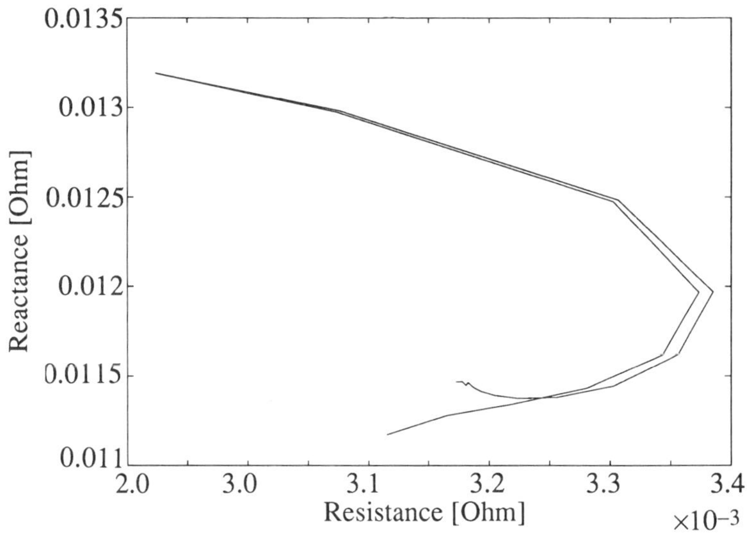

Figure 7.24 Impedance plane plot for the coil in Figure 7.22a. This plot is characteristic of the slot.

A similar solution method applies to a differential probe. A geometry of this type is shown in Figure 7.22b. Two coils, with currents in opposite directions are shown. Although the problem is rather similar in its description, the results are quite different. In particular we note that since the two coils move together, their impedance in the absence of a defect is zero. Similarly, if the coils are placed symmetrically about a symmetric defect, the impedance is also zero.

After the calculation of the magnetic vector potential, the impedance is calculated using Eq. (7.116) for each coil. The current density in coil 2 is negative therefore we can write

$$z = \frac{j\omega 2\pi J_{s1}}{I_{s1}^2}\sum_{i=1}^{k_1} \Delta_i A_{ci} r_{ci} - \frac{j\omega 2\pi J_{s2}}{I_{s2}^2}\sum_{j=1}^{k_2} \Delta_j A_{cj} r_{cj}$$

or, with currents of the same magnitude and with the same number of elements in each coil we can write

$$z = \frac{j\omega 2\pi J_s}{I_s^2}\left[\sum_{i=1}^{k} \Delta_i A_{ci} r_{ci} - \sum_{j=1}^{k} \Delta_j A_{cj} r_{cj}\right] \tag{7.117}$$

The impedance plot of the defect is shown in Figure 7.25 as registered by the differential probe shown in Figure 7.22b. Note in this case the more completely symmetric plot, due to the use of some *9,600* triangular elements and *4,961* nodes. The mesh is similar to that in Figure 7.23 but much larger both in dimensions and in number of elements. The outer geometry is now taken to 4 coil diameters away or to *30 cm*, and the length of the geometry is *16 cm*, effectively ensuring that the field is zero at the outer boundaries.

Figure 7.25 Impedance plane trajectory for the differential probe in Figure 7.22b.

In addition to impedance, contour plots of the magnetic vector potential can also be used to represent the solution. However, in the axisymmetric case, the correct variable to display is the magnitude of rA rather than of A. Plots of this type are shown in Figure 7.26a for the absolute probe in air and in Figure 7.26b for the differential probe inside the tube, but without the defect. Each of these plots gives information for a single probe position. While this may not seem to be very useful quantitatively, display of a sequence of such plots, perhaps in an animation sequence can reveal a great deal about the physics of the field interaction with materials and discontinuity.

As mentioned above, a number of parametric studies can be conducted including effects of frequency, permeability, defect size and location, foreign materials, etc.

In addition to the calculation of impedance, stored and dissipated energy in the system, input energy, and inductance are calculable from basic field relations. These are demonstrated for the three-dimensional calculation shown in the following section. However, the inductance of a coil can be immediately calculated from Eq. (7.117) by writing

$$z = R + j\omega L = -\frac{\omega 2\pi J_s}{I_s^2}\, Imag\left[\sum_{i=1}^{k} \Delta_i A_{ci} r_{ci}\right] + j\omega \frac{2\pi J_s}{I_s^2}\, Real\left[\sum_{i=1}^{k} \Delta_i A_{ci} r_{ci}\right]$$

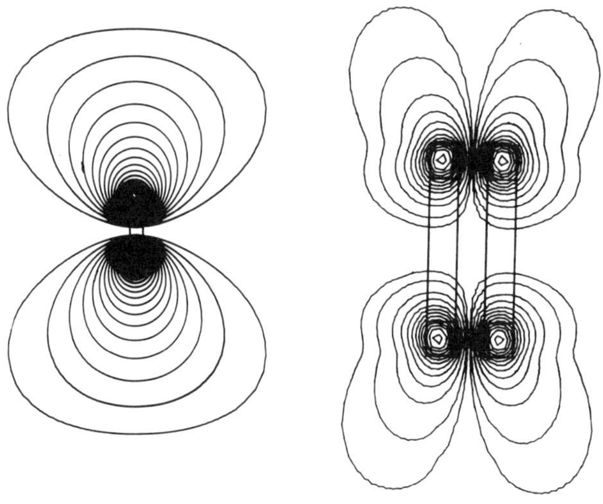

Figure 7.26 Contour plots showing the magnetic field of eddy current coils. a. Absolute probe in air. b. Differential probe inside nonmagnetic tube.

The inductance of the coil can therefore be written as

$$L = \frac{2\pi J_s}{I_s^2} Real\left[\sum_{i=1}^{k} \Delta_i A_{ci} r_{ci}\right] \tag{7.118}$$

In the two-dimensional case, a similar result can be obtained from Eq. (7.116) as

$$L = \frac{J_s}{I_s^2} Real\left[\sum_{i=1}^{k} \Delta_i A_{ci}\right] \tag{7.119}$$

and, in accordance with the definition of inductance in two dimensional geometries, this is an inductance per unit length.

7.7.3 The remote field effect

Calculation of testing parameters in the remote field effect is similar to those for the case of eddy current coils with a few changes:

1. The coils are much further apart and the corresponding fields are considerably lower in intensity.
2. Geometrical boundaries of the solution domain must be taken much further away from sources because even a small change in the flux distribution can affect the results. This is due to the very low field intensity present.
3. The quantities normally calculated are the magnetic flux density and induced voltages in the pickup coils.

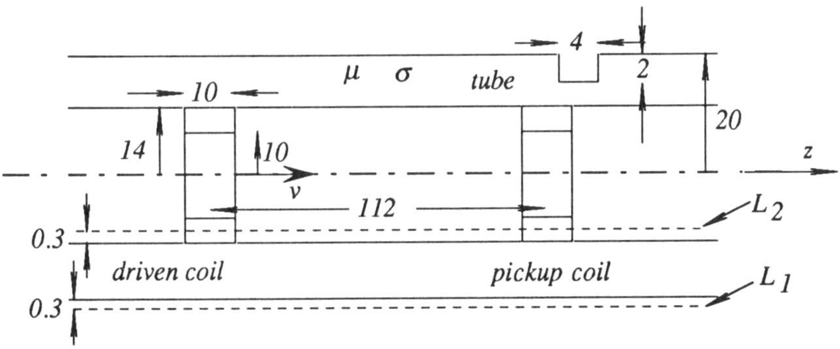

Figure 7.27 Geometry for a remote field effect test in ferromagnetic tubes.

A typical geometry is shown in Figure 7.27. The driving coil is *24 mm* in diameter (average) and the pickup coil is at a distance of *112 mm* in front of the driving coil. The object here is to calculate the induced voltage in the coil as well as the magnetic flux density below and above the tube surface. The tube itself is *6 mm* thick and made of steel. The magnetic flux density is calculated on lines L_1 and L_2 as shown in Figure 7.27. The induced voltage due to the axisymmetric, external defect is calculated. Each coil is *10 mm* long, contains *400* turns and is and *4 mm* thick. Current density in each coil is 3×10^6 A/m^2 at *50 Hz*. Conductivity of the tube is 5×10^7 S/m and relative permeability is *50*.

Using the formulation of the previous section, a single coil (the driving coil) is modeled. The mesh for this geometry is rather dense since low field intensities at relatively large distances must be modeled accurately. A total of *9,240* elements and *9,512* nodes are used to calculate the magnetic vector potential.

After calculation of the magnetic vector potential, the magnetic flux density is calculated based on the relations in Eq. (7.106) and (7.107). Using the definitions above for the centroidal value of the radial distance, the magnetic flux density in each triangular element is calculated as

$$B_r = -\frac{x_3 - x_2}{2\Delta} A_1 + \frac{x_3 - x_1}{2\Delta} A_2 - \frac{x_2 - x_1}{2\Delta} A_3 \qquad (7.120)$$

$$B_z = \frac{y_3 - y_2}{2\Delta} A_1 + \frac{y_3 - y_1}{2\Delta} A_2 - \frac{y_2 - y_1}{2\Delta} A_3 + \frac{A_1 + A_2 + A_3}{r_1 + r_2 + r_3} \qquad (7.121)$$

where the indices *1,2,3*, refer to the three nodes of a triangular element. These values are calculated for each element in the solution domain. It should be noted that these are constant within each element and are associated with the centroid of the element.

Plots of the flux densities along lines L_1 and L_2 are shown in Figure 7.28 for the radial flux density and in Figure 7.29 for the axial flux density. Both figures are calculated at zero velocity. To be noted is the fact that the axial flux density on the outer line is larger than on the inner line beyond a certain distance. If the pickup coil is placed beyond this distance the induced voltage due to outer surface defects is equal or larger than that due to internal surface defects. This crossover of the two curves is in fact the whole reason for the remote field effect.

In addition to flux densities, the induced voltage in the pickup coil can be calculated. A plot of the induced voltage as the pickup coil crosses under the defect is shown in Figure 7.30 in which zero distance is the center of the slot.

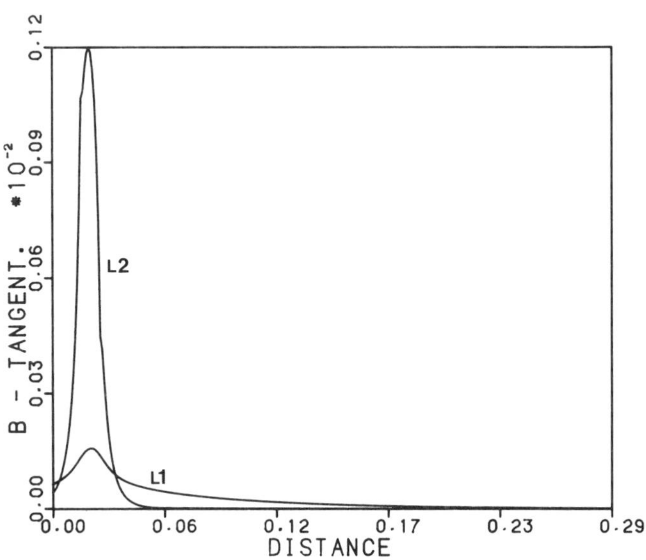

Figure 7.28 Radial flux density component for lines L_1 and L_2 in Figure 7.27.

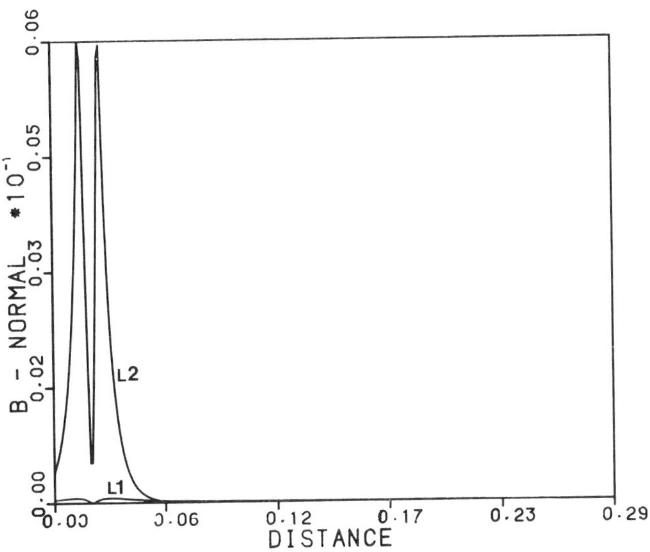

Figure 7.29 Axial flux density component for lines L_1 and L_2 in Figure 7.27.

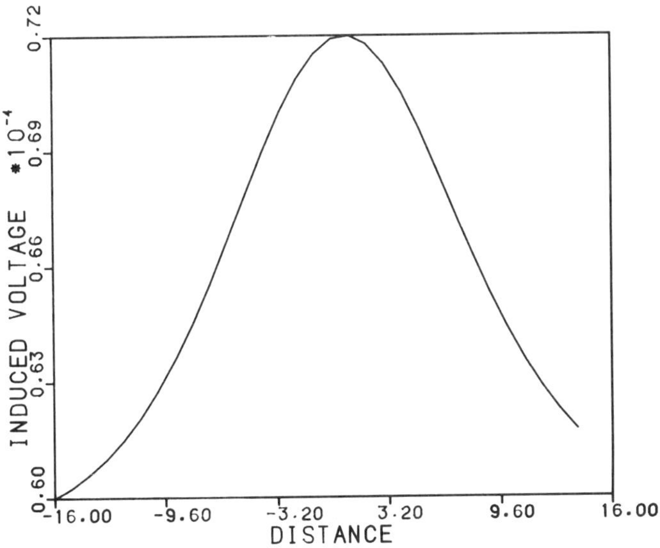

Figure 7.30 Induced voltage due to the outer axisymmetric slot as detected by the pickup coil.

7.8 EXAMPLES: THREE-DIMENSIONAL APPLICATIONS

Although the three-dimensional formulation is given in detail in section 7.3 of this chapter, modeling of *3-D* geometries is considerably more complicated than in two dimensions or axisymmetric geometries. There are many reasons for this. Not the least are problems with visualization and accuracy. Mesh data for the finite element program is quite extensive and the resulting system of equations extremely large. However, in spite of this, *3-D* solutions to *NDT* problems are possible and are demonstrated here.

As an example, we show mesh generation and solution to the problem of wear scar modeling in stainless steel tubing (Inconel 600) as occuring in steam generators. This particular problem is typical of the type of problems one may need to model as a three-dimensional application and also points out various considerations encountered in three-dimensional applications.

The geometry considered is shown in Figures 7.31a and 7.31b. It contains two identical wear scars on the opposite sides of the tube. Geometrical data and dimensions are given in these figures. Modeling is done at a frequency of *10 kHz*. The conducting tube has conductivity of 10^6 *S/m* and permeability of free space.

The first task is discretization into finite elements. This step is shown in Figures 7.32 through 7.36. Figure 7.32 shows the dimensions of the solution domain and choice of boundaries. While it is desirable to locate the boundary as far as possible from the coils, for reasons of economy they are located at about *40 mm* from the center of the coil. The coil itself has a radius of *9.5 mm*. Longitudinal dimensions are dictated both by the required distance from the coil and by the number of probe positions needed for modeling. The total tube length is *82 mm* with the probe moving within *32 mm* symmetrically about the scars. *25* probe positions are required to obtain a reasonably smooth impedance curve.

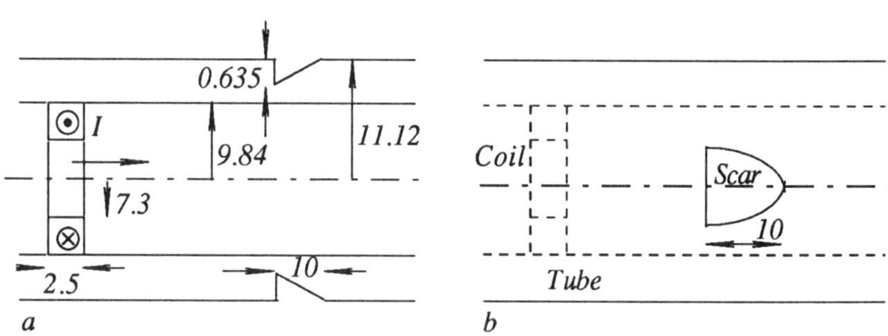

Figure 7.31 Geometry of a wear scar on the outer surface of an Inconel tube.

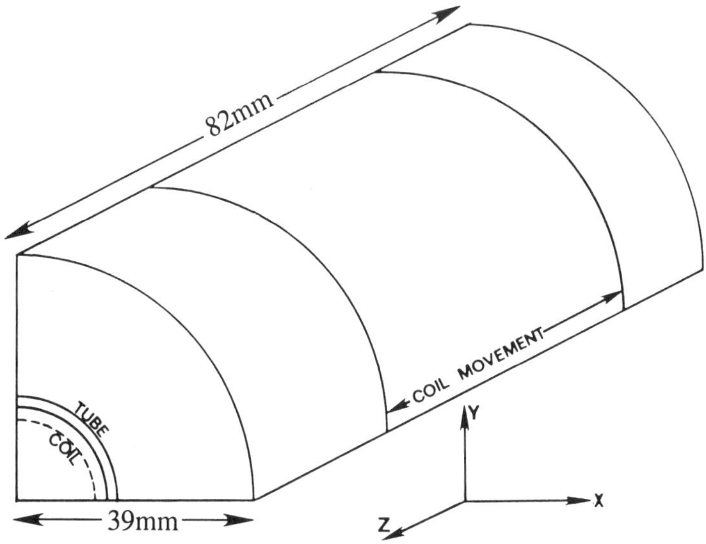

Figure 7.32 Basic geometry used for solution. A quarter of the actual tube is modeled.

To produce the mesh, the geometry is first divided into four quarters, one of which is modeled as shown in Figure 7.32. To simplify mesh generation, consider Figure 7.33. The end surface (that is, the quarter plane in Figure 7.32) is divided as shown in Figure 7.33. This basic structure is repeated at each axial layer of elements. This produces a mesh with a single layer of elements in the coil and two layers in the tube. In the circumferential direction there are eight elements. A *3-D* display of the mesh is shown in Figure 7.34. After the mesh is defined the wear scars are defined by specifying material properties. While the tube is defined in terms of the conductivity of Inconel, the wear scar is modeled as free space in the tube. The shape of the wear scar is obtained by moving the corresponding nodes in the tube. This is shown in Figure 7.35 where a few elements in the region of the scar are shown. In Figure 7.35a and 7.35c, the elements as generated by the mesh generator are shown. After adjustment of a few nodes, the elements in Figures 7.35b and 7.35d are obtained. Modifications can be done directly in the mesh data file or in the solution program after mesh data is read. It can also be done by writing a small program to read the data and rewrite it with the required modifications.

The result of the mesh generation step is a mesh with *4,032* hexahedral (eight-node) elements and *4,945* nodes. With the formulation in section 7.3 a total of *19,780* variables are calculated (*19,780* unknowns). If the formulation in section 7.4 is used, only three variables per node are needed and the total number of variables is *14,835*.

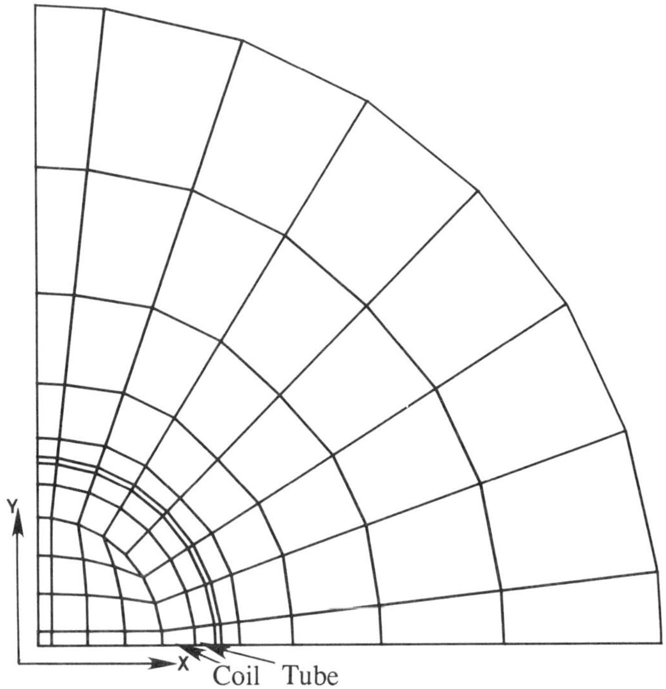

Figure 7.33 Division of the end surface into elements. The location of the coil and tube are shown.

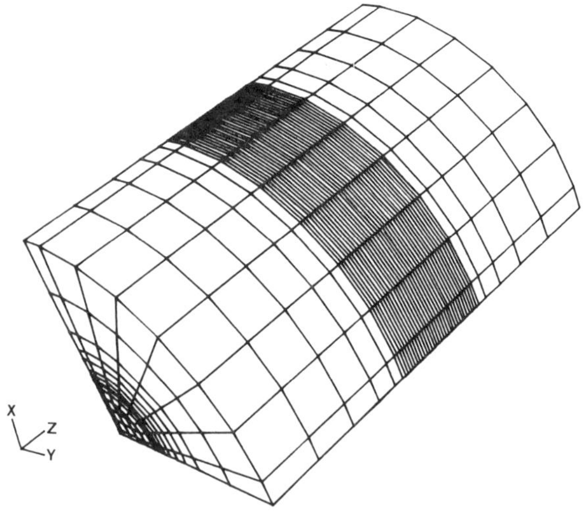

Figure 7.34 A 3-D display of the mesh with hidden lines removed. The uniform mesh area is the area of coil movement.

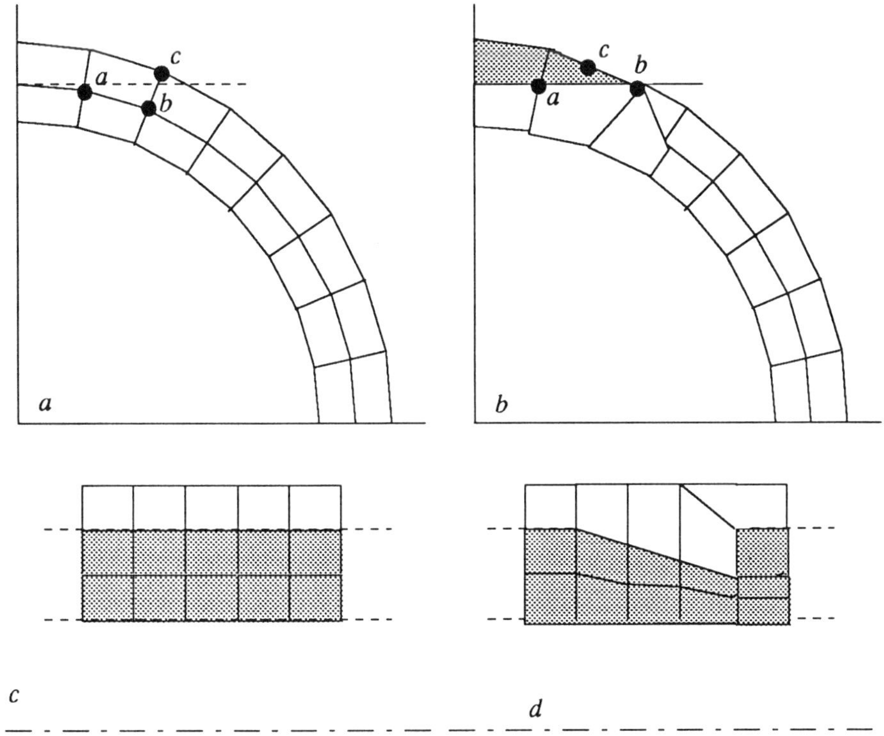

Figure 7.35 Adjustment of nodes to produce the wear scar. Two views of wear scar are given. On the left, the mesh as generated. On the right, adjusted nodes.

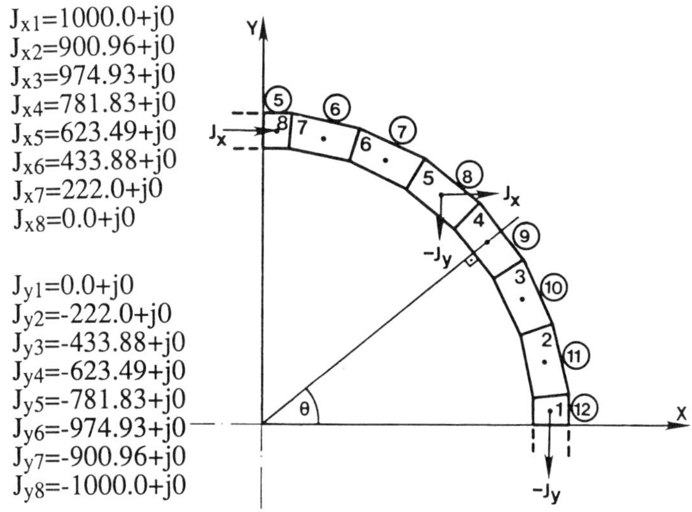

Figure 7.36 Modeling of current density in the coil.

Before the solution can commence, sources (current density in coils) must be defined. This step is shown in Figure 7.36 where the coil is divided into eight elements and the current densities calculated based on the direction of each element. The x and y components of the current density are given in the figure. These current densities are then entered as part of the input data to the finite element program and the resulting system of equations is solved in the normal fashion, using complex variables.

This problem, while certainly a large problem to solve, can be relatively easily solved using a conjugate gradient algorithm.

After solution, impedance of the coils (either absolute coil or differential coils) is calculated. However, unlike the axisymmetric solution in section 7.6 the impedance here must be calculated from energy considerations.

To calculate probe impedance we must first calculate the stored and dissipated energy. First we calculate the magnetic flux density and eddy current density. The magnetic flux density is calculated from the definition of the curl as

$$B_x = \frac{\partial A_z}{\partial y} - \frac{\partial A_y}{\partial z} \qquad B_y = \frac{\partial A_x}{\partial z} - \frac{\partial A_z}{\partial x} \qquad B_z = \frac{\partial A_y}{\partial x} - \frac{\partial A_x}{\partial y}$$

These are easily calculated in each element using the approximation for the finite element as

$$B_x = \sum_{i=1}^{8} \frac{\partial N_i}{\partial y} A_{zi} - \sum_{i=1}^{8} \frac{\partial N_i}{\partial z} A_{yi} \tag{7.122}$$

$$B_y = \sum_{i=1}^{8} \frac{\partial N_i}{\partial z} A_{xi} - \sum_{i=1}^{8} \frac{\partial N_i}{\partial x} A_{zi} \tag{7.123}$$

$$B_z = \sum_{i=1}^{8} \frac{\partial N_i}{\partial x} A_{yi} - \sum_{i=1}^{8} \frac{\partial N_i}{\partial y} A_{xi} \tag{7.124}$$

Eddy current densities in each element are similarly calculated as

$$J_x = -j\omega\sigma A_x = -\frac{1}{8}\sum_{i=1}^{8} j\omega\sigma A_{xi} \tag{7.125}$$

$$J_y = -j\omega\sigma A_y = -\frac{1}{8}\sum_{i=1}^{8} j\omega\sigma A_{yi} \tag{7.126}$$

$$J_z = -j\omega\sigma A_z = -\frac{1}{8}\sum_{i=1}^{8} j\omega\sigma A_{zi} \tag{7.127}$$

where A_{xi}, A_{yi} and A_{zi} indicate the three components of the magnetic vector potential at the eight nodes of the element. Both flux densities and current densities are assumed to be uniform in each finite element for each component. Eddy current density can also be calculated using the approximation for A in which case it varies linearly within a finite element but then integration over the element is normally required in evaluating the total current in an element. Now the stored and dissipated energies are calculated from the terms of the energy functional. The stored energy is

$$W_s = \frac{1}{2}\left(\nu_x B_x^2 + \nu_x B_y^2 + \nu_x B_z^2\right) v_i \tag{7.128}$$

and the dissipated energy is

$$P_d = \frac{v_i |J_e|^2}{\sigma} \tag{7.129}$$

or, in terms of the magnetic vector potential

$$P_d = v_i \omega^2 \sigma |A_i|^2 \tag{7.130}$$

where v_i is the volume of element i and A_i is the average magnitude of the magnetic vector potential. The resistance and inductance are now calculated from the expressions

$$W = \frac{1}{2} LI^2$$

and

$$P = RI^2$$

Thus, the impedance of the coil is

$$Z = R + j\omega L = \frac{1}{I^2}(P + j\omega 2W) \tag{7.131}$$

As with the axisymmetric case above, this impedance does not reflect the self resistance of the coil (dc resistance).

The impedance of the coil or coils is calculated at each probe position as in the axisymmetric case but using the expressions above and plotted to produce an impedance plane plot. The impedance plane plot for the wear scars in Figure 7.31 is given in Figure 7.37 for an absolute probe and in Figure 7.38 for a differential probe.

Figure 7.37 Absolute probe impedance plane trajectory for the wear scar in Figure 7.31.

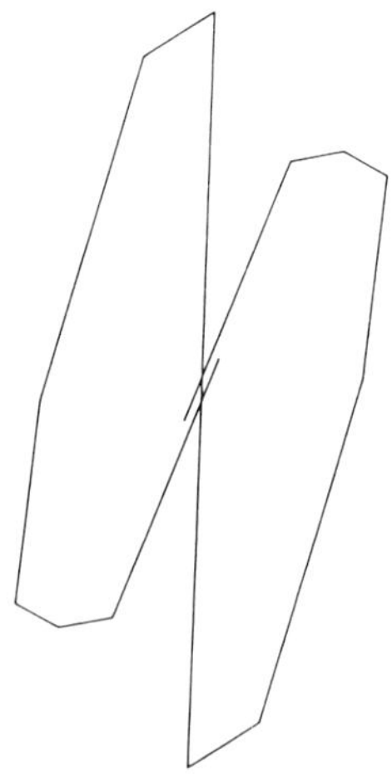

Figure 7.38 Differential probe impedance plane trajectory for the wear scar in Figure 7.31.

7.9 EXTENSIONS AND MODIFICATIONS

The first extension to the formulations given above is to time-harmonic wave propagation as used for solution of *TE* or *TM* modes in closed structures (waveguides and cavity resonators). The method is applied to the same types of problems as in chapter 6. The type of geometries considered here are shown in Figure 6.20. However, as we will see shortly, the finite element method is also modified to allow calculation for resonant frequencies in cavities without the need to solve an eigenvalue problem.

7.9.1 Solution of Helmholtz's equation

Formulation for *TM* modes starts with the following equation, given here in two dimensions as used for calculations in waveguides:

$$\frac{\partial^2 E_z}{\partial x^2} + \frac{\partial^2 E_z}{\partial y^2} + k_c^2 E_z = 0\,, \qquad k_c^2 = \gamma^2 + k^2 \tag{7.132}$$

with Dirichlet boundary conditions $(E_z=0)$ on the conducting surfaces of the cavity as shown in Figure 6.21. Using the same considerations as in section 6.5.1, we can write this equation as

$$\frac{1}{\mu_r}\left(\frac{\partial^2 E_z}{\partial x^2} + \frac{\partial^2 E_z}{\partial y^2}\right) + k_0^2\, \varepsilon_r E_z = 0 \tag{7.133}$$

where k_0 is the free space wavenumber and once calculated can be directly interpreted as the cutoff frequency of a particular mode of propagation using Eq. (6.34) or (7.132). From a formulation point of view, we must come up either with an energy related functional or use Galerkin's method. The latter is more convenient here, especially since we can apply the method for each term separately. This separation of terms allows the use of the already defined matrices for the Laplacian part of the equation and then add the matrices resulting from the second part of the equation. An appropriate approximation over any finite element in this case is

$$E_z(x,y) = \sum_{i=1}^{M} E_i N_i$$

where M is the number of nodes in the element. We use triangular elements in this case, therefore $M=3$. To simplify formulation, we note the following:

The first term in Eq. (1.133) is identical to that of Eq. (7.90) except there μ is permeability while μ_r in Eq. (7.133) is only relative permeability. Thus, after applying Galerkin's method we obtain the same general coefficients of the matrix as in Eq. (7.93):

$$[s] = \frac{1}{4\Delta\mu_r}\begin{bmatrix} (y_3-y_2)(y_3-y_2)+(x_3-x_2)(x_3-x_2) & (y_3-y_2)(y_3-y_1)+(x_3-x_2)(x_1-x_3) & (y_3-y_2)(y_1-y_2)+(x_3-x_2)(x_2-x_1) \\ (y_3-y_1)(y_3-y_2)+(x_1-x_3)(x_3-x_2) & (y_3-y_1)(y_3-y_1)+(x_1-x_3)(x_1-x_3) & (y_3-y_1)(y_1-y_2)+(x_1-x_3)(x_2-x_1) \\ (y_1-y_2)(y_3-y_2)+(x_2-x_1)(x_3-x_2) & (y_1-y_2)(y_3-y_1)+(x_2-x_1)(x_1-x_3) & (y_1-y_2)(y_1-y_2)+(x_2-x_1)(x_2-x_1) \end{bmatrix} \tag{7.134}$$

where isotropic materials were assumed.

The second term is again treated by comparison with previously defined matrices. The term is similar to the last term in Eq. (7.98). Thus, if we replace in Eq. (7.98) the constants $j\omega\mu_0\mu_r$ by $k_0^2\varepsilon_r$, we can use the results obtained in section 7.6.1. Using this approach on the second part of Eq. (7.133) we get the following system of equations

$$[s]\{E\} = k_0^2[q] \tag{7.135}$$

where $[s]$ is given in Eq. (7.134) while $[q]$ is

$$[q] = \frac{\varepsilon_r \Delta}{6}\begin{bmatrix} 2 & 1 & 1 \\ 1 & 2 & 1 \\ 1 & 1 & 2 \end{bmatrix} \tag{7.136}$$

This is an eigenvalue problem in complex variables since, in general, permittivity must be assumed to be complex. The eigenvalues represent the cutoff frequencies of the various modes and the eigenvectors, the electric field variation in the cross-section of a waveguide.

The system in Eq. (7.135) may be solved by any method of solution for eigenvalues or eigenvectors after boundary conditions are imposed.

The extension of this method to three dimensions is also straightforward. The equation required for solution of *3-D TM* modes is given in Eq. (1.160)

$$\frac{\partial^2 E_z}{\partial x^2} + \frac{\partial^2 E_z}{\partial y^2} + \frac{\partial^2 E_z}{\partial z^2} + k_c^2 E_z = 0\,, \qquad k_c^2 = \gamma^2 + k^2 \tag{7.137}$$

Again, without having to repeat the whole process, (see problem 7.10) we note that Eq. (7.137) can be considered to be scalar. The first term is the *3-D* Laplacian for which an approximation is found in section 7.5.3. Therefore, using the same generic element as in section 7.3.3, the $[s]$ and $[q]$ matrices for the three dimensional formulation are

$$s_{i,j} = \frac{1}{\mu_r}\int_{v_i}\left(\frac{\partial N_i}{\partial x}\frac{\partial N_j}{\partial x} + \frac{\partial N_i}{\partial y}\frac{\partial N_j}{\partial y} + \frac{\partial N_i}{\partial z}\frac{\partial N_j}{\partial z}\right)dxdydz \tag{7.138}$$

$$q_{i,j} = \varepsilon_r\int_{v_i} N_i N_j dxdydz \tag{7.139}$$

Thus, Eq. (7.135) remains unaltered but now the system is much larger because of the three dimensional nature of the problem it describes. Using the latter equation on the cavity in Figure 6.20b for the TM_{111} mode the resonant frequency is found to be *246.4 MHz*. This is within *1%* of the result obtained using finite differences for the same problem in section 6.5.5. Note also that the solution is actually closer to the exact solution (see Table 6.2). The finite element solution requires only a $5\times5\times5$ mesh of eight-node hexahedral elements with a total of *125* elements and *216* nodes. The system of equations is a 216×216 (sparse) system and can be solved relatively easily.

7.9.2 Solution of Helmholtz's equation as a deterministic problem.

It is possible to find the resonant frequencies of a cavity resonator by an indirect approach which does not require solution of an eigenvalue problem. The method involves solution of a sequence of deterministic problems from which the resonant frequencies are found by calculating the energy in the resonant cavity. The method uses the basic principle that energy in a resonant cavity increases considerably at and near resonance. This is equivalent to measuring the resonant frequency of a resonant cavity. To do so, the frequency of the cavity is incremented within a given range and resonance is detected. When resonance has been detected, the measured frequency is the resonant frequency of the cavity.

The advantage of a method of this sort is obvious: it is simple and does not require solution of eigenvalue problems and the method is physical in the sense that it closely resembles measurements. On the other hand, it requires solution of a *3-D* problem for a range of frequencies and this can be expensive.

To formulate the problem we start with the general time harmonic field equation as given in Eq. (7.6). However, in this case we do not neglect the displacement currents. Eq. (7.6), after performing all steps leading to Eq. (7.8) but without neglecting displacement currents yields

$$\frac{\partial^2 \mathbf{A}}{\partial x^2} + \frac{\partial^2 \mathbf{A}}{\partial y^2} + \frac{\partial^2 \mathbf{A}}{\partial z^2} = -\mu \mathbf{J}_s + j\omega\mu\sigma\mathbf{A} - \omega^2\mu\varepsilon\mathbf{A} \qquad (7.140)$$

This is the general, time harmonic eddy current equation. Comparing this equation with Eq. (7.8) and the formulation in section 7.3.2 reveals that formulation of Eq. (7.140) requires the addition of the term $\omega^2\mu\varepsilon\mathbf{A}$. We also note that the second and third terms on the right hand side are identical if we replace $j\omega\mu\sigma$ by $\omega^2\mu\varepsilon$. Thus, although not done here, we can expect that the matrix contributions of these two terms must also be identical except for the constants mentioned. Thus, following all steps in section 7.3.2 we get the following system of equations

$$\{[S] + j\omega\sigma[R] - \omega^2\mu\varepsilon[P]\}\{A\} = \{Q\} \qquad (7.141)$$

and the coefficients of the matrix $[P]$ are identical to those for the matrix $[R]$. These are

$$P_{3i-2,3j-2} = P_{3i-1,3j-1} = P_{3i,3j} = \int_{v_i} N_i N_j dxdydz \tag{7.142}$$

while the matrices [S], [R], and [Q] are given in Eqs. (7.42) through (7.51).

In effect, the method up to this point is an eddy current method which includes displacement currents. These currents are normally neglected in eddy current applications. Here however, the finite element method is applied at a series of frequencies and, given the source current density the system in Eq. (7.141) solved for each frequency. To detect resonance, the stored magnetic and stored electric energy in the solution domain is calculated. The stored magnetic energy is calculated using Eq. (7.128)

$$W_m = \frac{1}{2\mu}\left(B_x^2 + B_y^2 + B_z^2\right)v_i \tag{7.143}$$

where B_x, B_y, and B_z are given by Eqs. (7.122) through (7.124) and an isotropic material is assumed. The stored electric energy is calculated as

$$W_e = \frac{\varepsilon}{2}\left(B_x^2 + B_y^2 + B_z^2\right)v_i \tag{7.144}$$

and the three components of the electric field intensity are calculated as

$$E_x = -j\omega A_x = -\frac{1}{8}\sum_{i=1}^{8} j\omega A_{xi} \tag{7.145}$$

$$E_y = -j\omega A_y = -\frac{1}{8}\sum_{i=1}^{8} j\omega A_{yi} \tag{7.146}$$

$$E_z = -j\omega A_z = -\frac{1}{8}\sum_{i=1}^{8} j\omega A_{zi} \tag{7.147}$$

Summing up the terms in Eqs. (7.143) and (7.144) gives the total energy in the finite element solution domain. If this domain is the interior of a resonant cavity, this energy is the total energy in the cavity. A maximum in this energy occurs at resonance and the frequency at which this is achieved is the resonant frequency of the cavity.

7.9.3 Nonlinear applications

The need for nonlinear methods occurs often in modeling of nondestructive testing geometries. While ac applications are dominated by low intensity fields where nonlinearities are minimal, whenever ferromagnetic materials are magnetized, the behavior is nonlinear. In particular, active and residual leakage field methods of testing require nonlinear considerations. In the active case, magnetization of samples is performed by increasing the current in samples or by applying a magnetic field through a magnetic yoke. In either case, the field can be viewed as being increased from zero to a predetermined magnitude at which testing takes place.

In residual methods of testing, the opposite occurs. A sample is first magnetized and then the field is reduced gradually until the sample relaxes into a residual field state which is then used for flaw detection.

From a modeling point of view, either method requires two main considerations: one is the magnetization curve, the other the nonlinear iteration. Both of these and their implementation are discussed in detail in section 6.6 in conjunction with the application of the finite difference method to nonlinear calculations. All steps and considerations given there apply here without change. The only minor variation is that at each step of the nonlinear iteration a finite element rather than a finite difference solution must be obtained.

Because of the identical nature of modeling and the fact that nonlinear solutions only affect the nonlinear iteration (that is, the solution of the finite element discretization at each iteration remains linear) we do not discuss this aspect any further. However, the bibliography section provides references which expand on this aspect of calculation.

7.10 PROBLEMS

7.1 Consider the geometry described in problem 6.1. Use the formulation for electrostatic fields given in section 7.5.1 to solve the same problem. Show that the charge can be taken into account directly. Use as few elements as possible but in such a way that there is at least one node in the solution domain in addition to the node at which the charge q is located. Increase the number of nodes gradually and plot the solution at a selected node with respect to number of nodes (or elements) in the solution domain. Show that the solution converges to a final solution.

7.2 A two-dimensional application is shown in Figure 7.39. Use triangular elements to calculate the magnetic flux density everywhere in the solution domain. Use a reasonable number of elements and use symmetry boundary lines to keep the mesh small.

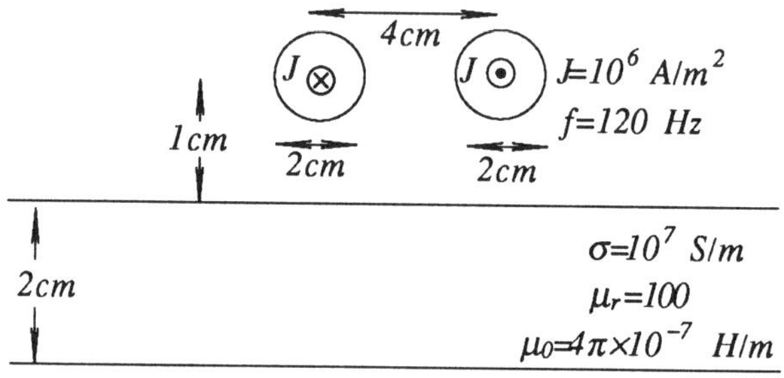

Figure 7.39.

7.3 An overhead transmission line is located over a conducting ground as shown in Figure 7.40. Assume the conductor is very thin and the earth to be flat and uniform. Calculate the magnetic field intensity at the surface of the earth. Compare this with an analytic solution obtained by assuming the ground to be perfectly conducting (see chapter 2).

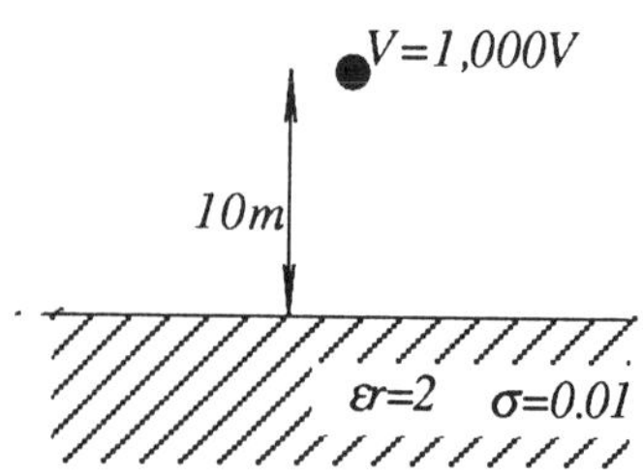

Figure 7.40.

7.4 A three-phase transmission line is shown in Figure 7.41. Using the currents given, calculate the magnetic flux density, electric field and potential everywhere in the solution domain. Plot the electric field and the magnetic field at ground level to find location of maximum field intensities.

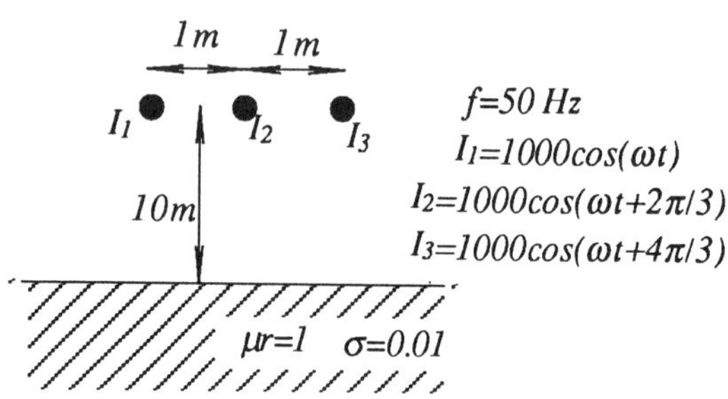

Figure 7.41.

7.5 A circular coil shown in Figure 7.42 is given. The coil contains *400* turns and carries a *1A* current. Frequency is *1 kHz*. Use the axisymmetric formulation given in section 7.7 to calculate the magnetic field everywhere in space. Pay particular attention to the location of boundaries and boundary conditions.

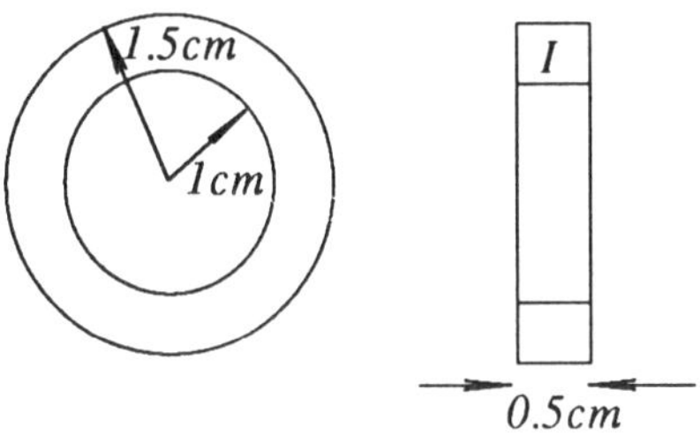

Figure 7.42.

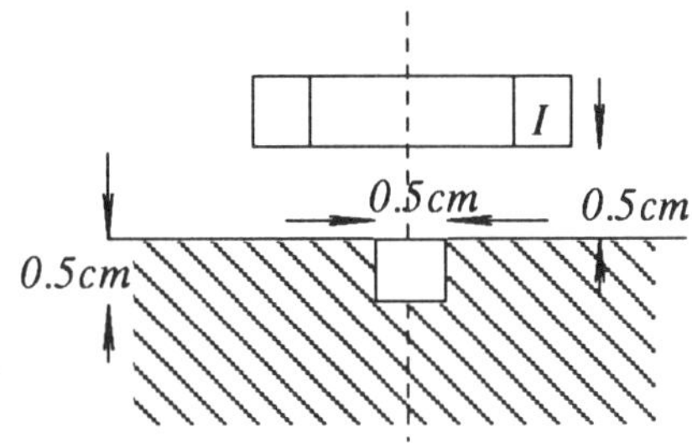

Figure 7.43.

7.6 The coil in Figure 7.42 is placed over a conductor, centered with a flat bottomed hole as shown in Figure 7.43. Define a solution domain that will allow you to move the coil in small increments vertically. Obtain an impedance plot as the coil moves away vertically from the hole. Comment on the use of this method to characterize defects. Change the hole diameter and depth to see how this changes the impedance curves.

7.7 A test for defects in circular rods is shown in Figure 7.44. Use the coil in Figure 7.42 in the configuration in Figure 7.44 but at a frequency of *100 kHz*. Move the coil past the axisymmetric slot and obtain an impedance plane trajectory plot for the defect.

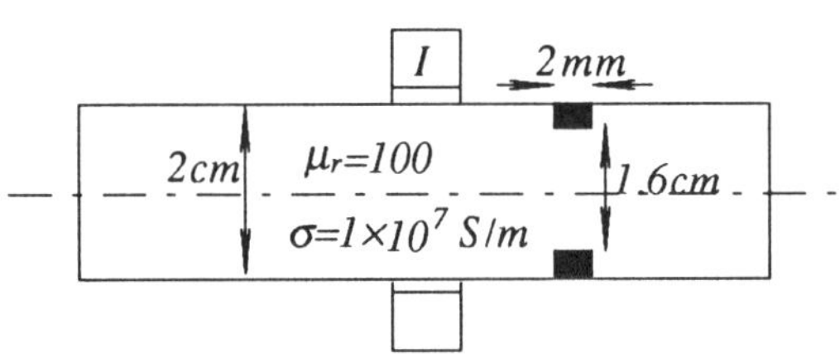

Figure 7.44.

7.8 Repeat problem 7.7 but with a differential probe as in Figure 7.45. The two coils are equal in all respects except that the currents are in opposite directions. Comment on this method in terms of sensitivity, amplitudes, sensitivity to noise, etc.

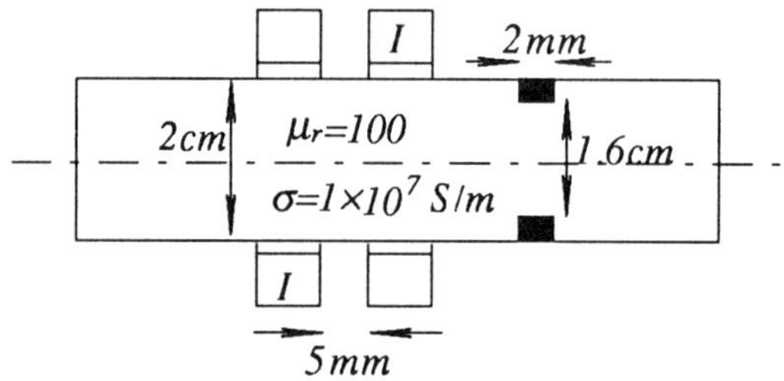

Figure 7.45.

7.9 Consider again the question of the coil over a bar but now there is no defect. Rather, the test is designed to test for thinning of the bar. To simulate thinning, the problem is solved by simulating a number of geometries, each with a slightly thinner bar (one situation is shown in dotted lines in Figure 7.46). Use a frequency of *100 kHz* and *10* thinning steps. Plot the impedance of the coil to show how a calibration curve for thinning may be obtained.

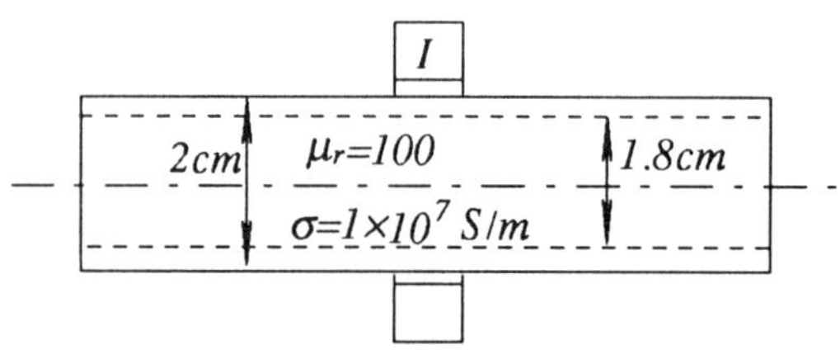

Figure 7.46.

7.10 Show that finite element formulation for *TE* modes leads to equations identical to those in Eqs. (7.138) and (7.139) and that the only difference in the formulation is in the use of the magnetic field intensity as the variable (H_z) and the need to impose Neumann boundary conditions for the magnetic field intensity on the boundaries. Explain how these are imposed considering the fact that Neumann boundary conditions are implicit in the formulation.

7.11 Calculate the first three resonant frequencies of a loaded resonant cavity (dimensions are $1 \times 1 \times 1$ *m*) and the cavity is loaded with a dielectric that fills half the cavity. The dielectric has relative permittivity of *80* and conductivity of 1×10^{-9} *S/m*. Use the formulation in section 7.9.2.

7.11 BIBLIOGRAPHY

Formulation in terms of finite elements has received considerable attention because of the effect formulations have on solutions and computation. Some formulations are more suited for particular applications while others are more accurate, more efficient or, perhaps, simpler to implement. The sources cited here are intended to expand on formulations given in this chapter. The number of sources is, however, very large and only representative examples are given here.

General discussion of finite elements, the finite element method and associated techniques can be found in references [1] and [2]. A more "field oriented" approach can be found in [3-5]. Special aspects of formulations, including Galerkin's method and the variational technique are discussed in [6-8] but also, in [1-5]. Linear static formulations and applications are listed in [9-15]. Eddy current calculations and formulations are given in references [16-27]. Time harmonic wave propagation is discused in [28-33]. It is interesting to note that some of the earliest finite element formulations [28,29] dealt with microwave applications. The modified eddy current equation given in section 7.9.2 is also discussed in [32,33]. Nonlinear formulations and applications can be found in [34-39], including applications in residual field testing.

[1] Huebner, K.H. (1975), *The Finite Element Method for Engineers*, Wiley-Interscience, John Wiley & Sons, Inc., New York.

[2] Zienkiewicz, O.C. (1977), *The Finite Element Method in Engineering*, third edition, McGraw-Hill Book Co., London.

[3] Sylvester, P.P. and Ferrari, R.P. (1990), *Finite Elements for Electrical Engineers*, Cambridge University Press, Cambridge.

[4] Ida, N. and Bastos, J.P.A. (1992), *Electromagnetics and Calculation of Fields*, Springer Verlag, New York.

[5] Hoole, S.R.H. (1989), *Computer-Aided Analysis and Design of Electromagnetic Devices*, Elsevier, New York.

[6] Finlayson, B.A. (1972), *The Method of Weighted Residuals and Variational Principles*, Academic Press, New York.

[7] Ferrari, R.L. (1985), Complementary variational formulation for eddy current problems using the variables E and H directly, *IEE Proceedings*, pt. A, Vol. 132, No. 4, pp. 157-164.

[8] Hajebe, S. and Kano, Y. (1988), Variational principles in 3-D magnetostatic fields, *IEEE Transactions on Magnetics*, Vol. 24, pp. 2940-2942.

[9] Andersen, O.W. (1973), Laplacian electrostatic field calculation by finite elements with automatic grid generation, *IEEE Transactions on Power Apparatus and Systems*, Vol. PAS-92, No. 5, pp. 1485-1492.

[10] Chari, M.V.K. Silvester, P.P. Konrad, A. Cendes, Z.J. and Palmo, M.A. (1981), Three dimensional magnetostatic field analysis of electrical

machinery by the finite element method, *IEEE Transactions on Power Apparatus and Systems*, Vol. PAS-100, No. 8, pp.4007-4019.

[11] Ida, N. and Lord, W. (1983), 3-D finite element predictions of magnetostatic leakage fields, *IEEE Transactions on Magnetics*, Vol. MAG-19, No. 5, pp. 2260-2265.

[12] Lord, W. and Hwang, J.H. (1975), Finite element modeling of magnetic field/defect interactions, *ASTM Journal of Testing and Evaluation*, Vol. 3, No. 1, pp. 21-25.

[13] Chari, M.V.K. Palmo, M.A. and Czendes, Z.J. (1979), Axisymmetric and three dimensional electrostatic field solution by the finite element method, *Electric Machines and Electromechanics*, Vol. 3, 1979, pp. 235-244.

[14] Chari, M.V.K. Silvester, P.P. and Konrad, A. (1981), Three-dimensional magnetostatic field analysis of electrical machinery by the finite element method, *IEEE Transactions on Power Apparatus and Systems*, Vol. PAS-100, No. 8, pp. 4007-4019.

[15] Cendes, Z.J. Weiss, J. and Hoole, S.R.H. (1982), Alternative vector potential formulations of 3-D magnetostatic problems, *IEEE Transactions on Magnetics*, Vol. MAG-18, No. 2, pp. 367-372.

[16] Kriezis, E.E. Tsiboukis, T.D. Panas, S.M. and Tegopoulos, J.A. (1992), Eddy currents: theory and applications, *Proceedings of the IEEE*, Vol. 80, No. 10, pp. 1559-1589.

[17] Biddlecombe, C.S. Heighway, E.A. Simkin, J. and Trowbridge, C.W. (1982), Methods for eddy current computation in three dimensions, *IEEE Transactions on Magnetics*, Vol. MAG-18, No. 2, pp. 492-497.

[18] Biro, O. Preis, K. Renhart, W. Richter, K.R. and Vrisk, G. (1990), Performance of different vector potential formulations in solving multiply connected 3-D eddy current problems, *IEEE Transactions on Magnetics*, Vol. MAG-26, pp. 438-441.

[19] Carpenter, C.J. (1977), Comparison of alternative formulations of 3-D magnetic field and eddy current problems at power frequencies, *IEE Proceedings*, Vol. 124, No. 11, pp. 1026-1034.

[20] Preston, T.W. and Reece, A.B.J. (1982), Solution of 3-dimensional eddy current problems: the T - Ω method, *IEEE Transactions on Magnetics*, Vol. MAG-18, No. 2, March 1982, pp. 486-491.

[21] Brown, M.L. (1982), Calculation of 3-dimensional eddy currents at power frequencies, *IEE Proceedings*, Vol. 129, Part A, No. 1, pp. 46-53.

[22] Ida, N. and Lord, W. (1985), A finite element model for 3-D eddy current NDT calculations, *IEEE Transactions on Magnetics*, Vol. MAG-21, No. 6, pp. 2635-2643.

[23] Hong, S. and Ida, N. (1992), Modeling of velocity terms in 3D eddy current problems, *IEEE Transactions on Magnetics*, Vol. 28, No. 2, pp. 1178-1181.

[24] Ida, N. Hoshikawa, H. and Lord, W. (1985), Finite element prediction of differential EC probe signals from Fe_3O_4 deposits in PWR steam generators, *NDT International*, Vol. 18, No. 6, pp. 331-338.

[25] Ida, N. Betzold, K. and Lord, W. (1983), Finite element modeling of absolute eddy current probe signals, *Journal of Nondestructive Evaluation*, Vol. 3, No. 3, pp. 147-154.

[26] Ida, N. Palanisamy, R. and Lord, W. (1983), Eddy current probe design using finite element analysis, *Materials Evaluation*, Vol. 41, No. 12, pp. 1389-1394.

[27] Song, H. and Ida, N. (1991), An eddy current constraint formulation for 3D electromagnetic field calculation, *IEEE Transactions on Magnetics*, Vol. 27, No. 5, pp. 4012-4015.

[28] Arlet P.L. et al. (1968), Application of finite elements to the solution of Helmholtz' equation, *IEE Proceedings*, Vol. 115, No. 12.

[29] Cendes, Z.J. and Silvester, P.P. (1970), Numerical solution of dielectric loaded waveguides: I - Finite element analysis, *IEEE Transactions on Microwave Theory and Techniques*, Vol. MTT-18, No. 12.

[30] Webb, J.P. (1985), The finite-element method for finding modes of dielectric-loaded cavities, *IEEE Transactions on Microwave Theory and Techniques*, Vol. MTT-33, No. 7, pp. 635-639.

[31] Barton, M.L. and Cendes, Z.J. (1987), New vector finite elements for three-dimensional magnetic field computation, *Journal of Applied Physics*, Vol. 61, No. 8, pp. 3919-3921.

[32] Brauer, J.R. Vander Heiden R.H. and Bruno, A.R. (1988), Finite element modeling of electromagnetic resonators and absorbers, *Journal of Applied Physics*, Vol. 63, No. 8, pp. 3197-3199.

[33] Ida, N. (1990), Computation of resonant frequencies for nonuniformly loaded cavities using a modified 3D eddy current formulation, *Proceedings of the International Seminar in Electromagnetic Field Analysis*, Oxford, pp. 267-274.

[34] Brauer, J.R. (1976), Saturated magnetic energy function for finite element analysis of electrical machines, *IEEE PES winter Meeting*, New York, paper C75, pp. 151-156.

[35] Kozakoff, D.J. and Simons, F.O. (1970), Three dimensional nonlinear magnetic field boundary value problem and its numerical solution, *IEEE Transactions on Magnetics*, Vol. MAG-6, pp. 828-833.

[36] Holzinger, C.S. (1970), Computation of magnetic fields within three-dimensional highly nonlinear media, *IEEE Transactions on Magnetics*, Vol. MAG-6, No. 1, pp. 60-65.

[36] Hoole, S.R.H. and Shin, Y. (1987), 3-D solution of sinusoidally excited saturated magnetic fields, *Journal of Applied Physics.*, Vol. 61, pp. 3928-3930.

[37] Lord, W. Bridges, J.M. Yen, W. and Palanisamy, R. (1978), Residual and active leakage fields around defects in ferromagnetic materials, *Materials Evaluation*, Vol. 36, No. 8, pp. 47-54.

[38] Heath, S.E. (1982) *Residual and active magnetostatic leakage field modeling*, M.S. Thesis, Colorado State University.

[39] Udpa, S.S. Lord, W. and Sun, Y. (1985), Numerical modeling of residual magnetic phenomena, *IEEE Transactions on Magnetics*, Vol. MAG-21, No. 6, pp. 2165-2168.

CHAPTER 8

Boundary integral, volume integral and combined formulations

8.1 INTRODUCTION

In chapter 1 we introduced Maxwell's equations in differential and integral forms. The preceding two chapters dealt exclusively with the formulation of Maxwell's equations in differential form. While no specific reason was given for this lopsided approach, we did use the integral form of Maxwell's equations for analytical solution of field problems in chapter 2.

Since the integral form is equivalent to the differential form it can be used for numerical solution of field problems as well. This is done in the present chapter.

There are two basic forms of integral equations that we use. One is the boundary integral form. In this form, the problem is formulated in terms of integral equations on the boundary of the geometry to be analyzed. The solution is in terms of equivalent surface sources: charge densities, current densities, and surface magnetization. The second is a volume integral equation whereby the approximation is in terms of equivalent volume sources, current densities, charge densities and volume magnetization.

Boundary integral equations are also combined with differential finite element approximations to obtain combined boundary integral/finite element solution throughout the required space.

8.2 BOUNDARY INTEGRAL METHODS

The methods described so far in chapters 6 and 7 are differential methods: they all depend on approximating a differential operator in well-defined boundaries and are subject to specified boundary conditions. Both the finite element and finite difference methods as presented, are of this type.

While the advantages of doing so are self-evident, there are a number of difficulties with the differential approach. One of the most important is the need to fix the boundary at an artificial location so as to have a finite space for solution. Inevitably, truncation of the boundary introduces errors in the solution and in some cases (such as in solution for waves), the problem becomes nonphysical. While it is possible to overcome most of these difficulties and still use a differential approach (infinite elements and radiation boundary conditions are often used for this purpose), and, in some cases, the exact boundary conditions of the problem can be satisfied (such as in closed systems, including waveguides and cavity resonators), there are a number of surprisingly simple problems that are almost impossible to solve using the differential approach. Alternatively, if a solution can be obtained it is expensive and complex. One simple example is the capacitance of a simple body: to solve it with any accuracy, we must solve Laplace's equation in the infinite domain, and satisfy the boundary conditions on the body and at infinity. Figure 8.1a shows this problem schematically.

If, for example, we chose to calculate the capacitance of a sphere of radius b, we could assume that a potential V_0 is connected to the sphere (with reference to a point at infinity) as shown in Figure 8.1b and proceed to calculate the capacitance from the basic relation

$$C = \frac{Q}{V}$$

The potential is known but we must calculate the charge on the sphere due to the potential or, if the charge is given, calculate the potential due to the charge. This is done by assuming a total charge Q on the sphere and, using Gauss' law, we find the electric potential due to the charge at any point in space as

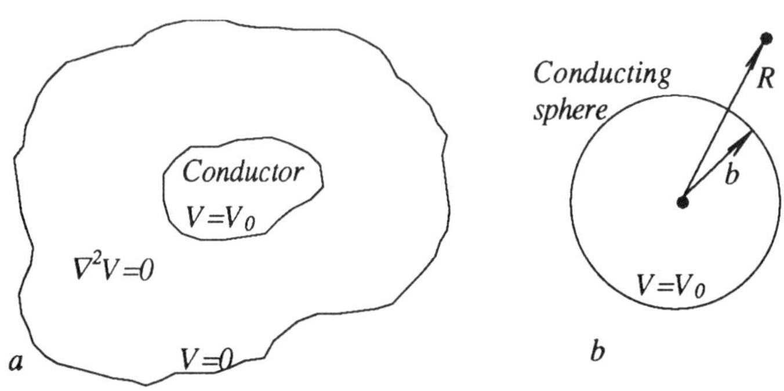

Figure 8.1 a. General electrostatic geometry. b. Spherical conductor in free space used for calculation of capacitance.

$$V(R) = \frac{Q}{4\pi\varepsilon R}$$

where ε is permittivity of the surrounding space and R the distance to the calculation point.

To satisfy the boundary condition on the surface of the sphere we set

$$V(R{=}b) = \frac{Q}{4\pi\varepsilon b} = V_0$$

or:

$$Q = 4\pi\varepsilon b V_0$$

and the capacitance of the sphere is

$$C = \frac{4\pi\varepsilon b V_0}{V_0} = 4\pi\varepsilon b$$

A very simple result. Since both Q and V_0 are now known, the electric field intensity and other parameters can be easily evaluated. If the body were more complex the solution would be more difficult but the principle remains the same.

However, the numerical solution to a problem of this type is much more involved, primarily because of the need to satisfy boundary conditions which enclose the whole of space.

A second class of applications for which the differential approach is ill-suited is that in which we are only interested in the effect of sources at a given location but not throughout the solution domain. A very simple example is an antenna. Viewing the antenna as a current source, we might wish to calculate the electric or magnetic field intensities at some distant point in space but cannot hope or wish to calculate the fields everywhere. This class of problem is shown in Figure 8.2. Again, as with the capacitance of a body, simple relations exist between the current in the antenna and the electric and magnetic fields throughout space.

From the discussion and example above it is also obvious what is necessary for an alternative solution. If the equivalent sources (charges or current densities) on the boundaries of the object (antenna or charged body) can be found, then simple equations, based on relations such as Ampere's or Gauss' laws can be used to find the fields.

This is in essence what boundary integral methods attempt to do: to calculate the equivalent sources on the surface of a volume, sources which produce the same fields outside the body. The process is shown schematically in Figures 8.3 and 8.4. The materials, including effects of inclusions, defects, etc., are replaced by equivalent surface and/or volume current densities and magnetization.

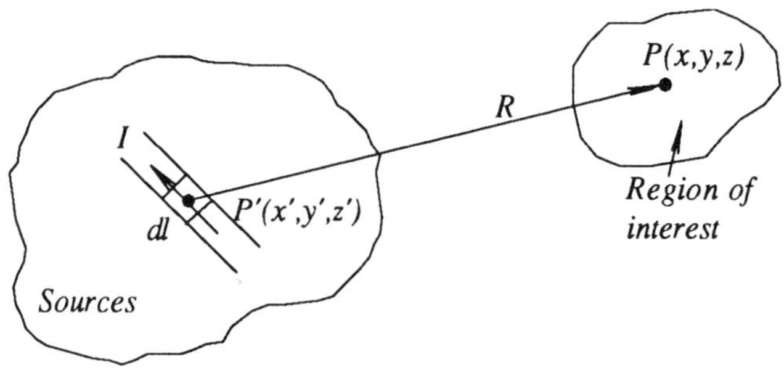

Figure 8.2 A far-field problem. The region of interest does not contain sources. The fields in this domain are generated by incident fields through the boundaries.

Because of this transformation, volume effects are replaced by surface sources and the dimensionality of the problem is reduced by one. Hence one of the most attractive aspects of the integral equations approach: only discretization of surfaces is required.

However, the integral formulation approach has significant disadvantages to the extent that at the present time, integral formulations are used less often and in fewer applications than differential formulations, especially for *NDT*.

The first, and perhaps the most significant disadvantage is that with very few exceptions only linear, isotropic applications can be handled. This is understood from Figures 8.3 and 8.4: for the sources on the surface to represent the volume uniquely, the relation between fields and sources must be linear.

Next, and equally significant for nondestructive testing is that the fields in the volume are not calculable while quite often, this is the domain in which defects may exist. Only fields outside the volume can be calculated.

For testing purposes, the more complex the geometry of the surface, the more difficult it would be to represent the sources. For awkward shaped objects the differential approach is often more useful.

Finally, in terms of the mechanics of solution, boundary integral methods are rather inefficient: the integration process is often slow and the resulting system of equations is represented by a full matrix. The system is slow and expensive to solve.

Nevertheless, the integral approach has significant applications and, when combined with a differential approach, can overcome many of the difficulties mentioned.

In the sections that follow some of the basic integral methods and their application to *NDT* are discussed.

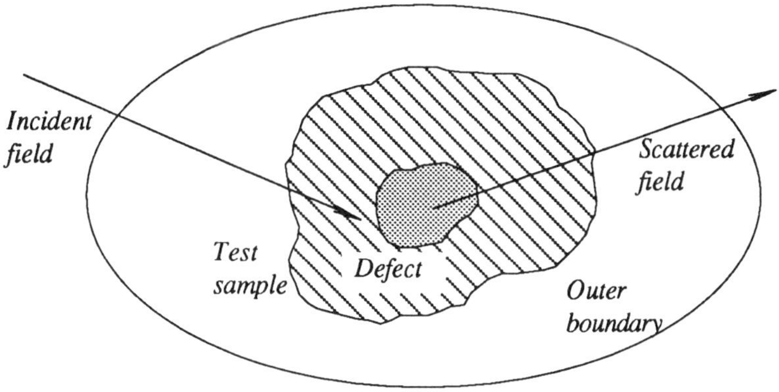

Figure 8.3 A testing geometry showing incident and scattered fields in a test sample and its surroundings.

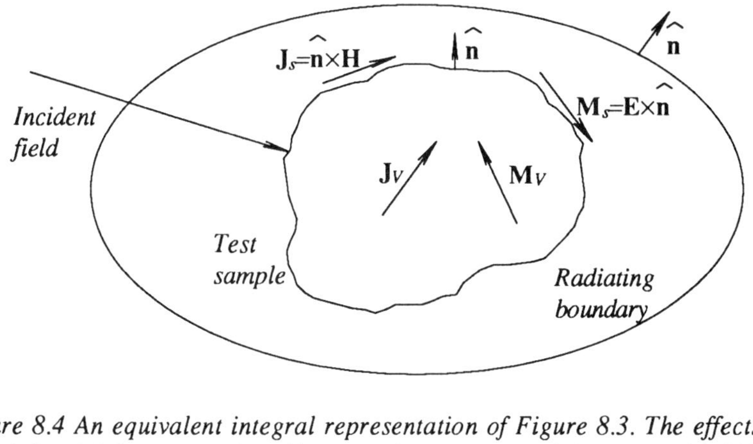

Figure 8.4 An equivalent integral representation of Figure 8.3. The effects of material and defects are replaced by equivalent sources and the scattered field by a radiating boundary.

8.3 THE METHOD OF MOMENTS: AN INTUITIVE APPROACH

To emphasize many of the points discussed above we consider now a very simple form of the method of moments (*MOM*) as used to calculate electrostatic fields. The method of moments is particularly effective for solution of integral equations. The method solves a general integral equation of the form:

$$V(z) = \int_{\Omega'} K(z,z')\rho(z')d\Omega' \tag{8.1}$$

where Ω is the domain over which the integration is performed and may be the whole solution domain or, more often, a subdomain. The function $K(z,z')$ is assumed to be known and is called the kernel of the integral or the Green's function for the problem. The unknown to be determined is $\rho(z')$ which in electrostatics is the charge density but, in general may be any source such as a current density or magnetization on the surface of a conductor.

The problem considered here is the calculation of the electric field or capacitance between two conducting strips intersecting at $45°$ as shown in Figure 8.5. The strips are assumed to be very long, on parallel planes, and separated a distance d apart. We do not actually calculate the electric field intensity, only the equivalent charge density on the two strips. From this, the electric field intensity can be immediately calculated by simple means.

To solve this problem, we use the potential at any point in space due to a charge density over a surface. This is given by

$$V(x,y,z) = \int_{S'} \frac{1}{4\pi\varepsilon R} \rho_s(x',y'z')dS' \tag{8.2}$$

where R is the distance between the observation point (x,y,z) to the source point (x',y',z'). This distance is given simply as

$$R = \sqrt{(x-x')^2 + (y-y')^2 + (z-z')^2} \tag{8.3}$$

The expression in Eq. (8.2) is merely the potential at (x,y,z) of a point charge of magnitude $\rho_s ds$ located at a point (x',y',z'). This expression implies that the charge density over a surface S can in fact be approximated by an equivalent point charge. That this is admissible, we only have to mention that the surface can be made as small as necessary and therefore, in the limit, this condition can always be satisfied, at least approximately.

The two strips in Figure 8.5 are charged, each with a total charge Q (one is positive and one is negative). We can assume that the upper plate is connected to a constant potential V_0 and the lower to a potential $-V_0$. The potentials are arbitrary and for purpose of calculation will be taken as $+1\ V$ and $-1\ V$. The strips are $5cm$ wide, separated $10\ cm$ and, although infinite in length, the actual length for computation is taken as $1m$. This is 20 times longer than the width and 10 times longer than the distance between the strips. Under these conditions, the error due to the finite length of the strip is minimal. However, the result can be improved by increasing the length of the strips.

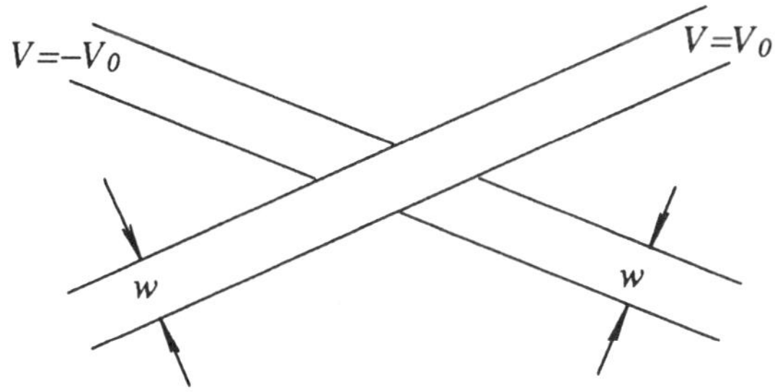

Figure 8.5 Two conducting strips, a distance d apart intersecting at 45°.

To solve the problem we divide each plate into an arbitrary number of subdomains which, in general, can be of arbitrary size. The division is shown schematically in Figure 8.6 where only a few domains are shown. The number of domains that should be used depends on accuracy required. Each subdomain is assigned an unknown charge density but the charge density is assumed to be uniform over the subdomain and associated with the center of the subdomain.

The potential due to each rectangle is calculated using Eq. (8.2). The potential on each plate is constant and therefore it is constant on each subdomain. However, to calculate the potential correctly, it is viewed as being generated due to all charge densities, including itself. The potential due to subdomain i at the center of subdomain j is calculated directly from Eq. (8.2) as

$$V(x_j,y_j,z_j)=\frac{\rho_{si}(x_i',y_i'z_i')\Delta S_i}{4\pi\varepsilon\sqrt{(x_j-x_i')^2+(y_j-y_i')^2+(z_j-z_i')^2}} \tag{8.4}$$

where ΔS_i is the area of subdomain i.

This is repeated for each subdomain but, in addition, the potential at the center of domain i due to the charge density on itself must be added. This is calculated by direct integration and is equal to

$$V(x_i',y_i',z_i')=\frac{\rho_{si}(x_i',y_i'z_i')}{4\pi\varepsilon}\left(2a\ln\frac{b+\sqrt{a^2+b^2}}{a}+2b\ln\frac{a+\sqrt{a^2+b^2}}{b}\right) \tag{8.5}$$

Thus, the potential on a subdomain i on the upper plate is equal to

$$V_i=\frac{\rho_{si}(x_i',y_i'z_i')}{4\pi\varepsilon}\left(2a\ln\frac{b+\sqrt{a^2+b^2}}{a}+2b\ln\frac{a+\sqrt{a^2+b^2}}{b}\right)+\frac{1}{4\pi\varepsilon}\sum_{\substack{j=1\\ j\neq i}}^{N}\frac{\rho_{sj}(x_j',y_j'z_j')\Delta S_j}{\sqrt{(x_j-x_i')^2+(y_j-y_i')^2+(z_j-z_i')^2}} \tag{8.6}$$

where N is the number of subdomains and a, b, are the dimensions of the subdomain. An identical equation is written for any subdomain, thus resulting in a system of N equations in N unknown charge densities. Assuming that there are M subdomains on the upper plate and N–M subdomains on the lower plate, and since $V_1, V_2, \ldots\ldots, V_n$, are known (each is equal to $+V_0$ or $-V_0$), we can write the following matrix equation

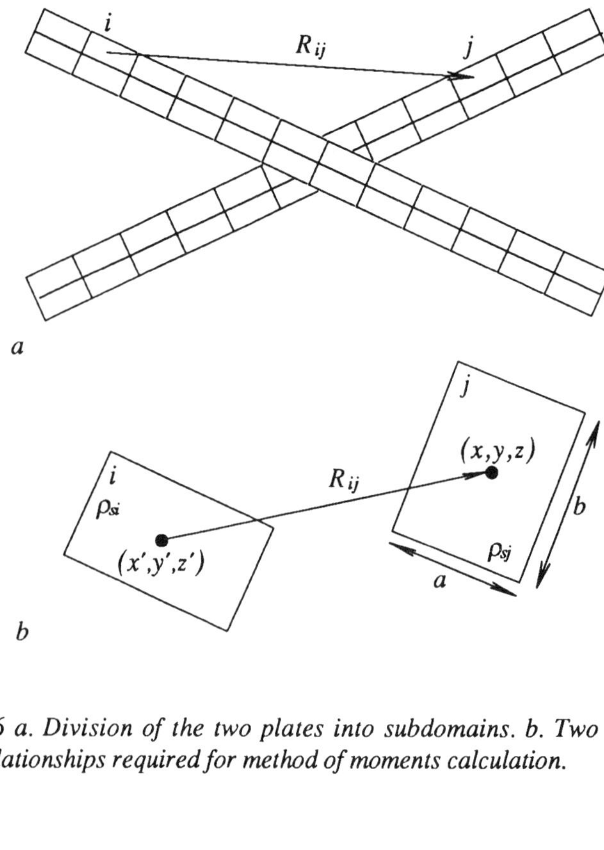

Figure 8.6 a. Division of the two plates into subdomains. b. Two subsections and the relationships required for method of moments calculation.

$$\begin{bmatrix} \frac{K_1}{4\pi\varepsilon} & \frac{\Delta_{s2}}{4\pi\varepsilon R_{21}} & \frac{\Delta_{s3}}{4\pi\varepsilon R_{31}} & \cdots & \frac{\Delta_{s(n-1)}}{4\pi\varepsilon R_{(n-1)1}} & \frac{\Delta_{sn}}{4\pi\varepsilon R_{n1}} \\ \frac{\rho_{s1}\Delta_{s1}}{4\pi\varepsilon R_{12}} & \frac{K_2}{4\pi\varepsilon} & \frac{\Delta_{s3}}{4\pi\varepsilon R_{32}} & \cdots & \frac{\Delta_{s(n-1)}}{4\pi\varepsilon R_{(n-1)2}} & \frac{\Delta_{sn}}{4\pi\varepsilon R_{n2}} \\ \vdots & \vdots & \vdots & \vdots & \vdots & \vdots \\ \frac{\Delta_{s1}}{4\pi\varepsilon R_{1(n-1)}} & \frac{\Delta_{s2}}{4\pi\varepsilon R_{2(n-1)}} & \frac{\Delta_{s3}}{4\pi\varepsilon R_{3(n-1)}} & \cdots & \frac{K_{n-1}}{4\pi\varepsilon} & \frac{\Delta_{sn}}{4\pi\varepsilon R_{(n-1)n}} \\ \frac{\Delta_{s1}}{4\pi\varepsilon R_{1n}} & \frac{\Delta_{s2}}{4\pi\varepsilon R_{2n}} & \frac{\Delta_{s3}}{4\pi\varepsilon R_{3n}} & \cdots & \frac{\Delta_{s(n-1))}}{4\pi\varepsilon R_{n(n-1)}} & \frac{K_n}{4\pi\varepsilon} \end{bmatrix} \begin{bmatrix} \rho_{s1} \\ \rho_{s2} \\ \vdots \\ \rho_{s(n-1)} \\ \rho_{sn} \end{bmatrix} = \begin{bmatrix} V_0 \\ V_0 \\ \vdots \\ -V_0 \\ -V_0 \end{bmatrix} \tag{8.7}$$

where the first M potentials are positive, the last $N-M$ are negative as required, and K_i represents the geometrical factor of Eq. (8.6) for each subdomain. Now the charge densities ρ_{s1} through ρ_{sN} are calculated. These charge densities represent the solution to the problem as posed, although often we calculate additional parameters such as capacitance or electric fields. We note here that the matrix although simple, (and symmetric) is a full matrix which for a large number of subdomains can be rather slow to solve.

Applying the division above and using the dimensions as used for discretization in Figure 8.6, with a total of *20* subdomains on each strip, the capacitance between the two strips is found to be *6.4 picoFarad*. This discretization is rather coarse. A larger number of subdomains will improve solution but also increase the size of the matrix that needs to be solved, particularly since the matrix is full.

The solution above although only a rudimentary method of moments solution indicates both the advantage of the method and the challenge involved. As a rule, we must be able to come up with a simple integral solution to the problem. In more general terms, we must find the Green's function corresponding to the geometry analyzed. This is often a very difficult task although, in the context of numerical methods it is possible to simplify the approximation such that an integral form can be obtained.

The same method can be used to analyze current distributions. We do not pursue this aspect of the calculation (see bibliography), but the principle is essentially the same. An integral form is obtained in terms of a kernel and an equivalent current distribution as

$$E(z) = \int_{\Omega'} K(z,z')I(z')d\Omega' \tag{8.8}$$

where E is the electric field intensity and, similar to V, is assumed to be known.

Finally, we point out that the method of moments can also be used with differential operators much the same as any weighted residual method.

Also to be noted is that we assumed that charge density (or more appropriately, the source) is distributed only on the surface. This is obviously not always the case. The integration in Eqs. (8.1) or (8.8) can be done on the volume as well in which case the charge densities are volume charge densities or the currents are volume currents. In fact both surface and volume sources may exist. As long as a solution for elemental source distributions (that is: the Green's function) can be found, the method can be applied to the given problem, including volume integral methods.

8.4 INTEGRAL EQUATIONS

Before continuing and discussing the general approach to boundary element and volume element methods we discuss here the general formulation from which all integral equations methods stem. The general scattering problem is first treated as this includes all aspects of modeling, including low frequency and high frequency applications. After this, special aspects of modeling at low frequencies, including static fields are given. Although some of the methods discussed here can be extended to time dependent application this aspect of modeling is not described here.

To formulate the problem we assume a two domain geometry as shown in Figure 8.7. Domain Ω_0 is the surrounding medium such as free space while domain Ω includes all conductors and dielectrics, such as materials under test. In domain Ω_0 there are sources such as current densities, charge densities and magnetization vectors. These sources produce the fields in the domain. Since we deal here with scattering, these fields are the incident fields. Materials in domain Ω interact with the incident fields to produce scattered fields. The total field, in any application, is the sum of the scattered and incident fields. Note that the use of magnetization as a source is somewhat unusual but it is a very convenient source, especially at low frequencies and static application since it allows us to account for magnetic materials in terms of equivalent magnetization.

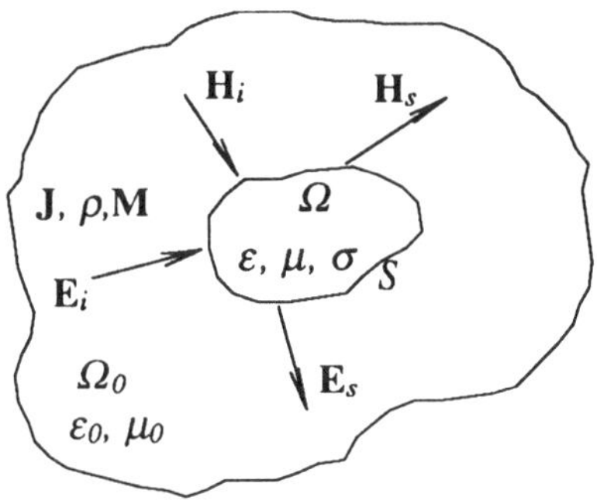

Figure 8.7 The general geometry used to formulate integral methods.

From the relations in Figure 8.7. we can write the total fields as

$$\mathbf{E}_t = \mathbf{E}_i + \mathbf{E}_s \tag{8.9}$$

$$\mathbf{H}_t = \mathbf{H}_i + \mathbf{H}_s \tag{8.10}$$

where i stands for incident, s for scattered and t for total fields.
Separating the two domains Ω_0 and Ω we can write a set of equations in each domain
In Ω_0:

$$\nabla\times\mathbf{E}_t = -\mathbf{M} - j\omega\mu_0\mathbf{H}_t \tag{8.11}$$

$$\nabla\times\mathbf{H}_t = \mathbf{J} + j\omega\varepsilon_0\mathbf{E}_t \tag{8.12}$$

In Ω:

$$\nabla\times\mathbf{H}_t = j\omega\varepsilon\mathbf{E}_t \tag{8.13}$$

$$\nabla\times\mathbf{E}_t = -j\omega\mu\mathbf{H}_t \tag{8.14}$$

In these equations, all sources are in domain Ω_0 while in domain Ω permittivity and permeability can be tensors and both can be complex variables. In particular permittivity ε is assumed to be of the form

$$\varepsilon = \varepsilon' + \frac{\sigma}{j\omega} \tag{8.15}$$

This is the familiar form of complex permittivity (see chapter 1) in which ε' is recognized as the constant or low frequency permittivity of dielectrics and σ is material conductivity. Thus, eddy currents can be viewed as being due to the second term in Eq. (8.15).

The equations in the two domains result in a set of interface conditions that must be satisfied. These are normally taken as the continuity of tangential components of the electric and magnetic field intensities on the interfaces between any two materials.

Finally, as in any well-defined boundary value problem we must include boundary conditions. Although both Dirichlet and Neumann boundary conditions may exist, the boundary condition that must be satisfied in an open domain is the radiation boundary condition or, more specifically, the Sommerfeld radiation condition. This condition guarantees that fields vanish at large distances from the interaction zone (that is, far from sources and scatterers). The radiation boundary condition is stated as

$$lim_{r\to\infty}\left(\nabla\times\mathbf{H}_s - jk_0\hat{\mathbf{r}}\times\mathbf{H}_s\right) = 0 \tag{8.16}$$

and

$$lim_{r\to\infty}\left(\nabla\times\mathbf{E}_s - jk_0\hat{\mathbf{r}}\times\mathbf{E}_s\right) = 0 \tag{8.17}$$

where

$$k_0 = \omega\sqrt{\mu_0\varepsilon_0}$$

is the free-space wavenumber. Note also that the radiation condition is written in terms of the scattered field alone, meaning that at the far boundary only the scattered field is cancelled by the radiation boundary condition.

At this point we have a set of differential equations with appropriate boundary conditions which is rather difficult to solve, although special cases of these equations are solved in the previous four chapters.

Now, we look at a general method to transform these equations into integral equations. First, we rewrite the two relations in Eqs. (8.11) and (8.12) in terms of a single variable, say **E**, by simple operations. By taking the curl of both sides of Eq. (8.11) we get

$$\nabla\times\nabla\times\mathbf{E}_t = -\ \nabla\times\mathbf{M} - j\omega\mu_0\nabla\times\mathbf{H}_t$$

Substitution of Eq. (8.12) for $\nabla\times\mathbf{H}_t$ gives

$$\nabla\times\nabla\times\mathbf{E}_t - k_0^2\mathbf{E}_t = -\ \nabla\times\mathbf{M} - j\omega\mu_0\mathbf{J} \tag{8.18}$$

Essentially identical steps result in an equation in **H** alone

$$\nabla\times\nabla\times\mathbf{H}_t - k_0^2\mathbf{H}_t = \nabla\times\mathbf{J} - j\omega\varepsilon_0\mathbf{M} \tag{8.19}$$

Similarly, in domain Ω we can write

$$\nabla\times\nabla\times\mathbf{E}_t - k^2\mathbf{E}_t = 0 \tag{8.20}$$

$$\nabla\times\nabla\times\mathbf{H}_t - k^2\mathbf{H}_t = 0 \tag{8.21}$$

where

$$k = \omega\sqrt{\mu\varepsilon}$$

is the (complex) wavenumber in any material in domain Ω.

Eqs. (8.18) through (8.21) can be transformed into integral equations by use of the vector form of Green's theorem. The latter is stated as

$$\int_v \left(\mathbf{A}\cdot(\nabla\times\nabla\times\mathbf{B}) - \mathbf{B}\cdot(\nabla\times\nabla\times\mathbf{A})\right)dv = \int_S \widehat{\mathbf{n}}\cdot(\mathbf{B}\times(\nabla\times\mathbf{A}) - \mathbf{A}\times(\nabla\times\mathbf{B}))ds \tag{8.22}$$

where **A** and **B** are two general vectors. Comparison of this with Eq. (8.18) or (8.19) reveals that **B** can be taken either as **E** or **H** but we must define **A**. We choose **A** to be either the scalar (free space) Green's function

$$G_0(\mathbf{R}) = \frac{e^{-jk_0R}}{|\mathbf{R}|} \tag{8.23}$$

or the free space dyadic Green's function

$$\overline{\overline{G}}_0(\mathbf{R}) = \overline{\overline{\mathbf{I}}}\frac{e^{-jk_0R}}{4\pi|\mathbf{R}|} \tag{8.24}$$

where

$$\overline{\overline{\mathbf{I}}} = \widehat{\mathbf{x}}\widehat{\mathbf{x}} + \widehat{\mathbf{y}}\widehat{\mathbf{y}} + \widehat{\mathbf{z}}\widehat{\mathbf{z}}$$

and $\hat{\mathbf{x}},\hat{\mathbf{y}},\hat{\mathbf{z}}$, are the unit vectors in the Cartesian system of coordinates. This expression is called the unit dyadic. It indicates that for a vector source each component of the source gives rise to a vector field, therefore the dyadic relation between source and field, or the Green's function (see chapter 2 for a simple, physical interpretation of the idea of a Green's function).

Returning now to Green's theorem we can substitute either of these results for the vector **A** in Eq. (8.18). If the scalar Green's function is chosen, it must be multiplied by an arbitrary, constant vector **b**. The dyadic Green's function is a vector and can be used directly. Thus, we choose

$$\mathbf{A} = \mathbf{b}G_0(\mathbf{R}) \qquad or: \qquad \mathbf{A} = \overline{\overline{G_0}}(\mathbf{R})$$

Both Green's functions satisfy the homogeneous equation in Ω_0 and the radiation boundary condition.

Substitution of **A** (using the scalar Green's function) gives the following relations:

Inside domain Ω_0:

$$\mathbf{E} = \int_{\Omega_0}\left(-j\omega\mu_0\mathbf{J}G_0 - \frac{\nabla\cdot\mathbf{J}}{j\omega\varepsilon_0}\nabla G_0 - \mathbf{M}\times\nabla G_0\right)dv' +$$
$$\int_S\left(-j\omega\mu_0(\hat{\mathbf{n}}\times\mathbf{H})G_0 + (\hat{\mathbf{n}}\cdot\mathbf{E})\nabla G_0 - (\mathbf{E}\times\hat{\mathbf{n}})\times\nabla G_0\right)ds' \tag{8.25}$$

$$\mathbf{H} = \int_{\Omega_0}\left(-j\omega\varepsilon_0\mathbf{M}G_0 - \frac{\nabla\cdot\mathbf{M}}{j\omega\mu_0}\nabla G_0 + \mathbf{J}\times\nabla G_0\right)dv' +$$
$$\int_S\left(-j\omega\varepsilon_0(\mathbf{E}\times\hat{\mathbf{n}})G_0 + (\hat{\mathbf{n}}\cdot\mathbf{H})\nabla G_0 + (\hat{\mathbf{n}}\times\mathbf{H})\times\nabla G_0\right)ds' \tag{8.26}$$

On the surface S between Ω_0 and Ω:

$$\mathbf{E} = 2\int_{\Omega_0}\left(-j\omega\mu_0\mathbf{J}G_0 - \frac{\nabla\cdot\mathbf{J}}{j\omega\varepsilon_0}\nabla G_0 - \mathbf{M}\times\nabla G_0\right)dv' +$$
$$2\int_S\left(-j\omega\mu_0(\hat{\mathbf{n}}\times\mathbf{H})G_0 + (\hat{\mathbf{n}}\cdot\mathbf{E})\nabla G_0 - (\mathbf{E}\times\hat{\mathbf{n}})\times\nabla G_0\right)ds' \tag{8.27}$$

$$\mathbf{H} = 2\int_{\Omega_0}\left(-j\omega\varepsilon_0\mathbf{M}G_0 - \frac{\nabla\cdot\mathbf{M}}{j\omega\mu_0}\nabla G_0 + \mathbf{J}\times\nabla G_0\right)dv' +$$

$$2\int_S\left(-j\omega\varepsilon_0(\mathbf{E}\times\hat{\mathbf{n}})G_0 + (\hat{\mathbf{n}}\cdot\mathbf{H})\nabla G_0 + (\hat{\mathbf{n}}\times\mathbf{H})\times\nabla G_0\right)ds' \qquad (8.28)$$

Inside Ω:

$$\mathbf{E} = 0 \qquad \mathbf{H} = 0$$

Note that the terms $\hat{\mathbf{n}}\cdot\mathbf{E}$ and $\hat{\mathbf{n}}\cdot\mathbf{H}$ are normal components of **E** and **H** while $\hat{\mathbf{n}}\times\mathbf{E}$ and $\hat{\mathbf{n}}\times\mathbf{H}$ are tangential components. The normal components can be eliminated and, therefore, only the tangential components of **E** or **H** on the surface of the scatterer and any sources outside the conductor are necessary for solution. The field outside the scatterer can then be determined from these fields and the Green's function in free space.

Equivalent relations are obtained using the dyadic Green's function. The main difference is that the expressions contain only tangential components of the electric and magnetic fields and there is no need to relate normal components to tangential components as for the scalar Green's function.

Finally we note that if there are no sources involved in the solution, the volume integrals in Eqs (8.25) through (8.28) disappear and the solution can be performed strictly on the surface. In particular, if we simply replace $\hat{\mathbf{n}}$ by $-\hat{\mathbf{n}}$ in the relations above, and replace G_0 by G (the Green's function in domain Ω), the relations above apply to the interior of domain Ω. However, in this case, the sources do not exist and only a surface integral exists. If sources do exist in domain Ω_0, these sources must be integrated as given (analytically) but the integration is only over the volumes that contain sources. In addition we also note here that the tangential components of the electric and magnetic field intensities are equivalent to:

$$\mathbf{J}_s = \hat{\mathbf{n}}\times\mathbf{H} \qquad (8.29)$$

$$\mathbf{M}_s = \mathbf{E}\times\hat{\mathbf{n}} \qquad (8.30)$$

These equivalent sources or the electric and magnetic charges that cause them are normally the variables we evaluate.

Using these relations the scattered electric and magnetic fields outside and inside domain Ω and the incident fields in domain Ω_0 are written. These are:
In Ω_0 (including the surface S):

$$\mathbf{E}_s = -j\omega\mathbf{A}_0 - \nabla V_0 - \frac{1}{\varepsilon_1}\nabla\times\mathbf{T}_0 \qquad (8.31)$$

$$\mathbf{H}_s = -j\omega\mathbf{T}_0 - \nabla\Phi_0 + \frac{1}{\mu_0}\nabla\times\mathbf{A}_0 \tag{8.32}$$

In Ω:

$$\mathbf{E}_s = j\omega\mathbf{A} + \nabla V + \frac{1}{\varepsilon_1}\nabla\times\mathbf{T} \tag{8.33}$$

$$\mathbf{H}_s = j\omega\mathbf{T} + \nabla\Phi - \frac{1}{\mu_0}\nabla\times\mathbf{A} \tag{8.34}$$

where the scalar functions Φ_0, V_0 and Φ, V are

$$\Phi_0 = -\frac{1}{4\pi\varepsilon_0}\int_s \frac{\nabla_s'\cdot\mathbf{M}}{j\omega} G_0 ds' = \frac{1}{4\pi\varepsilon_0}\int_s \rho_{sm} G_0 ds' \tag{8.35}$$

$$V_0 = -\frac{1}{4\pi\varepsilon_0}\int_s \frac{\nabla_s'\cdot\mathbf{J}}{j\omega} G_0 ds' = \frac{1}{4\pi\varepsilon_0}\int_s \rho_{se} G_0 ds' \tag{8.36}$$

$$\Phi = -\frac{1}{4\pi\varepsilon}\int_s \frac{\nabla_s'\cdot\mathbf{M}}{j\omega} G\, ds' = \frac{1}{4\pi\varepsilon}\int_s \rho_{sm} G\, ds' \tag{8.37}$$

$$V = -\frac{1}{4\pi\varepsilon}\int_s \frac{\nabla_s'\cdot\mathbf{J}}{j\omega} G\, ds' = \frac{1}{4\pi\varepsilon}\int_s \rho_{se} G\, ds' \tag{8.38}$$

The vector functions $\mathbf{A}_0$, $\mathbf{T}_0$, $\mathbf{A}$, and $\mathbf{T}$ are:

$$\mathbf{A}_0 = \frac{\mu_0}{4\pi}\int_S \mathbf{J}_s G_0 ds' \tag{8.39}$$

$$\mathbf{T}_0 = \frac{\varepsilon_0}{4\pi}\int_S \mathbf{M}_s G_0 ds' \tag{8.40}$$

$$\mathbf{A} = \frac{\mu}{4\pi}\int_S \mathbf{J}_s G\, ds' \tag{8.41}$$

$$\mathbf{T} = \frac{\varepsilon}{4\pi}\int_S \mathbf{M}_s G\, ds' \tag{8.42}$$

In Eqs. (8.35) through (8.38), ∇s indicates surface divergence and ρ_{se} and ρ_{sm} are surface electric and magnetic charge densities respectively. The four potentials are immediately recognized as the magnetic vector potential (**A**), the electric vector potential (**T**), the electric scalar potential (V) and the magnetic scalar potential (Φ). By requiring the continuity of the tangential component of the electric and magnetic field intensities on the surface of Ω we get using Eqs. (8.31) through (8.32) for the incident fields:

$$\mathbf{E}_{itan} = \left[j\omega(\mathbf{A}_0 + \mathbf{A}) + (\nabla V_0 + \nabla V) + \left(\frac{1}{\varepsilon_0}\nabla\times\mathbf{T}_0 + \frac{1}{\varepsilon}\nabla\times\mathbf{T} \right) \right]_{tan} \tag{8.43}$$

$$\mathbf{H}_{itan} = \left[j\omega(\mathbf{T}_0 + \mathbf{T}) + (\nabla\Phi_0 + \nabla\Phi) - \left(\frac{1}{\varepsilon_0}\nabla\times\mathbf{A}_0 + \frac{1}{\varepsilon}\nabla\times\mathbf{A} \right) \right]_{tan} \tag{8.44}$$

The first of these is called the electric field integral equation *(EFIE)* while the second is the magnetic field integral equation (*MFIE*).

These equations are rather general and depending on application may be simplified considerably. A simplification of this type is used in a subsequent section to treat conducting domains. On surfaces of conductors, the term $\mathbf{E}\times\hat{\mathbf{n}}$ in the scattered field is zero and the two equations above become

$$\mathbf{E}_{itan} = \left[j\omega\mathbf{A}_0 + \nabla V_0 \right]_{tan} \tag{8.45}$$

$$\mathbf{H}_{itan} = \frac{\mathbf{J}_s}{2} + \hat{\mathbf{n}}\times\nabla\times\mathbf{A}_0 \tag{8.46}$$

8.4.1 Boundary element methods

From the previous discussion it is evident that an electromagnetic problem can be described in integral form. The principles involved are:

1. Find an equivalent system of sources: charge densities and current densities that describe the problem. These may be surface or volume sources.
2. Find a solution for the required field due to these sources. The solutions are referred to as the Green's function for the problem. In analytic solutions, the Green's function refers to the whole system of sources. Therefore, it is rather difficult to find an appropriate Green's function. In numerical methods, the source distribution is often in terms of point or layer sources and the question of finding a solution is simplified.

The method above simply indicates that given a system of sources, the fields can be evaluated directly. However, in the boundary integral method, the sources are not known. These are, in fact the unknown values to be determined. The fields are determined indirectly after the equivalent sources are found. To determine the unknown sources, the given boundary conditions of the problem are used as was shown in the previous section.

In the following section we first formulate a number of problems in terms of boundary integral equations and show how these can be solved using finite element methods. Examples from scattering and testing of lossy dielectric materials are given.

8.4.2 Volume integral equations

In section 8.4 the problem is described in terms of equivalent surface sources. This leads to surface integral equations and consequently to surface or boundary integral equations.

Starting with the relations in Eqs. (8.9) and (8.10), we can view the electric and magnetic fields in domain Ω_0 as being generated by equivalent sources in volume Ω. Further, if Ω consists of a material with permittivity ε and permeability μ, the equivalent problem is to replace the material by free space and equivalent sources that account for magnetization (permeability effects) and polarization (permittivity effects) as shown in Figure 8.8.

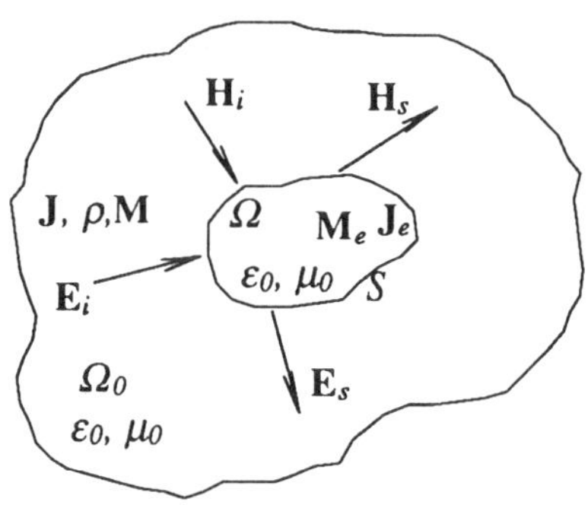

Figure 8.8 The use of equivalent volume sources to replace the material in domain Ω.

There are two types of sources possible in Ω. The first are applied sources while the second are induced sources. These can be written in terms of volume distributed charge densities and magnetization. The applied sources are $\mathbf{J}_0$ and $\mathbf{M}_0$ while the induced sources are

$$\mathbf{J}_e = j\omega\varepsilon_0\chi_e\mathbf{E} \qquad \mathbf{M}_e = j\omega\varepsilon_0\chi_m\mathbf{H} \tag{8.47}$$

where χ_e and χ_m are the electric and magnetic susceptibilities. Alternatively, we can write the induced sources in terms of equivalent volume electric and magnetic charge densities

$$\rho_e = -\,\varepsilon_0\cdot(\chi_e\mathbf{E}) \qquad \rho_m = -\,\mu_0\cdot(\chi_m\mathbf{H}) \tag{8.48}$$

and, following identical steps as above, the scattered fields due to Ω are

$$\mathbf{E}_s = -j\omega\mathbf{A} - \nabla V - \frac{1}{\varepsilon_0}\nabla\times\mathbf{T} \tag{8.49}$$

$$\mathbf{H}_s = -j\omega\mathbf{F} - \nabla\Phi + \frac{1}{\mu_0}\nabla\times\mathbf{A} \tag{8.50}$$

where the functions Φ, V, $\mathbf{A}$, and $\mathbf{T}$ are similar to those in Eqs. (8.35) through (8.38):

$$\Phi = \frac{1}{4\pi\varepsilon_0}\int_\Omega \rho_m G_0 dv \tag{8.51}$$

$$V = \frac{1}{4\pi\varepsilon_0}\int_\Omega \rho_e G_0 dv \tag{8.52}$$

$$\mathbf{A} = \frac{\mu_0}{4\pi}\int_\Omega \mathbf{J}_e G_0 dv \tag{8.53}$$

$$\mathbf{T} = \frac{\varepsilon_0}{4\pi}\int_\Omega \mathbf{M}_e G\, dv \tag{8.54}$$

Using Eqs. (8.9) and (8.10), the electric and magnetic field intensities at any point in space are

$$\mathbf{E}_t = \mathbf{E}_{inc} + \mathbf{E}_s = \mathbf{E}_{inc} - j\omega\mathbf{A} - \nabla V - \frac{1}{\varepsilon_0}\nabla\times\mathbf{T} \tag{8.55}$$

$$\mathbf{H}_t = \mathbf{H}_{inc} + \mathbf{H}_s = \mathbf{H}_{inc} - j\omega\mathbf{T} - \nabla\Phi + \frac{1}{\mu_0}\nabla\times\mathbf{A} \tag{8.56}$$

These equations are coupled in that **T** depends on magnetization and magnetization depends on **H**. However, it is often possible to separate the equations. As an example, if magnetization is zero (such as in dielectrics) the last term in Eq. (8.55) vanishes and the equation is in terms of **E** alone. Similarly, if polarization is zero (in magnetic or conducting materials), the last term in Eq. (8.56) vanishes and Eq. (8.56) is in terms of **H** alone.

8.4.3 Finite element/boundary element formulation

The boundary integral method has the distinct advantage that the equivalent surface sources are calculated and therefore, the problem is effectively reduced to that of solving the equivalent problem on the surface of conductors or scatterers. This solution is equivalent to a two-dimensional solution in that only surface variables are used. Without this transformation the problem would require a three-dimensional solution in an infinite domain. However, since only surface variables are used, the fields inside scatterers are not known, nor can they be calculated. This aspect of the formulation is sometimes detrimental to modeling of nondestructive testing of materials where inclusions and defects in scatterers are often the major aspect we are interested in.

An alternative formulation, one that maintains the advantages of boundary integral methods is to use boundary integral methods on surfaces of scatterers and standard finite element solutions inside the volume of scatterers. If the two solutions can be properly coupled then a solution for volume variables is obtained as in the finite element case while the boundary integral solution serves to limit the solution domain to the volume of scatterers. The question of location of boundary conditions for the problem is therefore solved by use of the boundary integral representation.

We now present a boundary element/finite element formulation (*FEM/BEM*) that uses this method, based on the boundary integral method discussed in section 8.4.1.

For the finite element representation in a given volume we can use any of the formulations given in chapter 7. However, for consistency with the type of results considered here, we assume that the Helmholtz equation for **H** and **E** applies in the volume of the scatterer and that sources are outside the scatterer as shown in Figure 8.9. Under these conditions, the equation for **E** and **H** inside domain Ω are

$$\nabla\times\nabla\times\mathbf{E} - k^2\mathbf{E} = 0 \tag{8.57}$$

and

$$\nabla\times\nabla\times\mathbf{H} - k^2\mathbf{H} = 0 \tag{8.58}$$

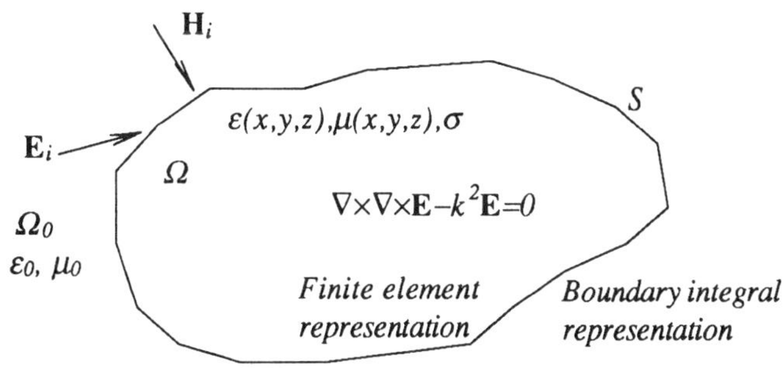

Figure 8.9 General geometry for a FEM/BEM solution. Finite elements are used for the interior and boundary elements on the surface S.

If materials are inhomogeneous, the following equivalent form is easier to use

$$\nabla\times\left(\frac{1}{j\omega\mu}\nabla\times\mathbf{E}\right) + j\omega\varepsilon\mathbf{E} = 0 \tag{8.59}$$

and

$$\nabla\times\left(\frac{1}{j\omega\varepsilon}\nabla\times\mathbf{H}\right) + j\omega\mu\mathbf{H} = 0 \tag{8.60}$$

To formulate these equations in terms of finite elements in the domain, we use Galerkin's method. A set of approximating functions is used for **E** or **H** and a set of weighting functions, equal to the shape functions N_m (see chapters 4 and 7) is used to weigh the approximation. For clarity, and to see how this formulation may be coupled to the boundary element method, we use **E** and **H** but it must be remembered that for implementation these must be replaced with the approximating functions. Integrating over the volume Ω (actually, the integration is done one element at a time) we get

$$\int_\Omega \nabla\times\left(\frac{1}{j\omega\mu}\nabla\times\mathbf{E}\right)N_m dv + \int_\Omega j\omega\varepsilon\mathbf{E}N_m dv = 0$$

Using the vector identity

$$\nabla\times(\nabla\times\mathbf{E})N_m = \nabla\times(N_m\nabla\times\mathbf{E}) - (\nabla N_m)\times(\nabla\times\mathbf{E})$$

Finally, using the relation

$$\int_{\Omega} \nabla\times(N_m \nabla\times\mathbf{E})\, dv = \int_S \hat{\mathbf{n}}\times(N_m \nabla\times\mathbf{E})\, ds$$

where S is the surface enclosing the volume we get

$$\int_{\Omega}\left(\frac{1}{j\omega\mu}\nabla\times\mathbf{E}\right)\times(\nabla N_m)dv + \int_{\Omega} j\omega\varepsilon\mathbf{E}N_m dv = \int_S (\hat{\mathbf{n}}\times\mathbf{H})N_m ds \quad (8.61)$$

This form, also known as the weak form, is advantageous because it only contains first order derivatives of the function to be evaluated. Similar steps lead to an equivalent formulation for **H** as

$$\int_{\Omega}\left(\frac{1}{j\omega\varepsilon}\nabla\times\mathbf{H}\right)\times(\nabla N_m)dv + \int_{\Omega} j\omega\mu\mathbf{H}N_m dv = \int_S (\mathbf{E}\times\hat{\mathbf{n}})N_m ds \quad (8.62)$$

While these formulations are similar to those given in chapter 7, they show one important difference: the two tangential terms $\hat{\mathbf{n}}\times\mathbf{H}$ and $\mathbf{E}\times\hat{\mathbf{n}}$ are defined only on the surface and they represent surface current densities and surface magnetization:

$$\mathbf{E}\times\hat{\mathbf{n}} = \mathbf{M} \qquad \hat{\mathbf{n}}\times\mathbf{H} = \mathbf{J} \quad (8.63)$$

Thus, the natural method of matching the finite element solution of the weak forms above is to use the boundary integral equations in Eqs. (8.27) and (8.28) and use the surface current density and surface magnetization to match the two. The boundary integral equations we obtain have the following terms for the surface

$$\mathbf{E} = \mathbf{E}_{inc} - \int_S \left(j\omega\mu_0(\hat{\mathbf{n}}\times\mathbf{H})G_0 + (\hat{\mathbf{n}}.\mathbf{E})\nabla' G_0 - (\mathbf{E}\times\hat{\mathbf{n}})\times\nabla' G_0\right)ds' \quad (8.64)$$

$$\mathbf{H} = \mathbf{H}_{inc} + \int_S \left(j\omega\mu_0(\hat{\mathbf{n}}\times\mathbf{E})G_0 - (\hat{\mathbf{n}}.\mathbf{H})\nabla' G_0 - (\hat{\mathbf{n}}\times\mathbf{H})\times\nabla' G_0\right)ds' \quad (8.65)$$

The latter equations are assumed to apply in domain Ω_0 (on the surface) and Eqs. (8.61) and (8.62) apply in domain Ω. Now it is obvious that Eqs. (8.64) and (8.65) should be used on the surface while the finite element equations should be used in Ω. The details of implementation are given next.

8.5 FINITE ELEMENT IMPLEMENTATION

To define a general implementation we use Figure 8.9. The general geometry contains two regions: a nonhomogeneous region Ω in which the fields are required and which contains conductors as well as dielectrics. The materials within this region can be anisotropic and nonhomogeneous but must be linear. The second domain, marked Ω_0 is homogeneous, and normally is taken as free space although this is not a requirement of the method. Domain Ω is completely contained within Ω_0.

To define a finite element/boundary element method (*FEM/BEM*), the finite element method is applied to the interior of domain Ω and the boundary integral method is applied to the interface between Ω and Ω_0. The two approximations are then coupled to provide a single result. This method is particularly advantageous in modeling of nondestructive evaluation of materials because the open boundary nature of the problem is eliminated

8.5.1 Finite element approximation in Ω

The weak form of the field equations are obtained in Eqs. (8.61) and (8.62) as

$$\int_\Omega \left(\frac{1}{j\omega\mu}\nabla\times\mathbf{E}\right)\times(\nabla N_m)dv + \int_\Omega j\omega\varepsilon\mathbf{E}N_m dv = \int_S (\hat{\mathbf{n}}\times\mathbf{H})N_m ds \quad (8.66)$$

for the electric field intensity and as

$$\int_\Omega \left(\frac{1}{j\omega\varepsilon}\nabla\times\mathbf{H}\right)\times(\nabla N_m)dv + \int_\Omega j\omega\mu\mathbf{H}N_m dv = \int_S (\mathbf{E}\times\hat{\mathbf{n}})N_m ds \quad (8.67)$$

for the magnetic field intensity, where S is the surface enclosing Ω. We also note that the right hand side represents current density $(\hat{\mathbf{n}}\times\mathbf{H})$ and magnetization $(\mathbf{E}\times\hat{\mathbf{n}})$. These are used to couple the finite element approximation with the boundary element approximation.

(a) The finite element

A finite element must first be chosen. Any three-dimensional element is appropriate. Since a number of finite elements that can be used here are defined in chapter 4, we assume that tetrahedral elements are used. Using the shape functions in section 4.4.2, the approximations within the finite element are:

$$\mathbf{E} = \sum_{j=1}^{m} \mathbf{E}_j N_j$$

and

$$\mathbf{J}_s = \widehat{\mathbf{n}} \times \mathbf{H} = \sum_{j=1}^{n} (\widehat{\mathbf{n}} \times \mathbf{H})_j N_j$$

where $\mathbf{J}_S$ is the surface current density and both $\mathbf{H}$ and $\mathbf{E}$ are vectors, m is the number of nodes in the finite element and n the number of nodes on the surface. Although any of the equations above can be used we use the electric field intensity. Thus, Eq. (8.66) can be written for a single element as

$$[S]_e\{E\}_e = [Q]_e\{J\}_e \tag{8.68}$$

Since the electric field is a vector, matrices $[S]$ and $[Q]$ are $3\,m \times 3\,m$ in size with each node corresponding to a 3×3 matrix. This is, in fact, similar to the matrices given in section (7.3). The coefficients of $[S]$ and $[Q]$ are

$$S_{3i-2,3j-2} = \int_{v_i} \left(\frac{1}{j\omega\mu} \frac{\partial N_i}{\partial y} \frac{\partial N_j}{\partial y} + \frac{1}{j\omega\mu} \frac{\partial N_i}{\partial z} \frac{\partial N_j}{\partial z} + j\omega\varepsilon N_i N_j \right) dv$$

$$S_{3i-1,3j-1} = \int_{v_i} \left(\frac{1}{j\omega\mu} \frac{\partial N_i}{\partial x} \frac{\partial N_j}{\partial x} + \frac{1}{j\omega\mu} \frac{\partial N_i}{\partial z} \frac{\partial N_j}{\partial z} + j\omega\varepsilon N_i N_j \right) dv$$

$$S_{3i,3j} = \int_{v_i} \left(\frac{1}{j\omega\mu} \frac{\partial N_i}{\partial x} \frac{\partial N_j}{\partial x} + \frac{1}{j\omega\mu} \frac{\partial N_i}{\partial y} \frac{\partial N_j}{\partial y} + j\omega\varepsilon N_i N_j \right) dv$$

$$S_{3i-1,3j-2} = -\frac{1}{j\omega\mu} \int_{v_i} \left(\frac{\partial N_i}{\partial x} \frac{\partial N_j}{\partial y} \right) dv$$

$$S_{3i,3j-1} = -\frac{1}{j\omega\mu} \int_{v_i} \left(\frac{\partial N_i}{\partial x} \frac{\partial N_j}{\partial z} \right) dv$$

$$S_{3i,3j-2} = -\frac{1}{j\omega\mu} \int_{v_i} \left(\frac{\partial N_i}{\partial y} \frac{\partial N_j}{\partial x} \right) dv$$

$$S_{3i-2,3j-1} = -\frac{1}{j\omega\mu}\int_{v_i}\left(\frac{\partial N_i}{\partial y}\frac{\partial N_j}{\partial x}\right)dv$$

$$S_{3i-1,3j} = -\frac{1}{j\omega\mu}\int_{v_i}\left(\frac{\partial N_i}{\partial z}\frac{\partial N_j}{\partial x}\right)dv$$

$$S_{3i-2,3j} = -\frac{1}{j\omega\mu}\int_{v_i}\left(\frac{\partial N_i}{\partial x}\frac{\partial N_j}{\partial y}\right)dv$$

$$Q_{3i-2,3j-2} = Q_{3i-1,3j-1} = Q_{3i,3j} = \int_{S_i} N_i N_j dS$$

and, as expected, the elemental matrix is symmetric. The nondiagonal terms of $[Q]$ are all zero. We also note that while the $[S]$ matrix is due to volume integration, the $[Q]$ matrix is due to surface contributions and these are nonzero only on the interface between Ω and Ω_0. We also note that both ε and μ may be space dependent but, within a single element, they are assumed to be constant.

Since Ω is nonhomogeneous we must assume that the electric field as well as the magnetic field has a jump at the interface between Ω and Ω_0. The jump condition for the electric field intensity is

$$\hat{\mathbf{n}}\cdot(\varepsilon_0\mathbf{E}_0 - \varepsilon\mathbf{E}) = 0 \tag{8.69}$$

$$\hat{\mathbf{n}}\times(\mathbf{E}_0 - \mathbf{E}) = 0 \tag{8.70}$$

where $\mathbf{E}$ is the field in Ω, $\mathbf{E}_0$ the field in Ω_0 and $\hat{\mathbf{n}}$ is the normal unit vector. To define the jump conditions we therefore need to calculate the normal components of $\mathbf{E}$. Using a unit vector as

$$\hat{\mathbf{n}} = \hat{\mathbf{x}}n_x + \hat{\mathbf{y}}n_y + \hat{\mathbf{z}}n_z$$

where

$$n_x = \frac{1}{n}\int_{V_i}\frac{\partial N_i}{\partial x}dv \qquad n_y = \frac{1}{n}\int_{V_i}\frac{\partial N_i}{\partial y}dv \qquad n_z = \frac{1}{n}\int_{V_i}\frac{\partial N_i}{\partial z}dv$$

$$n = \sqrt{\left(\int_{V_i} \frac{\partial N_i}{\partial x} dv\right)^2 + \left(\int_{V_i} \frac{\partial N_i}{\partial y} dv\right)^2 + \left(\int_{V_i} \frac{\partial N_i}{\partial z} dv\right)^2}$$

we can write the following relation between the components of the electric field intensity in the two domains

$$\begin{Bmatrix} E_x \\ E_y \\ E_z \end{Bmatrix} = \begin{bmatrix} a_{11} & a_{12} & a_{13} \\ a_{21} & a_{22} & a_{23} \\ a_{31} & a_{32} & a_{33} \end{bmatrix} \begin{Bmatrix} E_{x0} \\ E_{y0} \\ E_{z0} \end{Bmatrix}$$

where

$$a_{11} = n_x^2 \frac{\varepsilon_0}{\varepsilon} + n_y^2 + n_z^2$$

$$a_{22} = n_y^2 \frac{\varepsilon_0}{\varepsilon} + n_x^2 + n_z^2$$

$$a_{33} = n_z^2 \frac{\varepsilon_0}{\varepsilon} + n_x^2 + n_y^2$$

$$a_{12} = a_{21} = n_x n_y \left(\frac{\varepsilon_0}{\varepsilon} - 1\right)$$

$$a_{13} = a_{31} = n_x n_z \left(\frac{\varepsilon_0}{\varepsilon} - 1\right)$$

$$a_{23} = a_{32} = n_y n_z \left(\frac{\varepsilon_0}{\varepsilon} - 1\right)$$

8.5.2 Boundary element approximation on *S*

To perform the boundary element approximation we can start with Eq. (8.64) or Eq. (8.65). Since the finite element approximation is in terms of **E** we continue with the equation for **E**

$$\mathbf{E}_j = \mathbf{E}_j^{inc} - \frac{1}{4\pi}\int_S \left(j\omega\mu(\hat{\mathbf{n}}\times\mathbf{H})G_j + (\hat{\mathbf{n}}.\mathbf{E})\,\nabla' G_j - (\mathbf{E}\times\hat{\mathbf{n}})\times\nabla' G_j \right) ds$$

(a) The boundary element

To perform a finite element approximation on the interface between Ω and Ω_0, surface elements must be defined. The boundary element is essentially a two-dimensional element but must be consistent with the volume element used in the previous section. Since we choose, arbitrarily, to use a tetrahedral element for the finite element approximation, the boundary element must be a triangular element. If the volume element is hexahedral (see chapter 4), the boundary element must be quadrilateral.

It is also important that the boundary element and volume element be of the same order (linear in our case). The finite element approximations on the boundary are

$$\mathbf{E} = \sum_{j=1}^{p} \mathbf{E}_j M_j$$

$$\mathbf{J}_s = \widehat{\mathbf{n}}\times\mathbf{H} = \sum_{j=1}^{p} (\widehat{\mathbf{n}}\times\mathbf{H})_j M_j$$

where M_j are the shape functions for the boundary element and p is the number of nodes in the boundary element. These values are different from the shape functions N_j and number of nodes m used for the volume element.

8.5.3 Boundary element equations

Substitution of the approximations above into Eq. (8.64) provides the elemental equation for an element on the boundary

$$[B]_e\{E\}_e - \{E\}_e^{inc} = [C]_e\{J\}_e \tag{8.71}$$

where E^{inc} indicates the incident electric field at the nodes of the element. From Eq. (8.64), the coefficients of matrices $[B]$ and $[C]$ (each a $3n\times 3n$ matrix) are

$$B_{3i-2,3j-2} = B_{3i-1,3j-1} = B_{3i,3j} = \delta_{ij} + \int_{S_i}\left(n_x\frac{\partial G_i}{\partial x} + n_y\frac{\partial G_i}{\partial y} + n_z\frac{\partial G_i}{\partial z}\right)M_j ds$$

$$B_{3i-2,3j-1} = -B_{3i-1,3j-2} = \int_{S_i}\left(n_y\frac{\partial G_i}{\partial x} - n_x\frac{\partial G_i}{\partial y}\right)M_j ds$$

$$B_{3i-2,3j} = -B_{3i,3j-2} = \int_{S_i}\left(n_z\frac{\partial G_i}{\partial x} - n_x\frac{\partial G_i}{\partial z}\right)M_j ds$$

$$B_{3i-1,3j} = -B_{3i,3j-1} = \int_{S_i}\left(n_z\frac{\partial G_i}{\partial y} - n_y\frac{\partial G_i}{\partial z}\right)M_j ds$$

$$C_{3i-2,3j-2} = C_{3i-1,3j-1} = C_{3i,3j} = -j\omega\mu\int_{S_i} G_i M_i ds$$

The normals at the nodes of the element n_x, n_y, n_z, are calculated as for the finite elements but are defined at nodes i. These are given as

$$n_{xi} = \frac{1}{n}\int_{S_i} n_x M_i\, ds \qquad n_{yi} = \frac{1}{n}\int_{S_i} n_y M_i\, ds \qquad n_z = \frac{1}{n}\int_{S_i} n_z M_i\, ds$$

and

$$n = \sqrt{\left(\int_{S_i} n_x M_i\, ds\right)^2 + \left(\int_{S_i} n_y M_i\, ds\right)^2 + \left(\int_{S_i} n_z M_i\, ds\right)^2}$$

8.5.4 Coupling of the finite element and boundary element matrices

The two equations to be coupled at the interface are:

$$[S]_e\{E\}_e = [Q]_e\{J\}_e \tag{8.72}$$

$$[\mathrm{B}]_e\{\mathrm{E}\}_e - \{E\}_e^{inc} = [C]_e\{J\}_e \tag{8.73}$$

In Eq. (8.72), there are nonzero contributions to the right hand side due to current densities on the interface. For any internal node in the finite element mesh, the right hand side is zero. This allows the partitioning of the matrices $[S]$ and $[Q]$ as

$$\begin{bmatrix} s_{11} & s_{12} \\ s_{21} & s_{22} \end{bmatrix}_e \begin{Bmatrix} E \\ E_0 \end{Bmatrix}_e = \begin{bmatrix} 0 & 0 \\ 0 & q \end{bmatrix}_e \begin{Bmatrix} J \\ J_0 \end{Bmatrix}_e$$

where the index zero indicates quantities on the interface between Ω and Ω_0 and quantities without indices are defined only in the volume Ω.
Eq. (8.73) is only defined on the interface and, therefore, we may write

$$[B]_e\{E_0\}_e - \{E_0\}_e^{inc} = [C]_e\{J_0\}_e$$

Combining the two equations we get

$$\begin{bmatrix} s_{11} & s_{12} \\ s_{21} & s_{22} - qC^{-1}B \end{bmatrix}_e \begin{Bmatrix} E \\ E_0 \end{Bmatrix}_e = \begin{Bmatrix} 0 \\ qC^{-1}E_0^{inc} \end{Bmatrix}_e \qquad (8.74)$$

The coefficients of the matrices are all known. The incident electric field intensity is calculated on the surface of the domain Ω as in the previous sections and the system can then be solved by standard methods.

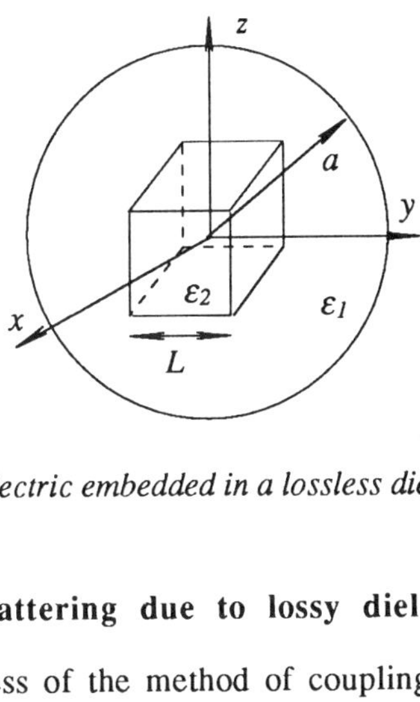

Figure 8.10 A lossy dielectric embedded in a lossless dielectric

8.5.5 Example: scattering due to lossy dielectrics

To show the usefulness of the method of coupling the finite element and boundary element methods, an example, consisting of a cubic inclusion inside a spherical lossless dielectric is presented. The geometry is shown in Figure 8.10. The dielectric constant of the dielectric sphere is *4* and that of the inclusion is either *4* or *80–j80*. The first case simulates a uniform sphere (no inclusion) while the second simulates an anisotropic material or an inclusion inside an otherwise uniform material. The shift in measured results (fields, scattering cross section) are then an indication of variation in material properties.

The model shown in Figure 8.10 is discretized using tetrahedral elements inside the volume and triangular elements on the surface of the sphere. The results obtained by solving Eq. (8.74) are presented in Figures 8.11 through

8.13. Figure 8.11 shows the scattering cross section of the sphere with and without the embedded cube, clearly showing the effect of the cubic lossy dielectric. Note that both frequency and dimensions are normalized. Also, the incident field has an x component of the electric field intensity (plane wave) and a y component of the magnetic field intensity.

Figure 8.12 shows the surface electric currents on the yz and xz planes while Figure 8.13 shows the surface magnetic current on the yz and xz planes both with and without the embedded cube.

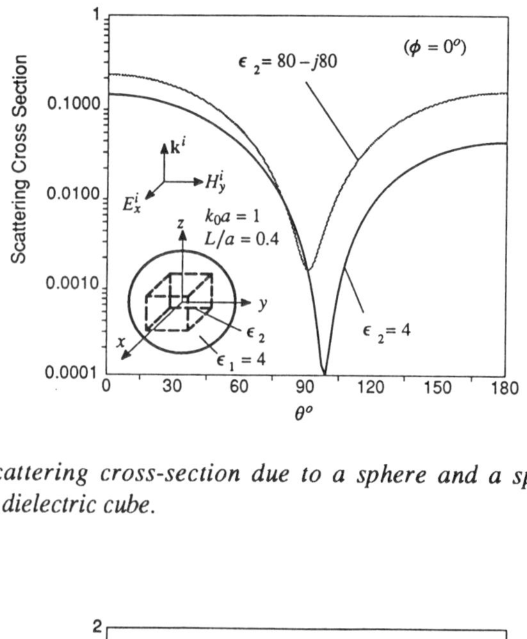

Figure 8.11 Scattering cross-section due to a sphere and a sphere with an embedded lossy dielectric cube.

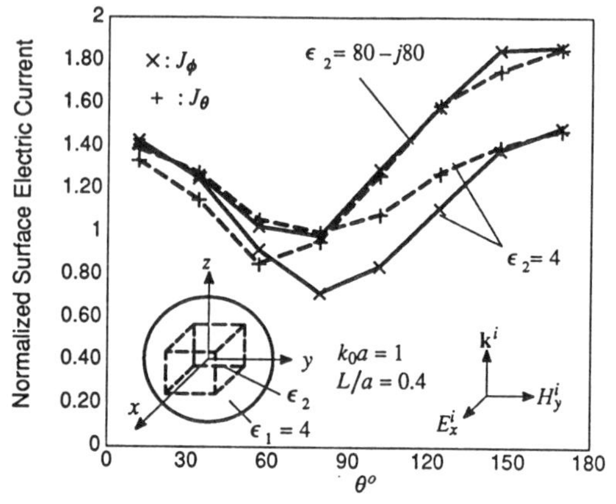

Figure 8.12 Normalized surface electric current on the yz and xz planes.

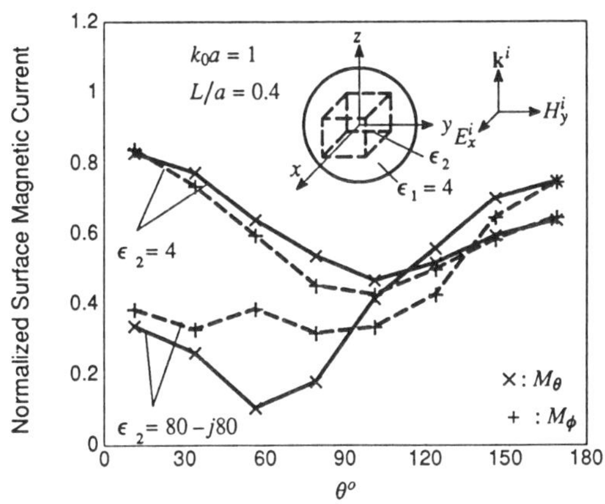

Figure 8.13 Normalized surface magnetic current on the yz and xz planes.

8.6 INTEGRAL EQUATIONS FOR STATIC FIELDS

8.6.1 Formulation

The volume integral equation given in Eq. (8.55) or (8.56) is general and, under the assumptions given applies to any field, including static electric and magnetic fields. However, for static fields, the electric and magnetic field equations decouple (**H** is zero for electrostatic fields and **E** is zero for magnetostatic fields). It is therefore possible to write the two equations as

$$\mathbf{E}_t = \mathbf{E}_i - \nabla V \tag{8.75}$$

$$\mathbf{H}_t = \mathbf{H}_i - \nabla \Phi \tag{8.76}$$

where the index i indicates the fields due to applied sources (current densities for **H** and charges for **E**) and the scalar potentials Φ and V are still the same as in Eqs. (8.51) and (8.52). However, the free space Green's function now becomes

$$G_0(\mathbf{R}) = \frac{1}{|\mathbf{R}|} \tag{8.77}$$

since $k_0=0$ in Eq. (8.23) at zero frequencies.

As an example, by substituting the potential Φ in Eq. (8.76), the magnetic field intensity can be written as

$$\mathbf{H}_t = \mathbf{H}_i - \nabla\left[\frac{1}{4\pi\varepsilon_0}\int_\Omega \rho_m G_0 dv\right] \tag{8.78}$$

This is not necessarily a good choice since the equivalent volume sources required in this equation are magnetic charge densities. Normally, volume magnetization is chosen to represent equivalent sources. To write the equation in terms of magnetization we use the relation

$$\mathbf{M} = \frac{\mathbf{B}}{\mu_0} - \mathbf{H} = \chi\mathbf{H}$$

where χ is the magnetic susceptibility of the material and, in general, is field dependent. Thus, Eq. (8.78) becomes

$$\mathbf{M} = \chi\mathbf{H}_i - \frac{\chi}{4\pi\varepsilon_0}\nabla\int_\Omega \rho_m G_0 dv$$

and, since

$$\rho_m = -\mu_0(\nabla.\mathbf{M})$$

we obtain

$$\mathbf{M}(\mathbf{r}) = \chi\mathbf{H}_i + \frac{\chi}{4\pi\varepsilon_0}\nabla\int_\Omega \mu_0(\nabla.\mathbf{M}(\mathbf{r}'))G_0 dv$$

To simplify this relation we use the following vector identity

$$(\nabla\cdot\mathbf{M})G = \nabla\cdot(G\mathbf{M}) - \mathbf{M}\cdot(\nabla G)$$

This gives

$$\mathbf{M}(\mathbf{r}) = \chi\mathbf{H}_i + \frac{\chi}{4\pi\varepsilon_0}\nabla\int_\Omega \mu_0\nabla\cdot(\mathbf{M}(\mathbf{r}')G_0)dv - \frac{\chi}{4\pi\varepsilon_0}\nabla\int_\Omega \mu_0\mathbf{M}(\mathbf{r}')\cdot\nabla(G_0)dv$$

By converting the first integral into a surface integral, it can be shown to be due to surface magnetic charge densities while the second integral is due to volume magnetic charge densities:

$$\mathbf{M}(\mathbf{r}) = \frac{\chi}{4\pi\varepsilon_0}\nabla\int_{\Omega}\mu_0\nabla\cdot(\mathbf{M}(\mathbf{r}')G_0)dv = \frac{\chi}{4\pi\varepsilon_0}\nabla\int_{S}\mu_0(\mathbf{M}(\mathbf{r}')\cdot\hat{\mathbf{n}}G_0)dS$$

$$= \frac{\chi}{4\pi\varepsilon_0}\nabla\int_{S}\mu_0\rho_{ms}G_0)dS$$

Since volume integral equations are normally implemented in the volume of a material, only volume magnetization is assumed and therefore the surface integral vanishes. The magnetization integral equation is written as

$$\mathbf{M}(\mathbf{r}) = \chi\mathbf{H}_i - \frac{\chi}{4\pi\varepsilon_0}\nabla\int_{\Omega}\mu_0\mathbf{M}(\mathbf{r}')\cdot\nabla(G_0)dv \qquad (8.79)$$

8.6.2 Method of moments implementation

We show here the method of moments implementation of this equation for static magnetic fields. Also, we assume for now that $\mathbf{H}_i$ is known from independent calculations. Under these conditions, the domain of interest (primarily magnetic materials) is discretized into any number of volume subdomains or elements. These elements are similar to finite elements but, for this purpose they are simply a small volume of some defined shape. Prisms and tetrahedra are often used for this purpose. In each volume we assume constant magnetization. This is the so-called pulse basis function. This magnetization is unknown and must be determined. The magnetization is associated with a convenient point inside the element, normally the centroid of the element. This point defines the point $\mathbf{r}'$ in calculations. These definitions are shown schematically in Figure 8.14.

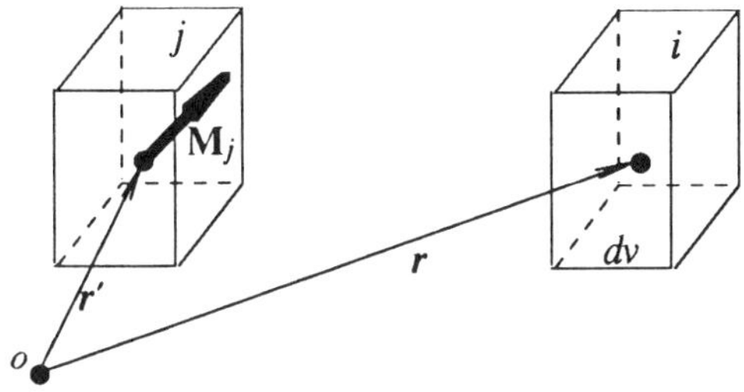

Figure 8.14 Calculation of magnetization at center of element i due to magnetization in element j.

Thus, subdomain j is characterized by magnetization $\mathbf{M}_j$ (or M_{jx}, M_{jy}, M_{jz}). Requiring that the field value be satisfied at the centroid of the subdomain we can write for each element in the volume:

$$\mathbf{M}_i = \chi_i \mathbf{H}_{ci} - \frac{\chi_i}{4\pi\varepsilon_0}\nabla \int_\Omega \sum_{j=1}^{N} \mu_0 \mathbf{M}_j \cdot \nabla\left(\frac{1}{|\mathbf{r} - \mathbf{r}'|}\right) dv \tag{8.80}$$

where we note that the summation is over the whole domain. This means that the field at the centroid of element i depends on the magnetization of all elements in the volume. This is typical of the method of moments and, indeed, of integral methods in general.

The equation above is easily implemented since $\mathbf{r}$ and $\mathbf{r}'$ are simply the position vectors of centroids i and j, and $\mathbf{M}_j$ is constant in each element. To simplify implementation, the following approximation is used:

$$\mathbf{M} = \sum_{j=1}^{N} \mathbf{M}_j \cdot w_{ij} \tag{8.81}$$

where w_{ij} is a weighting or testing function. $w_{ij}=1$ if $i=j$, $w_{ij}=0$ if $i \neq j$. With this, the equation is written as

$$\sum_{i=1}^{N} \frac{\mathbf{M}_i w_{ij}}{\chi_i} + \frac{1}{4\pi\varepsilon_0}\nabla \int_\Omega \sum_{j=1}^{N} \mu_0 \mathbf{M}_j w_{ij} \cdot \nabla\left(\frac{1}{|\mathbf{r} - \mathbf{r}'|}\right) dv = \mathbf{H}_{ci}$$

Since $\mathbf{M}$ is constant within each element, the magnitude of $\mathbf{M}$ can be taken out of the integration and the system written as

$$\sum_{j=1}^{N} \left[\mathbf{M}_j \frac{w_{ij}}{\chi_i} + \frac{1}{4\pi\varepsilon_0}\nabla \int_\Omega \mu_0 \widehat{\mathbf{M}}_j \cdot \nabla\left(\frac{1}{|\mathbf{r} - \mathbf{r}'|}\right) dv \right] = \mathbf{H}_{ci} \tag{8.82}$$

where $\widehat{\mathbf{M}}_j$ is the unit vector in the direction of $\mathbf{M}_j$.

The implementation now simply requires the coordinates of the centroids of the elements and integration of the Green's function over the element. For simple shapes such as prisms it is quite easy to evaluate the integrals. In other cases, numerical evaluation might be necessary. The implementation is identical to that for the simple method of moments in section 8.3, with the obvious changes due to the volumes of the elements.

As was noted before, the equations above can only be solved if the incident field (that is, the field due to current sources) is known apriori. The current density is calculated from the given sources in the absence of the scatterer. This is accomplished using the Biot-Savart or Ampere's laws to calculate the magnetic field intensity everywhere inside domain Ω.

The method as described above can be implemented in the context of finite elements since the approximation in Eq. (8.81) is identical with the approximation for finite element analysis if the weighting function w_{ij} is properly chosen as the shape function of the elements and the variable as the nodal values.

Also to be noted is that an essentially identical form can be obtained for electrostatic calculations in terms of **J**, using Eq. (8.75) as a starting point.

8.6.3 Formulation for magnetostatic fields

A formulation for magnetostatic applications is obtained by starting with the definition of the magnetic field in source-free regions. This approach is based on the Helmholtz theorem which states that any vector field can be separated into a curl-free and a divergence-free field. In our case, the field is separated into a term due to current sources $\mathbf{H}_s$ and a term due to magnetization sources $\mathbf{H}_m$. The equation for the magnetostatic field in a source-free region (from Eq. (1.2)) is:

$$\mathbf{H} = \mathbf{H}_s + \mathbf{H}_m \tag{8.83}$$

The second term is curl free and, therefore, can be defined as the gradient of a scalar potential which we call Φ. Physically, this scalar potential is a magnetic scalar potential

$$\mathbf{H}_m = -\nabla \Phi$$

If a region contains no current densities, the latter is sufficient for the calculation of the magnetic field intensity.

We wish now to formulate the magnetic field equations in terms of equivalent surface charges or currents under the current-free condition in the volume enclosed by the surface. To do so, we have two options: one is to use the general equations obtained in section 8.4. This is equivalent to the steps in the previous section except that surface integrals are used and the scalar potential functions Φ and V are evaluated on the surface of the domain Ω. However, it is more straightforward to use Green's (second) identity directly. The latter is stated for any two scalar functions Φ and Ψ as

$$\int_V \left(\Phi \nabla^2 \Psi - \Psi \nabla^2 \Phi\right) dV' = \int_S \left(\Psi \frac{\partial \Phi}{\partial n} - \Phi \frac{\partial \Psi}{\partial n}\right) ds' \tag{8.84}$$

where the surface integration is on the surface sources.

To make this expression useful we first substitute the scalar Green's function G_0 for Ψ

$$\int_V \left(\Phi \nabla^2 G_0 - G_0 \nabla^2 \Phi\right) dV' = \int_S \left(G_0 \frac{\partial \Phi}{\partial n} - \Phi \frac{\partial G_0}{\partial n}\right) ds' \tag{8.85}$$

while the scalar potential Φ is the same as defined in Eq. (8.51). Now we need to evaluate the two terms on the left hand side. From the definition of the Green's function:

$$\nabla^2 G_0 = -\delta(x - x')\delta(y - y')\delta(z - z')$$

where x',y',z' is the source point and x,y,z, the field point. Substitution of this in Eq. (8.84) gives for the first term

$$\int_V \Phi \nabla^2 G_0 dV' = -\int_V \Phi \delta(\mathbf{R}) dV' = -\Phi(\mathbf{r}) \tag{8.86}$$

where $\mathbf{r}$ is the field position vector, ($\mathbf{R}=\mathbf{r}-\mathbf{r}'$).

The second term is evaluated using the relation

$$\nabla \cdot \mathbf{H}_m = -\nabla \cdot \mathbf{M} = \frac{\rho_m}{\mu_0}$$

where $\mathbf{M}$ is the volume magnetization vector and ρ_m the equivalent magnetic volume charge density. This relation can be shown to be correct from the definition of magnetization

$$\mathbf{H} = \mathbf{H}_m + \mathbf{H}_s = \frac{\mathbf{B}}{\mu_0} - \mathbf{M}$$

and, since $\nabla \cdot \mathbf{H}_s = 0$ and $\nabla \cdot \mathbf{B} = 0$, the relation above follows. In terms of Φ the relation above is

$$\nabla \cdot \mathbf{H}_m = \nabla \cdot \nabla \Phi = \nabla^2 \Phi = -\nabla \cdot \mathbf{M} = \frac{\rho_m}{\mu_0}$$

Thus, the second term on the left hand side of Eq. (8.85) is:

$$\int_V G_0 \nabla^2 \Phi dV' = -\int_V G_0 \nabla \cdot \mathbf{M} dV' = \int_V G_0 \frac{\rho_m}{\mu_0} dV' \tag{8.87}$$

Substitution of Eqs. (8.86) and (8.87) in Eq. (8.85) gives

$$\Phi(\mathbf{r}) = -\int_V G_0 \nabla\cdot\mathbf{M} dV' - \int_S G_0 \frac{\partial \Phi}{\partial n} ds' + \int_S \Phi \frac{\partial G_0}{\partial n} ds' \tag{8.88}$$

or, in terms of an equivalent magnetic charge density

$$\Phi(\mathbf{r}) = \int_V G_0 \frac{\rho_m}{\mu_0} dV' - \int_S G_0 \frac{\partial \Phi}{\partial n} ds' + \int_S \Phi \frac{\partial G_0}{\partial n} ds' \tag{8.89}$$

In Eqs. (8.88) and (8.89) the volume integral is due to volume magnetization. For constant permeability, this term is equal to zero and we have an expression in terms of surface integrals alone.

$$\Phi(\mathbf{r}) = \int_S \Phi \frac{\partial G_0}{\partial n} ds' - \int_S G_0 \frac{\partial \Phi}{\partial n} ds' \tag{8.90}$$

Substituting the scalar Green's function from Eq. (8.77) we get

$$\Phi(\mathbf{r}) = \frac{1}{4\pi}\int_S \Phi \frac{\partial}{\partial n}\left(\frac{1}{|\mathbf{R}|}\right) ds' - \frac{1}{4\pi}\int_S \frac{1}{|\mathbf{R}|}\frac{\partial \Phi}{\partial n} ds' \tag{8.91}$$

While the Green's function is that of free space, we note that both the scalar potential Φ and its normal derivatives are, in general, unknown functions and must be determined.

8.6.4 Implementation

For implementation of the boundary integral method, all material surfaces are discretized either into finite elements or, by using the method of moments into subsections of some simple shape. Material surfaces can be either interfaces between two magnetic materials or between magnetic materials and free space. On surfaces bounded by free space, either Φ or $\partial\Phi/\partial n$ are known on the surface and the solution then can be defined in terms of the unknown values on the surface. On interfaces between two different materials, with different permeabilities, neither Φ or $\partial\Phi/\partial n$ are known and, therefore, the continuity relations at the surface must be used.

Continuity relations at the interface between two materials are:

$$\Phi_1 = \Phi_2$$

and

$$\mu_1 \frac{\partial \Phi_1}{\partial n} - \mu_2 \frac{\partial \Phi_2}{\partial n} = (\mu_1 - \mu_2) H_c$$

where H_c is now the normal component of the magnetic field intensity due to sources at the interface. This quantity can be calculated from independent relations and used to evaluate either Φ or $\partial\Phi/\partial n$ at the interface.

(a) Method of moments implementation

Assuming that the scalar potential and its normal derivatives are constant within a subsection, the method of moments implementation can be viewed as calculating the scalar potential due to each subsection at the center of a single subsection. That is, the scalar potential at the center of subsection i is evaluated as

$$\Phi = \sum_{j=1}^{N} \Phi_j . w_{ij}$$

where w_{ij} is the weighting or testing function. $w_{ij}=1$ if $i=j$, $w_{ij}=0$ if $i \neq j$. With this, the equation is written as

$$\Phi_i = \frac{1}{4\pi} \sum_{j=1}^{N} \Phi_j \int_S \frac{\partial}{\partial n}\left(\frac{1}{|\mathbf{r}_i - \mathbf{r}_j|}\right) ds_j - \frac{1}{4\pi} \sum_{j=1}^{N} \frac{\partial \Phi_j}{\partial n} \int_S \frac{1}{|\mathbf{r}_i - \mathbf{r}_j|} ds_j$$

and, as for the magnetization equation, the integration must be done over the subdomain before assembling the system of equations. The integration can be done analytically or numerically, depending on the shape of the subdomain. However, the calculation is rather lengthy because for each Φ_i the integration must be done over the whole surface. As mentioned above Φ_i must be known at the center of each subdomain and either Φ or $\partial\Phi/\partial n$ must be known on the surface. Under these conditions the solution is in terms of the unknown values of Φ.

(b) Finite element implementation

The finite element implementation is almost identical to the method of moments implementation. The differences between the two implementations are due to the approximation which, with few exceptions, uses a higher approximation than in the method of moments. The generic finite element approximation is

$$\Phi = \sum_{j=1}^{m} \Phi_j N_j$$

where the approximation is over an element and m indicates the number of nodes in the element and N_j are the shape functions of the element. The function Φ is then defined anywhere within the finite element, including at the nodes of the finite element. The normal derivative of the scalar potential is similarly defined as

$$\frac{\partial \Phi}{\partial n} = \sum_{j=1}^{m} \frac{\partial \Phi_j}{\partial n} N_j$$

Substituting the approximations for Φ and $\partial\Phi/\partial n$ in Eq. (8.91) we get

$$\Phi(\mathbf{r}) = \frac{1}{4\pi}\int_S \sum_{j=1}^{m} \Phi_j N_j \frac{\partial}{\partial n}\left(\frac{1}{|\mathbf{R}|}\right) ds' - \frac{1}{4\pi}\int_S \frac{1}{|\mathbf{R}|} \sum_{j=1}^{m} \frac{\partial \Phi_j}{\partial n} N_j ds'$$

As with the method of moments, this solution must be tested against known values of Φ. These values are at the nodes and therefore, for each node i we can write

$$\Phi_i = \frac{1}{4\pi}\int_S \sum_{j=1}^{m} \Phi_j N_j \frac{\partial}{\partial n}\left(\frac{1}{|\mathbf{r}_i - \mathbf{r}_j|}\right) ds_j - \frac{1}{4\pi}\int_S \frac{1}{|\mathbf{r}_i - \mathbf{r}_j|} \sum_{j=1}^{m} \frac{\partial \Phi_j}{\partial n} N_j ds_j$$

The integrals are evaluated either analytically or numerically, depending on the type of element. In most cases, the integration is more difficult than in the method of moments approach because the shape functions are space dependent.

The main difficulty in applying the finite element method is the evaluation of the integrals for $i=j$ since the integrals become singular. Thus, the diagonal terms in the equation above ($i=j$) cannot be calculated from direct integration. There are a number of techniques that can be used to evaluate singular integrals of this type. The simplest is to subdivide the area around the singular node and perform a sum of integrals of the small areas which is then equal (approximately) to that calculated above.

8.7 PROBLEMS

8.1 Calculate the capacitance of a single flat rectangular conducting plate, *2 m* by *1 m* in size. Assume the plate is thin and connected to an arbitrary known potential. For numerical calculation, this potential is given a numeric value. Use, *1, 2, 4, 8, 16*, and *64* subdomains. Show that the solution converges to a final value as the number of subdomains increases.

8.2 Repeat Problem 8.1 with the capacitor made of two plates, each $2\ m \times 1\ m$ in size, separated a distance of *1 m*. Use as many subdomains as are necessary to obtain a good solution. Calculate the capacitance and compare with the capacitance obtained with the formula for parallel plate capacitors.

8.3 Use the method of moments to calculate the total charge on a conducting sphere, *1 m* in diameter, and connected to a 10 *V* potential. Compare with the analytic solution.

8.4 Solve the problem described in Figure 8.10. Use the combined finite element/boundary element method in section 8.5 and compare your results with those given in Figures 8.12 and 8.13. From the equivalent current densities, calculate the electric and magnetic field intensities everywhere outside the sphere. Use any frequency and size of sphere, provided that the ratios given in Figure 8.11 are kept.

8.5 A ferromagnetic sphere, 10 *cm* in diameter is located in a uniform, dc magnetic field of *1 Tesla* as shown in Figure 8.15. Using the volume integral method, calculate the magneic field everywhere outside the sphere.

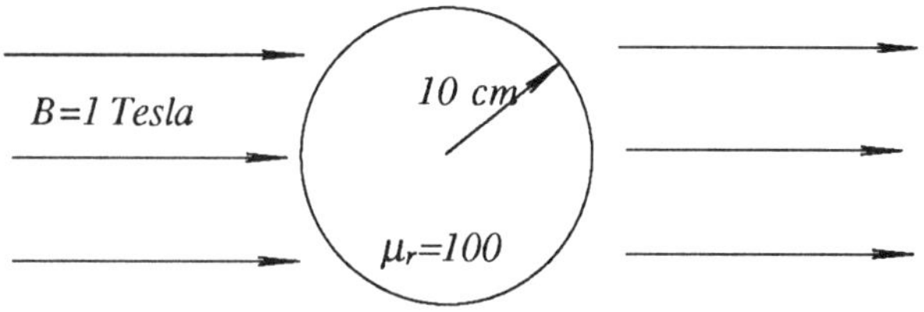

Figure 8.15.

8.6 A dielectric cube is located in a uniform dc electric field, 10 *V/m* in intensity. The cube and dimensions are given in Figure 8.16. Calculate the equivalent volume sources and the electric field everywhere outside the cube. Pay particular attention to treatment of the corners of the cube.

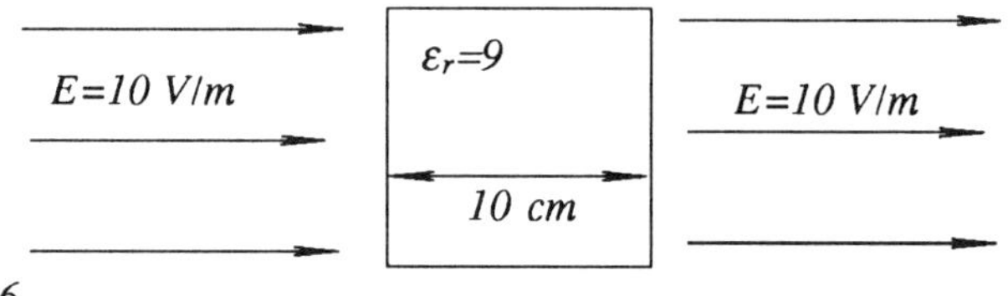

Figure 8.16.

8.8 BIBLIOGRAPHY

Integral equation methods have received considerable attention in electromagnetics because of their apparent efficiency in modeling. This was true in the early days of computers when large systems of equations resulting from differential methods could not be solved efficiently and is true today when larger and more complex models are required [1].

The method of moments has evolved first as a general mathematical tool [2,3]. Its application to electromagnetics dates back to the 1960's with Harrington's book [4] but the method has evolved before that as a method in applied mathematics [2]. A few applications of the method of moments are given in [5-16]. This list is very selective and does not cover a vast number of applications of the method in scattering and antenna calculations.

Formulation of field problems in terms of boundary integral methods and combined boundary integral/finite element methods are discussed in a number of sources. Some are given in references [17] through [23].

[1] Brebbia, C.A. (1984), *The Boundary Element Method for Engineers*, Pantech Press, London.

[2] Vorobyev, Y.V. (1965), *Method of Moments in Applied Mathematics*, Gordon and Breach Science Publishers, New York.

[3] Harrington, R.F. (1990), Origin and development of the method of moments for field computation, *IEEE Antenna and Propagation Society Magazine*, Vol. 32, No. 3, pp. 31-36.

[4] Harrington, R.F. (1967), *Field Computation by Moment Method*, The Macmillan Company, London.

[5] Moore, J. and Pizer, R. (eds.), (1977), *Moment Methods in Electromagnetics, Techniques and Applications*, John Wiley and Sons Inc., New York.

[6] Wilton, D.R. and Govind, S. (1977), Incorporation of edge conditions in moment method solutions, *IEEE Transactions on Antennas and Propagation*, Vol. AP-25, No. 6, pp. 845-850.

[7] Ney, M.M. (1985), Method of moments as applied to electromagnetic problems, *IEEE Transactions on Microwave Theory and Techniques*, Vol. MTT-33, No. 10, pp. 972-980.

[8] Taflove, A. (ed.), (1988), Special issue on numerical methods for electromagnetic wave interactions, *Wave Motion*, Vol. 10, No. 6, 1988.

[9] J.C. Toler, "Special issue on electromagnetic wave interactions with biological systems", IEEE Transactions on Microwave Theory and Techniques, Vol. MTT-32, No. 8,.

[10] Harrington, R.F. (1967), Matrix methods for field problems, *Proceedings of the IEEE*, Vol. 55, No. 2, pp. 136-149.

[11] Harrington, R.F. (1987), The method of moments in electromagnetics, *Journal of Electromagnetic Waves and Applications*, Vol. 1, No. 3, pp. 181-200.

[12] Miller, E.K. (1988), A selective survey of computational electromagnetics, *IEEE Transactions on Antennas and Propagation*, Vol. 36, No. 9, pp. 1281-1304.

[14] Reitan, D.K. and Higgins, T.H. (1957), Accurate determination of the capacitance matrix of a thin rectangular plate, *AIEE Transactions*, pp. 761-766

[15] Higgins, T.J. and Reitan, D.K. (1951), Calculation of the capacitance of a circular annulus by the method of subsurfaces, *AIEE Transactions*, Vol. 70, pp. 926-933.

[16] Storer, J.E. (1966), Impedance of thin wire loop antennas, *AIEE Transactions*, pp. 606-619.

[17] Sylvester, P.P. and Ferrari, R.P. (1990), *Finite Elements for Electrical Engineers*, Cambridge University Press, Cambridge.

[18] Ancelle, A. and Sabonadiere, J.C. (1980), Numerical solution of 3-D magnetic field problems using boundary integral equations, *IEEE Transactions on Magnetics*, Vol. MAG-16, No. 5, pp. 1089-1091.

[19] Bossavit, A. and Verite, J.C. (1982), A fixed FEM-BIEM method to solve 3D eddy current problems, *IEEE Transactions on Magnetics*, Vol. 18, No. 2, pp. 431-437.

[20] Davey, K.R. and Barnes, W.J. (1986), T-Ω technique applied to 3-D eddy current problems using the boundary interal equation, *IEE Proceedings, pt. B*, Vol. 138, No. 6, pp. 359-364.

[21] Enokinozo, M. and Todaka, T. (1990), Boundary element analysis for 3-D eddy current problem, *IEEE Transactions on Magnetics*, Vol. MAG-26, 1990, pp. 446-449.

[22] Poggio, A.J. and Miller, E.K. (1973), Integral equation solution of three-dimensional scattering problems, in Mittra, R. (ed.), *Computer Techniques for Electromagnetics*, Ch. 4, Pergamon Press, London.

[23] Albanese, R. and Rubinacci, R. (1988), Integral formulations for 3-D eddy currents computation using edge elements, *IEE Proceedings, Pt. A*, Vol. 135, pp. 457-461.

[24] Wang, J.S. (1992), *On edge based finite elements and method of moments solutions of electromagnetic scattering and coupling*, Ph. D. Dissertation, The University of Akron.

CHAPTER 9

Parabolic partial differential equations

9.1 INTRODUCTION

Following the pattern of chapter 5 where elliptic differential processes are discussed we discuss here parabolic processes and their relation to modeling in *NDT*. Parabolic processes are those that describe transient fields. These are essentially linear, transient eddy current applications. In terms of methods of solution, the finite element and finite difference methods are used and the finite difference time domain method (*FDTD*) is introduced. Time domain representation in finite element calculations is by finite difference methods. Examples from pulsed eddy current testing and scattering from inclusions in composite materials are given to highlight the type of problems that can be solved using these methods.

9.2 THE GENERAL PARABOLIC PARTIAL DIFFERENTIAL EQUATION

In chapter 5 we introduced the definition of the elliptic equation based on the coefficients of the general second order partial differential equation:

$$a\frac{\partial^2 F(u,v)}{\partial u^2} + b\frac{\partial^2 F(u,v)}{\partial u\,\partial v} + c\frac{\partial^2 F(u,v)}{\partial v^2} + d\frac{\partial F(u,v)}{\partial u} + e\frac{\partial F(u,v)}{\partial v} + gF(u,v) = f(u,v) \qquad (9.1)$$

where u,v may represent any two variables such as x,y in space, or x in space and t in time. This equation is said to be parabolic in form if the following condition is satisfied

$$b^2 - 4ac = 0 \tag{9.2}$$

The following is a parabolic equation

$$a\frac{\partial^2 F(u,v)}{\partial u^2} + e\frac{\partial F(u,v)}{\partial v} = f(u,v)$$

As examples, the following commonly encountered equations are parabolic equations

$$\frac{\partial^2 \mathbf{A}}{\partial x^2} = -\sigma\mu\frac{\partial \mathbf{A}}{\partial t}$$

$$\frac{\partial^2 \mathbf{A}}{\partial x^2} + \frac{\partial^2 \mathbf{A}}{\partial y^2} = \mu\mathbf{J}_s - \sigma\mu\frac{\partial \mathbf{A}}{\partial t}$$

The last equation, even though it is a function of three variables, is of the same form and is solved here as a parabolic equation. These equations have a number of properties in common. First, they are time dependent. Second, the solution is bounded in space but unbounded in time. Thus, the solution of any of these equations requires the definition of a solution domain with defined boundaries and boundary conditions but also requires the specification of initial conditions in time.

Both the equation and the boundary conditions must be satisfied in the solution process. This is shown schematically in Figure 9.1, where we assume in general that both Dirichlet and Neumann boundary conditions may exist in the solution domain. The domain is closed in all spatial variables (i.e. it is a closed surface for two-dimensional problems or a volume for three dimensional applications) but is open in time. The solution in time may progress indefinitely although, for practical implementation we stop this progression at some point.

9.3 TRANSIENT FINITE ELEMENT FORMULATION

The formulation of the transient eddy current (diffusion) equation is presented next. The formulation leads to a set of linear algebraic equations, which are solved as in the steady state formulation. In other words, a set of equations is solved at discrete time steps until the transient effects have disappeared. The inclusion of time is at a separate level, in that finite difference approximations in time are used while the spatial approximation is in terms of finite elements. Normally, a backward difference approximation is used to express the time derivative term of the finite element equation. The magnetic vector potential is used for formulation in axisymmetric and two-dimensional geometries although, from the discussion in chapters 4 and 7, other variables and dimensionality can be used. Unlike the eddy current formulation in chapter 7, the magnetic vector potential in transient applications is a single valued real quantity rather than a complex phasor.

A three-nodal point triangular element may be used to discretize the region. The element is chosen arbitrarily and, in fact, the formulation is general and an isoparametric triangular element can be used as well.

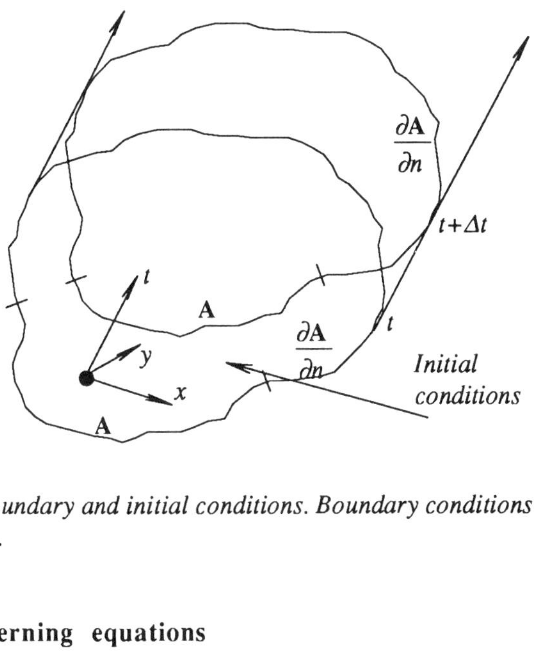

Figure 9.1 Boundary and initial conditions. Boundary conditions are satisfied at all time steps.

9.3.1 Governing equations

The governing equation is the time dependent form of Poisson's equation. We start with the general form of Maxwell's equations but also use the magnetic vector potential as defined in Eq. (1.63). Thus, we write Maxwell's first and second equations as

$$\nabla\times\mathbf{E} = -\nabla\times\left(\frac{\partial \mathbf{A}}{\partial t}\right) \tag{9.3}$$

$$\nabla\times\left(\frac{1}{\mu}\nabla\times\mathbf{A}\right) = \mathbf{J}_s - \sigma\frac{\partial \mathbf{A}}{\partial t} \tag{9.4}$$

where the following two relations were used:

$$\mathbf{B} = \nabla\times\mathbf{A} \qquad \text{and} \qquad \mathbf{E} = -\frac{\partial \mathbf{A}}{\partial t}$$

From Eq. (9.4) we write:

$$\frac{1}{\mu}\nabla^2\mathbf{A} = -\mathbf{J}_s + \sigma\frac{\partial \mathbf{A}}{\partial t} \tag{9.5}$$

This is the general time dependent equation where the displacement current as well as the static electric field are neglected (see section 7.3 for the complete equation).

In this section, only axisymmetric geometries will be considered. This means that the magnetic vector potential has only a ϕ component. Using the Laplacian in cylindrical coordinates under these conditions, yields the governing equation:

$$\frac{1}{\mu}\left[\frac{\partial^2 A_\phi}{\partial r^2} + \frac{1}{r}\frac{\partial A_\phi}{\partial r} + \frac{\partial^2 A_\phi}{\partial z^2} - \frac{A_\phi}{r^2}\right] = -J_s + \sigma\frac{\partial A_\phi}{\partial t} \tag{9.6}$$

where J_s is not a sinusoidal waveform and only the ϕ component is used.

An energy functional is used here in a manner similar to that adopted in chapter 7: the functional is assumed to be correct, while its correctness must be shown by deriving the Euler equation. However, we will view this energy functional to be correct at any given point in time and assume it is stationary for any time step without explicitly deriving the Euler equation. The functional is

$$F(A_\phi) = \int_v \left[\frac{1}{2\mu}\left[\left[\frac{\partial A_\phi}{\partial r} - \frac{A_\phi}{r}\right]^2 + \left[\frac{\partial A_\phi}{\partial z}\right]^2\right] - J_s \cdot A_\phi + \sigma\frac{\partial A_\phi}{\partial t}A_\phi\right]dv \tag{9.7}$$

To write this form we assume that the time derivative of A is independent of A. This assumption is necessary in deriving the variational form since then the time derivative is treated as a constant. The variation is,

$$\delta F(A_\phi) = \int_v \left[\frac{1}{2\mu}\left[\left[\delta\frac{\partial A_\phi}{\partial r} - \frac{\delta A_\phi}{r}\right]\left[\frac{\partial A_\phi}{\partial r} - \frac{A_\phi}{r}\right] + \left[\delta\frac{\partial A_\phi}{\partial z}\right]\left[\frac{\partial A_\phi}{\partial z}\right]\right]\right.$$
$$\left. - J_s \cdot \delta A_\phi + \sigma\frac{\partial A_\phi}{\partial t}.\delta A_\phi\right]dv \tag{9.8}$$

9.3.2 Approximation in space

The general approximations for A and $\partial A/\partial r$, $\partial A/\partial z$, *and* $\partial A/\partial t$ are

$$A(r,z,t) = \sum_{i=1}^{3} N_i(r,z)A_i(t) \tag{9.9}$$

$$\frac{\partial A(r,z,t)}{\partial r} = \sum_{i=1}^{3} \frac{\partial N_i(r,z)}{\partial r} A_i(t) \tag{9.10}$$

$$\frac{\partial A(r,z,t)}{\partial z} = \sum_{i=1}^{3} \frac{\partial N_i(r,z)}{\partial z} A_i(t) \tag{9.11}$$

$$\frac{\partial A(r,z,t)}{\partial t} = \sum_{i=1}^{3} N_i(r,z) \frac{\partial A_i(t)}{\partial t} \tag{9.12}$$

where the standard form of the finite element approximation is used. Eq. (9.12) indicates that the time derivatives are independent of the finite element approximation and must be treated separately. For a triangular element we get:

$$[S]_e \{A\} + [R]_e \left\{\frac{\partial A}{\partial t}\right\} = \{Q\}_e \tag{9.13}$$

where the elemental matrices $[S]$, $[R]$, and $\{q\}$ are

$$[S]_e = \begin{bmatrix} \int_{v_e} (a_1^2 + b_1^2)dv & \int_{v_e} (a_1a_2 + b_1b_2)dv & \int_{v_e} (a_1a_3 + b_1b_3)dv \\ \int_{v_e} (a_1a_2 + b_1b_2)dv & \int_{v_e} (a_2^2 + b_2^2)dv & \int_{v_e} (a_2a_3 + b_2b_3)dv \\ \int_{v_e} (a_1a_3 + b_1b_3)dv & \int_{v_e} (a_2a_3 + b_2b_3)dv & \int_{v_e} (a_3^2 + b_3^2)dv \end{bmatrix} \tag{9.14}$$

$$[R]_e = \begin{bmatrix} \int_{v_e} N_1^2 dv & \int_{v_e} N_1N_2 dv & \int_{v_e} N_1N_3 dv \\ \int_{v_e} N_1N_2 dv & \int_{v_e} N_2^2 dv & \int_{v_e} N_2N_3 dv \\ \int_{v_e} N_3N_1 dv & \int_{v_e} N_3N_2 dv & \int_{v_e} N_3^2 dv \end{bmatrix} \tag{9.15}$$

$$[Q]_e = \begin{bmatrix} \int_{v_e} N_1 dv \\ \int_{v_e} N_2 dv \\ \int_{v_e} N_3 dv \end{bmatrix} \tag{9.16}$$

where

$$a_1 = \frac{\partial N_1}{\partial r} + \frac{N_1}{r} \qquad a_2 = \frac{\partial N_2}{\partial r} + \frac{N_2}{r} \qquad a_3 = \frac{\partial N_3}{\partial r} + \frac{N_3}{r}$$

$$b_1 = \frac{\partial N_1}{\partial z} \qquad b_2 = \frac{\partial N_2}{\partial z} \qquad b_3 = \frac{\partial N_3}{\partial z}$$

and N_i are the shape functions of the element. If we choose a triangular element as in section 7.5.2, we obtain the matrices in section 7.7.1. The integration in the expressions above is performed analytically as in section 7.7.1. On the other hand, if an isoparametric element is used, numerical integration must be carried out as in chapter 11.

9.3.3 Approximation in time

To treat the time derivative, we first write Eq. (9.13) as

$$[S]_e \{A\} = \{Q\}_e - [R]_e \left\{\frac{\partial A}{\partial t}\right\} \tag{9.17}$$

To evaluate the time derivative we use a backward finite difference approximation as

$$\left.\frac{\partial A}{\partial t}\right|_{(t+\Delta t)} = \frac{A_{(t+\Delta t)} - A_{(t)}}{\Delta t} \tag{9.18}$$

where Δt is a time step to be chosen.
Substituting Eq. (9.18) into Eq. (9.17) gives

$$[S]\{A\}_{(t+\Delta t)} = \{Q\}_{(t+\Delta t)} - [R]\frac{\{A\}_{(t+\Delta t)} - \{A\}_{(t)}}{\Delta t}$$

The approximation of $\partial A/\partial t$ involves $A_{t+\Delta t}$, the quantity which we are solving for. Rearranging this equation results in,

$$[[S]+[R]]\{A\}_{(t+\Delta t)}=\{Q\}_{(t+\Delta t)}+\frac{1}{\Delta t}[R]\{A\}_{(t)} \tag{9.19}$$

Equation (9.19) is the form of the finite element equation which is assembled into the global matrices and eventually solved by any convenient method such as Gauss elimination. The next step is introduction of boundary conditions. These are specified as in chapter 4 and only Dirichlet boundary conditions need special treatment. Neumann boundary conditions are implicit in the formulation.

To solve the problem, initial values of A must be specified at every node at time zero as is evident from the right hand side of Eq. (9.19). Also, a time step must be defined and the value of J_s at each time increment given. The fact that J_s is specified at each time step allows the investigation of any sampled waveform. A finite element mesh is defined over the space of the solution (i.e. a two-dimensional plane or a three-dimensional volume). The same mesh is used at each time step and the boundary conditions remain constant throughout the solution. Thus, although the solution in time is explicit, time can be viewed as a third (or fourth) dimension in solution.

The solution of the system above must be done at each time step but this is a rather minor problem because the left hand side of Eq. (9.17) is independent of time and, therefore, the elimination is only performed once. The right hand side is updated at each time step and the system solved repeatedly.

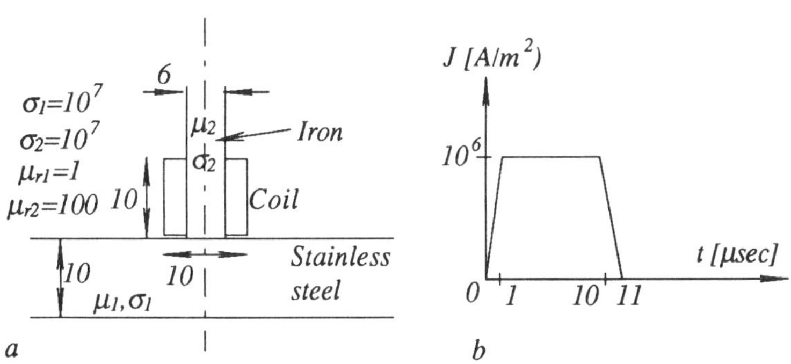

Figure 9.2 A probe over conducting sheet. a. The axisymmetric geometry. b. Current in the coil.

9.3.4 Example: transient field of a probe

As an example of the use of finite elements for solution of transient fields in materials, the geometry in Figure 9.2a is analyzed. It consists of a coil over a conducting sheet. The fields in the sheet and their variation with time are required. The main purpose of this example is to analyze the transient due to a pulse such as would be required in a multifrequency eddy current or pulsed eddy current test setup. Figure 9.2b shows the pulse used to drive the coil. For simplicity, it is assumed the ferromagnetic core of the coil is very long and the stainless steel sheet is large compared with the extent of the fields. The geometry is analyzed using triangular elements and, although the solution is transient, the mesh is conventional and in fact, it is designed to accommodate both the transient and the steady state solution. The mesh is shown in Figure 9.3 where the various materials are shown. Solution proceeds with a time step of $1\ \mu s$. The solution for each time step can be analyzed separately. We will however, only show 3 solution steps. These are given in Figures 9.4 through 9.6, at *2*, *10* and $30\ \mu s$. The solutions show penetration of fields into the conducting materials with time and also, that at $t{=}30\ \mu s$, the field is entirely due to induced fields since the current in the coil is zero. This can be clearly seen from the fact that the centers of contour lines in Figures 9.4 and 9.5 are in the coil while in Figure 9.6 the center has shifted into the magnetic core of the coil. If the transient is allowed to progress further, the field dissipates completely as expected.

The solution obtained here clearly indicates the flexibility in solution and the fact that the time step used is independent of the space discretization (unlike finite difference solutions). We also note that there is no need for any interface conditions.

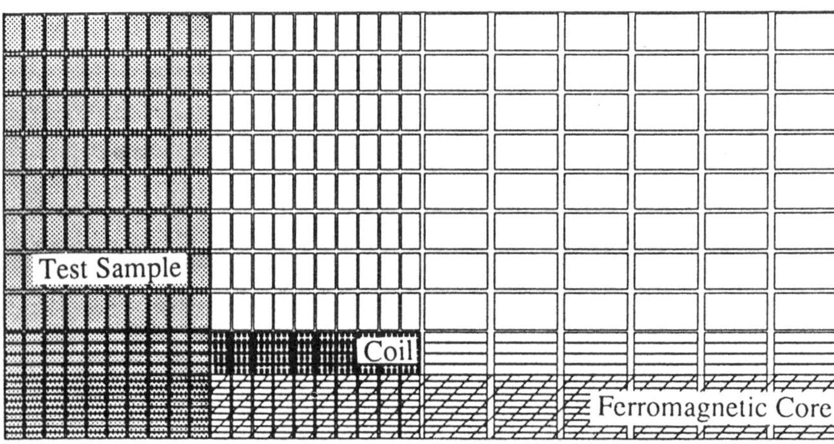

Figure 9.3 The mesh generated for Figure 9.2a.

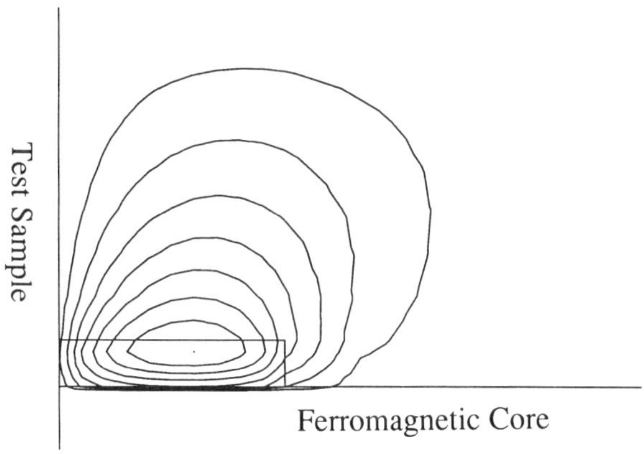

Figure 9.4 Solution for the magnetic field at t=2 μs.

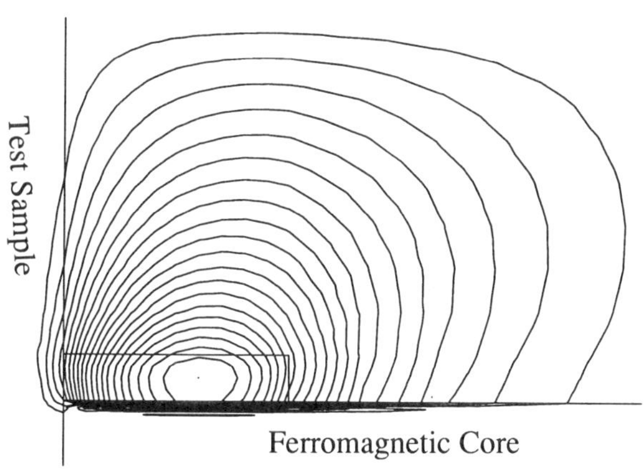

Figure 9.5 Solution for the magnetic field at t=10 μs.

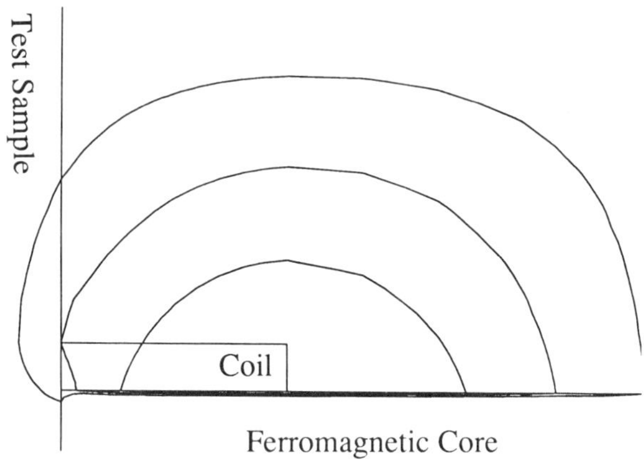

Figure 9.6 Solution for the magnetic field at t=30 μs.

9.4 TRANSIENT FINITE DIFFERENCE FORMULATION

The application of finite difference methods to solution of static and time-harmonic problems is discussed in chapter 6. Here we show how the same methods apply to solution of transient field problems. Although this can be done at any frequency, the discussion here is limited to low frequency applications. High frequency applications are discussed later on in this chapter in the context of *FDTD* methods. Much of the material given in chapter 6 applies here as well and this is not repeated. However, there are a few aspects of modeling that have to do with time and these are discussed in detail.

Rather than outline the method in general, we present an example. It consists of a busbar over a thick conducting plate. This is the same problem as in section 6.4 but here the solution is in the time domain rather than in the frequency domain. Thus the transient behavior of the problem is the goal of the calculation.

The geometry of the problem is shown in Figure 9.7 together with dimensions and material properties. The busbar carries a current as shown in Figure 9.8.

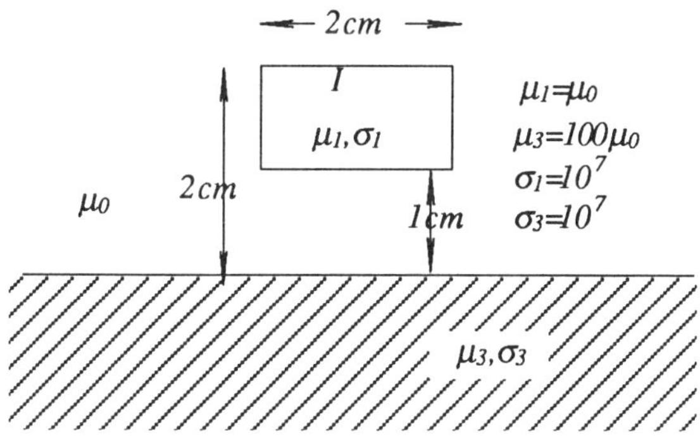

Figure 9.7 A busbar over ferromagnetic sheet, driven by a transient current source.

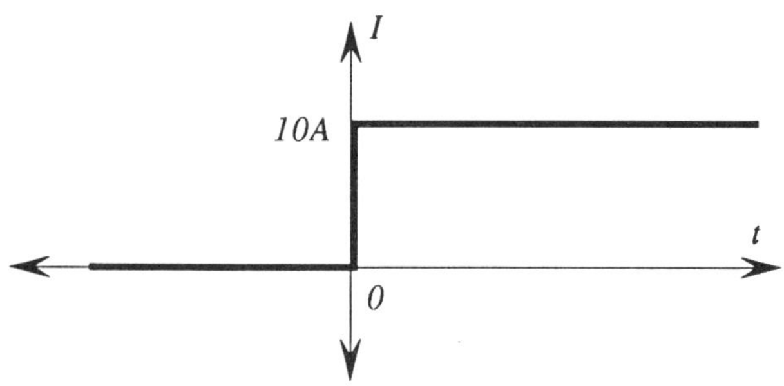

Figure 9.8 The current waveform used in Figure 9.7.

9.4.1 Governing equations

The governing equation is similar to that in Eq. (9.6) but now it is written as a two dimensional Poisson's equation:

$$\frac{1}{\mu}\left[\frac{\partial^2 A_z}{\partial x^2}+\frac{\partial^2 A_z}{\partial y^2}\right] = -J_s + \sigma\frac{\partial A_z}{\partial t} \tag{9.20}$$

For purposes of this formulation we rewrite it as

$$\frac{1}{\mu\sigma}\left[\frac{\partial^2 A_z}{\partial x^2}+\frac{\partial^2 A_z}{\partial y^2}\right] + \frac{J_s}{\sigma} = \frac{\partial A_z}{\partial t} \tag{9.21}$$

This is the general equation. There are three distinct materials we must consider as shown in Figure 9.9. The governing equations in the three materials are:

In material *(1)*:

$$\frac{1}{\mu_1\sigma_1}\left[\frac{\partial^2 A_z}{\partial x^2}+\frac{\partial^2 A_z}{\partial y^2}\right]+\frac{J_s}{\sigma_1}=\frac{\partial A_z}{\partial t} \tag{9.22}$$

In material *(2)*:

$$\left[\frac{\partial^2 A_z}{\partial x^2}+\frac{\partial^2 A_z}{\partial y^2}\right]=0 \tag{9.23}$$

and, in material *(3)*:

$$\frac{1}{\mu_3\sigma_3}\left[\frac{\partial^2 A_z}{\partial x^2}+\frac{\partial^2 A_z}{\partial y^2}\right]=\frac{\partial A_z}{\partial t} \tag{9.24}$$

Before continuing, we scale the variables as follows:

$$t = t^*\mu_1\sigma_1\times 10^{-4} = t^* 4\pi\times 10^{-4}$$

$$x = x^*\times 10^{-2}$$

$$A = A^*\times J_0\times\mu_1\times 10^{-4} = A^*\times 2\pi\times 10^{-6}$$

$$J_0 = \frac{J^*}{2\times 10^{-4}} = \frac{10}{2\times 10^{-4}}\ (Amp/m^2)$$

Now, the scaled equations are:

$$\frac{\partial A^*}{\partial t^*}=\left[\frac{\partial^2 A_z^*}{\partial x^2}+\frac{\partial^2 A_z^*}{\partial y^2}\right]+1 \qquad \textit{in material (1)} \tag{9.25}$$

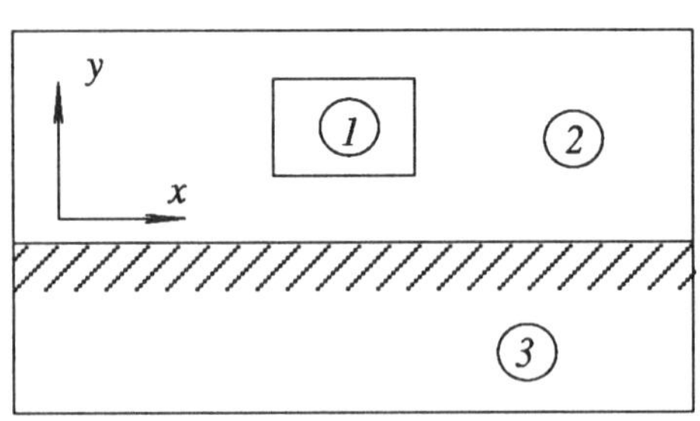

Figure 9.9 The three domains of the geometry in Figure 9.7.

$$\frac{\partial^2 A_z^*}{\partial x^2} + \frac{\partial^2 A_z^*}{\partial y^2} = 0 \qquad \text{in material (2)} \tag{9.26}$$

$$\frac{\partial A^*}{\partial t^*} = 10^{-2}\left[\frac{\partial^2 A_z^*}{\partial x^2} + \frac{\partial^2 A_z^*}{\partial y^2}\right] \qquad \text{in material (3)} \tag{9.27}$$

The next step is to define a grid. Although the grid is somewhat arbitrary, we must choose a fine enough grid to provide a realistic and accurate solution. For simplicity we use the same grid as in Figure 6.14. The grid is reproduced in Figure 9.10. For this grid, $\Delta x=0.1$, $\Delta y=0.1$ in materials *(1)* and *(2)* and $\Delta y=0.01$ in material *(3)*. Note that these are scaled values and the actual values are two orders of magnitude smaller ($\Delta x=\Delta y=0.001\ m$, in materials *(1)*, *(2)*, and *(3)* and $\Delta y=0.0001\ m$ in material *(3)*).

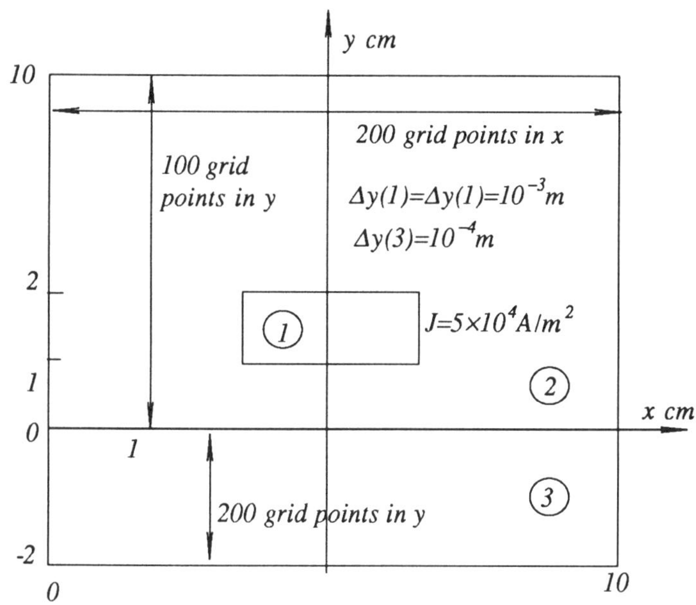

Figure 9.10 Finite difference grid definition. The left boundary is a Neumann boundary condition.

9.4.2 Finite difference formulae

The finite difference approximation is now written directly from Eqs. (9.25) through (9.27). To simplify notation the * symbol is dropped but the following relations are in terms of scaled quantities.

In material *(1)*:

$$A_{i,j}^{n+1} = A_{i,j}^{n} + \frac{\Delta t}{(\Delta x)^2}\left(A_{i-1,j}^{n} + A_{i+1,j}^{n} + A_{i,j-1}^{n} + A_{i,j+1}^{n} - 4A_{i,j}^{n}\right) + \Delta t \quad (9.28)$$

where a backward difference scheme is used for the time derivative (using Eq. (9.18)). The subscript indices indicate the grid location (i for nodes in the x direction, j for nodes in the y direction as in Figure 3.13. Superscript indices indicate the time step. Thus, the expression in Eq. (9.28) is explicit in time since only values of the magnetic vector potential at time step n are required for solution of the magnetic vector potential at time step $n+1$.
In material *(2)* the approximation is:

$$\left(A_{i-1,j}^{n+1} + A_{i+1,j}^{n+1} + A_{i,j-1}^{n+1} + A_{i,j+1}^{n+1} - 4A_{i,j}^{n+1}\right) = 0$$

and, in material *(3)*:

$$A_{i,j}^{n+1} = A_{i,j}^{n} + \frac{10^{-2}\Delta t}{(\Delta x)^2}\left(A_{i-1,j}^{n} - 2A_{i,j}^{n} + A_{i+1,j}^{n}\right) + \frac{\Delta t}{(\Delta y)^2}\left(A_{i,j-1}^{n} - 2A_{i,j}^{n} + A_{i,j+1}^{n}\right)$$

In the last expression the approximation of the derivatives with respect to x and y are separated since Δx and Δy are different in material *(3)*.

9.4.3 Boundary and interface conditions

Dirichlet boundary conditions are used on the outer boundary. It is also possible to cut the geometry in two and use Neumann boundary conditions on a vertical line through the center of the busbar but this is not used here for simplicity. The boundary conditions on geometry boundaries are $A=0$. This simply assumes that boundaries are far enough from sources to allow this simple condition.

The interface between material *(1)* and *(2)* needs no interface conditions since permeabilities of the two materials are the same. On the interface between materials *(2)* and *(3)* the following interface condition is applied (continuity of normal component of **B**).

$$A_0 = \frac{1}{3\left(1+\dfrac{\Delta y^{(3)}}{\Delta y^{(2)}}\dfrac{\mu^{(3)}}{\mu^{(2)}}\right)}\left[4A_{-1}^{(3)} - A_{-2}^{(3)} + \frac{\Delta y^{(3)}}{\Delta y^{(2)}}\frac{\mu^{(3)}}{\mu^{(2)}}\left(4A_1^{(2)} - A_2^{(2)}\right)\right]$$

$$= \frac{1}{33}\left[4A_{-1}^{(3)} - A_{-2}^{(3)} + 10\left(4A_1^{(2)} - A_2^{(2)}\right)\right]$$

where the superscript indicates material number and the subscript refers to the nodes of Figure 6.13.

This interface condition is discussed in section 6.3.3 and the fact that time dependency is involved does not change the interface condition since the continuity of normal components of **B** is independent of time.

While the choice of Δx and Δy is arbitrary, Δt must be chosen to satisfy the stability criterion as defined in Eq. (3.27):

$$\frac{\Delta t}{(\Delta x)^2} \le 0.25$$

By choosing $\Delta t = 0.2\times(\Delta x)^2$ the stability criterion is satisfied everywhere, including both terms of the approximation for material *(3)*.

9.4.4 Initial conditions

Initial conditions must be specified for the solution. Normally, the initial distribution for A is specified. The conditions are specified in terms of known values when the solution begins. It may be necessary to assume a solution, at least in part of the domain. In the problem we solve here, the current density at time t^- is zero. It is therefore reasonable to use a zero initial condition for the solution. If however, the reverse were done and the current were switched off, the initial solution would be the steady state solution. In any case, the initial solution must be specified at every node of the mesh.

9.4.5 Solution

Now the solution can proceed while the only aspect remaining is the question of how to stop it. This is normally a simple decision: we either let the solution progress for a given number of time steps, a given length of time, or let it continue until a steady state solution is obtained. Here we choose to solve for *5,000* time steps or a total actual time of $t=5000\times\Delta t\times 4\pi\times 10^{-4}=13\ ms$.

As examples of the type of results we can obtain, the solution (magnetic vector potential) is shown at four times in Figures 9.11 through 9.14. Figures 9.11 through 9.14 show the magnetic field distribution after *500* time steps (*1.256 ms*), *2,000* time steps (*5.02 ms*), *3,500* time steps (*8.79 ms*) and *5,000* time steps (*13 ms*) clearly showing the increase in field penetration as the solution progresses towards steady state. The solution at steady state is shown in Figure 9.15 for reference. From this and from the plots in Figures 9.11 through 9.14 it is clear that the solution after *13 ms* has not reached steady state.

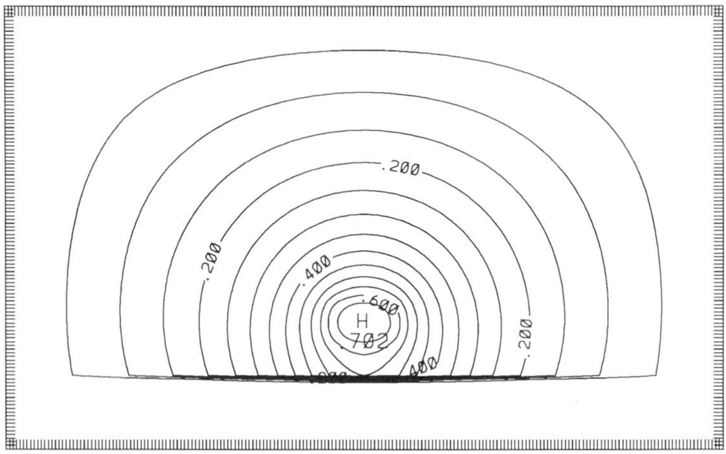

Figure 9.11 Transient field in Figure 9.7 after 500 time steps (1.256 ms).

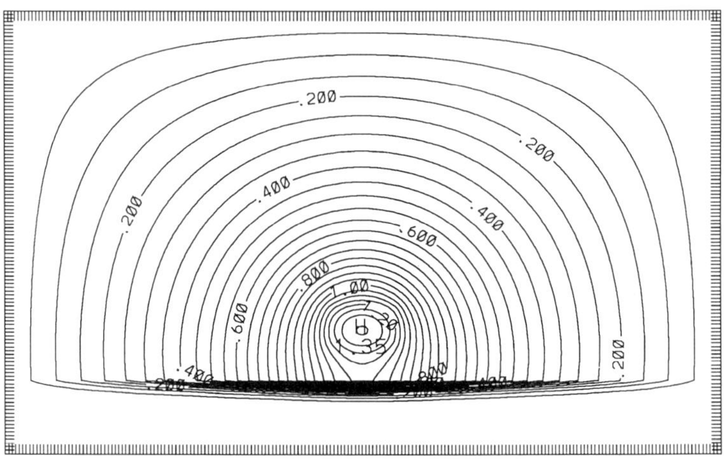

Figure 9.12 Transient field in Figure 9.7 after 2,000 time steps (5.02 ms).

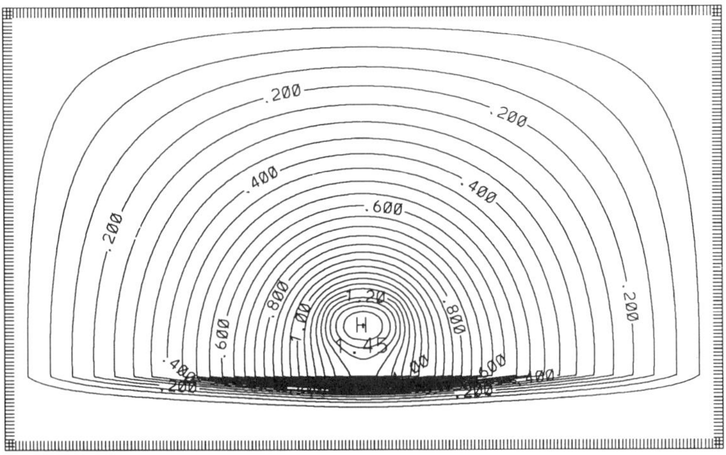

Figure 9.13 Transient field in Figure 9.7 after 3,500 time steps (8.79 ms).

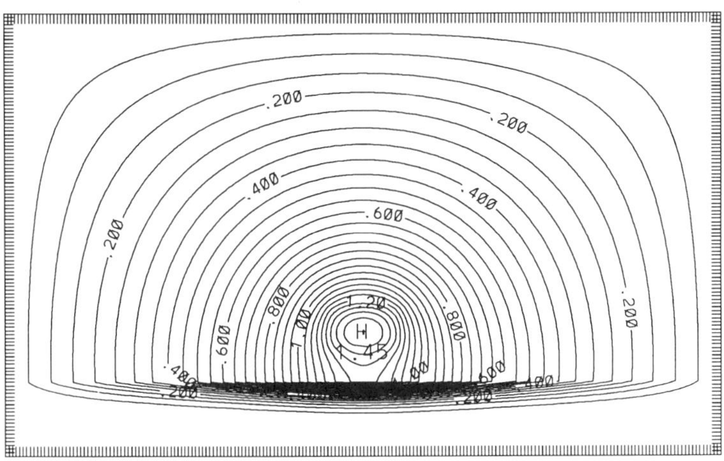

Figure 9.14 Transient field in Figure 9.7 after 5,000 time steps (13 ms).

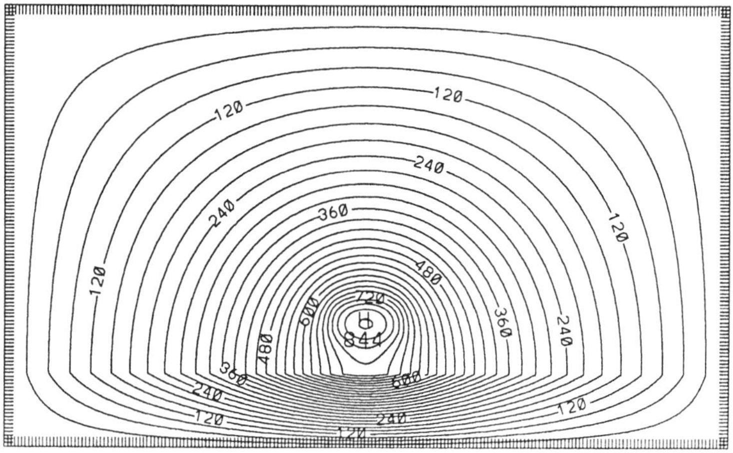

Figure 9.15 Steady state field in Figure 9.7. This solution is the dc solution for the geometry analyzed here and was calculated in section 6.3.4.

9.5 THREE-DIMENSIONAL SOLUTIONS

The extension of the finite difference method given above to *3-D* is straightforward: the only difference is in the equation we start with which has three components for the magnetic vector potential. In three dimensions, the magnetic field intensity, magnetic flux density or the electric field intensity can be used directly without the need to use a potential function. However, this only has to do with the spatial part of the equation. Modeling of the time derivative is identical to that given above. Thus, the type of equation to solve is

$$\frac{1}{\mu}\left[\frac{\partial^2 \mathbf{A}}{\partial x^2}+\frac{\partial^2 \mathbf{A}}{\partial y^2}+\frac{\partial^2 \mathbf{A}}{\partial z^2}\right] = -\mathbf{J}_s + \sigma\frac{\partial \mathbf{A}}{\partial t} \tag{9.29}$$

where the vector notation for the magnetic vector potential indicates the three components of the vector potential at each node of the finite difference mesh. The one difficulty with three-dimensional solutions is the size of the problem. In effect, the problem is four-dimensional as shown in Figure 3.6.

Implementation of the finite difference approximation follows the steps in sections 9.4.1 and 9.4.2. Equation (9.27) is separated into three parts as

$$\hat{\mathbf{x}}\frac{1}{\mu}\left[\frac{\partial^2 A_x}{\partial x^2}+\frac{\partial^2 A_x}{\partial y^2}+\frac{\partial^2 A_x}{\partial z^2}\right]+\hat{\mathbf{y}}\frac{1}{\mu}\left[\frac{\partial^2 A_y}{\partial x^2}+\frac{\partial^2 A_y}{\partial y^2}+\frac{\partial^2 A_y}{\partial z^2}\right]$$
$$+\hat{\mathbf{z}}\frac{1}{\mu}\left[\frac{\partial^2 A_z}{\partial x^2}+\frac{\partial^2 A_z}{\partial y^2}+\frac{\partial^2 A_z}{\partial z^2}\right]$$
$$=-\hat{\mathbf{x}}J_{sx}+\hat{\mathbf{x}}\sigma\frac{\partial A_x}{\partial t}-\hat{\mathbf{y}}J_{sy}+\hat{\mathbf{y}}\sigma\frac{\partial A_y}{\partial t}-\hat{\mathbf{z}}J_{sz}+\hat{\mathbf{z}}\sigma\frac{\partial A_z}{\partial t}$$

and, using the finite difference formulae in section 9.4.2 for each component of the equation separately, we obtain an approximation in each material. A more efficient discretization process for finite difference solution is discussed shortly in the context of the finite difference time domain method however, the method above is much simpler in concept.

In addition, anisotropic and nonlinear solutions can be obtained by modifying the equation as is done in chapter 6. The only unique aspect of modeling of the time dependent equation is approximation of time and this does not change with modification of materials, or dimensionality.

The finite element method described in section 9.3 can be extended to three dimensions. All aspects of formulation are identical except, of course, the use of a *3-D* element and the resulting finite element matrices. Matrices $[S]$, $[R]$, and $[Q]$ are built in the same way as above but for the *3-D* element and these are discused in section 7.3. The vector of unknown magnetic vector potential values has now three components per node. The general form of the solution is

$$\left[[S]+[R]\right]\{\mathbf{A}\}_{(t+\Delta t)}=\{Q\}_{(t+\Delta t)}+\frac{1}{\Delta t}[R]\{\mathbf{A}\}_{(t)}$$

where the vector notation for **A** indicates the three components necessary.

The four-component solution of section 7.4 can be used in an identical fashion, where again, the only change is the addition of the time derivative which is the same as for the two-dimensional case.

9.6 FINITE DIFFERENCE TIME DOMAIN METHODS

The finite difference time domain (*FDTD*) method was originally developed for solution of time dependent applications, especially at high frequencies and for scattering problems. The method uses, traditionally, the field variables since *3-D* scattering applications were originally treated. However, the method applies equally well at low frequencies and can be easily adapted for solution in terms of the magnetic vector potential which, as shown in chapter 1 is a single component vector in two-dimensional and axisymmetric applications. We discuss first the formulation of *2-D* applications in low frequency eddy currents in terms of the *FDTD* method, followed by axisymmetric and then, separately, by *3-D* applications. The magnetic vector potential is used for *2-D* and axisymmetric applications and the field variables (**E** or **H**) for *3-D* applications.

9.6.1 Formulation: two-dimensional applications

The general, time dependent equation in terms of the magnetic vector potential (see section 1.13) is

$$\frac{\partial^2 \mathbf{A}}{\partial x^2} + \frac{\partial^2 \mathbf{A}}{\partial y^2} + \frac{\partial^2 \mathbf{A}}{\partial z^2} = - \mathbf{J}_s + \sigma_c \mu_c \frac{\partial \mathbf{A}}{\partial t} + \varepsilon_c \mu_c \frac{\partial^2 \mathbf{A}}{\partial t^2} \tag{9.30}$$

where the first term on the right hand side is the applied current density, the second is the induced current density (eddy current term) and the third is the displacement current density. The last term is not normally associated with eddy current applications and, in fact, this term makes the equation a hyperbolic equation. We, however, retain this term for now because of the implications on specification of boundary conditions. The index c stands for lossy materials (conductors).

The type of problems we treat here are assumed to have sources outside the solution domain and, since we solve the problem in the plane, only one component of the magnetic vector potential exists. Using Figure 9.16, the equation can be rewritten as

$$\frac{\partial^2 A_z}{\partial x^2} + \frac{\partial^2 A_z}{\partial y^2} = \sigma_c \mu_c \frac{\partial A_z}{\partial t} + \varepsilon_c \mu_c \frac{\partial^2 A_z}{\partial t^2} \tag{9.31}$$

The general problem we are solving is shown in Figure 9.16 where Ω_a is the exterior domain (normally free space but not necessarily so) and Ω_c is, in general, a lossy dielectric or conductor. We view this material as the scatterer or the material in which eddy currents are induced. Sources are specified either through the external boundary conditions or by internal boundary conditions. The latter means that the source may be replaced by the equivalent magnetic vector potential distribution on or near the boundaries of sources as shown in Figure 9.17. In low frequency applications, the first term is dominant and the problem is parabolic. At higher frequencies, the second term is dominant and the problem is dominated by a hyperbolic term. However, these definitions are rather loose since the terms depend on material properties as well and, for example, in a highly conductive material the propagation term is normally negligible at almost all frequencies.

The two domains in Figure 9.16 are now defined through the corresponding equations that apply in each.

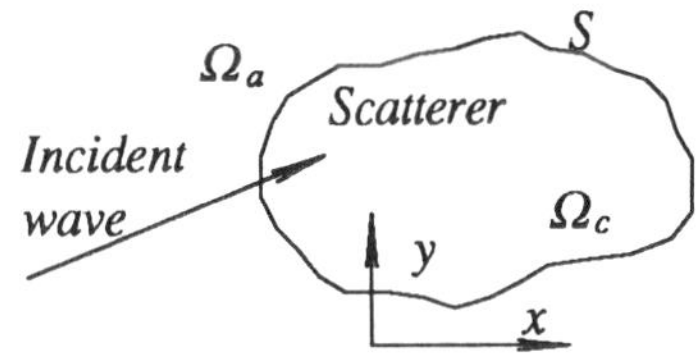

Figure 9.16 The general solution domain.

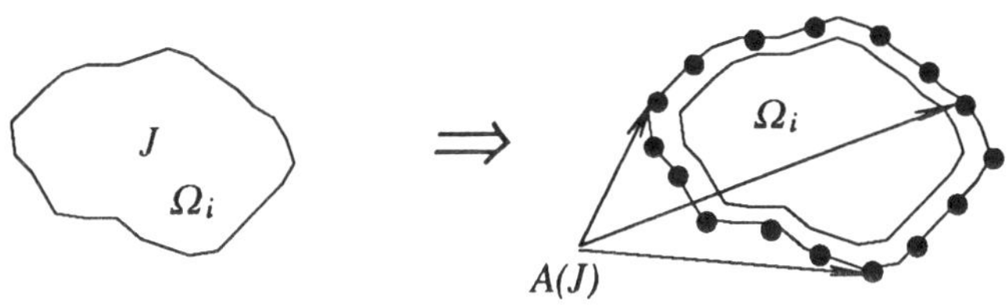

Figure 9.17 Method of accounting for sources through use of equivalent potential values on or around the boundaries of sources.

In Ω_a:

$$\frac{\partial^2 A_z}{\partial x^2} + \frac{\partial^2 A_z}{\partial y^2} = \sigma_a \mu_a \frac{\partial A_z}{\partial t} + \varepsilon_a \mu_a \frac{\partial^2 A_z}{\partial t^2} \tag{9.32}$$

In Ω_c:

$$\frac{\partial^2 A_z}{\partial x^2} + \frac{\partial^2 A_z}{\partial y^2} = \sigma_c \mu_c \frac{\partial A_z}{\partial t} + \varepsilon_c \mu_c \frac{\partial^2 A_z}{\partial t^2} \tag{9.33}$$

In addition, the interface conditions on the boundary between Ω_a ands Ω_c are:

$$\left[\frac{\partial A_z}{\partial t}\right]^- = \left[\frac{\partial A_z}{\partial t}\right]^+ \tag{9.34}$$

$$\mu_c \left[\frac{\partial A_z}{\partial n}\right]^- = \mu_a \left[\frac{\partial A_z}{\partial n}\right]^+ \tag{9.35}$$

where the negative sign represents the exterior field, the positive sign the interior field, μ_a is the relative permeability in Ω_a and μ_c the relative permeability in Ω_c. In addition, the following initial conditions are needed

$$A_z(x,y,0) = A_{zi}(x,y,0) \tag{9.36}$$

$$\frac{\partial A_z(x,y,0)}{\partial t} = \frac{\partial A_{zi}(x,y,0)}{\partial t} \tag{9.37}$$

where A_i indicates the incident field. Now, all that we need to define is an appropriate set of boundary conditions.

Although not specifically taken into account, permeability, permittivity, or conductivity need not be constant or isotropic, allowing the application of this method to analysis of composite materials.

9.6.2 Scaling

Before discussing the boundary conditions for this problem, we mention a number of difficulties that need to be addressed before a solution can be attempted. These are:

1. Implementation of the radiation boundary condition.
2. Implementation of interface conditions.
3. The large number of grid points required to capture the eddy current phenomenon adequately.
4. Appropriate local boundary conditions for the low frequency case where the wavelengths are much longer than the dimensions of the scatterers.

To address some of these problems, a scaling of the equations is required and performed as follows:

1. Divide A by A_0, where A_0 is the amplitude of the incident wave.
2. Divide the x and y coordinates by L, where L is an arbitrary length chosen such that it is of the order of the body's (scatterer) size (diameter, width, etc.).
3. Divide t by $1/\omega$, where ω is the angular frequency of the incident wave A_0.
4. Divide the scaled time ωt by $\omega L\sqrt{\mu_a \varepsilon_a}$

All scaled values are denoted with a bar.

After scaling, the problem can be defined in terms of the following equations:

In Ω_a:

$$\frac{\partial^2 \overline{A}_z}{\partial x^2} + \frac{\partial^2 \overline{A}_z}{\partial y^2} = L\sigma_a \sqrt{\frac{\mu_a}{\varepsilon_a}} \frac{\partial \overline{A}_z}{\partial t} + \frac{\partial^2 \overline{A}_z}{\partial t^2} \tag{9.38}$$

In Ω_c:

$$\frac{\partial^2 \overline{A}_z}{\partial x^2}+\frac{\partial^2 \overline{A}_z}{\partial y^2}=L\sigma_c\sqrt{\frac{\mu_c\varepsilon_c}{\mu_a\varepsilon_a}}\frac{\partial \overline{A}_z}{\partial t}+\frac{\mu_c\varepsilon_c}{\mu_a\varepsilon_a}\frac{\partial^2 \overline{A}_z}{\partial t^2} \tag{9.39}$$

The interface conditions on the boundary between Ω_a ands Ω_c are now:

$$\overline{A}_z^- = \overline{A}_z^+ \tag{9.40}$$

$$\mu_c\left[\frac{\partial \overline{A}_z}{\partial n}\right]^- = \mu_a\left[\frac{\partial \overline{A}_z}{\partial n}\right]^+ \tag{9.41}$$

where the notation is the same as in Eqs. (9.34) and (9.35). Initial conditions are:

$$\overline{A}_z(\overline{x},\overline{y},0) = \overline{A}_{zi}(\overline{x},\overline{y},0) \tag{9.42}$$

$$\frac{\partial \overline{A}_z(\overline{x},\overline{y},0)}{\partial t}=\frac{\partial A_{zi}(\overline{x},\overline{y},0)}{\partial t} \tag{9.43}$$

9.6.3 Boundary conditions

The boundary condition we choose is a radiation boundary condition. However, a general boundary condition cannot be obtained for the whole frequency spectrum. A radiation boundary condition is defined such that none of the wave propagating towards a boundary is reflected back into the solution domain. The Sommerfeld radiation boundary condition:

$$\lim_{r\to\infty} r^{1/2}\left(\frac{\partial A_{zs}}{\partial r}+\frac{\partial A_{zs}}{\partial t}\right)=0 \tag{9.44}$$

can be used for high frequency applications. For low frequency applications, separate radiation boundary conditions must be used. The low frequency radiation boundary condition is discussed next.

(a) Radiation boundary condition at low frequencies

The idea of a radiation boundary condition is to ensure either that the waves propagating outwards are cancelled or that reflections from artificially introduced boundaries are minimized. In either case, the net effect is that artificial boundaries may be placed near or on the boundaries of the required solution domain. This is important in two respects. First, the solution domain is not unnecessarily extended for the purpose of satisfying the mathematical requirements of modeling. Second, it ensures that the solution is correct in that nonphysical reflections from the artificial boundaries do not influence the results. The latter is particularly important in high frequency applications, especially in low loss dielectrics because waves can travel with little attenuation for large distances. Reflections can introduce large errors in solutions. At low frequencies this is less of a problem but the first reason for radiation boundary conditions is still valid: the closer the boundary can be placed, the lower the number of discretization points necessary and the more efficient the solution. We also note that although strictly speaking wave solutions belong to the class of hyperbolic processes which are discussed in chapter 10, they exist in all time dependent fields and for proper modeling of eddy current applications in the time domain must be taken into account.

We will not present here the derivation of radiation boundary conditions in detail. Instead we give some useful boundary conditions for the time dependent and the time-harmonic equations. Before doing so, we point out the following:

1. The equations used here to describe the eddy current phenomena are, in effect, low frequency wave equations. We call these parabolic equations because they behave like parabolic equations. That is, the wave term is small compared to the other terms of the equations. This in fact is the whole idea of the eddy current equation. In most other applications, the wave term (actually displacement currents) are neglected completely. However, retaining the propagation term has some benefits, not the least of which is the use of the radiation boundary conditions given below.
2. Because the equation is a low frequency equation, the solution must contain two terms: one is a term that satisfies Laplace's equation. The second must satisfy the wave equation and must, therefore, take into account the scattered field in the solution domain.

The wave on the boundary of the solution domain is shown in Figure 9.18. It consists of an incident wave on the boundary (A_i) and a scattered wave (A_s) from the boundary. This can be written as

$$A_t = A_i + A_s$$

If the scattered wave is cancelled (or at least minimized) by the radiation boundary condition, the total wave at the boundary is due to the incident wave alone and no reflections back into the solution domain can occur.

An appropriate radiation boundary condition is the following:

$$\frac{\partial A_s}{\partial r} + \frac{\partial A_s}{\partial t} + C(r)A_s = 0 \tag{9.45}$$

where this is written as a function of the r coordinate and $C(r)$ is some function of r to be determined. We discuss this function in the implementation stage but now we simply state that this has the form

$$C(r) = \frac{1}{r}\left(- \frac{b}{a + b\ln r} \right)$$

where a and b are constants.

In two dimensional applications, the boundary condition takes the form

$$\frac{\partial A_s}{\partial x}\cos\phi + \frac{\partial A_s}{\partial y}\sin\phi + \frac{\partial A_s}{\partial t} + C(r)A_s = 0 \qquad (9.46)$$

which is identical to Eq. (9.45) if the coordinate transformation is taken into account. Again, the choice of $C(r)$ and implementation of the cosine and sine terms are discussed shortly.

For low frequency time-harmonic applications we can still use the boundary condition in Eq. (9.45) or (9.46) However, the term $C(r)$ for time harmonic applications is

$$C(r) = \frac{1}{r}\left[- \frac{1}{\gamma + \ln \omega L\sqrt{\mu_a \varepsilon_a} - \ln 2 + \ln r} \right]$$

The finite difference time domain method can be used with Dirichlet and Neumann boundary conditions as well, if these are known. In these special cases, the problem of modeling is simplified since the approximation for radiation boundary conditions is not required. Treatment of this type of boundary condition is the same as for any other finite difference method and is done as in chapter 6. An example of *FDTD* solution with Dirichlet boundary conditions is given in section 9.7.2.

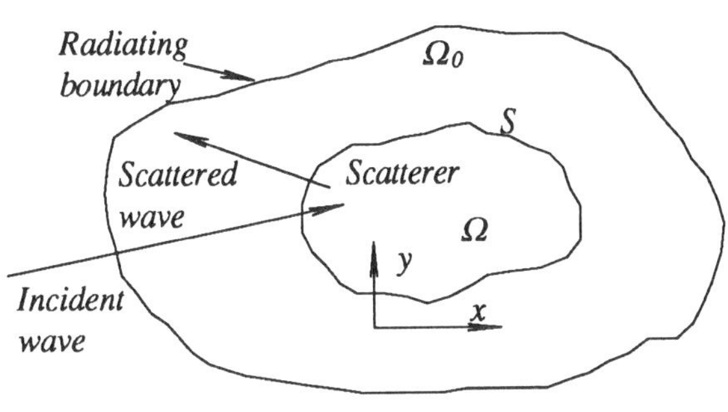

Figure 9.18 Generic geometry and the incident and scattered fields. The incident field can originate inside or outside Ω_0.

9.6.4 Formulation for axisymmetric geometries

The solution in axisymmetric geometries is similar with that in two dimensions with the obvious change in the system of coordinates. From the discussion in sections 9.3.1, and 9.6.1, we can write the field equations in terms of the magnetic vector potential as:
In Ω_a:

$$\frac{\partial^2 A_\phi}{\partial z^2}+\frac{\partial^2 A_\phi}{\partial r^2}+\frac{1}{r}\frac{\partial A_\phi}{\partial r}-\frac{A_\phi}{r^2}=\sigma_a\mu_a\frac{\partial A_\phi}{\partial t}+\mu_a\varepsilon_a\frac{\partial^2 A_\phi}{\partial t^2} \tag{9.47}$$

In Ω_c:

$$\frac{\partial^2 A_\phi}{\partial z^2}+\frac{\partial^2 A_\phi}{\partial r^2}+\frac{1}{r}\frac{\partial A_\phi}{\partial r}-\frac{A_\phi}{r^2}=\sigma_c\mu_c\frac{\partial A_\phi}{\partial t}+\mu_c\varepsilon_c\frac{\partial^2 A_\phi}{\partial t^2} \tag{9.48}$$

In this formulation, there are no free current density sources within the solution domain. Any currents that may be used are assumed to be known before hand and incorporated in the solution domain in terms of nodal point potentials. An example of how this might be accomplished is given later. Even though current densities are replaced by equivalent nodal potentials, we still assume that any equivalent current densities are in domain Ω_a. In addition to the equations in each material we must define the interface conditions between materials. The interface conditions take the form:

$$\left[\frac{\partial A_\phi}{\partial t}\right]^- = \left[\frac{\partial A_\phi}{\partial t}\right]^+ \tag{9.49}$$

$$\mu_c\left[\frac{\partial A_\phi}{\partial n}\right]^- = \mu_a\left[\frac{\partial A_\phi}{\partial n}\right]^+ \tag{9.50}$$

where again, the notation is as in Eqs. (9.34) and (9.35). For simplicity, we assume that the interface is parallel to the r coordinate as shown in Figure 9.19. Under this condition, the normal component is axial, in the z direction. Thus:

$$\mu_c\left[\frac{\partial A_\phi}{\partial z}\right]^- = \mu_a\left[\frac{\partial A_\phi}{\partial z}\right]^+ \tag{9.51}$$

where μ_c is the relative permeability of the lossy medium and μ_a that of the surrounding material. Using the same scaling techniques as in section 9.6.2 we get the following:

In Ω_a:

$$\frac{\partial^2 \overline{A}_\phi}{\partial z^2} + \frac{\partial^2 \overline{A}_\phi}{\partial r^2} + \frac{1}{r}\frac{\partial \overline{A}_\phi}{\partial r} - \frac{\overline{A}_\phi}{r^2} = L\sigma_a \sqrt{\frac{\mu_a}{\varepsilon_a}}\frac{\partial \overline{A}_\phi}{\partial t} + \frac{\partial^2 \overline{A}_\phi}{\partial t^2} \tag{9.52}$$

In Ω_c:

$$\frac{\partial^2 \overline{A}_\phi}{\partial z^2} + \frac{\partial^2 \overline{A}_\phi}{\partial r^2} + \frac{1}{r}\frac{\partial \overline{A}_\phi}{\partial r} - \frac{\overline{A}_\phi}{r^2} = L\sigma_c \sqrt{\frac{\mu_c\varepsilon_c}{\mu_a\varepsilon_a}}\frac{\partial \overline{A}_\phi}{\partial t} + \frac{\mu_c\varepsilon_c}{\mu_a\varepsilon_a}\frac{\partial^2 \overline{A}_\phi}{\partial t^2} \tag{9.53}$$

The interface conditions on the boundary between Ω_a ands Ω_c are now:

$$\overline{A}_\phi^{\,-} = \overline{A}_\phi^{\,+} \tag{9.54}$$

$$\mu_c\left[\frac{\partial \overline{A}_\phi}{\partial z}\right]^- = \mu_a\left[\frac{\partial \overline{A}_\phi}{\partial z}\right]^+ \tag{9.55}$$

In addition, the following initial conditions are needed

$$\overline{A}_\phi(\overline{r},\overline{z},0) = \overline{A}_{\phi i}(\overline{r},\overline{z},0) \tag{9.56}$$

$$\frac{\partial \overline{A}_\phi(\overline{r},\overline{z},0)}{\partial t} = \frac{\partial A_{\phi i}(\overline{r},\overline{z},0)}{\partial t} \tag{9.57}$$

where A_i is the incident magnetic vector potential and the bar indicates scaled values as described in section 9.6.2.

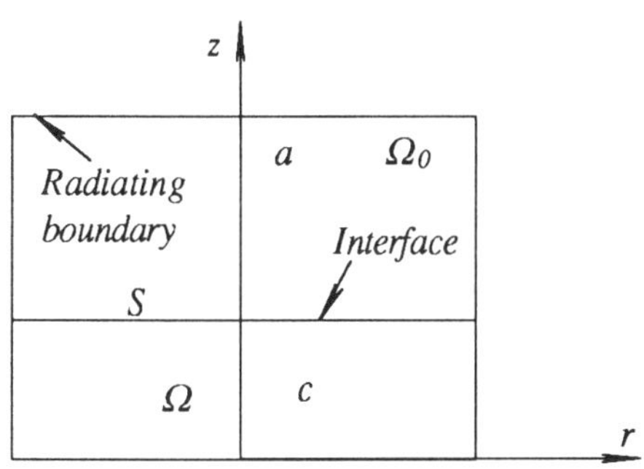

Figure 9.19 Simplified solution domain for axisymmetric geometries.

9.6.5 Boundary conditions in axisymmetric geometries

Unlike the two-dimensional situation where we assumed that the surrounding medium Ω_a extends to the boundaries everywhere, in the generic geometry we discuss here, there is a need for two radiation boundary conditions: one in domain Ω_a and one in domain Ω_c. The main difference in these boundary conditions is that the far field solution must be three-dimensional.

We assume the outgoing wave is to be cancelled. A_0 is the outgoing component of the magnetic vector potential. The radiation boundary condition in domain Ω_a is

$$\frac{\partial A_0}{\partial r} + \frac{\partial A_0}{\partial t} + \frac{1}{r}A_0 = 0 \tag{9.58}$$

In lossy media, the radiation condition is obtained as:

$$\frac{\partial A_0}{\partial r} + \frac{\mu_c \varepsilon_c}{\sqrt{\mu_a \varepsilon_a}} \frac{\partial A_0}{\partial t} + \left[\frac{L\mu_c \sigma_c}{2\sqrt{\mu_a \varepsilon_a}} + \left(\frac{1}{r}\right)\right] A_0 = 0 \tag{9.59}$$

In low loss materials, $L\mu_c\sigma_c << 2(\mu_a\varepsilon_a)^{1/2}$ and the first term in the bracket disappears. In low loss dielectrics and perfect dielectrics, the radiation boundary condition is therefore simplified to

$$\frac{\partial A_0}{\partial r} + \frac{\mu_c \varepsilon_c}{\sqrt{\mu_a \varepsilon_a}} \frac{\partial A_0}{\partial t} + \frac{1}{r}A_0 = 0 \tag{9.60}$$

On the other hand, if the material is highly conductive, outgoing waves are attenuated rapidly and the boundary condition is reduced to the Dirichlet boundary condition $A_0 = 0$.

9.6.6 Implementation of field equations: 2-D geometries

Equations (9.38) and (9.39) are represented in a standard form using central difference schemes for spatial as well as time derivatives. As is normally done, Δx is chosen to be equal to Δy and Δt is then chosen such that the Courant stability criterion $((\Delta t/\Delta x)^2 \le 0.5)$ is satisfied. The radiation condition is then discretized and included in the approximation. The finite difference mesh is the same as used for other finite difference approximations such as the one in section 9.3.4 or those given in chapter 6. The difference formulae used in solving this problem are as follows:
In domain Ω_a:

$$A_{i,j}^{k+1} = aA_{i,j}^{k} - bA_{i,j}^{k-1} + c(A_{i+1,j}^{k} + A_{i-1,j}^{k} + A_{i,j+1}^{k} + A_{i,j-1}^{k} - 4A_{i,j}^{k}) \tag{9.61}$$

where the constants are:

$$a = \frac{2}{1 + \frac{l_c^2}{2k_c^2}\Delta t} \qquad b = \frac{1 - \frac{l_c^2}{2k_c^2}\Delta t}{1 + \frac{l_c^2}{2k_c^2}\Delta t} \qquad c = \frac{\frac{1}{k_c^2}\left(\frac{\Delta t}{\Delta x}\right)^2}{1 + \frac{l_m^2}{2k_c^2}\Delta t}$$

where

$$l_c = \frac{L\sqrt{\omega\mu_m\sigma_m}}{\sqrt{k_a}} \qquad k_c = \frac{L\omega\sqrt{\mu_m\sigma_m}}{k_a}$$

and

$$k_a = \omega L\sqrt{\mu_a\sigma_a}$$

In most cases, this domain is taken as free space, or at the least, to be free of lossy materials. In this case, the approximation is simpler and is given by

$$A_{i,j}^{k+1} = 2A_{i,j}^{k} - A_{i,j}^{k-1} + \left(\frac{\Delta t}{\Delta x}\right)^2(A_{i+1,j}^{k} + A_{i-1,j}^{k} + A_{i,j+1}^{k} + A_{i,j-1}^{k} - 4A_{i,j}^{k}) \quad (9.62)$$

In domain Ω_C:

$$A_{i,j}^{k+1} = a'A_{i,j}^{k} - b'A_{i,j}^{k-1} + c'(A_{i+1,j}^{k} + A_{i-1,j}^{k} + A_{i,j+1}^{k} + A_{i,j-1}^{k} - 4A_{i,j}^{k}) \quad (9.63)$$

where

$$a' = \frac{2}{1 + \frac{l_c^2}{2k_c^2}\Delta t} \qquad b' = \frac{1 - \frac{l_c^2}{2k_c^2}\Delta t}{1 + \frac{l_c^2}{2k_c^2}\Delta t} \qquad c' = \frac{\frac{1}{k_c^2}\left(\frac{\Delta t}{\Delta x}\right)^2}{1 + \frac{l_c^2}{2k_c^2}\Delta t}$$

9.6.7 Implementation of radiation boundary condition: 2-D applications

Inclusion of the radiation boundary conditions is more complicated, primarily because the term $C(r)$ must be chosen and also because the approximation for the radiation boundary varies depending on the shape of boundary and location on the boundary. The boundary condition we use for two-dimensional applications is given in Eq. (9.44).

As expected, radiation boundary conditions are dependent on the geometry modeled. Thus, it is not possible to give a general representation of boundary conditions. However, there are a number of generic considerations that can be made at this point.

1. At corner nodes, the radiation boundary condition is represented by backward and forward finite difference formulae to avoid difficulties at corners.
2. On any boundary leading towards a corner implementation is in terms of the same backward/forward formulae used at corners.
3. At the central node on a boundary, either $sin\phi$ or $cos\phi$ is zero (in Eq. 9.46) since the angle ϕ is the angle between the horizontal (say x axis) and the position vector **r**. r is the distance from the origin to the point on the boundary at which boundary conditions are implemented. For ease of implementation, the origin is located at the center of the geometry and the solution domain chosen is some convenient shape such as a circle or a square.
4. The radiation boundary conditions are only enforced for the scattered field as indicated in Eq. (9.46). The incident field at the boundary is not affected and the total field is the sum of the two.

If we consider a generic rectangular boundary as in Figure 9.20, the radiation boundary condition is approximated on this boundary as follows:
At point (*1*):

$$(A_s)_{i,j}^{k+1} = a\,(A_s)_{i,j}^{k} - b\left[(A_s)_{i+1,j}^{\;k} - (A_s)_{i,j}^{\;k}\right] \tag{9.64}$$

On the boundary between points (*1*) and (2]:

$$(A_s)_{i,j}^{k+1} = a\,(A_s)_{i,j}^{k} - b\left[(A_s)_{i+1,j}^{\;k} - (A_s)_{i,j}^{\;k}\right] - c\left[(A_s)_{i,j}^{\;k} - (A_s)_{i,j-1}^{\;k}\right] \tag{9.65}$$

At point (*2*):

$$(A_s)_{i,j}^{k+1} = a\,(A_s)_{i,j}^{k} - c\left[(A_s)_{i,j}^{\;k} - (A_s)_{i,j-1}^{\;k}\right] \tag{9.66}$$

On the boundary between point (*2*) and [*3*):

$$(A_s)_{i,j}^{k+1} = a\,(A_s)_{i,j}^{k} - b\left[(A_s)_{i,j}^{\;k} - (A_s)_{i-1,j}^{\;k}\right] - c\left[(A_s)_{i,j}^{\;k} - (A_s)_{i,j-1}^{\;k}\right] \tag{9.67}$$

At point [*3*]:

$$(A_s)_{i,j}^{k+1} = a\,(A_s)_{i,j}^{k} - b\left[(A_s)_{i,j}^{k} - (A_s)_{i-1,j}^{k}\right] \tag{9.68}$$

Equation. (9.66) is used on the bottom boundary while Eqs. (9.65) and (9.67) are used on the respective boundaries on the lower part of the geometry. To any of these nodal values, we must add the incident wave. The total field is therefore

$$A_{i,j}^{k+1} = (A_i)_{i,j}^{k+1} + (A_s)_{i,j}^{k+1} \tag{9.69}$$

The coefficients a, b, and c are obtained from Eq. (9.46) in the following way:

$$a = 1 - C(r)\frac{\Delta t}{\Delta x} \qquad b = \cos\theta\frac{\Delta t}{\Delta x} \qquad c = \sin\theta\frac{\Delta t}{\Delta x} \tag{9.70}$$

where $C(r)$ is given in Eq. (9.46) or (9.45) and θ is the angle between r and the horizontal axis. At high frequencies, $C(r)=1/2r$. This condition is used in the following chapter.

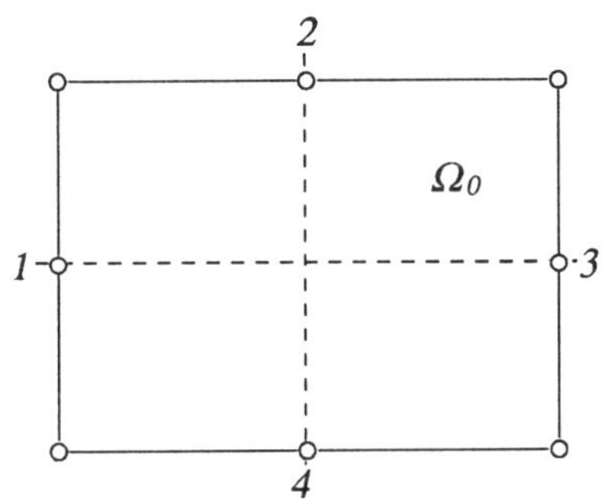

Figure 9.20 Implementation of boundary conditions.

We should note at this point that the reason for Eq. (9.69) is the fact that our basic assumption was that the sources of the electric field must be outside the solution domain (such as a plane wave impinging on the boundaries). If, however, there are sources in the solution domain, both scattered and incident waves must be canceled at the boundary. In such a case, Eq. (9.69) is dropped and Eqs. (9.64) through (9.68) are used in exactly the same form but the field now is the total field rather than the scattered field.

9.6.8 Implementation of field equations on interfaces: 2-D applications

Implementation of interface conditions can be done in one of two ways. First, the forward/backward method used in section 9.4 and in chapter 6 can be used here as well. This is particularly true if the interface is a simple planar interface and the interface condition need only be applied on one component of the field as was the case in section 9.4. A second method, which is much more general, is to assume an arbitrarily shaped boundary and impose the continuity on the tangential and normal components.

The method we adopt here is a compromise between the two that allows simple implementation of interface conditions. First we note from Figure 9.21 that the interface is approximated as a staircase approximation as is done in chapter 3. Second, we assume that all nodes on the boundary can be assumed to be in one material or the other. This is justified based on the fact that a grid can always be refined. Four possible situations are shown in Figure 9.21.

For the situation shown in Figure 9.21a we can write using a backward difference formula in material *(a)* and a forward formula in material *(c)*:

$$\mu_a(A_1 - A_3) = \mu_c(A_1 - A_2) \tag{9.71}$$

From this we can write for point *(2)*

$$A_2 = A_3 \frac{\mu_a}{\mu_c} + A_1\left(1 - \frac{\mu_a}{\mu_c}\right) \tag{9.72}$$

Substitution of this in the field approximation gives the approximation of a point on the boundary in which only nodes on one side of the boundary are used. The approximation in Eq. (9.62) is

$$A_1^{k+1} = 2A_1^k - A_1^{k-1} + \left(\frac{\Delta t}{\Delta x}\right)^2\left[A_5^k + A_4^k + A_3\left(1 + \frac{\mu_a}{\mu_c}\right) + A_1^k\left(-3 - \frac{\mu_a}{\mu_c}\right)\right] \tag{9.73}$$

Alternatively, we could have related all values to nodes in domain c and find a similar expression.

As for the situation in Figure 9.21b, we simply invoke the first interface condition and the node on the boundary is essentially shifted into material *(a)*. Therefore, Eq. (9.61), (9.62) or (9.63) can be used without change. Assuming that node *(1)* is in material *(a)* we use Eq. (9.61) directly.

In Figure 9.21c, we invoke the same approximation as for Figure 9.21a since nodes *(1)* and *(5)* (used for the approximation in the x direction) and nodes *(1)* and *(2)* (used for the approximation of the y component) are in material *(c)*. Thus we can relate the remaining two nodes to these three nodes using Eq. (9.71) for each component. As for the expression for the situation in Figures 9.21d, nodes *(2)* and *(5)* are shifted into material *(a)* and there is no need for special treatment. If four out of the five nodes are in material *(a)*, we simply use Eq. (9.61) or (9.62) and if four out of the five nodes are in material *(c)* we use Eq. (9.63).

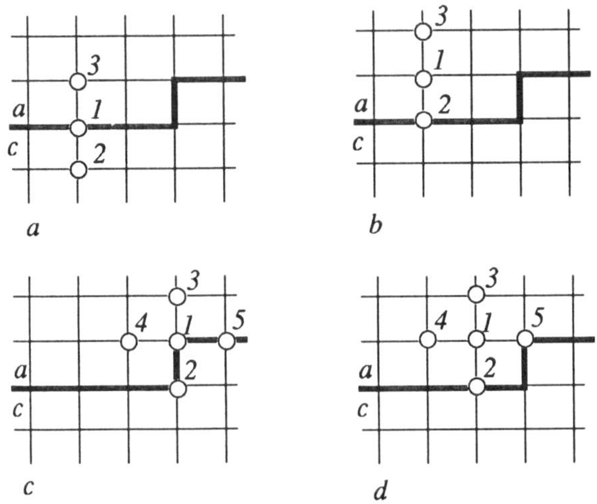

Figure 9.21 Treatment of nodes on interfaces. a. Node (1) on interface is treated as interface node. b. Node (2) is shifted to domain (a). c. Node (1) is treated as interface node. d. Nodes (5) and (2) are shifted to domain (a).

9.6.9 Implementation of field equations: axisymmetric geometries

Comparison of the *2-D* field equations in Eqs. (9.38) through (9.46) and those for the axisymmetric case in Eqs. (9.52) through (9.60), reveals that there are only two differences between the two systems: One is the Laplacian in Eqs. (9.52) and (9.53) and the second is the radiation boundary condition given by Eq. (9.60). Therefore, we will use all approximations as given above except for the changes mentioned. To define the problem we use Figure 9.19. The general assumptions are the same as for the two-dimensional applications: sources are in material *(a)* and scatterers in material *(c)*. Thus, using the equations from section 9.6.5 we get, in domain Ω_a:

$$A_{i,j}^{k+1} = a\, A_{i,j}^{k} - b\, A_{i,j}^{k-1} + c\left(A_{i+1,j}^{k} + \frac{i-1}{i} A_{i-1,j}^{k} + A_{i,j+1}^{k} + A_{i,j-1}^{k} - \frac{4i^2 - i + 1}{i^2} A_{i,j}^{k} \right) \tag{9.74}$$

where

$$a = \frac{2 + \frac{l_c^2}{k_c^2}\Delta t}{1 + \frac{l_c^2}{k_c^2}\Delta t} \qquad b = \frac{1}{1 + \frac{l_c^2}{k_c^2}\Delta t} \qquad c = \frac{\frac{1}{k_c^2}\left(\frac{\Delta t}{\Delta z}\right)^2}{1 + \frac{l_c^2}{k_c^2}\Delta t}$$

In most cases, this domain is taken as free space. In this case, the approximation is simpler and is given by

$$A_{i,j}^{k+1} = 2A_{i,j}^{k} - b\,A_{i,j}^{k-1} + \left(\frac{\Delta t}{\Delta z}\right)^2\left(A_{i+1,j}^{k} + \frac{i-1}{i}A_{i-1,j}^{k} + A_{i,j+1}^{k} + A_{i,j-1}^{k} - \frac{4i^2 - i+1}{i^2}A_{i,j}^{k}\right) \tag{9.75}$$

In domain Ω_c:

$$A_{i,j}^{k+1} = a'A_{i,j}^{k} - b'A_{i,j}^{k-1} + c'\left(A_{i+1,j}^{k} + \frac{i-1}{i}A_{i-1,j}^{k} + A_{i,j+1}^{k} + A_{i,j-1}^{k} - \frac{4i^2 - i+1}{i^2}A_{i,j}^{k}\right) \tag{9.76}$$

where

$$a' = \frac{2 + \frac{l_c^2}{k_c^2}\Delta t}{1 + \frac{l_c^2}{k_c^2}\Delta t} \qquad b' = \frac{1}{1 + \frac{l_c^2}{k_c^2}\Delta t} \qquad c' = \frac{\frac{1}{k_c^2}\left(\frac{\Delta t}{\Delta z}\right)^2}{1 + \frac{l_c^2}{k_c^2}\Delta t}$$

Implementation of the radiation boundary condition is again similar but two changes must be incorporated:

1. The radiation boundary condition in Eq. (9.59) must be used.
2. Because of the axis of symmetry, only the boundaries to the right (or left) of the axis are implemented. On the axis, because of symmetry, the magnetic vector potential is zero and there is no need for a radiation boundary condition. Refering to Figure 9.22, the radiation boundary condition is implemented as follows:

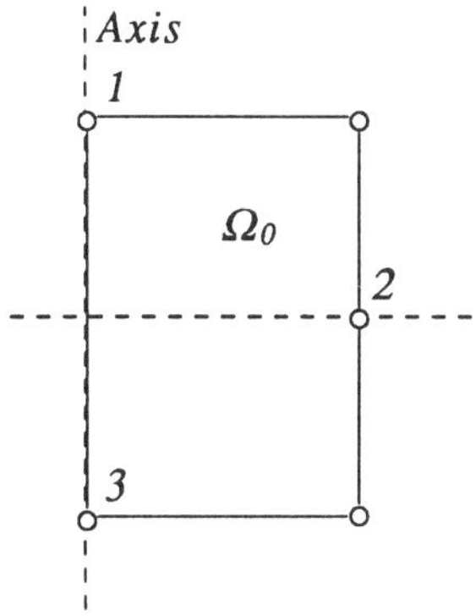

Figure 9.22 Implementation of radiation boundary condition in axisymmetric geometries. The condition on the axis is A=0 from symmetry considerations.

At node 2:

$$(A_t)_{i,j}^{k+1} = a\,(A_t)_{i,j}^{k} - b\left[(A_t)_{i,j}^{k} - (A_t)_{i-1,j}^{k}\right] \tag{9.77}$$

Between node *1* and node 2:

$$(A_t)_{i,j}^{k+1} = a\,(A_t)_{i,j}^{k} - b\left[(A_t)_{i,j}^{k} - (A_t)_{i-1,j}^{k}\right] - c\left[(A_t)_{i,j}^{k} - (A_t)_{i,j-1}^{k}\right] \tag{9.78}$$

At nodes *1* and *3*:

$$(A_t)_{i,j}^{k+1} = 0 \tag{9.79}$$

Between node 2 and node *3*:

$$(A_t)_{i,j}^{k+1} = a'(A_t)_{i,j}^{k} - b\left[(A_t)_{i,j}^{k} - (A_t)_{i-1,j}^{k}\right] - c\left[(A_t)_{i,j+1}^{k} - (A_t)_{i,j}^{k}\right] \tag{9.80}$$

where

$$a = 1 - \frac{\Delta t}{r\Delta z} \qquad a' = 1 - C(r)\frac{\Delta t}{r\Delta z} \qquad b = \sin\theta\frac{\Delta t}{\Delta z} \qquad c = \cos\theta\frac{\Delta t}{\Delta z} \tag{9.81}$$

and $C(r)$ is given as

$$C(r) = \frac{l_c^2}{2k_c^2} + \frac{1}{r}$$

where l_c and k_c are the the same as in section 9.6.6.

In these formulations we assume that the sources of the field are within the solution domain and therefore the magnetic vector potential at the boundary represents the total field. All other aspects of implementation are identical to those for the two-dimensional calculations and will not be repeated here.

9.7 EXAMPLES

9.7.1 Two-dimensional solutions

As an example to the use of the method for low frequency eddy current problems, the geometry shown in Figure 9.23 is considered. It consists of two cylindrical conductors and a plane wave impinging from the left of the geometry. The incident field is at *6 kHz* (uniform field) and, although the problem can be easily solved in the frequency domain, it is solved here as a transient problem using the *FDTD* method. Material properties are as follows:

$$\varepsilon_c = \varepsilon_a = \varepsilon_0 = 8.85\times10^{-12} \ [F/m] \qquad \mu_c = \mu_a = \mu_0 = 4\pi\times10^{-7} \ [H/m]$$

The length for scaling is $L=0.05m$. Also: $k_a=6.287468\times10^{-6}$, $l_m^2=7.534687\times10^5$, $k_m^2=1$, $\Delta t=0.09256006$ (actual step 1.543719×10^{-11} s), and $\Delta x=0.1308997$ (actual step is 6.544985×10^{-3} m). Conductivity of cylinders is $\sigma_c=4\times10^4$ S/m and cylinders radii are $a=6.544985\times10^{-2}$ m.

The equations used are: in Ω_a, there are no lossy materials and we can write:

$$\frac{\partial^2 \overline{A_z}}{\partial x^2} + \frac{\partial^2 \overline{A_z}}{\partial y^2} = \frac{\partial^2 \overline{A_z}}{\partial t^2}$$

In Ω_c:

$$\frac{\partial^2 \overline{A_z}}{\partial x^2} + \frac{\partial^2 \overline{A_z}}{\partial y^2} = L\sigma_c \sqrt{\frac{\mu_c\varepsilon_c}{\mu_a\varepsilon_a}} \frac{\partial \overline{A_z}}{\partial t} + \frac{\mu_c\varepsilon_c}{\mu_a\varepsilon_a} \frac{\partial^2 \overline{A_z}}{\partial t^2}$$

The interface and boundary conditions are as discussed above. Implementation in terms of finite differences is given in Eq. (9.61) for domain Ω_a, in Eq. (9.63) for Ω_c and in Eqs. (9.64) through (9.75) for the interface and boundary conditions. Under these conditions the solution is shown in Figure 9.24. It clearly shows the skin effect in the conductors and the low frequency nature of the solution (that is, there are no propagation effects visible in spite of the use of a wave equation for solution). Note also that the boundaries are very close to the conducting cylinders.

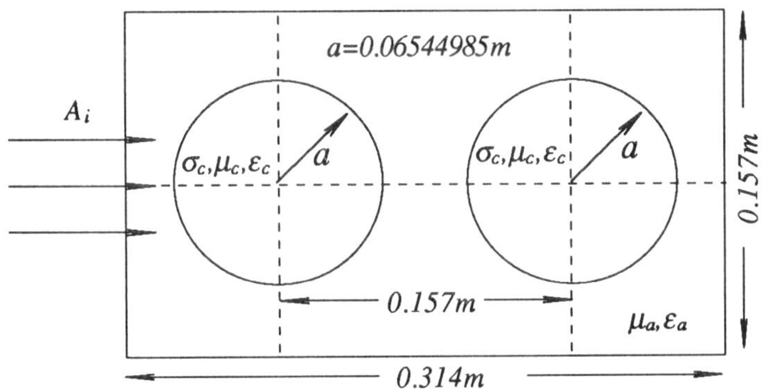

Figure 9.23 Two cylinders in a plane wave field at 6,000 Hz.

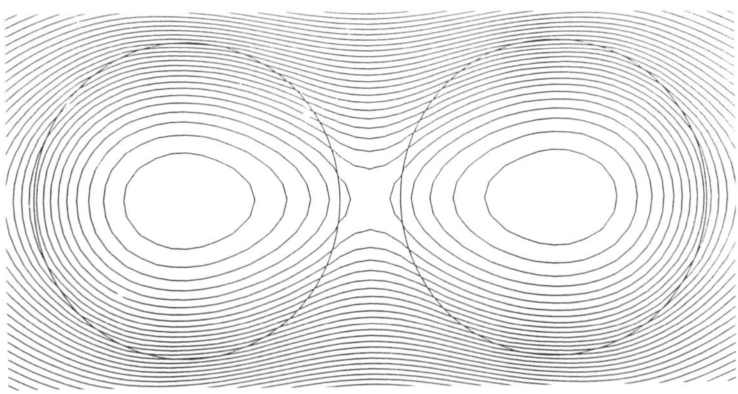

Figure 9.24 Contour plot of the field in the vicinity of the two cylinders shown in Figure 9.23.

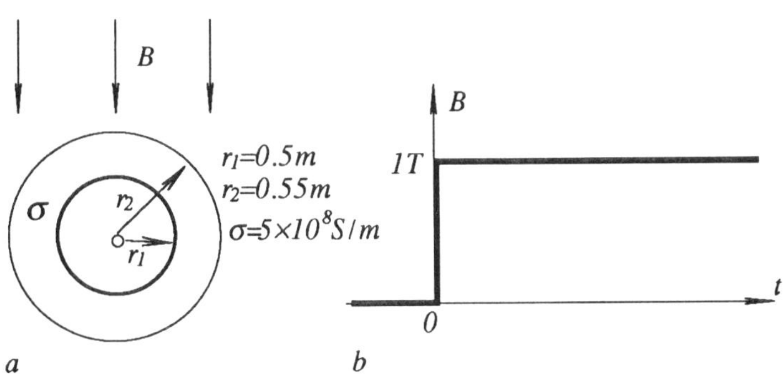

Figure 9.25 a. Conducting sphere in a transient magnetic field. b. Shape of the magnetic flux density.

9.7.2 Axisymmetric solutions

As an example to the use of the axisymmetric formulation, the problem of a hollow conducting sphere in a transient electromagnetic field is solved here. Quantities of interest are the magnetic flux density inside and outside the sphere and the shielding effect in the interior of the sphere during the transient phase of solution. Other parameters to calculate are the eddy current density in the sphere and the power dissipated.

The geometry considered is shown in Figure 9.25a. It consists of a conducting sphere with conductivity $\sigma = 5\times 10^7$ $[S/m]$ and dimensions as shown. The magnetic field is zero until time $t=0$ and then switched on instantaneously to $B=1T$. This gives a step field as shown in Figure 9.25b. The field far from the sphere is assumed to be uniform in space and directed in the z direction.

Considering the geometry in Figure 9.25, domain Ω_a is free space and therefore we can write for domain Ω_a:

$$\frac{\partial^2 A_\phi}{\partial z^2} + \frac{\partial^2 A_\phi}{\partial r^2} + \frac{1}{r}\frac{\partial A_\phi}{\partial r} - \frac{A_\phi}{r^2} = 0 \tag{9.82}$$

In Ω_c, only a diffusion term may exist since the wave propagation term is negligible. Thus the equation of interest in this domain is

$$\frac{\partial^2 A_\phi}{\partial z^2} + \frac{\partial^2 A_\phi}{\partial r^2} + \frac{1}{r}\frac{\partial A_\phi}{\partial r} - \frac{A_\phi}{r^2} = \mu_c \sigma_a \frac{\partial A_\phi}{\partial t} \tag{9.83}$$

The interface conditions are defined to guarantee continuity of the tangential components of the magnetic field and electric field on the interfaces between conductor and free space. These are given as

$$\left[\frac{\partial A_\phi}{\partial t}\right]^- = \left[\frac{\partial A_\phi}{\partial t}\right]^+$$

$$\mu_c \left[\frac{\partial r A_\phi}{\partial r}\right]^- = \left[\frac{\partial r A_\phi}{\partial r}\right]^+$$

where r is the radial distance of points on the interface surface. Permittivity of free space is taken as equal to ε_0, and μ_c is the relative permeability of the sphere. The equations above do not contain a wave term, therefore only one initial condition for the magnetic vector potential is needed namely, $A(r,z,0)$. This initial condition must be known before solution commences and is determined from the given boundary field. This is done shortly but we must first define the boundary conditions.

In this application, the radiation boundary condition is not used since the magnetic flux density on the boundary of the solution domain is given and is constant. For a given constant flux density, we can choose the magnetic vector potential from the relation $\nabla\times\mathbf{A}=\mathbf{B}$. One choice is

$$A_b = \frac{B_0 r}{2} \tag{9.84}$$

This gives the magnetic vector potential on the boundary which results in a constant flux density B_0 on the same boundary and also satisfies Eq. (9.82) in domain Ω_a as can be verified by direct substitution.

Also from the physical behavior of the problem we know that initially, as the field is switched on, the field inside the sphere is zero while the field outside is the superposition of the boundary field and an equivalent dipole field produced by the sphere. This dipole field is given as

$$A_d = \frac{\mu M \sin\theta}{4\pi r^2} \tag{9.85}$$

where M is the equivalent dipole moment. From the fact that $A=0$ in the interior of the sphere, (at $t=0$), including inside the conductor we get the equivalent dipole moment by equating the expression above with that of the boundary vector potential at the surface of the sphere. This gives

$$\frac{B_0 R}{2} + \frac{\mu M \sin\theta}{4\pi R^2} = 0$$

where R is the outer radius of the sphere as shown in Figure 9.25a. The dipole moment is, therefore,

$$M = -\frac{2\pi R^3 B_0}{\mu} \tag{9.86}$$

Substitution of this in Eq. (9.85) gives the magnetic vector potential due to the dipole as

$$A_d = -\frac{R^3 B_0 \sin\theta}{2r^2}$$

and, the total field anywhere outside the sphere at time $t=0^+$ is

$$A = A_d + A_b = \frac{B_0 r}{2} - \frac{R^3 B_0 \sin\theta}{2r^2}$$

while inside the sphere the total magnetic vector potential is

$$A = 0$$

These are therefore the initial conditions of the problem. We note that even though this initial condition imposes an additional, negative potential on the boundary, it is minimal since it decays as $1/r^2$ and, if the boundary is taken at a reasonable distance from the sphere, it affects the solution very little.

The problem is solved with the center of the sphere at the origin. This approach requires specification of boundary conditions on the four surfaces shown in Figure 9.26.

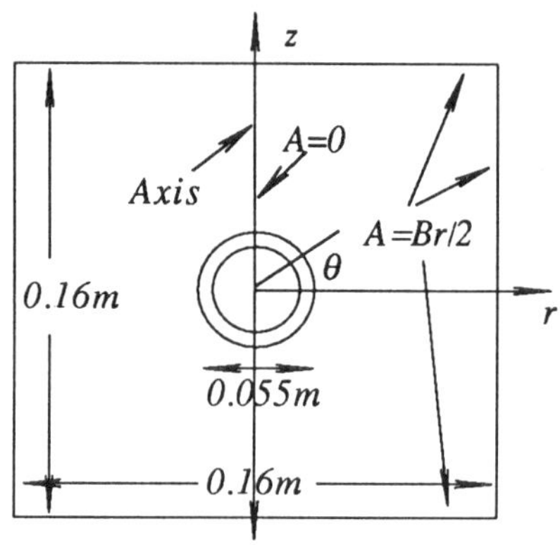

Figure 9.26 Solution domain and boundary condition. Only the domain to the right of the symmetry line is actually modeled.

The choice given above as $A=B_0r/2$ provides an approximation on all four boundaries. On the top boundary, $A=B_0r/2$ (that is: $B=B_0$). On the lower boundary, we can write

$$B_z = \frac{A}{r} + \frac{\partial A}{\partial r}$$

from which we have at $y=0$

$$B_z|_{r\to 0} = 2\frac{\partial A}{\partial r}$$

At $r=0$, $A=0$ because this is the axis of symmetry. On the right boundary r is constant and the boundary condition can be written as $B_0r_0/2$. These boundary conditions are shown in Figure 9.26. Since the whole boundary is defined in terms of the magnetic vector potential, only Dirichlet boundary conditions need be considered. The finite difference formulae for this application are given in Eqs.

(9.74) and (9.76). The interface conditions are specified as in Eq. (9.69) or (9.70) on the surfaces of the sphere (internal or external).

A finite difference grid must be defined. For good accuracy, a *160 mm* by *160 mm* domain is discretized with a 129×129 grid such that $\Delta r = \Delta z = 0.00125\ m$. With a time interval of $1\ \mu s$, a solution encompassing *20 ms* requires 20,000 time steps. However, because the approximation is explicit, this does not require excessive computer time. As with previous solutions, the grid can be modified to produce a more accurate result, provided that the stability criterion is satisfied.

Sample results are given in Figures 9.27 through 9.29. The magnetic flux density is shown in Figures 9.27 and 9.28 while the induced current density is shown in Figure 9.29.

As expected, the flux density goes up, tending towards the boundary condition while the overshoot in Figure 9.27 is reduced with time. The current density is also going down with time. After *20 ms* the current density is still about $1.7 \times 10^8\ A/m^2$, clearly indicating that the transient is far from dissipated. Also to be noted from Figure 9.27 is the fact that at very short time intervals after switching on the field, the solution is rather inaccurate. This only happens at short times because of the explicit nature of the solution. It takes a certain number of time steps for the solution to propagate through the mesh. However, after that, the solution is quite smooth.

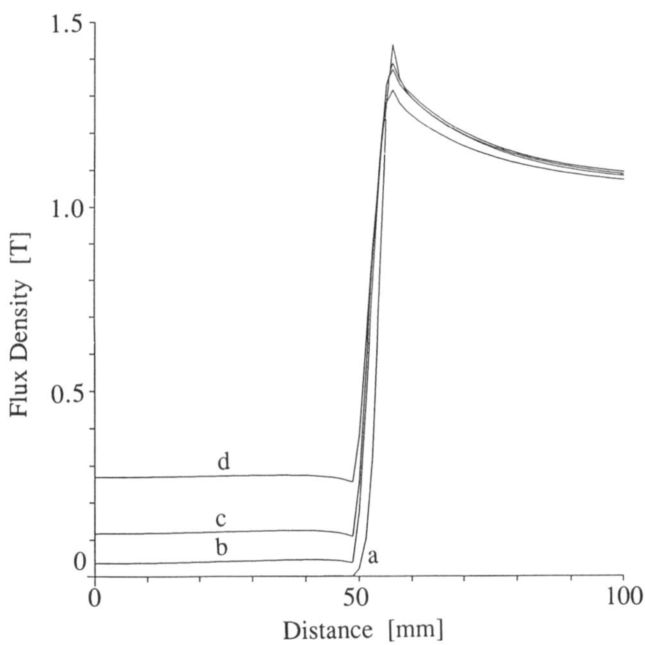

Figure 9.27 Magnetic flux density (B_z) in the radial direction at $\phi=0$, $z=0$. a. At $t=1$ ms, b. At $t=5$ ms. c. At $t=10$ ms. d. At $t=20$ ms.

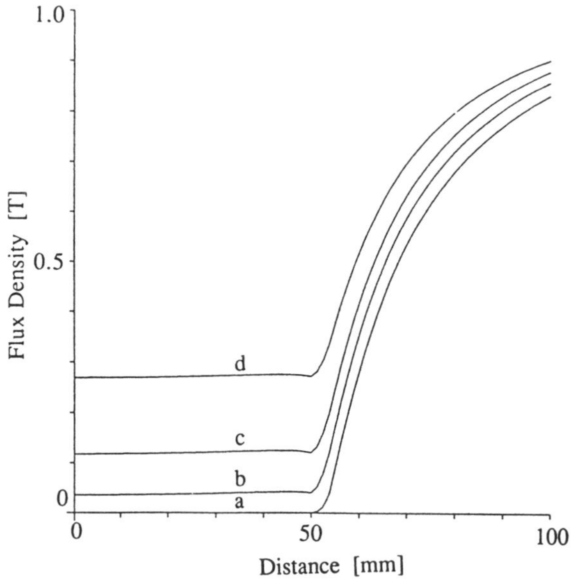

Figure 9.28 Magnetic flux density (B_z) in the axial direction at ϕ=0, r=0. a. At t=1 ms, b. At t=5 ms. c. At t=10 ms. d. At t=20 ms.

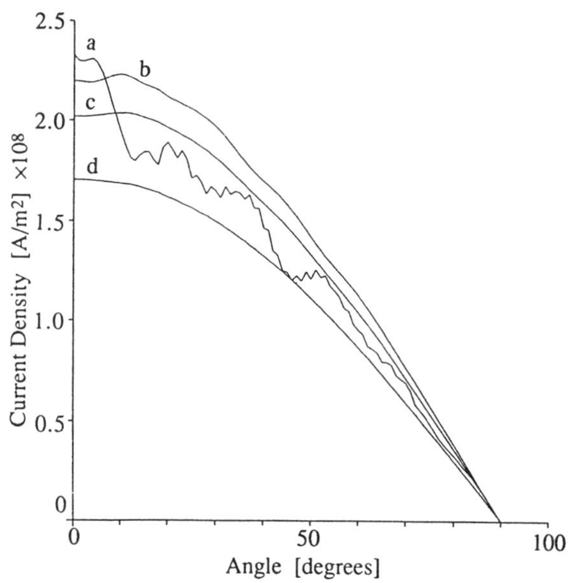

Figure 9.29 Current density in the sphere. ϕ component of the current density is shown at ϕ=0. θ is the angle between the r axis and location in the sphere. a. At t=1 ms, b. At t=5 ms. c. At t=10 ms. d. At t=20 ms.

9.8 PROBLEMS

9.1 Consider Figure 9.30. Using the finite element method, calculate the magnetic field intensity everywhere in the solution domain. Assume a ferromagnetic material with relative permeability of 100 is tested. Use the formulation in section 9.3.

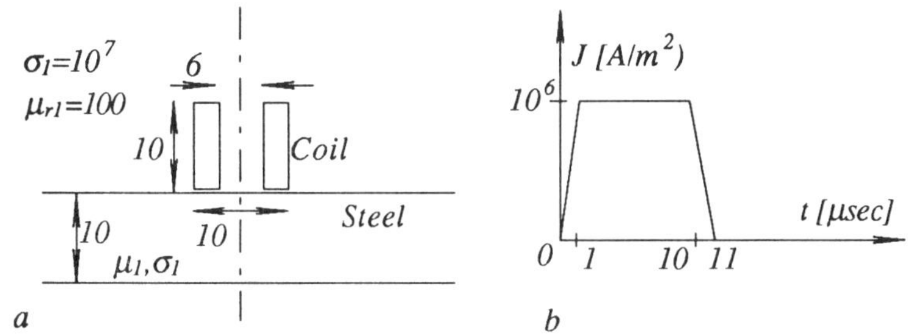

Figure 9.30.

9.2 Solve the same problem assuming the steel is replaced with a nonferromagnetic material ($\mu_{rl}=1$). Compare penetration of fields in materials and comment on the effect of this penetration on testing in materials.

9.3 Solve problems 9.1 and 9.2 using the pulse shown in Figure 9.31.

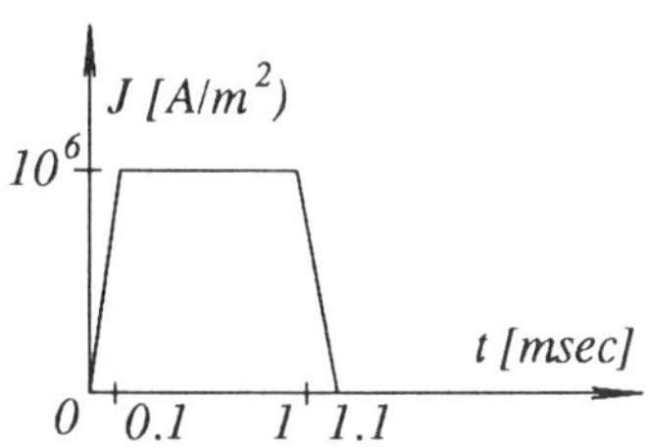

Figure 9.31.

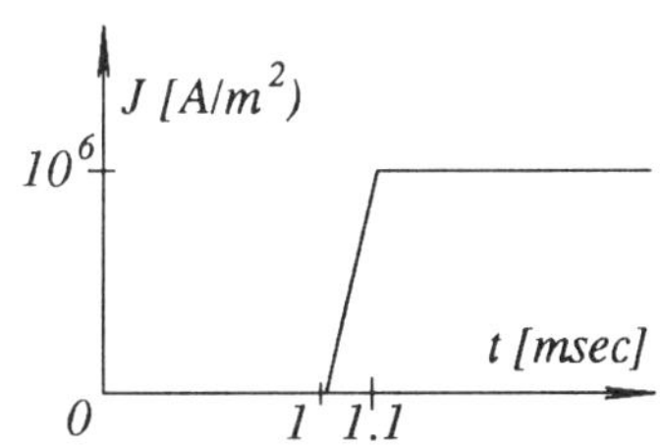

Figure 9.32.

9.4 Solve problem 9.2 using the pulse in Figure 9.32. Comment on the differences between solutions in the previous problems and this in terms of the steady state solution.

9.5 As an exercise solve the problem in Figure 9.7 using the finite element method of section 9.3 and compare results with the finite difference method solutions in section 9.4. Comment on accuracy, number of nodes/elements, speed of solution and in particular on the time steps required in the two solutions. Use the pulse given in Figure 9.8. Note: it is not normally possible to model a perfect pulse. For this reason, the vertical line is assumed to rise in a given time say, $1\ \mu s$.

9.6 Using the method in section 9.3 extend the axisymmetric formulations to three dimensions using any finite element. Obtain the matrices $[S]$, $[R]$ and $\{Q\}$ (see Eqs. (9.14) through (9.16)).

9.7 Solve problem 9.1 using the finite difference method of section 9.4. Assume a hole, *6 mm* in diameter, extending through the plate is present. Compare the solution with that in problem 9.1.

9.8 The test arrangement in Figure 9.33 is given. The lower coil is identical in size to the upper driving coil. The pulse is applied to the upper coil and the induced voltage in the lower (pickup) coil is measured. Calculate the induced voltage in the coil if the driving coil (upper coil) is at a lift-off of *0, 1, 2,* and *5 mm* above the sheet.

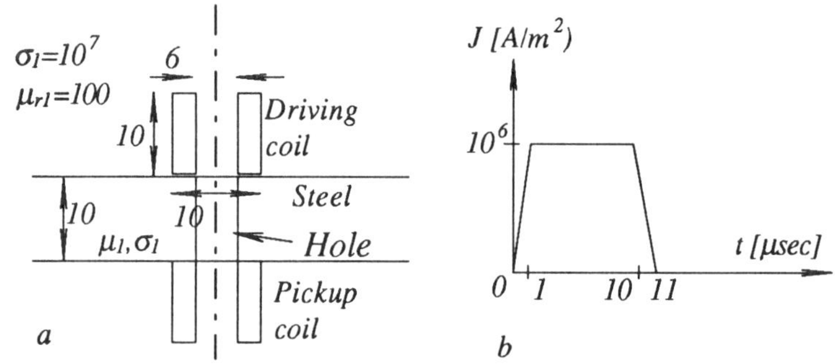

Figure 9.33.

9.9 Repeat problem 9.8 but leave the driving coil stationary and calculate the induced voltage in the pickup coil at a lift-off of the pickup coil of *0, 1, 2,* and *5 mm*. What are the main differences between the two solutions in terms of solution methodology and results?

9.10 A conducting cylinder of radius *1 cm* is located at the center of a large sheet of a dielectric material, *10 cm* thick. An electromagnetic field of unit strength (either flux density or, for convenience assume $A=1$) as shown in Figure 9.34 is applied. Calculate the magnetic field intensity at a distance *1 mm* from the surface of the dielectric (on the line shown), with and without the cylinder. Use a frequency of *1 kHz* and the *FDTD* method. Use conductivity of cylinder as $10^7 S/m$, that of dielectric as $\sigma=0$. Permeabilities are those of free space and relative permittivity of dielectric is *3*. Assume the incident field is time-harmonic (sinusoidal) but solve the problem as a transient problem.

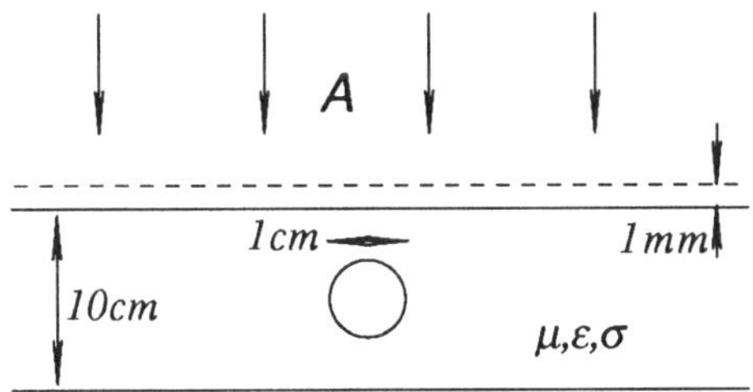

Figure 9.34.

9.11 Repeat problem 9.9 but with free space instead of the conducting cylinder. Comment on the results and their bearing on the use of this method for testing.

9.12 Replace the cylinder in Figure 9.34 by a lossy dielectric with conductivity $\sigma=10^2$ *S/m* and relative permittivity of *9*. Solve for the field at a frequency of *1 GHz*. Comment on the implication for testing at microwave frequencies.

9.9 BIBLIOGRAPHY

The references given here deal primarily with the aspects of computation presented in this chapter rather than with the general transient electromagnetic problem. In [1-2], the formulation in section 9.3 is given in more detail. References [3-4] give some more information on finite element formulation of transient solutions. Various aspects of parabolic equations and their solutions in terms of finite differences can be found in [5-9]. References [10-11] discuss in detail the finite difference time domain method in terms of potential functions. The question of radiating boundary conditions is addressed in many sources, some of which are references [12-16]. Some additional application related to *NDT* are given in [17-20].

[1] Allen, B. Ida, N. and Lord, W. (1985), Finite element modeling of pulsed eddy current NDT phenomena, *IEEE Transactions on Magnetics*, Vol. MAG-21, No. 6, pp. 2250-2253.

[2] Allen, B.L. (1983), *Finite Element Modeling of Pulsed Electromagnetic NDT Phenomena*, M.Sc. Thesis, Colorado State University.

[3] Tandon, S.C. and Chari, M.V.K. (1981), Transient solution of the diffusion equation by the finite element method, *Journal of Applied Physics*, Vol. 52, No. 3, pp. 2431-2432.

[4] Leonard, P.J. and Rodger, D. (1988), Finite element scheme for transient 3D eddy currents, *IEEE Transactions on Magnetics*, Vol. 24, No. 1, Jan. pp. 90-93.

[5] Dufort, E.C. and Frankel, S.P. (1953), Stability conditions in the numerical treatment of parabolic differential equations, *Mathematical Tables and Aids to Computation*, Vol. 7, pp. 135-152.

[6] Richtmyer, R.D. and Morton, K.W. (1967), *Differential Methods for Initial Value Problems*, John Wiley.

[7] Forsythe, G.E. and Wasow, W.R. (1964), *Finite Difference Methods of Partial Differential Equations*, Wiley.

[8] Southwell, R.V. (1946), *Relaxation Methods in Theoretical Physics*, Oxford University Press, London.

[9] James, M.L. Smith, G.M. and Walford, J.C. (1977), *Applied Numerical Methods for Digital Computation*, Harper and Row, New York.

[10] Lee, M.E. (1989), *Potential Finite Difference Time-Domain Methods for Electromagnetic Interface Problems*, Ph.D. Dissertation, College of Engineering, The University of Akron, Akron, OH.

[11] Lee, M.E. Hariharan, S.I. and Ida, N. (1990), Solving General Time Dependent Two Dimensional Eddy Current Problems, *Journal of Computational Physics*, Vol. 89, No. 2, pp. 319-348.

[12] Bayliss, A. and Turkel, E. (1980), Radiation boundary conditions for wave-like equations, *Communications of Pure and Applied Mathematics*, Vol. 33, No. 6, pp. 707-725.

[13] Engquist, B. and Majda, A. (1977), Absorbing boundary conditions for the numerical simulation of waves, *Mathematics of Computation*, Vol. 31, pp. 629-651.

[14] Friedlander, F.G. (1962), On the radiation field of pulse solutions of the wave equation, Proceedings of the Royal Society, London, Ser. A, Vol. 269, pp. 53-65.

[15] Mur, G. (1981), Absorbing boundary conditions for finite-difference approximation of the time-domain electromagnetic field equations, *IEEE Transactions on Electromagnetic Compatibility*, Vol. EMC-24, pp. 1073-1077.

[16] Atkinson, F.V. (1949), On Sommerfeld's radiation condition, *Philosophical Magazine*, Vol. XL, pp. 645-651.

[17] Wang, J.S. Ida, N. and Hariharan, S.I. (1989), Numerical modeling of transient wave propagation for high frequency NDT, *Review of Progress in Quantitative Nondestructive Evaluation*, D. O. Thompson and D. E. Chimenti, Eds., Plenum Press, Vol. 8A, New York, pp. 259-266.

[18] Lee, M.E. and Ida, N. Hollow sphere in a step field, *COMPEL, Vol. 9, Supplement A*, 1990, pp. 278-280.

[19] Taflove, A. (1980), Application of the finite difference time-domain method to sinusoidal steady state electromagnetic penetration problems, *IEEE Transactions on Electromagnetic Compatibility*, Vol. EMC-22, No.3, pp. 192-202.

[20] Wang, J.S. Ida, N. and Hariharan, S.I. (1989), Numerical modeling of transient wave propagation for high frequency NDT, *Review of Progress in Quantitative Nondestructive Evaluation*, D. O. Thompson and D. E. Chimenti, Eds., Plenum Press, Vol. 8A, New York, pp. 259-266.

[21] Lee M.E. and Ida, N. (1990), Axisymmetric electromagnetic fields coupled to lossy media, *IEEE Transactions on Magnetics*, Vol. 26, No. 2, pp. 575-578.

CHAPTER 10

Hyperbolic partial differential equations

10.1 INTRODUCTION

After discussing the static and quasistatic field equations in chapters 6 and 7 and the transient field equation in chapter 9, the extension of the methods described in those chapters to the solution of the wave equation are outlined here. The solution to the wave equation is discussed briefly in chapter 9 in that the finite difference time domain method and the associated radiation boundary conditions included a wave term. Normally, this term is neglected in eddy current applications because it is rather small. However, all time-dependent electromagnetic phenomena can be viewed as wave propagation phenomena. The method of solution is based on how dominant a particular term is. If the wave term (that is, displacement currents) are dominant, the problem must be solved as a wave equation. If conduction currents are dominant (that is, losses are relatively high and the wave is highly damped) a solution as a transient eddy current problem is appropriate. There are cases in which both terms must be included.

In this chapter, the time dependent wave equation is treated. This is a hyperbolic equation and the methods of solution are somewhat unique. Finite difference methods are more commonly used for their solution because of the need for time discretization but finite element methods are also used.

The need for radiation boundary conditions is more critical in solution of the wave equation since waves may propagate large distances, may be guided by test samples, and are reflected from any material interface. Thus, proper handling of boundaries is critical and artificial Dirichlet boundary conditions cannot be used.

10.2 THE GENERAL HYPERBOLIC PARTIAL DIFFERENTIAL EQUATION

Following the procedures in chapters 5 and 9 we introduce the definition of the hyperbolic partial differential equation based on the coefficients of the general second order partial differential equation

$$a\frac{\partial^2 F(u,v)}{\partial u^2}+b\frac{\partial^2 F(u,v)}{\partial u\,\partial v}+c\frac{\partial^2 F(u,v)}{\partial v^2}+d\frac{\partial F(u,v)}{\partial u}+e\frac{\partial F(u,v)}{\partial v} + gF(u,v)=f(u,v) \tag{10.1}$$

where u,v, represent any two variables such as x,y in space, or x in space and t in time. This equation is said to be hyperbolic in form if the following condition is satisfied

$$b^2-4ac>0$$

The following is, therefore, a hyperbolic equation:

$$a\frac{\partial^2 F(u,v)}{\partial u^2}-c\frac{\partial^2 F(u,v)}{\partial v^2}=0$$

However, we are interested here in time dependent equations of the form:

$$\frac{\partial^2 F(x,t)}{\partial x^2}=k^2\frac{\partial^2 F(x,t)}{\partial t^2} \tag{10.2}$$

This is a one-dimensional wave equation. Similar equations in two- or three-dimensions will be considered:

$$\frac{\partial^2 \mathbf{E}(x,y,t)}{\partial x^2}+\frac{\partial^2 \mathbf{E}(x,y,t)}{\partial y^2}=k^2\frac{\partial^2 \mathbf{E}(x,y,t)}{\partial t^2} \tag{10.3}$$

$$\frac{\partial^2 \mathbf{E}(x,y,z,t)}{\partial x^2}+\frac{\partial^2 \mathbf{E}(x,y,z,t)}{\partial y^2}+\frac{\partial^2 \mathbf{E}(x,y,z,t)}{\partial z^2}=k^2\frac{\partial^2 \mathbf{E}(x,y,z,t)}{\partial t^2} \tag{10.4}$$

In addition, the following type of equations will be treated here as well:

$$\frac{\partial^2 \mathbf{E}(x,y,z,t)}{\partial x^2} + \frac{\partial^2 \mathbf{E}(x,y,z,t)}{\partial y^2} + \frac{\partial^2 \mathbf{E}(x,y,z,t)}{\partial z^2} = k^2 \frac{\partial^2 \mathbf{E}(x,y,z,t)}{\partial t^2} + p \frac{\partial \mathbf{E}(x,y,z,t)}{\partial t} \tag{10.5}$$

$$\frac{\partial^2 \mathbf{E}(x,y,z,t)}{\partial x^2} + \frac{\partial^2 E(x,y,z,t)}{\partial y^2} + \frac{\partial^2 \mathbf{E}(x,y,z,t)}{\partial z^2} = k^2 \frac{\partial^2 \mathbf{E}(x,y,z,t)}{\partial t^2} + p \frac{\partial \mathbf{E}(x,y,z,t)}{\partial t} + q\mathbf{J} \tag{10.6}$$

Equations (10.3) and (10.4) are two- and three-dimensional wave equations respectively. Eq. (10.5) is a transient wave equation where the last term is an eddy current term. This term is responsible for losses. Eq. (10.6) is a nonhomogeneous wave equation where **J** is the source term in the equation.

These equations are different than parabolic equations in that although a first order time derivative may exist, the solution is dominated by a propagation term which is due to the second order time derivative in the equation. The solutions to these equations are time dependent. The solution is bounded in space but unbounded in time. The equation and its boundary conditions must be satisfied throughout the solution domain. This is shown schematically in Figure 9.1 where we assume in general that both Dirichlet and Neumann boundary conditions as well as radiation boundary conditions may exist in the solution domain. The domain is closed in all spatial variables (i.e. it is a closed surface for two-dimensional problems or a volume for three-dimensional applications) but is open in time. Again, as for parabolic processes, the solution in time may progress indefinitely although, for practical implementation we stop this progression at some point.

The time-harmonic form of the various wave equations is not discussed here except in the sense that a time dependent solution to the wave equation can be obtained by transform methods. This was also the case for parabolic equations but the solution to the time harmonic wave equation was discussed in some detail in chapters 6 and 7 and is not be repeated here. The time harmonic forms of the wave equations are discussed in chapter 1.

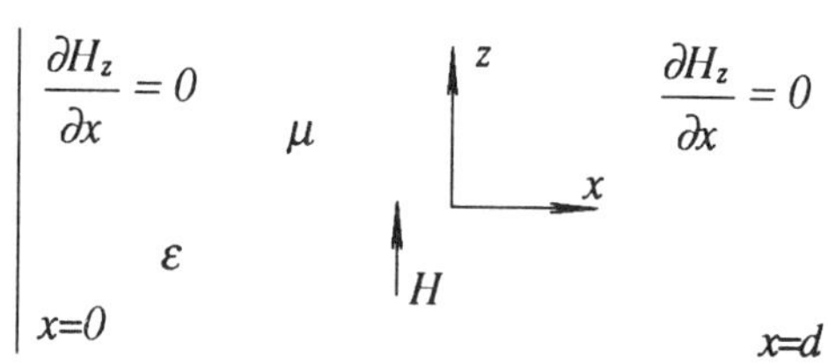

Figure 10.1 A one-dimensional geometry. The plates are infinite in extent.

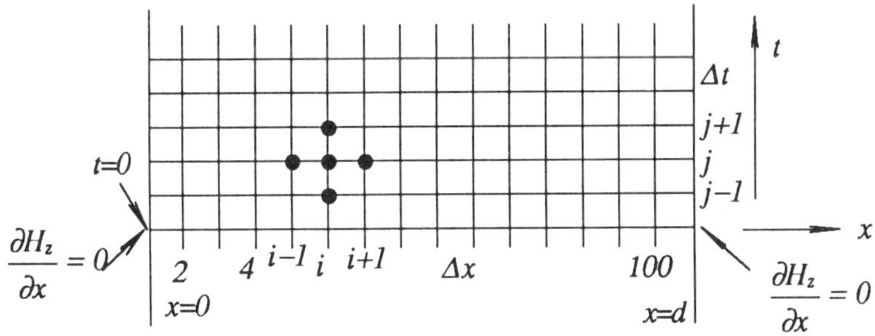

Figure 10.2 Finite difference grid over the solution domain.

10.2.1 Finite difference modeling for standing waves

As an example of modeling of wave propagation we solve here a one dimensional wave equation in a bounded domain. Although this is a rather simple application, it exhibits all aspects of wave propagation and therefore contains all the elements needed for modeling. Because it is a one-dimensional wave problem it can be easily solved. The problem is bounded in space but not in time. Handling of boundary conditions is simplified considerably because of this constraint.

The wave equation to be solved is

$$\frac{\partial^2 H_z}{\partial t^2} = \frac{1}{\mu\varepsilon}\frac{\partial^2 H_z}{\partial x^2} \tag{10.7}$$

This is solved in the domain shown in Figure 10.1. The domain between the two plates is a dielectric with permittivity ε and permeability μ. For simplicity, and to avoid the need to handle interface conditions between different materials, the dielectric is assumed to be isotropic and a single dielectric fills the whole solution space.

Eq. (10.7) indicates that the magnetic field intensity has a single component in the x direction which varies in the z direction. Boundary conditions are $\partial H_z/\partial x=0$ on both boundaries as shown. This type of boundary condition is not typical in wave applications, except for propagation in guiding structures such as waveguides and cavity resonators.

In addition to boundary conditions we must specify a correct set of initial conditions for the wave. The initial conditions give the electric field intensity and its first time derivative with respect to time at $t=0$. These can be written in general terms as

$$H_z(x,0) = f(x) \tag{10.8}$$

$$\frac{\partial H_z}{\partial t}(x,0) = g(x) \tag{10.9}$$

The initial conditions are related to the particular problem being solved as will be shown shortly. Starting with the initial condition, the wave propagates between the two boundaries of the solution domain, is reflected repeatedly and, depending on conditions, interferes with itself to produce standing waves, or a combination of standing and propagating waves, depending on conditions in the solution domain.

To solve the problem we define a finite difference grid as shown in Figure 10.2. The grid includes the boundary conditions, but not the initial conditions. These have to be defined separately. Grid spacing is arbitrary at this point but, in general, we must choose it judiciously since the convergence criterion in Eq. (3.29) must be satisfied. This means that once Δx is chosen, the following relation is used to calculate Δt.

$$\frac{(\Delta t)^2}{\mu\varepsilon(\Delta x)^2} \leq 1 \tag{10.10}$$

For best results, the choice should be very close to unity. Thus, the time step must be

$$\Delta t \approx \sqrt{\mu\varepsilon(\Delta x)^2}$$

This choice will prove to be difficult to satisfy, especially at lower frequencies. To get a better feeling for the value of the time step, we assume here that $\mu=\mu_0=4\pi\times10^{-07}$ and $\varepsilon=\varepsilon_0=8.85\times10^{-12}$ and that $\Delta z=1\ mm$. Under these conditions, $\Delta t<3.335\times10^{-12}$. If we were to assume that the fundamental frequency of the wave is $1\ mHz$, the number of time steps per cycle of the wave is almost $300{,}000$. This is a very large number which is dictated not by resolution requirements but by stability requirements. On the other hand, if we could choose $\Delta x=10\ mm$, the number of time steps would be $30{,}000$ and for $\Delta x=0.1\ m$, the number of time steps would be $3{,}000$ for the same conditions. Note also that at much higher frequencies, the number of time steps per cycle is lower. As an example, at $1\ GHz$, we would only need 300 time steps per cycle using $\Delta x=1\ mm$.

After the finite difference mesh is defined, the finite difference approximation is written.

First, the wave equation is discretized using central difference formulae as shown in chapter 6. Using the notation and dimensions in Figure 10.2, this takes the following form

$$H_{i,j+1} = 2H_{i,j} - H_{i,j-1} + \frac{(\Delta t)^2}{\mu\varepsilon(\Delta x)^2}\left[H_{i-1,j} - 2H_{i,j} + H_{i+1,j}\right] \quad (10.11)$$

While this approximation is rather straightforward we note that to calculate the magnetic field intensity at time step $j+1$ (present time step for an explicit solution) the magnetic field intensity at time steps j and $j-1$ must be known. The values at time step j $(t=0)$ are known from the first initial condition in Eq. (10.8). The second initial condition can be used to calculate the field at time step $j-1$ $(t-\Delta t)$ by using a central difference scheme for the second initial condition. The central difference approximation for Eq. (10.9) is

$$H_{x,\,t+\Delta t} - H_{x,\,t-\Delta t} = g(x)$$

or, in short form notation:

$$H_{i,j-1} = H_{i,j+1} + g(x) \quad (10.12)$$

Once $g(x)$ is known (and, in most cases it will be zero), the approximation in Eqs. (10.12) can be substituted in Eq. (10.11) to obtain an expression which only contains values at time step i and $i+1$. From this, a new solution for time step $i+1$ is obtained. Any subsequent time step can then be calculated directly from Eq. (10.11).

In addition to the discretization in the domain using Eq. (10.11) and the initial condition in Eq. (10.12) we must also implement boundary conditions. This is done as in section 6.2.6 since both boundaries can be considered to be symmetry boundaries. Note that using this concept, we in fact assume implicitly that the geoemtry in Figure 10.1 is repetitive both to the left of the left boundary and to the right of the right boundary. This, however, does not alter the solution inside the geometry given. Thus, on the boundary:

At node *(1)*:

$$H_{i,j+1} = 2H_{i,j} - H_{i,j-1} + \frac{(\Delta t)^2}{\mu\varepsilon(\Delta x)^2}\left[2H_{i+1,j} - 2H_{i,j}\right] \quad (10.13)$$

At node *(101)*:

$$H_{i,j+1} = 2H_{i,j} - H_{i,j-1} + \frac{(\Delta t)^2}{\mu\varepsilon(\Delta x)^2}\left[2H_{i-1,j} - 2H_{i,j}\right] \quad (10.14)$$

As an example of a particular solution consider the initial solution given in Figure 10.3. This represents the field at time $t=0$. The initial conditions are, therefore

$$H_z(x,0) = 1 \cos\left(\frac{\pi}{d}x\right) \tag{10.15}$$

and

$$\frac{\partial H_z}{\partial t}(x,0) = 0 \tag{10.16}$$

Thus, the finite difference approximation for Eq. (10.12) is

$$H_{i,j-1} = H_{i,j+1} \tag{10.17}$$

Substitution of this into Eq. (10.11) gives

$$H_{i,j+1} = 2H_{i,j} - H_{i,j+1} + \frac{(\Delta t)^2}{\mu\varepsilon(\Delta x)^2}\left[H_{i-1,j} - 2H_{i,j} + H_{i+1,j}\right]$$

or, after rearranging:

$$H_{i,j+1} = H_{i,j} + \frac{(\Delta t)^2}{2\mu\varepsilon(\Delta x)^2}\left[H_{i-1,j} - 2H_{i,j} + H_{i+1,j}\right] \tag{10.18}$$

Similar modification must be done at node *(1)* and *(101)* for the first time step.

This expression allows calculation of the field intensity at time $t+\Delta t$. Together with the expression in Eq. (10.16), the solutions at times t and $t+\Delta t$ are now known and the solution can proceed using Eq. (10.11) to calculate any subsequent time step.

For the geometry in Figure 10.2, with a total distance of *1m* between boundaries, and *100* divisions in the x direction we get:

$$\Delta x = 0.01\ m, \qquad \Delta t = 3.3\times 10^{-11}\ s \tag{10.19}$$

The solutions after *303*, *606*, *757*, *909*, *1,212*, and *1,515* time steps (corresponding to *10*, *20*, *25*, *30*, *40*, and *50* ns) are shown in Figure 10.4. It is obvious that the solution is a standing wave pattern with maximum amplitude twice the amplitude of the original field (initial condition) and minimum zero.

For other initial conditions, the solution will be different but the method remains the same.

This method of solution for bounded problems can be extended to two or three dimensions. A number of related problems can be solved using this method, including fields in waveguides, although these are usually calculated using the time-harmonic form of the equation. There is little to be gained in pursuing examples in two and three dimensions as these are discussed in chapter

6 for bounded space. The addition of time is a relatively minor problem and does not modify the method except for the need to satisfy the stability criterion.

For unbounded domain applications we have to define appropriate radiation boundary conditions to be able to solve the problem. This aspect of modeling is described next.

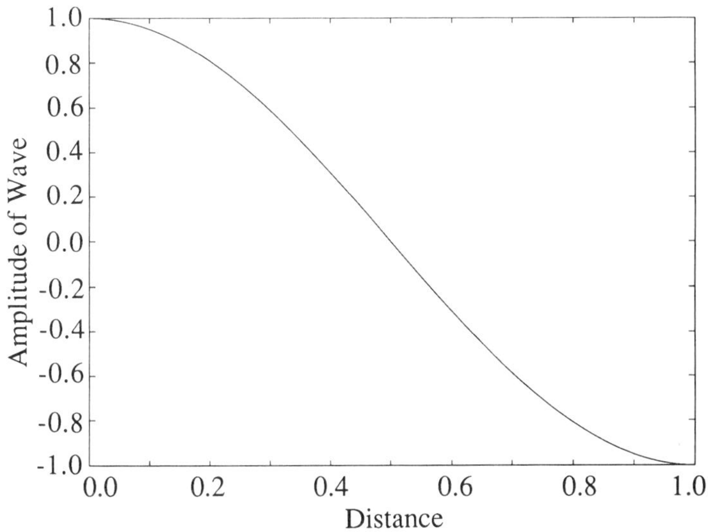

Figure 10.3 Initial conditions for standing wave solution.

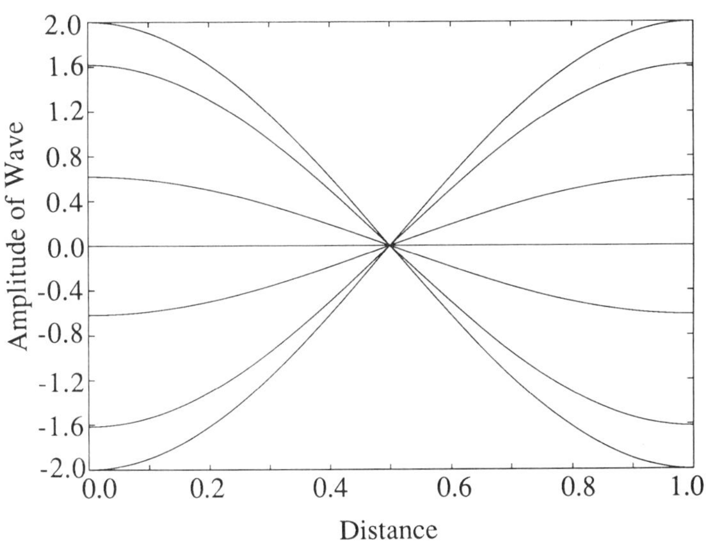

Figure 10.4 Standing waves at 10, 20, 25, 30, 40, and 50 ns.

10.3 THE FINITE DIFFERENCE TIME DOMAIN METHOD

The finite difference time domain (*FDTD*) method is introduced in the previous chapter in sections 9.6.1 through 9.6.9. The application there is geared towards transient applications, with emphasis on low frequency eddy current applications. However, the wave nature of the equations was retained in most developments since it was a requirement for proper definition of the radiation boundary condition. We therefore use here most of the results obtained in sections 9.6.1 through 9.6.9 with minor changes. The radiation boundary condition is modified to apply to high frequency applications since most of the interest in wave propagation in *NDT* is at high frequencies.

The generic problem we are solving is shown in Figures 9.16 and 9.18. These figures show two domains, Ω_a and Ω_c. Ω_a contains any sources that may exist while Ω_c contains scatterers such as defects, various materials, etc. Although the material in Ω_a is taken as a completely general material for the purpose of formulation, in practical applications this is often free space and therefore lossless.

10.3.1 The field equations

The type of equation solved here is

$$\frac{\partial^2 A_z}{\partial x^2} + \frac{\partial^2 A_z}{\partial y^2} = \sigma\mu\frac{\partial A_z}{\partial t} + \varepsilon\mu\frac{\partial^2 A_z}{\partial t^2} \tag{10.20}$$

This is a two-dimensional homogeneous wave equation with a damping term due to conductivity of materials in the solution domain. Thus, sources are not represented in the solution, except indirectly. In other words, while a current source is not included in the formulation, any source can be represented by its equivalent magnetic field distribution through the magnetic vector potential.

This equation has the following form in the two materials.
In Ω_a:

$$\frac{\partial^2 A_z}{\partial x^2} + \frac{\partial^2 A_z}{\partial y^2} = \sigma_a\mu_a\frac{\partial A_z}{\partial t} + \varepsilon_a\mu_a\frac{\partial^2 A_z}{\partial t^2} \tag{10.21}$$

In Ω_c:

$$\frac{\partial^2 A_z}{\partial x^2} + \frac{\partial^2 A_z}{\partial y^2} = \sigma_c\mu_c\frac{\partial A_z}{\partial t} + \varepsilon_c\mu_c\frac{\partial^2 A_z}{\partial t^2} \tag{10.22}$$

In addition, the interface conditions on the boundary between Ω_a ands Ω_c are:

$$\left[\frac{\partial A_z}{\partial t}\right]^- = \left[\frac{\partial A_z}{\partial t}\right]^+ \tag{10.23}$$

$$\mu_c\left[\frac{\partial A_z}{\partial n}\right]^- = \mu_a\left[\frac{\partial A_z}{\partial n}\right]^+ \tag{10.24}$$

where the negative sign represents the exterior field, the positive sign the interior field, μ_a is the relative permeability in Ω_a and μ_c the permeability in Ω_c. In addition, the following initial conditions are needed

$$A_z(x,y,0) = A_{zi}(x,y,0) \tag{10.25}$$

$$\frac{\partial A_z(x,y,0)}{\partial t} = \frac{\partial A_{zi}(x,y,0)}{\partial t} \tag{10.26}$$

where A_i indicates the incident field. In addition to these we will also have to specify a set of boundary conditions.

The relations above are recognized as the same relations we obtained in section 9.6.1.

10.3.2 The scaled equations

Again, as in section 9.6.2 we also scale the problem as described and obtain a final system to be solved. Governing equations are:
In Ω_a:

$$\frac{\partial^2 \overline{A}_z}{\partial x^2} + \frac{\partial^2 \overline{A}_z}{\partial y^2} = L\sigma_c\sqrt{\frac{\mu_a}{\varepsilon_a}}\frac{\partial \overline{A}_z}{\partial t} + \frac{\partial^2 \overline{A}_z}{\partial t^2} \tag{10.27}$$

In Ω_a:

$$\frac{\partial^2 \overline{A}_z}{\partial x^2} + \frac{\partial^2 \overline{A}_z}{\partial y^2} = L\sigma_a\sqrt{\frac{\mu_c\varepsilon_c}{\mu_a\varepsilon_a}}\frac{\partial \overline{A}_z}{\partial t} + \frac{\mu_c\varepsilon_c}{\mu_a\varepsilon_a}\frac{\partial^2 \overline{A}_z}{\partial t^2} \tag{10.28}$$

Interface conditions on the boundary between Ω_a and Ω_c:

$$\overline{A}_z^- = \overline{A}_z^+ \tag{10.29}$$

$$\mu_c\left[\frac{\partial \overline{A}_z}{\partial n}\right]^- = \mu_a\left[\frac{\partial \overline{A}_z}{\partial n}\right]^+ \tag{10.30}$$

Initial conditions are:

$$\overline{A}_z(\overline{x},\overline{y},0) = \overline{A}_{zi}(\overline{x},\overline{y},0) \tag{10.31}$$

$$\frac{\partial \overline{A}_z(\overline{x},\overline{y},0)}{\partial \overline{t}} = \frac{\partial A_{zi}(\overline{x},\overline{y},0)}{\partial t} \tag{10.32}$$

10.3.3 Boundary conditions

There are a number of radiation boundary conditions that can be used for high frequency applications. One of these is given in Eq. (9.44) and is known as the Sommerfeld radiation condition:

$$\lim_{r\to\infty} r^{1/2}\left(\frac{\partial A_{zs}}{\partial r} + \sqrt{\mu_a \varepsilon_a}\,\frac{\partial A_{zs}}{\partial t}\right) = 0 \tag{10.33}$$

or, after using the same scaling procedure as for the equations above, the radiation boundary condition is

$$\lim_{r\to\infty} \overline{r}^{-1/2}\left(\frac{\partial \overline{A}_{zs}}{\partial \overline{r}} + \frac{\partial \overline{A}_{zs}}{\partial \overline{t}}\right) = 0 \tag{10.34}$$

A better choice, one that is used here, is

$$\frac{\partial A_{zs}}{\partial r} + \frac{\partial A_{zs}}{\partial t} + \frac{1}{2r} A_{zs} = 0 \tag{10.35}$$

It should be noted that these expressions are written for the scaled potentials and assume that only free space extends to the radiation boundary. A similar expression is obtained for radiation boundary conditions if the material is lossy. This condition is given as

$$\left(\frac{\partial}{\partial r} + \frac{k_c}{k_a}\frac{\partial}{\partial t} + \frac{1}{2r} + \frac{l_c^2}{2k_c}\right) A_s = 0 \tag{10.36}$$

where k_m, k_a, l_m, were defined in section 9.6.6.

This radiation boundary condition can be used for low frequencies as well but is not well suited for this purpose. However, at high frequencies it yields good results while being simpler and easier to implement than the Sommerfeld radiation condition.

In addition to radiation boundary conditions there are three other types of boundary conditions that may be used. These are the magnetic wall, the electric wall and the impedance boundary conditions. Also, in some instances, simple Dirichlet boundary conditions may be used. An example of the latter is given in section 10.2 where the magnetic flux density on boundaries is given.

(a) Electric wall boundary condition

A condition that is often encountered, especially in high frequency applications is the interaction of waves with a perfectly conducting material. In this case, the tangential component of the electric field intensity must vanish on the surface. This condition is often written as

$$\hat{\mathbf{n}} \times \mathrm{E} = 0 \tag{10.37}$$

Electric wall boundary conditions are encountered in waveguides and cavity resonators and in other situations in which high conductivity materials are present. In still other geometries, the electric wall condition may be specified on part of the boundary with other conditions on other parts of the boundaries of the geometry.

(b) Magnetic wall boundary condition

An equivalent condition for the magnetic field intensity also exists. It is a magnetic wall boundary condition:

$$\hat{\mathbf{n}} \times \mathrm{H} = 0 \tag{10.38}$$

The magnetic wall boundary condition is often applied to model apertures in cavities as well as symmetries in structures

(c) Impedance boundary condition

A very simple boundary condition can sometimes be affected by use of a resistive layer at the location of the boundary. The layer, if properly matched to the surrounding medium, may represent physical conditions such as a lossy dielectric or may be used to absorb all outgoing energy of the wave, thus allowing no backward propagating waves. It is therefore similar to a radiation boundary condition in this sense. An appropriate choice is to assume a surface resistivity given as

$$R_s = \frac{1}{\sigma\delta} \tag{10.39}$$

where σ is the layer conductivity and δ the skin depth at the given frequency. The layer does not necessarily have to be thin and, in fact, the boundary may be replaced by a number of discretization layers (finite differences or finite elements) of a given high conductivity and a Dirichlet boundary condition applied to the outer boundary of the layer. The resistivity boundary condition is particularly useful in applications of boundary integral methods.

10.3.4 Finite difference approximation

The finite difference formulae for general two-dimensional applications are given in Eqs. (9.61) and (9.63). The only difference is in the way the radiation boundary condition is modeled. The boundary conditions are implemented in Eqs. (9.64) through (9.70). However, since this is a high frequency condition, the term $C(r)$ in Eq. (9.70) is replaced by $1/2r$. The constants in Eq. (9.70) are therefore

$$a = 1 - \frac{1}{2r}\frac{\Delta t}{\Delta x} \qquad b = \cos\theta \frac{\Delta t}{\Delta x} \qquad c = \sin\theta \frac{\Delta t}{\Delta x} \qquad (10.40)$$

The constants b and c are the same as for the low frequency boundary condition.

10.3.5 Axisymmetric applications

As with the two-dimensional applications discussed above, axisymmetric applications follow the same pattern as in section 9.6.4 and 9.6.5. There is little point in discussing these here since Eqs. (9.47) through (9.57) are used directly with the boundary conditions in Eqs. (9.60) and (9.61). Finite difference implementation is given in Eqs. (9.61) through (9.63). Therefore, we limit ourselves here to examples of implementation in high frequency *NDT*. This method is particularly useful in testing of dielectric materials (including lossy dielectrics) such as plastics and composites at microwave frequencies.

10.3.6 Three-dimensional applications

The use of the magnetic vector potential as the unknown variable dominates most applications in two dimensions and in axisymmetric geometries. The reason for this is mainly due to the fact that under these conditions the field equations are essentially scalar equations as is amply shown in chapter 1.

In three dimensions however, the three components of the magnetic vector potential must be used and sometimes, a fourth, scalar quantity must be added (see sections 7.3 and 7.4). There is therefore little incentive to using potential functions in three dimensions. In fact, some of the first FDTD applications were in three dimensions, in terms of both the magnetic field intensity and the electric field intensity.

For purposes of completeness we show here the so-called Yee algorithm for *3-D FDTD* applications although we show no examples to its application. Implementation is rather straightforward and there are many examples to its application in the literature. Some of these are given in the bibliography section. The basis of the method is to write the general field equations and separate these into the three components. The three components are then implemented directly on a standard cell of the finite difference grid. The two curl equations needed for implementation are:

$$\nabla\times\mathbf{E} = -\mu\frac{\partial\mathbf{H}}{\partial t} \tag{10.41}$$

$$\nabla\times\mathbf{H} = \sigma\mathbf{E} + \varepsilon\frac{\partial\mathbf{E}}{\partial t} \tag{10.42}$$

In finite element and finite difference applications, these equations are modified by defining vector potentials which could then be used to remove the two variables and write the equation in terms of a single vector potential function. If, instead, we wish to solve Eqs. (10.41) and (10.42) directly, we must solve for three components of **E** and three components of **H** or a total of six unknown values. This is accomplished by writing the equations explicitly in terms of their components. From Eq. (10.41):

$$\begin{aligned}&\hat{\mathbf{x}}\left(\frac{\partial E_z}{\partial y}-\frac{\partial E_y}{\partial z}\right)+\hat{\mathbf{y}}\left(\frac{\partial E_x}{\partial z}-\frac{\partial E_z}{\partial x}\right)+\hat{\mathbf{z}}\left(\frac{\partial E_y}{\partial x}-\frac{\partial E_x}{\partial y}\right)\\&=-\left(\hat{\mathbf{x}}\mu\frac{\partial H_x}{\partial t}+\hat{\mathbf{y}}\mu\frac{\partial H_y}{\partial t}+\hat{\mathbf{z}}\mu\frac{\partial H_z}{\partial t}\right)\end{aligned} \tag{10.43}$$

$$\begin{aligned}&\hat{\mathbf{x}}\left(\frac{\partial H_z}{\partial y}-\frac{\partial H_y}{\partial z}\right)+\hat{\mathbf{y}}\left(\frac{\partial H_x}{\partial z}-\frac{\partial H_z}{\partial x}\right)+\hat{\mathbf{z}}\left(\frac{\partial H_y}{\partial x}-\frac{\partial H_x}{\partial y}\right)\\&=\hat{\mathbf{x}}\sigma E_x+\hat{\mathbf{y}}\sigma E_y+\hat{\mathbf{z}}\sigma E_z+\hat{\mathbf{x}}\varepsilon\frac{\partial E_x}{\partial t}+\hat{\mathbf{y}}\varepsilon\frac{\partial E_y}{\partial t}+\hat{\mathbf{z}}\varepsilon\frac{\partial E_z}{\partial t}\end{aligned} \tag{10.44}$$

Separating components (and assuming that ε, μ and σ, are not time dependent) we get the required six equations

$$\frac{\partial H_x}{\partial t} = \frac{1}{\mu}\left(\frac{\partial E_y}{\partial z}-\frac{\partial E_z}{\partial y}\right) \tag{10.45}$$

$$\frac{\partial H_y}{\partial t} = \frac{1}{\mu}\left(\frac{\partial E_z}{\partial x}-\frac{\partial E_x}{\partial z}\right) \tag{10.46}$$

$$\frac{\partial H_z}{\partial t} = \frac{1}{\mu}\left(\frac{\partial E_x}{\partial y} - \frac{\partial E_y}{\partial x}\right) \tag{10.47}$$

$$\frac{\partial E_x}{\partial t} = \frac{1}{\varepsilon}\left(\frac{\partial H_z}{\partial y} - \frac{\partial H_y}{\partial z} - \sigma E_x\right) \tag{10.48}$$

$$\frac{\partial E_y}{\partial t} = \frac{1}{\varepsilon}\left(\frac{\partial H_x}{\partial z} - \frac{\partial H_z}{\partial x} - \sigma E_y\right) \tag{10.49}$$

$$\frac{\partial E_z}{\partial t} = \frac{1}{\varepsilon}\left(\frac{\partial H_y}{\partial x} - \frac{\partial H_x}{\partial y} - \sigma E_z\right) \tag{10.50}$$

A typical finite difference cell used for implementation of this algorithm is shown in Figure 10.5. The spatial grid spacings are assumed to be equal: $\Delta x = \Delta y = \Delta z = h$. The time increment is Δt. Perhaps the most unique feature of implementation is that the cell is actually subdivided and use is made of half-cell distances. Also, as shown in Figure 10.5, on each side of the cell only those components that are related through one of the equations above are used. Thus, for example, on the top surface of the cell, the z component of the magnetic field intensity is related to the y and x components of the electric field intensity. This defines Eq. (10.47). The same is true of the lower surface. Thus, Eqs. (10.45) through (10.47) are defined on the outer surfaces of the cell while Eqs. (10.48) through (10.50) are defined on the cross sectional planes through the center of the cells.

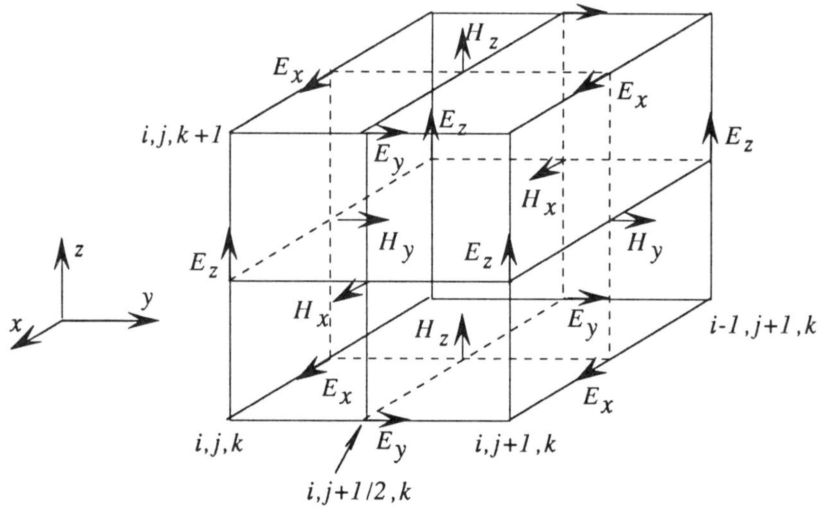

Figure 10.5 The basic cell for finite difference time domain implementation in three dimensions.

Using the cell in Figure 10.5, and the idea of the half spacing distance, approximation for derivatives are:

$$\frac{\partial E^m_{i,j,k}}{\partial x} = \frac{E^m_{i+1/2,j,k} - E^m_{i-1/2,j,k}}{h} \tag{10.51}$$

for the electric field and

$$\frac{\partial H^m_{i,j,k}}{\partial x} = \frac{H^m_{i+1/2,j,k} - H^m_{i-1/2,j,k}}{h} \tag{10.52}$$

with similar forms for other derivatives. In these relations, i, j, k, represent a point on the grid with coordinates x, y, z. m representing a single point in time. The cell shown in Figure 10.5 represents one point in time. Time derivatives are represented as

$$\frac{\partial E^m_{i,j,k}}{\partial t} = \frac{E^{m+1/2}_{i,j,k} - E^{m-1/2}_{i,j,k}}{\Delta t} \tag{10.53}$$

with similar forms for the magnetic field intensity. Using these expressions, Eqs. (10.45) through (10.50) are represented in the finite difference notation as

$$\begin{aligned} E^{m+1/2}_{xi+1/2,j,k} = E^m_{xi+1/2,j,k} + \frac{\Delta t}{\varepsilon_{i+1/2,j,k}h}\Big[H^{m+1/2}_{zi+1/2,j+1/2,k} - H^{m+1/2}_{zi+1/2,j-1/2,k} + \\ H^{m+1/2}_{yi+1/2,j,k-1/2} - H^{m+1/2}_{yi+1/2,j,k+1/2}\Big] - \frac{\sigma_{i+1/2,j,k}\Delta t}{\varepsilon_{i+1/2,j,k}} E^m_{xi+1/2,j,k} \end{aligned} \tag{10.54}$$

$$\begin{aligned} E^{m+1/2}_{yi,j+1/2,k} = E^m_{yi,j+1/2,k} + \frac{\Delta t}{\varepsilon_{i,j+1/2,k}h}\Big[H^{m+1/2}_{xi,j+1/2,k+1/2} - H^{m+1/2}_{xi,j+1/2,k-1/2} + \\ H^{m+1/2}_{zi-1/2,j+1/2,k} - H^{m+1/2}_{zi+1/2,j+1/2,k}\Big] - \frac{\sigma_{i,j+1/2,k}\Delta t}{\varepsilon_{i,j+1/2,k}} E^m_{yi,j+1/2,k} \end{aligned} \tag{10.55}$$

$$\begin{aligned} E^{m+1/2}_{zi,j,k+1/2} = E^m_{zi,j,k+1/2} + \frac{\Delta t}{\varepsilon_{i,j,k+1/2}h}\Big[H^{m+1/2}_{yi+1/2,j,k+1/2} - H^{m+1/2}_{yi-1/2,j,k-1/2} + \\ H^{m+1/2}_{xi,j-1/2,k+1/2} - H^{m+1/2}_{xi,j+1/2,k+1/2} - \frac{\sigma_{i,j,k+1/2}\Delta t}{\varepsilon_{i,j,k+1/2}} E^m_{zi,j,k+1/2} \end{aligned} \tag{10.56}$$

$$\begin{aligned} H^{m+1/2}_{xi,j+1/2,k+1/2} = H^{m-1/2}_{xi,j+1/2,k+1/2} + \frac{\Delta t}{\mu_{i,j+1/2,k+1/2}h}\Big[E^m_{yi,j+1/2,k+1} - E^m_{yi,j+1/2,k} + \\ E^m_{zi,j,k+1/2} - E^m_{zi,j+1,k+1/2}\Big] \end{aligned} \tag{10.57}$$

$$H_{yi+1/2,\,j,k+1/2}^{m+1/2} = H_{yi+1/2,\,j,k+1/2}^{m-1/2} + \frac{\Delta t}{\mu_{i+1/2,\,j,k+1/2}h}\Big[E_{zi+1,\,j,k+1/2}^{m} - E_{zi,\,j,k+1/2}^{m} + E_{xi+1/2,\,j,k}^{m} - E_{xi+1/2,\,j,k+1}^{m}\Big] \tag{10.58}$$

$$H_{zi+1/2,\,j+1/2,k}^{m+1/2} = H_{zi+1/2,\,j+1/2,k}^{m-1/2} + \frac{\Delta t}{\mu_{i+1/2,\,j+1/2,k}h}\Big[E_{xi+1/2,\,j+1,k}^{m} - E_{xi+1/2,\,j,k}^{m} + E_{yi,\,j+1/2,k}^{m} - E_{yi+1,\,j+1/2,k}^{m}\Big] \tag{10.59}$$

As with most previous finite difference approximations, these are explicit in time. Material properties are treated as in other approximations, as those at the point at which the approximation is evaluated.

Finally we mention that boundary conditions are implemented as for other *FDTD* methods but these must be written in terms of **H** or **E**, whichever is appropriate.

10.4 EXAMPLES

10.4.1 Scattering from conductors

The general *NDT* model in Figure 10.6 is used to demonstrate the two-dimensional formulation. The medium below the interface is a lossy dielectric ($\mu_o,\varepsilon_o,\sigma_c = 0.05\ S/m$). The foreign materials inside the test material are good conductors ($\mu_o,\varepsilon_o,\sigma_c = 10^7\ S/m$). The diameter of the cylinders is $L = 0.2\ m$. There are two cylinders inside medium 2. The numbers in brackets at the corners of the geometry represent mesh divisions. This application, although the scale is large, can be viewed as similar to that of modeling conducting fibers in a lossy matrix at microwave frequencies.

An oblique plane wave is incident on the test sample from free space. The incident wave penetrates into and reflects from the material surface. Figure 10.7 is the field contour plotting for $45°$ incidence with respect to the y-axis, with frequency $f = 4\ GHz$ and $\sigma = 0.05\ S/m$, without the scatterers. Figure 10.8 shows the wave surface plot for two conductors in the absence of the lossy dielectric. The scatterers are now introduced into the material. Figure 10.9 shows the wave surface with cylinders and lossy dielectric (cylinders embedded in the lossy dielectric), where the conductivity of the conducting cylinders is $10^7\ S/m$, and the incidence is the same as above. The wave is scattered from the surface and then propagates through free-space, after a refraction at the interface. The field distortion, due to the presence of foreign materials is detectable on the surface of the material. The corresponding field surface is given in Figure 10.10 where the scattered field due to the dielectric is subtracted from the total field in Figure 10.9. The solutions given here are typical of composite materials with fiber reinforcement. The scatterers modeled here are good conductors. This is not a necessary condition in order to apply the formulation.

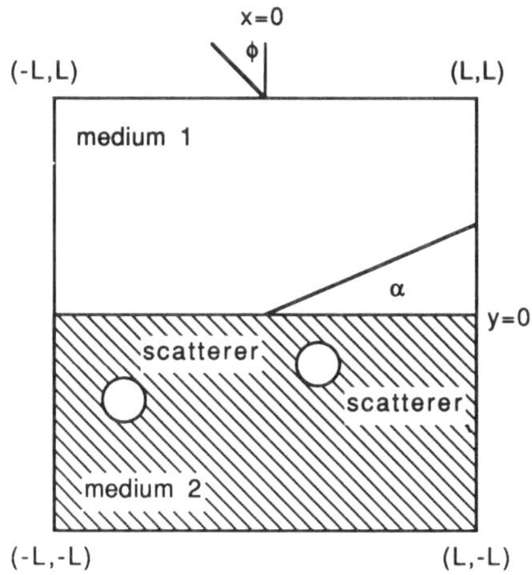

Figure 10.6 General geometry used for computation. The material above the interface is free space. Inclusions in the lossy dielectric below the interface are evaluated.

Figure 10.7 Wave penetration in the lossy dielectric in the absence of scatterers.

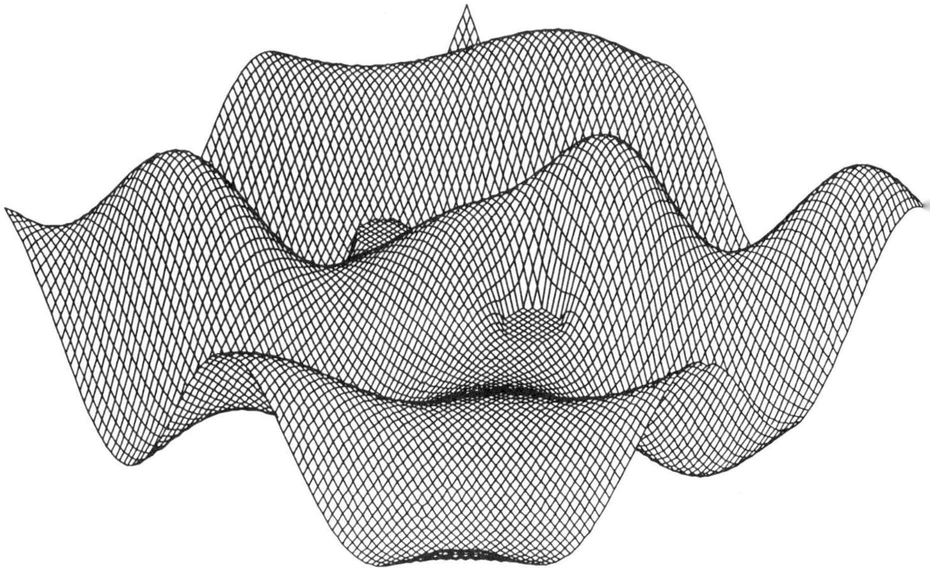

Figure 10.8 Wave propagation in free space. There are no scatterers or materials in the solution domain.

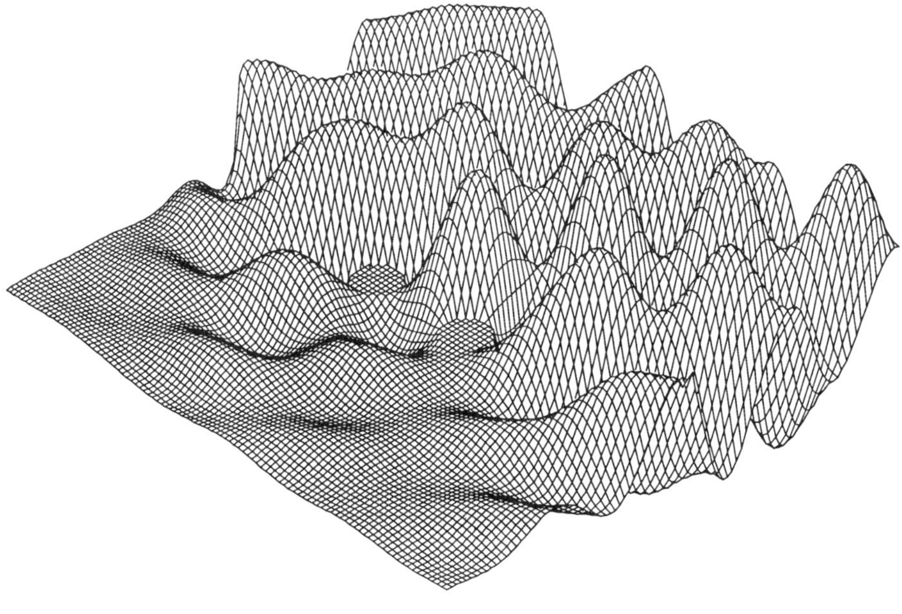

Figure 10.9 Wave surface plot in the presence of the dielectric half-space with four cylinders as shown in Figure 10.6.

Figure 10.10 Field due to the scatterers alone.

10.4.2 Testing of composite materials

Another example of a significant *NDT* problem at high frequencies is the interaction of electromagnetic fields produced by a current excited small loop with lossy and lossless materials. The problem is solved in the time domain in order to show the generality of the approach even though in most cases, the fields are time harmonic. The loop is considered to be very small so that the current distribution in the loop is uniform.

The problem considered here is axisymmetric in nature with the source in the solution domain. The source produces the incident field component and is modeled by the incident components of the magnetic vector potential at a group of grid points, which are near to, and enclose the source. This type of source specification is referred to at the beginning of the chapter as internal boundary conditions. For time-harmonic fields, A is a complex phasor. The incident magnetic vector potential produced by a small loop, carrying a spatially uniform current I, under time-harmonic excitation is given as

$$A_i \rightarrow \frac{I\pi a^2}{4\pi} e^{i(k_a r' - \omega t)} \left(\frac{jk_a}{r'} + \frac{1}{(r')^2} \right) sin\theta \qquad (10.60)$$

where r is the radius of the loop, r' is the distance of the observation point from the loop's center, ω is the angular frequency of the exciting current and θ is the angle which the position vector **r'** makes with the z-axis (that is, the inclination of the loop with respect to the surface of the test sample). On the axis, $\theta=0$ and therefore the condition $A=0$ is imposed. Figure 10.11 shows the various parameters involved in solution including grid divisions (shown at the corners of the domain). The composite material is assumed to be flat, large, and is $25mm$ thick. Although losses in the dielectric are assumed to be small (10^{-10} S/m), high losses can be analyzed equally well.

An example of the use of this geometry is testing of flat composite materials or any other lossy dielectric. A small loop carrying a current is located over a flat composite material. Application of the finite difference formulae given in Eqs. (9.74) through (9.80), provides the means of solution over this mesh.

The field distribution in Figure 10.12, clearly shows the effect of the slab on the field, including the guiding effect of the composite. The field is shown in an area of $75 \times 75cm$ while the blank area is the near field and is shown separately in Figure 10.13. Figures 10.14 shows changes in propagation as the relative permittivity of the material is changed from 3 to 6. Figure 10.15 shows propagation of waves generated by the small loop in free space.

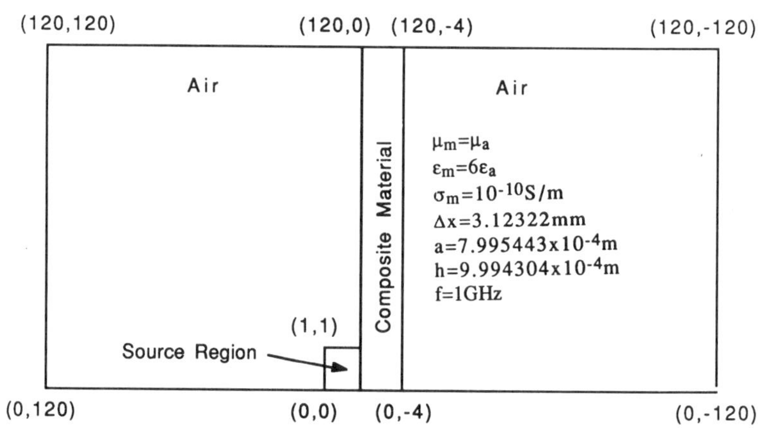

Figure 10.11 The axisymmetric geometry used for testing in composite materials using a small loop. Loop is shown in the inset.

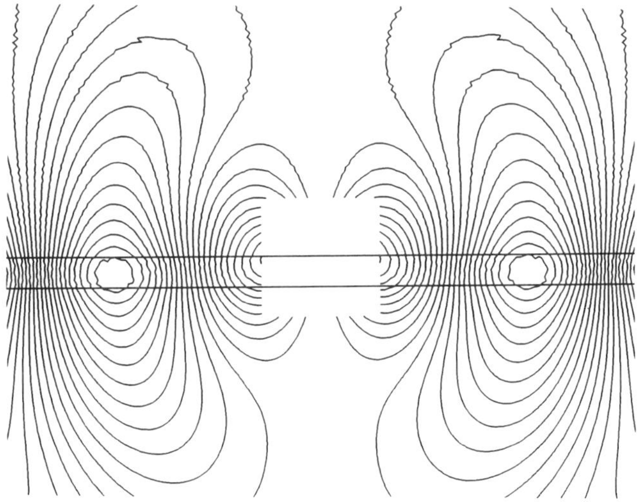

Figure 10.12 Field distribution over the slab (εr=3) showing propagation of waves along the slab. The blank area is the near field and is shown separately.

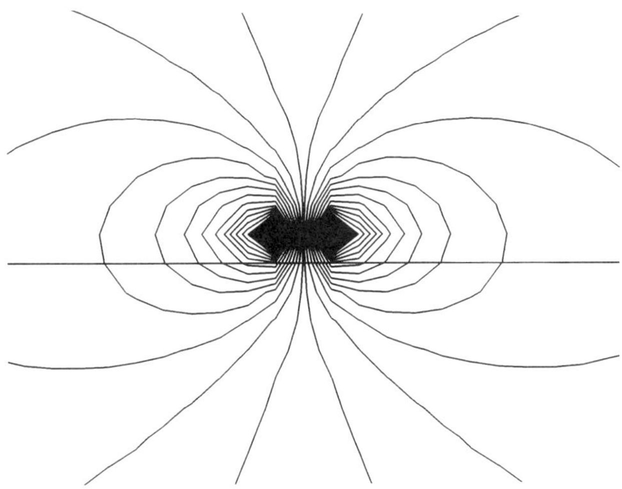

Figure 10.13 The near field for the solution in Figure 10.12. Note that propagation effects are not visible here.

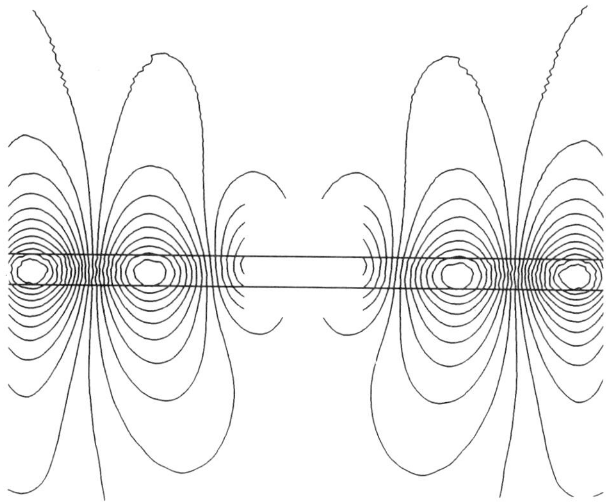

Figure 10.14 Field distribution over the slab in Figure 10.12 but with relative permittivity of 6.

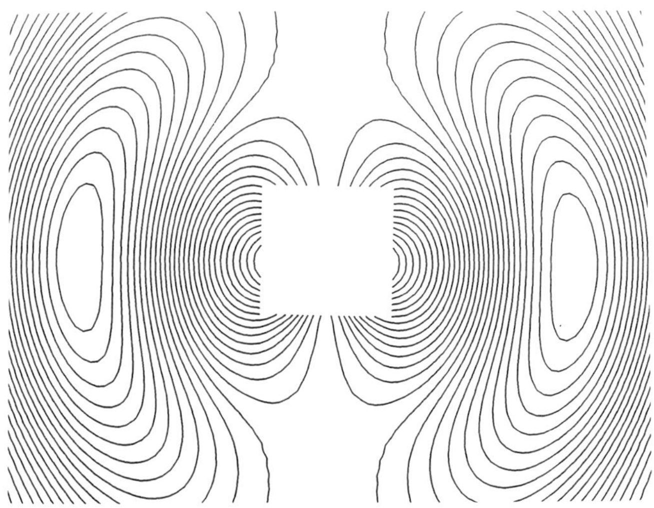

Figure 10.15 Field distribution over the slab in Figure 10.12 but with relative permittivity of 9.

10.5 PROBLEMS

10.1 Consider the one-dimensional problem in Figures 10.1 and 10.2. Assume that the initial condition is as shown in Figure 10.16. Calculate the field inside the solution domain and show that standing waves are obtained. For simplicity assume the solution domain is divided into *10* domains (*11* points).

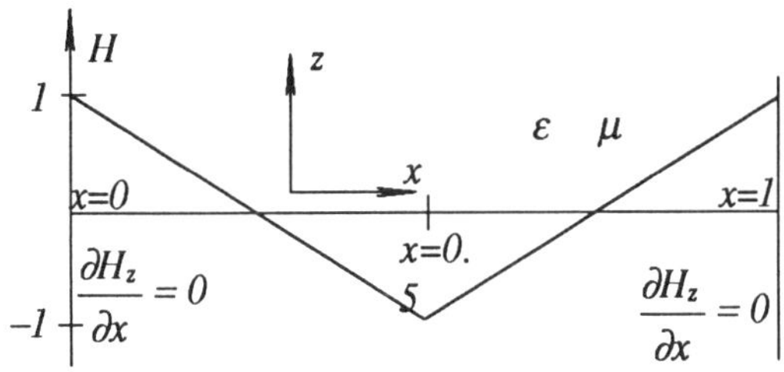

Figure 10.16.

10.2 Repeat problem 10.1 with the initial condition in Figure 10.17. How does the field inside the solution domain behave now?

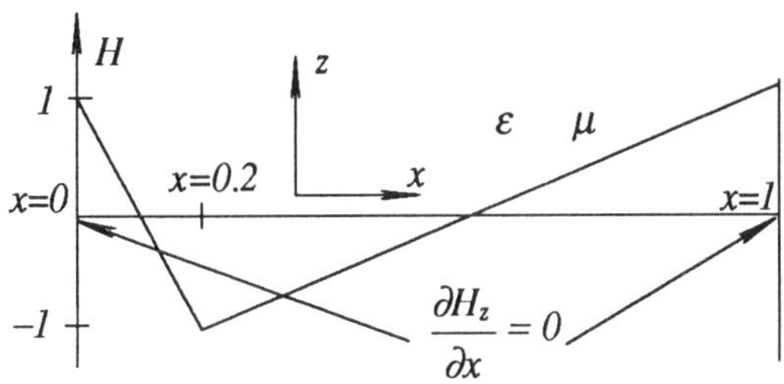

Figure 10.17.

10.3 Use the principle in section 10.2 to solve for the electric field rather than the magnetic field. The correct boundary conditions now are $E_z = 0$ on the boundary. Use the initial condition in Figure 10.18 to solve for the electric field everywhere. Plot the solution at a number of points in time to see that standing waves occur.

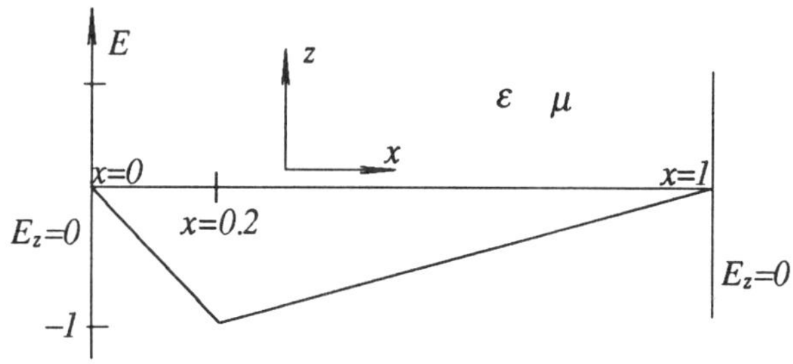

Figure 10.18.

10.4 Show that the method of section 10.2 can be easily extended to two dimensions. Assume a wave propagating in the z direction and varying in the x and y direction (such as Eq. (10.3)). Define a two-dimensional grid in space and one dimension in time and find the finite difference expression for solution. Perform the formulation for *TM* and for *TE* modes.

10.5 A rectangular waveguide, $10cm$ by $5cm$ is shown in Figure 10.19. Assuming that the initial condition is given as

$$E_z = 10 \sin(3\pi x/a) \sin(\pi y/b)$$

Calculate the electric field everywhere (this is a *TM* mode solution) in the cross section of the waveguide.

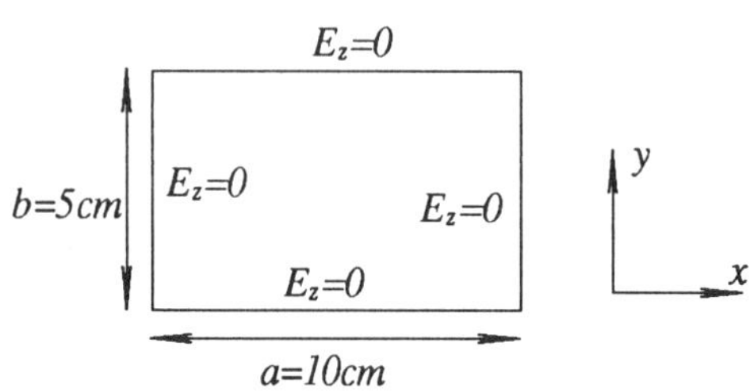

Figure 10.19.

10.6 Using steps similar to those in problem 10.4 extend the finite difference solution to three dimensions. Use the *TM* and *TE* equations (see section 1.20).

10.7 A current line carrying a sinusoidal current at *1 GHz* and amplitude *1 A* is located above a perfectly conducting plane as shown in Figure 10.20. The current line is thin and in free space. Using the *FDTD* method calculate the electric and magnetic fields everywhere in the vicinity of the line. Show (by plotting the field lines) that the wave from the line impinges on the surface of the conductor and is reflected from the surface.

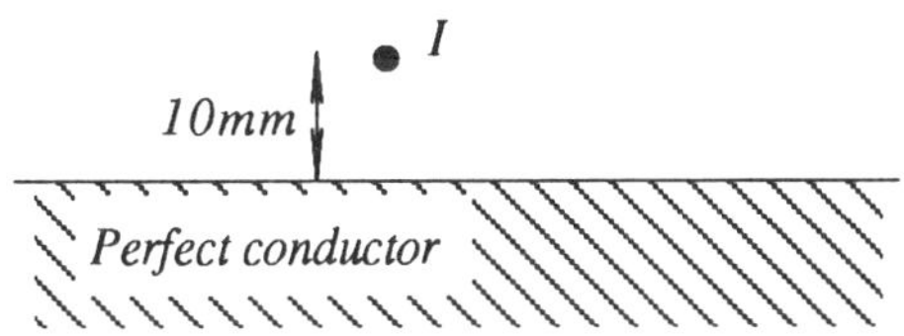

Figure 10.20.

10.8 In the geometry of problem 10.7, a slot is cut as shown in Figure 10.21. The conductor is changed to a dielectric with relative permittivity $\varepsilon_r = 4$. Calculate the field immediately above the surface of the dielectric (at dotted line) with and without the slot.

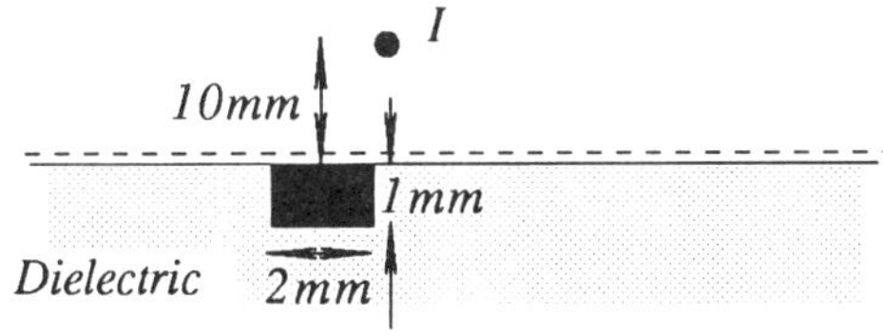

Figure 10.21.

10.6 BIBLIOGRAPHY

Finite difference time domain modeling based on potential functions is discussed extensively in references [1-5] including boundary conditions, implementation, and applications to *NDT*. Discussion of boundary conditions in general can be found in [6-10]. The finite difference time domain in *3-D* geometries can be found in [11-15]. The formula for the field of a small loop over a layer can be found in [16].

[1] Lee, M.E. (1989), *Potential Finite-Difference Time-Domain Methods for Electromagnetic Interface Problems*, Ph.D. Dissertation, The University of Akron.

[2] Lee M.E. and Ida, N. (1990), Axisymmetric electromagnetic fields coupled to lossy media, *IEEE Transactions on Magnetics*, Vol. 26, No. 2, pp. 575-578.

[3] Lee, M.E. Hariharan, S.I. and Ida, N. (1989), Transient calculations of two-dimensional eddy current problems, IEEE Transactions on Magnetics, Vol. 25, No. 4, pp. 3140-3144.

[4] Lee, M.E. and Ida, N. (1990), A transient axisymmetric formulation for modeling of composite materials, *Review of Progress in Quantitative Nondestructive Evaluation*, D. O. Thompson and D. E. Chimenti, Eds., Plenum Press, Vol. 9A, New York, pp. 295-302.

[5] Wang, J.S. Ida, N. and Hariharan, S.I. (1989), Numerical modeling of transient wave propagation for high frequency NDT, *Review of Progress in Quantitative Nondestructive Evaluation*, D. O. Thompson and D. E. Chimenti, Eds., Plenum Press, Vol. 8A, New York, pp. 259-266.

[6] Engquist, B. and Majda, A. (1977), Absorbing boundary conditions for the numerical simulation of waves, *Mathematics of Computation*, Vol. 31, pp. 629-651.

[7] Bayliss, A. and Turkel, E. (1980), Radiation boundary conditions for wave-like equations, *Communications of Pure and Applied Mathematics*, Vol. 33, No. 6, pp. 707-725.

[8] Friedlander, F.G. (1962), On the radiation field of pulse solutions of the wave equation, *Proceedings of the Royal Society*, London, Ser. A, Vol. 269, pp. 53-65.

[9] Mur, G. (1981), Absorbing boundary conditions for finite-difference approximation of the time-domain electromagnetic field equations, *IEEE Transactions on Electromagnetic Compatibility*, Vol. EMC-24, pp. 1073-1077.

[10] Atkinson, F.V. (1949), On Sommerfeld's radiation condition, *Philosophical Magazine*, Vol. XL, pp. 645-651.

[11] Wilcox, C.H. (1956) An expansion theorem for electromagnetic fields, *Communications of Pure and Applied Mathematics*, Vol. 9, pp. 115-134.

[12] Yee, K.S. (1966), Numerical solution of initial boundary value problems involving Maxwell's equations in isotropic media, *IEEE Transactions on Antenna and Propagation*, Vol. AP-14, No. 3, pp. 302-307.

[13] Taflove, A. (1980), Application of the finite difference time-domain method to sinusoidal steady state electromagnetic penetration problems, *IEEE Transactions on Electromagnetic Compatibility*, Vol. EMC-22, No.3, pp. 192-202.

[14] Taflove, A. and Umashankar, K.R. (1990), The finite-difference time-domain method for numerical modeling of electromagnetic wave interactions, *Electromagnetics*, Vol. 10, pp. 105-126.

[15] Christ, A. and Hartangel, H.L. (1987), Three-dimensional finite-difference method for the analysis of microwave-device embedding, *IEEE Transactions on Microwave Theory and Techniques*, Vol. MTT-35, No. 8, pp. 688-696.

[16] Wait, J.R. and Spies, K.R. (1972), Subsurface electromagnetic fields of a circular loop of current located above ground, *IEEE Transactions on Antennas and Propagation*, Vol. AP-20, pp. 520-522.

CHAPTER 11

Miscellaneous numerical methods

11.1 INTRODUCTION

This chapter is a collection of numerical methods required in the process of numerical solution of partial differential equations. Most of the methods were mentioned in the previous chapters but discussion of the methods has been left for this final, overview chapter. The main reason for this is that in many cases these rather fundamental methods are known. Also, while the methods are simple, there are many methods and techniques that can be used interchangeably and therefore, it is not possible to discuss all methods and variants possible.

The approach here is to discuss one or two fundamental methods for each numerical technique, to give its general algorithm, and to point to methods in order of their importance. Thus, some techniques are emphasized more than others, pointing to their perceived importance in conjunction with finite element and finite difference methods of solution.

The sequence of methods is also somewhat unique in that methods are described in order of their use in formulations. Thus, for example, numerical integration techniques are described first, followed by solution of systems of equations and then by eigenvalue problems. In most cases, direct methods are given first, followed by iterative methods.

11.2 NUMERICAL INTEGRATION

11.2.1 Gaussian quadrature: Gauss-Legendre integration

The need for numerical integration to find the elemental matrix contributions to the global matrix is introduced earlier in chapter 4. From the formulation of the isoparametric elements, integrals of the form

$$\int_{-1}^{+1}\int_{-1}^{+1}\int_{-1}^{+1} f(u,v,w)dudvdw \tag{11.1}$$

or

$$\int_{0}^{+1}\int_{0}^{+1}\int_{0}^{+1} f(u,v,w)dudvdw \tag{11.2}$$

arise, where the limits depend on the type of local coordinates used and, more often than not, are a matter of choice. These volume integrals are evaluated using Gaussian quadrature, which can be written as

$$Y = \int_{a}^{b} f(x)dx = K\sum_{\iota=1}^{n} W_i f(x_i) \tag{11.3}$$

where w_i are a set of special weights attached to the quadrature points x_i and K is a normalization constant. The weights and points are found in special tables but, for Gauss-Legendre quadrature are calculated from integrals of the form

$$\int_{-1}^{+1} f(x)dx$$

In order to use the Gauss quadrature with arbitrary limits (such as $a=0$ and $b=1$ in Eq. (11.2)) a new coordinate can be defined

$$\chi = \frac{2x-(b+a)}{b-1} \tag{11.4}$$

such that χ varies between -1 and $+1$ for x varying between a and b. Rewriting Eq. (11.4) for x:

$$x = \frac{1}{2}(b-a)\chi + \frac{1}{2}(b+a) \tag{11.5}$$

and therefore dx in Eq. (11.3) is

$$dx = \frac{1}{2}(b-a)d\chi$$

Substituting this into Eq. (11.3) gives

$$Y = \int_a^b f(x)\,\frac{b-a}{2}\,d\chi = \int_{-1}^{+1} F(\chi)\,\frac{b-a}{2}\,d\chi \tag{11.6}$$

Eq. (11.3) becomes:

$$\int_a^b f(x)dx = \frac{b-a}{2}\sum_{i=1}^{n} W_i f'(x_i) \tag{11.7}$$

where n is the number of quadrature points chosen for the integration. Note that for limits of integration $a=-1$, $b=+1$, the integration reduces to

$$\int_{-1}^{1} f'(x)dx = \sum_{i=1}^{n} W_i' f'(x_i') \tag{11.8}$$

while for limits $a=0$, $b=1$, the integration is

$$\int_{0}^{1} f'(x)dx = \sum_{i=1}^{n} W_i'' f'(x_i'') \tag{11.9}$$

and the weights and points are different in the three integrations. The weights and integration points in Eq. (11.8) are normally considered to be standard and are normally listed in tables. The weights and points in Eqs. (11.7) (for a general limit of integration) or in Eq. (11.9) are obtained from these standard values. Tables 11.1 and 11.2 show the weights and integration points for two values of n and for the limits of integration in Eq. (11.2). For the limits in Eq. (11.2) standard tables are available (see bibliography).

Table 11.1 Quadrature points and corresponding weights for two-point quadrature (in each direction in space).

Quadrature point number	Quadrature point x_i	Quadrature weight w_i
1	0.2113249	1.0
2	0.7886751	1.0

Table 11.2 Quadrature points and corresponding weights for three-point quadrature (in each direction in space).

Quadrature point number	Quadrature point x_i	Quadrature weight w_i
1	0.1127017	0.5555555556
2	0.5	0.8888888889
3	0.8872983	0.5555555556

This method, for a given number of quadrature points n, can integrate a polynomial of order $(2n-1)$ exactly. Thus for a linear element the minimum number of quadrature points is one and for a second order element at least two quadrature points are necessary. Both two and three point quadrature are used here but the best suited method in terms of economy and accuracy for field problems is the two-point quadrature (for linear and second order elements).

For two-point quadrature the integration over a three-dimensional hexahedral element is written as

$$\int_{-1}^{+1}\int_{-1}^{+1}\int_{-1}^{+1} f(u,v,w)dudvdw = \sum_{i=1}^{2}\sum_{j=1}^{2}\sum_{k=1}^{2} W_iW_jW_k f(u_i,v_j,w_k) \qquad (11.10)$$

where u_i are the quadrature points in the u direction, v_i are the quadrature points in the v direction and w_i are the quadrature points in the w direction. Similarly, for the three-point quadrature, the integral can be written as

$$\int_{-1}^{+1}\int_{-1}^{+1}\int_{-1}^{+1} f(u,v,w)dudvdw = \sum_{i=1}^{3}\sum_{j=1}^{3}\sum_{k=1}^{3} W_iW_jW_k f(u_i,v_j,w_k) \qquad (11.11)$$

The two-point quadrature requires a total of eight quadrature points in each element (two in each spatial direction) as shown schematically in Figure 11.1a. The three-point quadrature requires *27* quadrature points and is shown schematically in Figure 11.1b.

Integration between *0* and *1* is done in a similar manner but now the weights and points are taken from Table 11.2.

The quadrature points as used here and as listed in Table 11.1 are roots of Legendre polynomials. Thus the name Gauss-Legendre quadrature. The standard limits of integration between -1 and $+1$ are due to the Legendre polynomials which are defined in the range $[-1,+1]$. It is also possible to build other quadrature methods based on the same basic technique but with other polynomials. For example, Gauss-Laguerre quadrature allows integration between *0* and ∞, which is convenient in some instances.

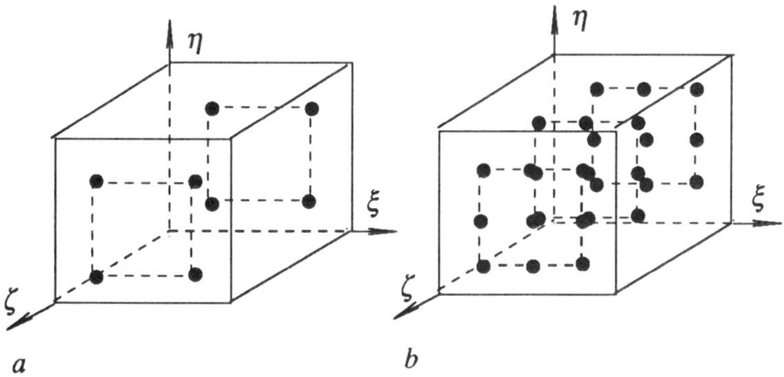

Figure 11.1 Quadrature points for a three-dimensional element. a. Two point quadrature. b. Three point quadrature.

11.3 NUMERICAL DIFFERENTIATION

Numerical differentiation is sometimes necessary to evaluate derived functions. An example is the calculation of the electric field from the potential in electrostatic calculations or the magnetic flux density from the magnetic vector potential in formulations that use the magnetic vector potential.

Normally only first order derivatives are necessary but sometimes second order derivatives need to be evaluated. For these purposes the finite difference techniques presented in chapter 3 may be used. Thus, for example, if the solution is known at a number of points in the solution domain, and derivatives with respect to a variable are needed, we can calculate the first order derivative as

$$\frac{dy_i}{dx} \approx \frac{y_{i+1} - y_{i-1}}{2h} \tag{11.12}$$

and the second order derivative as

$$\frac{d^2 y_i}{dx^2} \approx \frac{y_{i+1} - 2y_i + y_{i-1}}{h^2} \tag{11.13}$$

where central difference schemes are used in both cases. The values of the function, y, are viewed here as generic. If the derivatives with respect to x are required, h becomes Δx and y_{i+1}, y_{i-1} are values of the function in the x direction. Sometimes, especially at boundaries of a domain, only forward or backward difference formulae are practical. These are discussed in chapter 3.

This approach is most often associated with finite difference methods or whenever only the function at discrete points is defined. The calculations are straightforward but suffer from all problems associated with calculating derivatives of discrete functions including large errors in the calculation.

In conjunction with finite element calculation, in addition to knowledge of the nodal values, the approximation for the function and its derivatives is also given. This can then be used to calculate derivatives. As an example, the approximation of a function over a finite element (tetrahedral element in this case) is given (from Eq. (4.63)) as

$$\phi(x,y,z) = \sum_{i=1}^{4} L_i \phi_i$$

where ϕ_i is the value of the (unknown) function at node i and L_i is the shape function at the same node. The derivatives of the approximation with respect to x, y, and z, are written as

$$\frac{\partial \phi(x,y,z)}{\partial x} = \sum_{i=1}^{4} \frac{\partial L_i}{\partial x} \phi_i$$

$$\frac{\partial \phi(x,y,z)}{\partial y} = \sum_{i=1}^{4} \frac{\partial L_i}{\partial y} \phi_i$$

$$\frac{\partial \phi(x,y,z)}{\partial z} = \sum_{i=1}^{4} \frac{\partial L_i}{\partial z} \phi_i$$

where ϕ_i are the nodal values at the nodes of the element.

By substituting the nodal values for each element, the derivatives of the function are obtained. However, in the case of the tetrahedral element, the approximation for the function is linear and therefore the first order derivatives are constant within the element. Second order derivatives cannot be calculated. Second order derivatives can be calculated for quadratic or second order elements. This particular form of calculating derivatives has the advantage of defining the derivative everywhere within the element.

11.4 SOLUTION OF LINEAR SYSTEMS OF EQUATIONS

The finite element as well as the implicit finite difference formulations presented in previous chapters result in systems of linear equations of the form

$$[G]\{x\} = \{Q\} \tag{11.14}$$

where for the general eddy current case the system is complex and may be nonsymmetric. The system is also banded and diagonally dominant. The solution therefore follows simple, well-established procedures, at least in principle. One only has to apply an algorithm such as Gauss elimination and solve for $\{x\}$.

The problems with this straightforward approach are obvious when the size of the system of equations necessary for a *3-D* solution is considered. A medium sized finite element discretization could have some *2,000* nodes in the solution region, resulting in a system of *2,000* equations for a scalar potential formulation, *6,000* equations for a vector potential or field formulation or *8,000* equations for a combined vector potential/scalar potential formulation. Although the system is banded, the bandwidth is quite large and may easily exceed *1,000*. If the system is to be stored on a computer in the form of Eq. (11.14), between about *6* megawords to *36* megawords of memory would be needed for the storage of $[G]$ alone. Even though some computers can store systems of this type, it quickly becomes an unrealistic task. Assuming that only the coefficients in the band of a symmetric matrix ($N\times(BW+1)$), where N is the number of equations and BW is the semibandwidth - that is, the part of the bandwidth below or above the diagonal) are stored, the storage requirements for the above problem are reduced to between *3* to *18* megawords of memory (depending on bandwidth). This is small enough to solve on some computers but, in general, especially with systems much larger than the one described above, there is a real need to reduce memory requirements not only because of storage but also because of speed of execution. The following are the principal points that need consideration in the application of a computer to the solution of systems of equations resulting from finite element discretization of three dimensional applications. These considerations apply in two dimensions as well but the incentive to minimization of storage and execution time is much weaker in the two dimensional case because the problem size is considerably smaller.

1. Input data structure: The input data set of a finite element solution is large. It contains as a minimum nodes and elements data. Unless properly organized and minimized, large amounts of computer resources could be spent on this aspect alone. Proper input to the solution algorithm may also improve the speed such as is the case in reordering the nodes in the mesh to obtain a minimum bandwidth.
2. Storage: In the solution process, various data arrays have to be stored and manipulated. The fastest access to data is achieved if the data is available in memory. In most three dimensional cases it is impossible to store the finite element matrices in memory and it becomes necessary to use secondary storage such as magnetic disks and tapes. Within the secondary storage devices, one can use direct access or sequential files when both are applicable

(disks), to minimize access time (direct access files) or storage (sequential files). It is extremely important to use the computer memory as much as possible and to consider the tradeoff between solution time and storage. One more tradeoff in this context is between direct and iterative methods and is based on the fact that iterative methods require considerably less storage, especially for sparse matrices.

3. Solution time: For the most part this is a problem of tradeoff. If "infinite memory space" is available (in the sense that the problems foreseen can be accommodated) one can minimize the solution time independently. On computers with virtual memory it is usually possible to define arrays large enough to accommodate the problem at hand. This however does not necessarily improve matters as paging to and from the disk may not only slow the process but, in extreme cases, slow the computer to a standstill. Improvements of solution time must be considered in conjunction with storage schemes, the overall algorithm and the software of the available computer.
4. Algorithms: In many cases, more than one algorithm or combination of algorithms can be used. The choice is not always obvious as one may choose a faster algorithm if memory is available or a slower one to improve storage. Moreover, one algorithm may be faster on one class of computers but significantly slower on others. It is here however that most of the improvement can be made by choosing a proper algorithm and modifying it to take advantage of the particular computer and the properties of the problems to be solved. For example, one may want to reduce time spent on numerical integration by lowering the order of integration if the elements in the mesh are not significantly distorted, or one may want to consider only diagonal pivoting if the diagonal is known to be dominant.
5. Type of equations: The equations resulting from the finite element formulation can be linear or nonlinear, symmetric or nonsymmetric, and in some cases ill-conditioned. The treatment of each type of equation is different and must be considered not only for the type of equations solved but also for possible expansion in the future. Thus for example, optimization of an algorithm, suitable for solution of magnetostatic problems has little value if it cannot be expanded to accommodate eddy current or nonlinear problems if this is a possible future application. These considerations, in addition to ill-conditioning (either implicit in the formulation or by applying constraints to the system of equations) may well imply the use of a certain algorithm, overriding previous considerations.

In the light of the foregoing arguments it is clear that the solution of equations resulting from finite element formulations is a problem, and to a certain extent, computer dependent. A general solution routine for many classes of problems is certainly possible to develop but, when large problems need to be solved the efficiency of the program may need to be optimized even at the expense of generality and transportability. It must be remembered however that it is not the relative efficiency of different solution routines that is considered here but the ability to solve the problem. In other words, a program which is computer dependent is preferable to a universal program that cannot accommodate the size of problems that need to be solved.

The system of equations in Eq. (11.14) is very general and applicable to many situations in engineering and mathematics. The solution of sparse matrices, either symmetric or nonsymmetric, has received considerable attention both as a mathematical problem and as part of the finite element process. Many methods and variants for the solution of Eq. (11.14) exist but these can be classified in two major groups: direct and iterative methods. We describe next a number of the most commonly used direct and iterative methods but we emphasize here that iterative methods are almost always more efficient for large systems of equations (especially for sparse matrices) while for small systems, direct methods are usually faster. This division is not as clear-cut as it may seem because, for example, ill-conditioned systems are more readily solved using direct methods.

11.4.1 Solution of linear systems of equations: direct methods

Direct methods of solution for linear systems of equations lead to solution in a single step: after a prescribed and well-defined sequence of operations, a solution to the system of equations is obtained. This normally involves two general steps: an elimination step and a backsubstitution or solution step. Direct methods of solution are based on the Gauss elimination algorithm or its variants. The Gauss elimination algorithm is described next to form the basis for direct methods.

(a) The Gauss elimination method

To describe the Gauss elimination method we start with a general system of equations which may be symmetric or nonsymmetric, sparse or full, or banded.

$$\begin{bmatrix} a_{11}\, a_{12}\, a_{13}\, a_{14}\, a_{15} \cdots \\ a_{21}\, a_{22}\, a_{23}\, a_{24}\, a_{25} \cdots \\ a_{31}\, a_{32}\, a_{33}\, a_{34}\, a_{35} \cdots \\ a_{41}\, a_{42}\, a_{43}\, a_{44}\, a_{45} \cdots \\ \vdots \quad \vdots \quad \ddots \end{bmatrix} \begin{bmatrix} x_1 \\ x_2 \\ x_3 \\ x_4 \\ \vdots \end{bmatrix} = \begin{bmatrix} b_1 \\ b_2 \\ b_3 \\ b_4 \\ \vdots \end{bmatrix} \qquad (11.15)$$

At the first elimination stage the first line is multiplied by a_{i1}/a_{11} and subtracted from the corresponding terms of the ith row, where $i=1,2,3....n$. After this step the system is

$$\begin{bmatrix} a_{11} & a_{12} & a_{13} & a_{14} & a_{15} & \\ & a_{22}' & a_{23}' & a_{24}' & a_{25}' & \\ & a_{32}' & a_{33}' & a_{34}' & a_{35}' & \\ & a_{42}' & a_{43}' & a_{44}' & a_{45}' & \cdots \\ & & & \vdots & & \ddots \end{bmatrix} \begin{bmatrix} x_1 \\ x_2 \\ x_3 \\ x_4 \\ \vdots \end{bmatrix} = \begin{bmatrix} b_1 \\ b_2' \\ b_3' \\ b_4' \\ \vdots \end{bmatrix}$$

where the terms indicated by $(')$ are modified during the first elimination step as, for example,

$$a_{32}' = a_{32} - (a_{31}/a_{11})a_{12} \qquad\qquad b_2' = b_2 - (a_{21}/a_{11})b_1$$

The latter is quite important since it shows that elimination of the left hand side and that of the right hand side can be done independently. In applications in which repeated solutions such as for different probe positions or frequencies, the elimination of the matrix $[a]$ is done only once while the right hand side vector is eliminated for each probe position. Repeating the process, starting with the second row, the matrix, after the second elimination step looks like

$$\begin{bmatrix} a_{11} & a_{12} & a_{13} & a_{14} & a_{15} & \\ & a_{22}' & a_{23}' & a_{24}' & a_{24}' & \\ & & a_{33}'' & a_{34}'' & a_{35}'' & \\ & & a_{43}'' & a_{44}'' & a_{45}'' & \cdots \\ & & & \vdots & & \ddots \end{bmatrix} \begin{bmatrix} x_1 \\ x_2 \\ x_3 \\ x_4 \\ \vdots \end{bmatrix} = \begin{bmatrix} b_1 \\ b_2' \\ b_3'' \\ b_4'' \\ \vdots \end{bmatrix}$$

where, for example,

$$a_{33}'' = a_{33}' - (a_{32}'/a_{22}')a_{23}'$$

$$b_3'' = b_3' - (a_{32}'/a_{22}')b_2'$$

and the superscripts show that these coefficients have been modified for the second time.

Continuing this procedure up to the last row, we obtain an upper triangular system. This modified system is denoted $[U]\{x\}=\{c\}$ in which the modified coefficients on the right hand side are denoted $\{c\}$ to distinguish them from the original right hand side vector.

$$\begin{bmatrix} u_{11} & u_{12} & u_{13} & & & & \\ & u_{22} & u_{23} & u_{24} & & & \\ & & u_{33} & u_{34} & u_{35} & & \\ & & & \cdot & & & \\ & & & & \cdot & & \\ & & & & & \cdot & \\ & & & & & u_{n-1,n-1} & u_{n-1,n} \\ & & & & & & u_{n,n} \end{bmatrix} \begin{Bmatrix} x_1 \\ x_2 \\ x_3 \\ \cdot \\ \cdot \\ \cdot \\ x_{n-1} \\ x_n \end{Bmatrix} = \begin{Bmatrix} c_1 \\ c_2 \\ c_3 \\ \cdot \\ \cdot \\ \cdot \\ c_{n-1} \\ c_n \end{Bmatrix} \quad (11.16)$$

The system is now solved, starting from the last row. First $x_n=c_n/u_{n,n}$ is calculated. With this the unknown x_{n-1} is obtained as

$$x_{n-1} = \frac{c_{n-1} - u_{n-1,n}x_n}{u_{n-1,n-1}}$$

The values for x_n and x_{n-1} are now substituted in the unknown vector and x_{n-2} obtained in the same manner. This is the back-substitution step in the Gauss elimination method. The Gauss elimination method can be summarized in the following algorithm:

Elimination:

$$a_{ij}^{k+1} = a_{ij}^{k} - a_{ik}^{k}\left[a_{kk}^{k}\right]^{-1} a_{kj}^{k} \qquad i,j>k \qquad (11.17)$$

$$b_i^{k-1} = b_i^{k} - a_{ik}^{k}\left[a_{kk}^{k}\right]^{-1} b_k^{k} \qquad i>k \qquad (11.18)$$

which expresses the operations performed at the kth step, $k=1,2,...N-1$. After this step, known as the elimination step, the system of equations is triangular and the solution is obtained by backsubstitution. The general step in backsubstitution is:

$$x_k = \left[\overset{k}{a}_{kk}\right]^{-1}\left(\overset{k}{b}_k - \sum_{j=k+1}^{N} \overset{k}{a}_{kj} x_j\right) \qquad k=N, N-1, \ldots\ldots, 1 \qquad (11.19)$$

A few general points are in order here since these apply to most direct methods:

1. It is important to note that direct methods modify the original matrix *[A]*. If the original matrix is required, it must be stored separately.
2. The number of operations required for elimination is of the order of N^3 where N is the number of equations.
3. If the matrix is banded, the limits on the algorithm in Eq. (11.17) is the bandwidth rather than the number of equations N. This is self-understood because beyond the bandwidth the coefficients are zero and are not modified in the elimination process.
4. For sparse matrices, most coefficients that are zero may become nonzero in the process.

(b) The Gauss-Jordan elimination method

A simple modification to the method in the previous section allows solution without the need for backsubstitution. This may not seem important since the backsubstitution process is rather simple. However in some parallel solution algorithms, the backsubstitution process is very slow and methods that do not require backsubstitution are more efficient.

The Gauss-Jordan elimination method is based on the simple fact that the elimination steps, which eliminates all coefficients below the diagonal, can be used equally well to eliminate the coefficients in the row (that is, above the diagonal). At the end of the first elimination step, the matrix becomes:

$$\begin{bmatrix} a_{11} & & & & & \\ & a_{22}' & a_{23}' & a_{24}' & a_{25}' & \\ & a_{32}' & a_{33}' & a_{34}' & a_{35}' & \\ & a_{42}' & a_{43}' & a_{44}' & a_{45}' & \cdots \\ & & & \vdots & & \ddots \end{bmatrix} \begin{bmatrix} x_1 \\ x_2 \\ x_3 \\ x_4 \\ \vdots \end{bmatrix} = \begin{bmatrix} b_1 \\ b_2' \\ b_3' \\ b_4' \\ \vdots \end{bmatrix}$$

while at the end of the elimination process, the system is:

$$\begin{bmatrix} a_{11} & & & & \\ & a_{22}' & & & \\ & & a_{33}'' & & \\ & & & a_{44}''' & \cdots \\ & & & \vdots & \ddots \end{bmatrix} \begin{bmatrix} x_1 \\ x_2 \\ x_3 \\ x_4 \\ \vdots \end{bmatrix} = \begin{bmatrix} b_1 \\ b_2' \\ b_3' \\ b_4' \\ \vdots \end{bmatrix} \tag{11.20}$$

The solution process now is simply one of dividing the right hand side by the diagonal or, if the diagonals have been normalized, the right hand side is the solution.

(c) The *LU* decomposition

Another general method of solution is based on decomposition of the matrix into equivalent systems. One such method is the *LU* decomposition. The method is based on the fact that a system of equations can be decomposed into the following

$$[A] = [L][U] \tag{11.21}$$

where $[L]$ is the lower triangular matrix and $[U]$ is an upper triangular matrix. These matrices have the form:

$$[L] = \begin{bmatrix} l_{11} & 0 & 0 & \cdots \\ l_{21} & l_{22} & 0 & \\ l_{31} & l_{32} & l_{33} & \cdots \\ & \vdots & & \ddots \end{bmatrix} \qquad [U] = \begin{bmatrix} 1 & u_{12} & u_{13} & \cdots \\ 0 & 1 & u_{23} & \\ 0 & 0 & 1 & \cdots \\ & \vdots & & \ddots \end{bmatrix}$$

The solution based on this method is in two steps. Writing the decomposed system:

$$[A]\{x\} = [L][U]\{x\} = \{b\} \tag{11.22}$$

The first step calculates an intermediate result $\{c\}$:

$$[L]\{c\} = \{b\} \tag{11.23}$$

In the second step, $\{c\}$ is used to calculate $\{x\}$:

$$[U]\{x\} = \{c\} \tag{11.24}$$

To perform the decomposition, the following algorithms may be used

$$l_{ij} = a_{ij} - \sum_{m=i}^{j-1} l_{jm}u_{im} \qquad j=2,....,n; \quad i=j,j+1,....,n$$

$$u_{ij} = \frac{a_{ij} - \sum_{m=1}^{i-1} l_{jm}u_{mj}}{l_{ii}} \qquad i=2,....,n; \quad j=i+1,i+2....,n$$

and, for the first row/column:

$$l_{i1} = a_{i1} \qquad i=1,2,....,n; \quad j=1$$

$$u_{ij} = \frac{a_{1j}}{a_{11}} \qquad j=1,2,....,n; \quad i=1$$

The same operations as for the l coefficients are performed on the right hand side of the system $\{b\}$.

(d) Choleski's method

One of the most commonly employed methods is the Choleski decomposition method, for symmetric matrices. It is based on the principle of the $[L][U]$ decomposition above but, because the matrix $[A]$ is symmetric, the upper matrix $[U]$ is the transpose of the lower matrix and therefore can be replaced by $[L]^T$:

$$[A] = [L][L]^T \tag{11.25}$$

where $[L]$ is a lower triangular matrix (including the diagonals).

The system $[A]\{x\}=\{b\}$ is now written in the form:

$$[L][L]^T\{x\} = \{b\} \tag{11.26}$$

The two steps given above are now:

$$[L]\{c\} = \{b\} \tag{11.27}$$

$$[L]^T\{x\} = \{c\} \tag{11.28}$$

Eq. (11.27) has the following form:

$$\begin{bmatrix} l_{11} & & & 0 \\ l_{21} & l_{22} & & \\ l_{31} & l_{32} & l_{33} & \\ & \vdots & & \\ & \vdots & & \end{bmatrix} \begin{bmatrix} c_1 \\ c_2 \\ c_3 \\ \vdots \\ \vdots \end{bmatrix} = \begin{bmatrix} b_1 \\ b_2 \\ b_3 \\ \vdots \\ \vdots \end{bmatrix}$$

from which the values of $\{c\}$ are obtained by what might be called a forward solution as $c_1=b_1/l_{11}$; and similarly, $c_2=(b_2-l_{21}c_1)/l_{22}$, etc. The rest of the terms are calculated similarly.

After calculating vector $\{c\}$, Eq. (11.28) is used to solve for $\{x\}$:

$$\begin{bmatrix} l_{11} & l_{21} & l_{31} & & & \\ & l_{22} & l_{32} & l_{42} & & \\ & & \cdot & & & \\ & & & \cdot & & \\ & & & & \cdot & \\ & & & & l_{n-1,n-1} & l_{n-1,n} \\ & & & & & l_{n,n} \end{bmatrix} \begin{bmatrix} x_1 \\ x_2 \\ x_3 \\ \cdot \\ \cdot \\ x_{n-1} \\ x_n \end{bmatrix} = \begin{bmatrix} c_1 \\ c_2 \\ c_3 \\ \cdot \\ \cdot \\ c_{n-1} \\ c_n \end{bmatrix}$$

from this $\{x\}$ is obtained, using a process similar to the back-substitution process in the Gauss elimination method.

The terms l_{ij} of matrix $[L]$ are given by the expressions:

$$l_{11} = (a_{11})^{1/2} \qquad l_{ii} = \left(a_{ii} - \sum_{j=i}^{i-1} l_{ij}^2 \right)^{1/2} \qquad i=2,\ldots..,n \qquad (11.29)$$

$$l_{ij} = \frac{1}{l_{jj}} \left(a_{ij} - \sum_{m=i}^{j-1} l_{jm} l_{im} \right) \qquad i=2,\ldots..,n; \quad j=i,i+1,\ldots..,n \qquad (11.30)$$

and, for all coefficients below the diagonal, we have

$$l_{ij} = 0 \qquad i<j$$

The Choleski method requires fewer operations than the Gauss elimination method and therefore is faster but it is not applicable to non-symmetric matrices. For nonsymmetric matrices the $[L][U]$ decomposition method must be used.

There are other direct methods and they are generally, derived from the methods described above. These methods have specific advantages for specific problems. However, the methods presented here are most commonly used in finite element programs because of their simplicity and good performance.

As with the Gauss elimination, when used in conjunction with banded or sparse matrices, the limits on elimination must be changed accordingly. However this is a minor point that only affects speed of execution and storage requirements and, although a very important point it is not fundamental to the understanding of the method.

11.4.2 Solution of linear systems of equations: iterative methods

Iterative methods are very useful for the solution of linear systems of equations whenever the number of equations is very large and the storage of the matrix as a sparsely populated matrix (either a matrix with a relatively small bandwidth or otherwise a sparse matrix) is possible. Iterative methods can also be used with dense systems but the advantage in storage of coefficients is lost in this case and often, direct methods are faster for solution of dense matrices.

Iterative methods have two basic advantages for solution of systems of equations of the type discussed in this work. First, only nonzero coefficients of the system are stored, creating a significant advantage for application of the method to sparse systems of equations. Second, the iteration can be used to improve results and minimize the effects of computation errors such as roundoff.

However, iterative methods have significant disadvantages. The number of iterations may in some cases be very large resulting in a slow solution. Also, and perhaps more importantly, convergence of the iterative process is related to the condition of the matrix. Ill-conditioned matrices may not converge to a solution while the worse the condition of the matrix the slower the solution.

We describe here a number of iterative methods including the general iterations of Jacobi and Seidel as well as the conjugate gradient method. The latter is emphasized because it is an efficient method for solution of systems resulting from finite element applications.

A general iterative technique can be defined as follows, assuming that Eq. (11.14) has an exact solution:

$$\{x\} = [G]^{-1}\{Q\} \tag{11.31}$$

A general iterative method is defined as

$$\{x\}^{k+1} = \{x\}^k + \phi\{\Delta x\}^k \tag{11.32}$$

where k is the iteration number. An iterative algorithm can be built through any method that allows evaluation of a new value for $\{x\}$ based on the old value for $\{x\}$. It is also possible to show a general method of deriving iterative methods based on decomposition of matrices. Since we are only interested here in simple methods we do not follow this general path but rather, give the algorithms for the Gauss-Jacobi and Gauss-Seidel methods.

(a) Gauss-Jacobi and Gauss-Seidel methods

These two methods are the simplest iterative methods and may be used in many applications. They are presented here mainly as an introduction to iterative methods.

The solution starts by assuming an initial solution to the system of equations in Eq. (11.14). The initial solution can be either an estimate to the solution or, if an estimate is not available, an assumed solution. The iteration starts by assuming all values of x are known, except the current value being calculated. Thus, for example, using Eq. (11.31) we can write for x_2:

$$x_2^{1} = \frac{b_2 - a_{21}x_1^{0} - a_{23}x_3^{0} - a_{24}x_4^{0} - \ldots\ldots - a_{2n}x_n^{0}}{a_{22}} \tag{11.33}$$

In this relation, the superscript indicates the iteration number. In Eq. (11.33), x_2 at iteration number *1* is calculated using values, $x_1, x_3, x_4, \ldots\ldots, x_n$ at iteration number *0*. Thus, generalizing the operation in Eq. (11.33) we can write for x_j at the $k+1$ iteration

$$x_j^{k+1} = \frac{b_j - \sum_{i=1}^{j-1} a_{ji}x_i^{k} - \sum_{i=j+1}^{n} a_{ji}x_i^{k}}{a_{jj}} \tag{11.34}$$

This is the general iteration for the Gauss-Jacobi method. The main point is that all values of x on the right hand side are values calculated at the previous iteration while on the left hand side, all values are at the current iteration.

The Gauss-Seidel method is a slight modification of the Gauss-Jacobi method. Since all values of x with subscripts lower than j have been calculated in the current iteration, and since these are better estimates to the final solution, they can be used in the calculation of x_j. The Gauss-Seidel iteration is therefore

$$x_j^{k+1} = \frac{b_j - \sum_{i=1}^{j-1} a_{ji}x_i^{k+1} - \sum_{i=j+1}^{n} a_{ji}x_i^{k}}{a_{jj}} \tag{11.35}$$

A few properties of these algorithms that are important for applications in the type of systems of equations encountered in this work are:

1. The coefficients of the system of equations are not modified. This means that these are available at any stage for other purposes.
2. Ill-conditioned matrices cannot in general, be solved unless modified.
3. The solution usually converges to the required solution in a finite number of iterations. The number of iterations may be large and also depends on how accurate the final solution need be.
4. In Eqs. (11.34) and (11.35), the sums have nonzero contributions only if the corresponding coefficients are nonzero. This has far-reaching consequences for solution. It means that only nonzero coefficients of the system need be stored, in contrast to direct methods in which all coefficients in the bandwidth of the matrix must be stored.

In addition to the algorithms for solution we must also include a method for stopping the solution. This was also discussed in chapter 3 in conjunction with explicit methods of solution of finite difference equations. Again, there are a number of methods that can be used. One method is to calculate the average change in solution and require that this be smaller than a preset tolerance:

$$\frac{1}{n}\sum_{i=1}^{n}\left|x_i^k - x_i^{k-1}\right| \le \varepsilon \qquad (11.36)$$

The tolerance in Eq. (11.36) must be chosen carefully. A tolerance that is too large will allow fast convergence but the resulting solution vector may not be accurate enough. If ε is too small, too many iterations may be required or the solution may not converge.

Finally, we note that these methods apply equally well to nonsymmetric matrices but, depending on the condition of the matrix, may not converge at all. As a rule, this does not happen with finite difference and finite element matrices.

(b) Relaxation methods

Relaxation methods are methods of accelerating the solutions obtained by any of the iterative methods. The principle is to weigh the solution so that a faster solution process is obtained. To see how a relaxation method works, assume that a solution at iteration $k+1$ has been obtained and that the solution at iteration k is also known. Using these two solutions we can write the change in solution as

$$\Delta x_i^{k+1} = x_i^{k+1} - x_i^k \qquad (11.37)$$

This change is multiplied by a weighting factor $(\Omega-1)$ and added to the original value of x_i^{k+1} as

$$x_i^{k+1} = x_i^k + (\Omega - 1)\Delta x_i^{k+1} \tag{11.38}$$

If $\Omega > 1$ the procedure is called "over-relaxation"; if $\Omega = 0$ there is no relaxation and if $\Omega < 1$ we have "under-relaxation". Generally, we use $1.5 < \Omega < 2$, for this type of matrix, although the choice of Ω depends on the specific problem. The use of over-relaxation or what is often called successive over-relaxation, is very simple and can be used with either method described above. Even though the relaxation factor is not fixed and must be found by experimentation it can increase solution speed considerably.

(c) The conjugate gradient method

To solve the system $[A]\{x\}=\{b\}$ the conjugate gradient algorithm (CG) can be used. This method is relatively new but has evolved into a highly useful method for solution of symmetric and nonsymmetric systems of equations. We only show here the basic CG method and its more useful preconditioned counterpart. However, there are many variants to the method, including algorithms for nonsymmetric and complex matrices.

The conjugate gradient method is based on minimization of a function in an n-dimensional space. In formal terms, an error function is created and this error is minimized in a number of steps. In gradient methods, minimization is performed using direction vectors in the direction of maximum change (that is, the gradient) in the function. In the conjugate gradient method, two vectors are chosen but, in addition to the requirement that they approach the exact solution (minimum in the error function) along the gradient, the two vectors must also be orthogonal to each other. The following is the CG algorithm.
Initialization:

$$r_0 = b - Ax_0 \qquad p_0 = r_0 \tag{11.39}$$

This defines the two direction vectors needed for progression of the iteration. Note that r_0 is the residual since x_0 is assumed to be known.
Conjugate gradient iteration: for iteration k:

$$\alpha_k = \frac{r_k \cdot r_k}{(p_k A p_k)} \tag{11.40}$$

Now update the solution vector and calculate the updated direction vectors

$$x_{k+1} = x_k + \alpha_k A p_k \tag{11.41}$$

We note that this equation represents a line that connects the previous solution vector to the current solution vector and therefore, α_k represents the slope, or the gradient in solution. Thus, the new direction vector is defined as

$$r_{k+1} = r_k - \alpha_k A p_k \qquad (11.42)$$

$$\beta_k = \frac{r_{k+1} \cdot r_{i+1}}{r_k \cdot r_i} \qquad (11.43)$$

$$p_{k+1} = p_{k+1} + \beta_k p_k \qquad (11.44)$$

In addition, we must insert a check to identify the convergence of the solution. This is done between Eq. (11.41) and Eq. (11.42). Test of convergence is done as for any iterative method, normally by comparing the current solution and previous solution (see for example Eq. (11.36).

The solution of a system of equations using the conjugate gradient method converges to the exact solution in a maximum of N steps where N is the number of equations in the system. Because of roundoff errors, the number of iterations may be higher. In practice, and depending on the condition of the matrix, the number of iterations is much smaller than N. For finite element systems, the number of iterations is about $0.1N$ to $0.2N$.

(d) The Incomplete Choleski conjugate gradient method

The incomplete Choleski conjugate gradient (*ICCG*) method, uses the conjugate gradient (*CG*) method with preconditioning of the matrix $[A]$. Preconditioning is based on the fact that if the matrix $[A]$ and the vector $\{b\}$ are multiplied by a matrix $[C]$, the solution of the system is not altered provided $[C]$ is of the same order as $[A]$. For a system:

$$[A]\{x\} = \{b\}$$

the preconditioned matrix is

$$[C][A]\{x\} = [C]\{b\} \qquad (11.45)$$

If the matrix $[C]$, when multiplied by $[A]$, generates a product $[C][A]$ close to the unity matrix, the preconditioned matrix is better adapted to the solution by an iterative method since in the limit, if the preconditioning operation produces a unity matrix, the solution is obtained in one iteration. Therefore, if the matrix $[C]$ is equal to $[A]^{-1}$, preconditioning results in a unity matrix and the solution converges in one iteration. However, finding the inverse of $[A]$ is equivalent to solving the system and this preconditioning matrix is of little use.

A widely used preconditioning matrix is the product $[[L][L]^T]^{-1}$ where $[L]$ is obtained from the Choleski method, presented above. However, only the terms of $[[L][L]^T]^{-1}$ of the existent nonzero coefficients of $[A]$ are calculated. This results in an incomplete approximation of $[[L][L]^T]^{-1}$, and hence the name of the method. The method is valid for any symmetrical matrix $[A]$.

The incomplete Choleski decomposition is calculated using Eqs. (11.29) and (11.30). After the matrix $[[L][L]^T]^{-1}$ is obtained, the following algorithm is applied. This is essentially the same algorithm as in Eqs. (11.39) through (11.44) but with modifications to the direction vector p as shown.
Initialization:

$$r_0 = b - Ax_0 \qquad p_0 = (LL^T)^{-1} r_0 \tag{11.46}$$

Conjugate gradient iteration. For iteration k:

$$\alpha_k = \frac{(r_k(LL^T)^{-1} r_k)}{(p_k A p_k)} \tag{11.47}$$

$$x_{k+1} = x_k + \alpha_k p_k \tag{11.48}$$

$$r_{k+1} = r_k - \alpha_k A p_k \tag{11.49}$$

$$\beta_k = \frac{(r_{k+1}(LL^T)^{-1} r_{k+1})}{(r_k(LL^T)^{-1} r_k)} \tag{11.50}$$

$$p_{k+1} = (LL^T)^{-1} r_{k+1} + \beta_k p_k \tag{11.51}$$

This method (or one of its many variations) is considered to be the most efficient iterative method for solving matrix systems generated by the finite element method for electromagnetic problems. The increase in memory needs compared to relaxation methods is largely justified by its efficiency. However, it must be noted that the calculation of $[[L][L]^T]^{-1}$, even though it is an incomplete decomposition is lengthy and consumes most of the time needed for solution.

11.4.3 Solution of linear systems of equations: other methods

The subject of solution of large systems of equations has received considerable attention because of its importance in numerical methods. While the methods described above are perhaps most often used, there are other methods which are used and which have special properties that make them suitable for specific applications. However, it is not reasonable to summarize here all methods and to pick a few would indicate a preference. Therefore this large number of methods are left out (see bibliography section).

11.5 SOLUTION OF NONLINEAR SYSTEMS OF EQUATIONS

To solve a system of nonlinear equations, we must establish an iterative procedure that allows approximation of the solution based on known values of the solution.

Consider the following system of equations:

$$\begin{bmatrix} a_{11}(x_1,x_2) & a_{12}(x_1,x_2) \\ a_{21}(x_1,x_2) & a_{22}(x_1,x_2) \end{bmatrix} \begin{Bmatrix} x_1 \\ x_2 \end{Bmatrix} = \begin{Bmatrix} q_1 \\ q_2 \end{Bmatrix} \tag{11.52}$$

This is a nonlinear system of equations since, to assemble the matrix $[a]$, we must first know the solution vector $\{x\}$. This of course cannot be done by standard methods of solution. Rather, the process is iterative. In this process we assume that an initial solution is known, say $\{x_0\}$. This initial solution is then used to assemble the matrix $\{a\}$. The system of equations now is:

$$\begin{bmatrix} a_{11}(x_{10},x_{20}) & a_{12}(x_{10},x_{20}) \\ a_{21}(x_{10},x_{20}) & a_{22}(x_{10},x_{20}) \end{bmatrix} \begin{Bmatrix} x_1 \\ x_2 \end{Bmatrix} = \begin{Bmatrix} q_1 \\ q_2 \end{Bmatrix} \tag{11.53}$$

From this, a new solution vector $\{x_1\}$ is obtained and the process repeated until the new solution $\{x_k\}$ and the previous solution $\{x_{k-1}\}$ are within a required tolerance.

The method just described is a successive approximation method and may be summarized as follows:

1. Assume an initial solution vector (or, if an approximate solution is known, this may be used instead).
2. Form the system matrix based on this approximate solution. This includes matrix $[a]$ and vector $\{q\}$ if both are dependent on the solution vector $\{x\}$.
3. Solve the system of equations.
4. Compare the solution vector with the previous solution vector to see if the solution has converged to the final solution.
5. Repeat steps 2 to 4 until the required tolerance in solution is obtained.

This process indicates two important aspects of the solution of nonlinear equations:

1. The solution at every iteration is that of a linear system of equations since the dependencies in the matrix are known.
2. Although we assumed that the correct final solution is eventually obtained, this is not always the case. As we might suspect, the solution may not

converge to the correct solution and in fact, convergence may be dependent on our initial solution.

We also note that the particular iteration we chose here is equivalent to the Gauss-Jacobi iteration described in section 11.4.2.1.

11.5.1 The Newton-Raphson method

A faster method for solution of nonlinear equations is based on the Newton algorithm for nonlinear equations. The method can be best understood referring to Figure 11.2. Suppose we wish to find a root of the function $f(x)=0$ plotted in this figure. To do so we must find the intersection of the curve with the x axis. At point p_1, the slope to the function is given as

$$\frac{\partial f(x_1)}{\partial x} = \frac{f(x_1)}{x_1 - x_2}$$

or, rewriting, we get

$$x_2 = x_1 - \frac{f(x_1)}{\partial f(x_1)/\partial x}$$

Thus, if we view x_1 as an approximation to the solution, then the second term is the correction estimate needed to get closer to the correct solution. Generalizing the process gives

$$x_{n+1} = x_n - \frac{f(x_n)}{\partial f(x_n)/\partial x}$$

or

$$x_{n+1} = x_n - \left[\frac{\partial f(x_n)}{\partial x}\right]^{-1} f(x_n) \qquad (11.54)$$

This can be done for any number of equations, including a system of equations. To do so, we first rewrite the system in the form required above namely, $f(x)=0$. For a matrix system, this is:

$$[A]\{x\} - \{Q\} = 0$$

Substituting this in Eq. (11.54) gives:

$$\{x\}_{n+1} = \{x\}_n - \left[\frac{\partial [A]_n\{x\}_n}{\partial x}\right]^{-1} [[A]_n\{x\}_n - \{Q\}] \qquad (11.55)$$

where we have assumed that $\{Q\}$ does not depend on the solution vector $\{x\}$.

The first term in square brackets in Eq. (11.55) is immediately recognized as the Jacobian and can normally be evaluated exactly. The second term is simply the residual obtained by substituting the approximate solution $\{x\}_n$ into the original system of equations.

The steps involved in the method are as follows:

1. Assume an initial solution $\{x\}_0$.
2. Form the matrix $[A]$ and the residual $[A]\{x\}-\{Q\}$ using the current solution for $\{x\}$.
3. Calculate the Jacobian matrix by calculating the derivatives of the term $[A]\{x\}$ with respect to each unknown value of $\{x\}$.
4. Calculate the new solution for $\{x\}$ using Eq. (11.55).
5. Check if the new solution is within the required tolerance with respect to the previous vector $\{x\}$. If so the current solution is the correct solution.
6. Repeat steps 2 through 5 until convergence has been achieved.

Under normal conditions, the solution converges to the correct solution although, in general, there is no guarantee that this will be the case or, if it is, that the solution converges within a reasonable number of iterations.

As a rule however, the Newton-Raphson method converges faster than most other methods of solution for nonlinear equations.

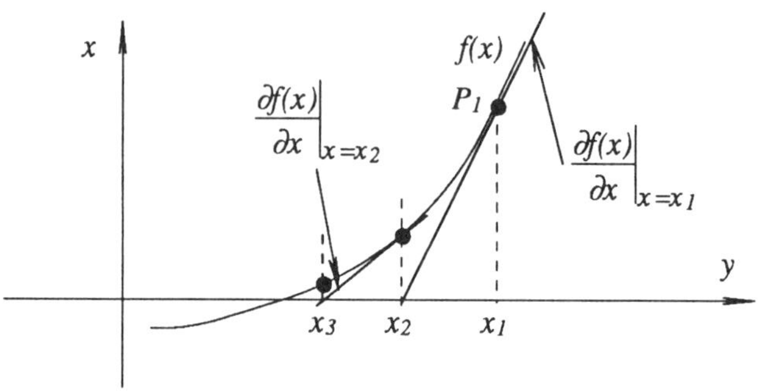

Figure 11.2 The use of Newton's method to find the root of a function.

11.6 METHODS OF SOLUTION FOR EIGENVALUES AND EIGENVECTORS

Electromagnetic eigenvalue problems result in systems of equations of the form

$$[S\,S]\{x\} = \lambda[M]\{x\} \tag{11.56}$$

where $[M]$ and $[SS]$ are normally symmetric, positive definite matrices and can be real or complex, depending on the physical conditions. The term λ is called an eigenvalue or characteristic value while the solution vector $\{x\}$ is the eigenvector or characteristic vector. The system only has solutions for discrete values of λ and then, the solution vector is arbitrarily scaled in the sense that only the relations between values of the eigenvector are correct, not their absolute values.

The generalized eigenvalue problem in Eq. (11.56) can be reduced to a standard eigenvalue problem by writing

$$[C]\{x\} = \lambda\{x\} \tag{11.57}$$

where

$$[C] = [M]^{-1}[S\,S]$$

and $[M]^{-1}$ is the inverse of $[M]$.

A number of methods of solving the standard eigenvalue problem exist and we discuss some of these next. While there are many methods suitable for the systems normally encountered, the most commonly used method for complex systems of equations is the so-called *QZ* algorithm. This, and the equivalent *QR* algorithm for real matrices is outlined briefly below. The Jacobi method for eigenvalues and eigenvectors is also presented since it is a simple iterative method that illustrates the steps involved in solution. Givens transformation is presented as an extension of Jacobi's method and also since it is often used in conjunction with the *QR* and *QZ* methods. Other methods, like the Householder method or the iteration (power) method are not given for lack of space. Also, we mention that while the methods given here are useful, there are more sophisticated algorithms for eigenvalue problems, especially for very large systems of equations. While solution of real systems of equations is rather straightforward, complex systems are rather difficult to solve and the existing algorithms are not well adapted to this purpose.

11.6.1 The Jacobi and Givens transformations

The principle of all transformation methods is in a systematic modification of the matrix until its diagonals or some other part of the matrix are the only nonzero coefficients. For a matrix with all zero coefficients except for the diagonal, the diagonals are the eigenvalues. Thus, reduction to a diagonal form provides the eigenvalues of the original matrix, if, in the transformation process, the properties of the matrix have not been changed. This is the case if the transformation is done using so-called similarity transformations

$$[C]' = [T]^{-1}[C][T]$$

where $[T]$ is the transformation matrix. If in addition, the transformation matrix is orthogonal, $[T]^{-1} = [T]^T$, and we may write

$$[C]' = [T]^T [C][T] \tag{11.58}$$

In the Jacobi method, the transformation matrix is of the form

$$[T] = \begin{bmatrix} 1 & 0 & 0 & 0 & 0 & 0 \\ 0 & \cos\alpha & 0 & -\sin\alpha & 0 & 0 \\ 0 & 0 & 1 & 0 & 0 & 0 \\ 0 & \sin\alpha & 0 & \cos\alpha & 0 & 0 \\ 0 & 0 & 0 & 0 & 1 & 0 \\ 0 & 0 & 0 & 0 & 0 & 1 \end{bmatrix} \tag{11.59}$$

where $[T][T]^{-1}=1$, indicating an orthogonal transformation matrix. The only nonzero coefficients are in columns i, j and rows i, j. All diagonal values are equal to 1 except for diagonals i, i and j, j as shown.

The principle is to apply this transformation in an iterative fashion. In each iteration, the value of a is calculated such that the off diagonal terms of the modified matrix $[C]'$ at locations (i, j) and (j, i), become zero. This process is repeated until all off-diagonal terms are zero. If any of the previously zeroed coefficients become nonzero due to the modifications of other coefficients in the transformation in Eq. (10.10), the process is repeated until all off-diagonal terms are within a tolerance parameter from zero. Since the orthogonal transformation retains the symmetry in the matrix only the upper part of the matrix is considered. The algorithm for one Jacobi iteration is:

1. Zero the off-diagonal terms from the condition (by expanding Eq. (11.59))

$$c'_{i,j} = 0 = (-c_{i,i} + c_{j,j})\cos\alpha\sin\alpha + c_{i,j}(\cos^2\alpha - \sin^2\alpha)$$

From this:

$$cos^2\alpha = \frac{1}{2} + \frac{c_{i,i} - c_{j,j}}{2\sqrt{(c_{i,i} - c_{j,j})^2 + 4c_{i,j}^2}}$$

$$sin^2\alpha = \frac{1}{2} - \frac{c_{i,i} - c_{j,j}}{2\sqrt{(c_{i,i} - c_{j,j})^2 + 4c_{i,j}^2}}$$

$$sin\,\alpha cos\,\alpha = \frac{c_{i,j}}{\sqrt{(c_{i,i} - c_{j,j})^2 + 4c_{i,j}^2}}$$

After these calculations, the off-diagonal coefficients $c_{i,j}$ and $c_{j,i}$ are set to zero.
2. The diagonal coefficients $c_{i,i}$ and $c_{j,j}$ are modified as:

$$c'_{i,i} = c_{i,i}cos^2\alpha + c_{j,j}sin^2\alpha + 2c_{i,j}sin\,\alpha cos\,\alpha$$

$$c'_{j,j} = c_{i,i}sin^2\alpha + c_{j,j}cos^2\alpha - 2c_{i,j}sin\,\alpha cos\,\alpha$$

Similarly, all coefficients in rows *i* and *j* and columns *i* and *j* are modified as:

$$c'_{k,i} = c_{k,i}cos\alpha + c_{k,i}sin\alpha \qquad k=1,N \qquad k\neq i$$

$$c'_{k,j} = -\,c_{k,i}sin\alpha + c_{k,j}cos\alpha \qquad k=1,N \qquad k\neq i$$

and

$$c'_{k,i} = c'_{i,k} \qquad and \qquad c'_{k,j} = c'_{j,k}$$

The only remaining difficulty is to decide on the signs of $sin\alpha$ and $cos\alpha$. If we arbitrarily choose $sin\alpha$ to be positive, then $cos\alpha$ must be chosen to have the same sign as $c_{k,j}$ This procedure leads to a modified coefficient matrix

$$[C]'^{\,k+1} = [T]^T_k\,[C]_k\,[T]_k$$

After n iterations, the matrix $[C]'$ is diagonal and the eigenvalues are the diagonal values. If the transformation matrices are saved and multiplied together, or if they are multiplied as they are obtained we can write the eigenvectors of the original matrix $[C]$ as the columns of the matrix

$$[X] = [T]_1\,[T]_2\,[T]_3 \ldots [T]_n$$

where each column is one solution $\{x\}_i$, corresponding to the eigenvector l_i.

The foregoing steps constitute the Jacobi method and, while simple, it is not the most efficient method for determination of eigenvalues and eigenvectors. Givens' method is a modification of the basic Jacobi method, and, unlike Jacobi's method which is an iterative method, it is a single sweep method (requires the elimination of off-diagonal terms only once, because terms that have been zeroed do not change after elimination). However, it does not produce a diagonal system but, rather, a tridiagonal system which must be solved using, for example, the *QR* or *QZ* algorithms. The main purpose of discussing it here is that it is often used in conjunction with the *QR* or *QZ* methods. The same is true of other methods for eigenvalue solution such as the Householder method which also reduces the original system to a tridiagonal system and which then must be solved for eigenvalues and eigenvectors.

Using the same transformation as for the Jacobi method, each step eliminates the off-diagonal terms $c_{i-1,j}$ and $c_{i,j-1}$. In all other respects the algorithm is the same. The coefficients of the transformation matrix ($sin\alpha$ and $cos\alpha$) are found from the condition

$$c'_{i-1,j} = 0 = -c_{i-1,i} sin\alpha + c_{i-1,j} cos\alpha$$

From this we write:

$$cos\alpha = \frac{c_{i-1,i}}{\sqrt{c_{i-1,i}^2 + c_{i-1,j}^2}}$$

$$sin\alpha = \frac{c_{i-1,j}}{\sqrt{c_{i-1,i}^2 + c_{i-1,j}^2}}$$

The remaining steps in the procedure are the multiplication of this matrix with the original matrix as in the Jacobi method. After all possible coefficients have been zeroed, the modified matrix is

$$[C]' = \begin{bmatrix} * & * & 0 & 0 & 0 & 0 \\ * & * & * & 0 & 0 & 0 \\ 0 & * & * & * & 0 & 0 \\ 0 & 0 & * & * & * & 0 \\ 0 & 0 & 0 & * & * & * \\ 0 & 0 & 0 & 0 & * & * \end{bmatrix}$$

where * indicates a nonzero coefficient.

Since the eigenvalues and eigenvectors are yet to be found we only note here that the eigenvalues of the tridiagonal system above are the same as the eigenvalues of the original matrix and the eigenvectors of the original matrix can be found from those of the tridiagonal system by their product with the product of the transformations used for the Givens method

$$x_i = [T]_1 [T]_2 [T]_3 \ldots [T]_n x''_i$$

where x_i denotes an eigenvector of the original system and x_i'' denotes the corresponding eigenvector (for the same eigenvalue) of the tridiagonal system.

11.6.2 The QR and QZ methods

The basic QR algorithm uses the following decomposition for step k of the iteration:

$$[C]_k = [Q]_k [R]_k \qquad (11.60)$$

where $[Q]^k$ is an orthogonal matrix and $[R]^k$ is an upper triangular matrix. The modified matrix at step $k+1$ is now written as

$$[C]_{k+1} = [R]_k [Q]_k \qquad (11.61)$$

and $[R]^k$ is related directly to the Givens transformations above. Assuming that n Givens transformations are required to obtain the tridiagonal system, we have:

$$[R]_k = [T]_n^t [T]_{n-1}^t [T]_{n-2}^t \ldots [T]_1^t [C]_k$$

and

$$[C]_{k+1} = [R]_k [T]_1 [T]_2 [T]_3 \ldots [T]_n$$

where we use the fact that

$$[R]_k = [Q]_k^t [C]_k$$

Obviously, the Givens transformation is not the only one that can be used and, in fact, the Householder transformation is often used for this purpose.

The QZ algorithm is more often used in conjunction with complex matrices of the type we encounter in field computation. The algorithm is based on the following principles:

For a generalized eigenvalue problem:

$$[S\,S]\{x\} = \lambda [M]\{x\}$$

we find a decomposition

$$[Q][SS][Z]\{y\} = \lambda[Q][M][Z]\{y\} \tag{11.62}$$

such that $[Q][SS][Z]$ and $[Q][M][Z]$ are both upper triangular matrices. Then the eigenvalues of the original matrix are related to the eigenvalues of the transformed matrix by

$$\{x\} = [Z]\{y\} \tag{11.63}$$

and the eigenvectors are the diagonals of $[Q][A][Z]$.

In the first step, the matrix $[SS]$ is reduced to a complex upper Hessenberg matrix using transformation methods (i.e. by Givens or Householder methods). This can always be done for a general complex matrix. The matrix $[M]$ is triangularized while modifying $[A]$ so that the eigenvalues are retained. These matrices are as follows

$$[SS]' = \begin{bmatrix} * & * & * & * & * & * \\ * & * & * & * & * & * \\ 0 & * & * & * & * & * \\ 0 & 0 & * & * & * & * \\ 0 & 0 & 0 & * & * & * \\ 0 & 0 & 0 & 0 & * & * \end{bmatrix} \qquad [M]' = \begin{bmatrix} * & * & * & * & * & * \\ 0 & * & * & * & * & * \\ 0 & 0 & * & * & * & * \\ 0 & 0 & 0 & * & * & * \\ 0 & 0 & 0 & 0 & * & * \\ 0 & 0 & 0 & 0 & 0 & * \end{bmatrix}$$

The iteration now becomes

$$[SS]_{k+1} = [Q]_k [SS]_k [Z]_k$$

$$[M]_{k+1} = [Q]_k [M]_k [Z]_k$$

After $[SS]$ has been triangularized, the transformation in Eqs. (11.62) and (11.63) are satisfied and the eigenvectors and eigenvalues are obtained.

The discussion above is brief and only includes a few aspects of eigenvalue solution. Many other methods, some much more efficient than those presented exist. The bibliography section points to some of these. There are also special computer libraries such as *EISPACK* dedicated to this aspect of computation. Eigenvalue solvers also exist in all other major computer packages such as *LINPACK*, *IMSL*, *NAG* to mention only a few.

11.7 INSERTION OF DIRICHLET BOUNDARY CONDITIONS

In conjunction with finite element matrices (and, to a lesser extent, in finite difference applications), we are often faced with the need to impose Dirichlet boundary conditions to an assembled matrix. From chapter *4* or *7*, we recall that these matrices are assembled from individual finite elements and, because treatment of each finite element is identical, boundary conditions cannot be included in the definition of the finite element. After matrix $[SS]$ and vector $\{Q\}$ are assembled, and before the solution step can begin, it is necessary to insert the imposed potentials at the boundary nodes on which Dirichlet conditions are specified.

This is a rather simple task but, because the systems of equations can be large it needs some attention. Before describing the process used we discuss briefly the standard method of reducing the system of equations to take into account known values of the solution vector. If, for example, in the system $[A]\{x\}=\{b\}$ with four equations, the variable x_2 has a specified value $x_2=P$, the system is modified as follows:

$$\begin{bmatrix} a_{11} & a_{12} & a_{13} & a_{14} \\ 0 & 1 & 0 & 0 \\ a_{31} & a_{32} & a_{33} & a_{34} \\ a_{41} & a_{42} & a_{43} & a_{44} \end{bmatrix} \begin{bmatrix} x_1 \\ x_2 \\ x_3 \\ x_4 \end{bmatrix} = \begin{bmatrix} b_1 \\ P \\ b_3 \\ b_4 \end{bmatrix} \tag{11.64}$$

The product of the second line by the unknown vector gives $x_2=P$. This system can be solved under the above form or it can be reduced to:

$$\begin{bmatrix} a_{11} & a_{13} & a_{14} \\ a_{31} & a_{33} & a_{34} \\ a_{41} & a_{43} & a_{44} \end{bmatrix} \begin{bmatrix} x_1 \\ x_3 \\ x_4 \end{bmatrix} = \begin{bmatrix} b_1 - a_{12}P \\ b_3 - a_{32}P \\ b_4 - a_{42}P \end{bmatrix} \tag{11.65}$$

It is tempting to reduce the matrix to the form in Eq. (11.65) because the number of unknowns is reduced. However, unless the number of known boundary points is extremely large in proportion to the total number of points in a solution domain, it is unlikely that the system can be reduced significantly. On the other hand, reduction requires a relatively large number of operations and, it is likely that the reduction in the matrix size together with the increase in operations due to the reduction step will result in little or no gain. There is also the question of keeping track of the unknowns which are in different locations in the vector after reduction. Thus, while in Eq. (11.64) the unknowns are ordered and their indices are their location in the vector this is not the case in Eq. (11.65). The solution of Eq. (11.64) is faster but, because of the arguments above and because the coefficients that have been zeroed are not available for further operations, this matrix cannot be used as is. We must either reduce it to that in Eq. (11.65) or use a different technique for insertion of boundary conditions.

To avoid this inconvenience, we adopt a modified method: the diagonal term is replaced by a large number K. Typical values could be $K=1.0E^{+14}$. In the right hand side vector, the value KP is placed while the nondiagonal terms are not modified. This makes the off-diagonal coefficients negligible relative to the large diagonal and the effect is as if the off-diagonal terms are zeroed. If K is large compared to the off-diagonal terms, the error due to this method is negligible.

Using this method we get

$$\begin{bmatrix} a_{11} & a_{12} & a_{13} & a_{14} \\ a_{21} & K & a_{23} & a_{24} \\ a_{31} & a_{32} & a_{33} & a_{34} \\ a_{41} & a_{42} & a_{43} & a_{44} \end{bmatrix} \begin{bmatrix} x_1 \\ x_2 \\ x_3 \\ x_4 \end{bmatrix} = \begin{bmatrix} b_1 \\ KP \\ b_3 \\ b_4 \end{bmatrix} \tag{11.66}$$

The advantage of this is that the form of the matrix, including size and relative location of coefficients is not modified, but because K is very large compared to all other coefficients in row *(2)* we get the same result as above since performing the product in row *(2)* and neglecting all small terms we get

$$Kx_2 = KP$$

which enforces the condition $x_2=P$. This method is commonly used in finite element applications because it is simple, accurate, and requires very few operations. The value of K can also be chosen by finding the value of the largest diagonal and ensuring that K is a few orders of magnitude larger than that diagonal, but for practical purposes it is often sufficient to choose an adequately large value for K.

11.8 BIBLIOGRAPHY

The subjects treated in this chapter are in effect a number of appendices. The references given below are divided into the various subjects treated here. Because the material is covered more as an overview, only a few general references are given for each subject.

Numerical integration in conjunction with finite elements is discussed in almost all texts on finite elements. Reference [1] is given here because it gives tables of integration points and weights for Gauss-Legendre, Gauss Laguerre and other methods of integration. A general discussion of the solution of systems of equations, especially as these relate to finite element solutions can be found in [3-10] while [7,8] discuss relaxation methods. These subjects are also treated in almost any text on linear algebra. Some other methods of solution, including multigrid methods are given in [11-24]. The conjugate gradient method is discused in [25-31]

[1] Carnahan, B. Luther, H.A, and Wilkes, J.O. (1969), *Applied Numerical Methods*, John Wiley, New York.

[2] Irons, B.M. (1966), Engineering applications of numerical integration in stiffness method, *Journal A.I.A.A.*, Vol. 14, pp. 2035-2037.

[3] Reid, J.K. (1976), Solution of linear systems of equations: direct methods (general), in *Lecture Notes in Mathematics*, V.A. Barker, Ed., Springer-Verlag, Berlin, No. 572, pp. 102-130.

[4] Varga, R.R. (1962), *Matrix Iterative Analysis*, Series in Automatic Computation, Prentice-Hall, Englewood Cliffs, NJ.

[5] Gustavson, F.G. (1972), Some basic techniques for solving sparse systems of equations, *Sparse Matrices and Their Applications*, D.J. Rose, and Willoughby, R. A., Ed., Plenum Press, New York, pp. 41-53.

[6] George, J.A. (1976) Sparse matrix aspects of the finite element method, *Proceedings of the Second International Symposium on Computing Methods in Applied Science and Engineering*, Springer-Verlag, 1976.

[7] De Allen, D.N. and Dennis, S.C.R. (1951), The application of relaxation methods to the solution of differential equations in three dimensions", *Qarterly Journal of Mechanics and Applied Mathematics*, Vol. IV, part 2.

[8] Stoll, R.L. (1970), Solution of linear steady state eddy current problems by complex successive over-relaxation, *IEE Proceedings*, Pt. A, Vol. 117, pp. 1317-1323.

[9] Axelson, O. (1976), Solution of linear systems of equations: iterative methods, in *Lecture Notes in Mathematics*, Barker, V. A., Ed., Springer-Verlag, Berlin, No. 572, pp. 1-51.

[10] Meyer, C. (1973), Solution of linear equations - State-of-the-art, *Journal of the Structure Div., ASCE*, Proceedings Paper 9861, pp. 1507-1527.

[11] Brandt, A. (1982), Guide to multigrid development, in *Lecture Notes in Mathematics*, No. 960, Hackbush, W. and Trottenberg, U., Eds., Springer-Verlag, Berlin.

[12] Brandt, A. (1979), Multi-level adaptive finite-element methods. I. variational methods, *Special Topics of Applied Mathematics*, Frehse, J., Pallasche, U. and Trottenberg, U., Eds., North-Holland Publishing Co., Amsterdam, pp. 91-128.

[13] Hackbush, W. (1978), A fast method for solving Poisson's equation in a general region, in *Lecture Notes in Mathematics*, No. 631, R. Bulirsch, R.D. Grigorieff, and J. Schroder, Eds., Springer-Verlag, Berlin, pp. 51-62.

[14] Mansfield, L. (1981), On the multi-grid solution of finite element equations with isoparametric elements, *Numerical Mathematics*, Vol. 37, pp. 423-432.

[15] George, J.A. (1975), Nested dissection of a regular finite element mesh, *SIAM Journal of Numerical Analysis*, Vol. 10, No. 2, pp. 345-363.

[16] Fellipa, C.A. (1975), Solution of linear equations with skyline-stored symmetric matrix, *Computers and Structures*, Vol. 5, pp.13-29.

[17] George, J.A. (1972), On block elimination of finite element systems of equations, *Sparse Matrices and Their Applications*, Rose, D.J. and Willowghby, R.A., ed., Plenum Press, New York, pp. 101-114.

[18] Gupta, S.K. and Tanji, K. K. (1977), Computer program for solution of large, sparse, unsymmetric systems of linear equations, *International Journal for Numerical Methods in Engineering*, Vol. 11, pp. 1251-1259.

[19] Hood, P. (1976), Frontal solution program for unsymmetric matrices, *International Journal for Numerical Methods in Engineering*", Vol. 10, pp. 379-399.

[20] Irons, B.M. A frontal solution program for finite element analysis, *International Journal for Numerical Methods in Engineering*, 2, pp. 5-32.

[21] Meijerink, J.A. and van der Vorst, H. A., An iterative solution method for linear systems of which the coefficient matrix is a symmetric M-matrix, *Mathematics of Computation*, Vol. 31, No.137, pp. 148-162.

[22] Pao, Y.C. (1978), Algorithms for direct access Gaussian solution of structural stiffness matrix equations, *International Journal for Numerical Methods in Engineering*, Vol. 12, pp. 751-764.

[23] Stabrowski, M.M. (1982), Block-indexed approach to solution of very large linear equation systems with DBDF coefficient matrix, *IEEE Transactions on Magnetics*, Vol. MAG-18, No. 2, pp. 605-608.

[24] Young, D.M. (1971), *Iterative solution of large linear systems*, Computer Science and Applied Mathematics, Academic Press, New York.

[25] Hestenes, M.R. and Stiefel, E. (1952), Methods of conjugate gradients for solving linear systems, *J. Res. Natl. Bureau Standards*, Vol. 49, pp. 409-436.

[26] Jennings, A. and Malik, G.M. (1978), The solution of sparse linear equations by the conjugate gradient method, *International Journal for Numerical Methods in Engineering*, Vol. 12, No. 1, pp. 141-158.

[27] Jacobs, D.A.H. (1980), Generalisations of the conjugate gradient method for solving non-symmetric and complex systems of algebraic equations, *RD/L/N 70/80, August 1980*, Central Electricity Research Laboratories, Leatherhead, Surrey, U. K.

[28] Daniel, J.W. (1967), The conjugate gradient method for linear and non-linear operator equations, *SIAM Journal of Numerical Analysis*, Vol. 4, pp. 10-26.

[29] Fletcher, R. and Reeves, C. M. (1964), Function minimization by conjugate gradients, *Computer Journal*, Vol. 7, pp. 149-154.

[30] Reid, J.K. (1972), The use of conjugate gradients for systems of linear equations possessing 'property A', *SIAM Journal of Numerical Analysis*, Vol. 9, No. 2, pp. 325-332.

[31] Reid, J.K. (1971), On the method of conjugate gradients for the solution of large sparse systems of linear equations, in *Large Sparse Sets of Linear Equations*, Reid, J. K., Ed., Academic Press, pp. 231-252.

Index